Table I (Cont.)

z	0	1	2	3	4	5	6	7	8	9
.0	.5000	.5040	.5080	.5120	.5160	.5199	.5239	.5279	.5319	.5359
.1	.5398	.5438	.5478	.5517	.5557	.5596	.5636	.5675	.5714	.5753
.2	.5793	.5832	.5871	.5910	.5948	.5987	.6026	.6064	.6103	.6141
.3	.6179	.6217	.6255	.6293	.6331	.6368	.6406	.6443	.6480	.6517
.4	.6554	.6591	.6628	.6664	.6700	.6736	.6772	.6808	.6844	.6879
.5	.6915	.6950	.6985	.7019	.7054	.7088	.7123	.7157	.7190	.7224
.6	.7257	.7291	.7324	.7357	.7389	.7422	.7454	.7486	.7517	.7549
.7	.7580	.7611	.7642	.7673	.7704	.7734	.7764	.7794	.7823	.7852
.8	.7881	.7910	.7939	.7967	.7995	.8023	.8051	.8078	.8106	.8133
.9	.8159	.8186	.8212	.8238	.8264	.8289	.8315	.8340	.8365	.8389
1.0	.8413	.8438	.8461	.8485	.8508	.8531	.8554	.8577	.8599	.8621
1.1	.8643	.8665	.8686	.8708	.8729	.8749	.8770	.8790	.8810	.8830
1.2	.8849	.8869	.8888	.8907	.8925	.8944	.8962	.8980	.8997	.9015
1.3	.9032	.9049	.9066	.9082	.9099	.9115	.9131	.9147	.9162	.9177
1.4	.9192	.9207	.9222	.9236	.9251	.9265	.9279	.9292	.9306	.9319
1.5	.9332	.9345	.9357	.9370	.9382	.9394	.9406	.9418	.9429	.9441
1.6	.9452	.9463	.9474	.9484	.9495	.9505	.9515	.9525	.9535	.9545
1.7	.9554	.9564	.9573	.9582	.9591	.9599	.9608	.9616	.9625	.9633
1.8	.9641	.9649	.9656	.9664	.9671	.9678	.9686	.9693	.9699	.9706
1.9	.9713	.9719	.9726	.9732	.9738	.9744	.9750	.9756	.9761	.9767
2.0	.9772	.9778	.9783	.9788	.9793	.9798	.9803	.9808	.9812	.9817
2.1	.9821	.9826	.9830	.9834	.9838	.9842	.9846	.9850	.9854	.9857
2.2	.9861	.9864	.9868	.9871	.9875	.9878	.9881	.9884	.9887	.9890
2.3	.9893	.9896	.9898	.9901	.9904	.9906	.9909	.9911	.9913	.9916
2.4	.9918	.9920	.9922	.9925	.9927	.9929	.9931	.9932	.9934	.9936
2.5	.9938	.9940	.9941	.9943	.9945	.9946	.9948	.9949	.9951	.9952
2.6	.9953	.9955	.9956	.9957	.9959	.9960	.9961	.9962	.9963	.9964
2.7	.9965	.9966	.9967	.9968	.9969	.9970	.9971	.9972	.9973	.9974
2.8	.9974	.9975	.9976	.9977	.9977	.9978	.9979	.9979	.9980	.9981
2.9	.9981	.9982	.9982	.9983	.9984	.9984	.9985	.9985	.9986	.9986
3.0†	.9987	.9987	.9987	.9988	.9988	.9989	.9989	.9989	.9990	.9990

† For $z \geq 4$ the areas are 1 to four decimal places.

Adapted from *Probability with Statistical Applications*, second edition, by F. Mosteller, R. E. K. Rourke, and G. B. Thomas, Jr. Reading, Mass.: Addison-Wesley, 1970, p. 473.

INTRODUCTORY
Statistics

INTRODUCTORY
Statistics

NEIL WEISS AND MATTHEW HASSETT

Arizona State University

 ADDISON-WESLEY

PUBLISHING COMPANY

Reading, Massachusetts • Menlo Park, California • London • Amsterdam • Don Mills, Ontario • Sydney

SPONSORING EDITOR: Ron Hill
PRODUCTION EDITOR: Mary Cafarella
DESIGNER: Marshall Henrichs
ART COORDINATOR: Joseph Vetere
ILLUSTRATOR: ANCO (Boston)
COVER DESIGN: Ann Scrimgeour Rose
CHAPTER OPENING PHOTOGRAPHS: Marshall Henrichs

Library of Congress Cataloging in Publication Data

Weiss, Neil A.
  Introductory statistics.

  Includes bibliographies and index.
  1. Statistics. I. Hassett, Matthew J., joint
author. II. Title.
QA276.12.W45   519.5   80-23520
ISBN 0-201-09507-6

ISBN 0-201-09507-6
FGHIJ-HA-8987654321

Preface

This book is intended for use in introductory statistics courses. It has been written for the students who typically take such courses on most campuses—potential users of statistics who will need to analyze data in the life, social, or management sciences. The mathematical prerequisite is a working knowledge of introductory high school algebra.

In order to better serve the intended readership we have included the following features:

1. *Emphasis on application* We have concentrated our exposition on the application of statistical techniques to the analysis of data. Although the statistical theory has been kept to a minimum we have made every attempt to give a thorough development of the rationale for using each statistical test.

2. *Real data* Real data sets have been used wherever possible. Even in those cases where real data sets were far too complex for pedagogical purposes, we have been guided by real data in the construction of our simplified data sets.

3. *Detailed and careful explanation* We have attempted to include every step of explanation that a typical reader might need. We have worked on the principle that

whenever the omission of a page might cause a substantial portion of our readers to labor unduly for an extra hour, that page should *not* be omitted. We hope that the inclusion of detailed and careful explanation will result in better understanding with a more efficient use of time.

4. *Extensive examples and exercise sets* It is our feeling that the vast majority of students learn by reading examples and doing exercises. Therefore we have included a large number of examples and extensive exercise sets. A special feature of this book is the inclusion of routine exercises directly following text examples for immediate *reinforcement* of learning and/or as a check of the student's understanding. A student who tries to compute a standard deviation immediately after reading the first example of such a computation will know rather quickly whether he or she has read carefully enough.

Since our readers will have rather diverse mathematical backgrounds we have included a wide variety of exercises, and labelled them by level of difficulty. Each exercise set contains a number of problems that are *routine applications* of basic material that every student should master. In most exercise sets there are also included *intermediate* and *advanced* problems. The intermediate problems are underscored with a "▬" and contain supplementary material that is not covered in the text but that may be of interest to some of the more highly motivated students. The advanced exercises (underscored with a "■") introduce more abstract concepts or lead the student through algebraic derivations and are intended for the student with special mathematical background and aptitude.

5. *Computer packages and supplementary aids* Many students in introductory statistics courses will, in the near future, analyze their own data. Regardless of their current level of sophistication, they will use computer packages and attempt to master more complicated techniques than are covered in this book. We cannot hope to teach the use of computer packages or more advanced techniques here, but we *have* included descriptions and examples of computer analyses and lists of recommended reference books for further study.

6. *Objectives, summaries, and practice tests.* Frequently students in introductory statistics courses feel a certain amount of anxiety and confusion about how their work should be structured. The instructional aids provided in the text—chapter objectives, summaries, and review tests—are designed to aid the student by providing a definite structure for study.

We have attempted to write a text that offers a great deal of flexibility in the choice of material to be covered. Chapters 1 through 8 contain the material that we consider to form the core of an introductory statistics course: descriptive statistics, basic probability, the normal distribution, the sampling distribution of the mean, and inferences concerning means (large and small samples). In Chapter 9 we introduce the χ^2- and F-statistics in the context of inferences concerning variances. However, instructors who wish to omit inferences concerning variances may do so by simply concentrating on those parts of the chapter that discuss how to use the χ^2 and F tables. The remainder of the chapters on inferential statistics can be taken in any combination and in any order.

It is our pleasure to thank the following reviewers whose comments were invaluable to the "fine tuning" of the book:

Larry Griffey (Florida Junior College, Jacksonville)
Larry Haugh (University of Vermont)
William Stines (North Carolina State University)
David Lund (University of Wisconsin)
J. Marie Haley (Embry Riddle Aeronautical University, Prescott, Arizona)
William Tomhave (University of Minnesota)
Jay Devore (California State Polytechnic University)
Abraham Weinstein (Nassau Community College, Garden City, N.Y.)
William Topp (University of the Pacific, Stockton)
Francisco J. Samaniego (University of California, Davis)
Richard S. Kleber (St. Olaf College, Northfield, Minnesota)
Margaret Lial (American River College, Sacramento)

We are especially grateful to Professors Larry Griffey and Larry Haugh, who reviewed the manuscript several times and helped us to preserve essential elements when we were tightening up the manuscript to avoid excessive length.

We also thank Professors Mikel Aickin, Paul Dunlap, Michael Driscoll, Greg Nielson, Donald Stewart, Dennis Young, and Lt. Colonel Jon Epperson, Lt. Colonel Ray Mitchell, and Major Sam Thompson for several useful consultations. In addition, we are grateful to Professor J. Marie Haley for providing the answers to the exercises.

Our special thanks go as well to Peggy Hassett who spent countless hours researching data sets for examples and exercises, solving problems, and purging the text of errors. Also, we should like to express our appreciation to Carol Weiss for her careful proofreading of the book.

Ethel Bauer and Juanita Moore did a superb job of typing. Finally, we thank everyone at Addison-Wesley for their help in the publication of the book.

Tempe, Arizona   N.W.
July 1981   M.H.

Contents

CHAPTER

The nature of statistics

1.1 Two kinds of statistics

You probably feel that you already know something about statistics. If you read newspapers, watch the news on television, or follow sports, you hear the word "statistics" frequently. In this section we will use such familiar examples as baseball statistics and voter polls to introduce you to the two major kinds of statistics: **descriptive statistics** and **inferential statistics**.

Each spring in the late 1940's the major-league baseball season was officially opened when President Harry S Truman threw out the "first ball" of the season at the opening game of the Washington Senators. Both President Truman and the Washington Senators had reason to be interested in statistics—statisticians were interested in them. Consider, for example, the year of 1948.

Example 1 ◖ In 1948 the Washington Senators played 153 games, winning 56 and losing 97. They finished seventh in the American League and were led in hitting by B. Stewart whose batting average was .279. This information was compiled by baseball statisticians who took the complete records for each game of the season and organized and simplified the great mass of information contained in them. Although baseball fans take this information for granted, it requires a great deal of time and effort to compile. Without it baseball would be much harder to understand: Picture yourself trying to pick the best hitter in the American League if you had only the official score sheets for each game played. (There were more than 600 games played that year. The best hitter was Ted Williams, who led the league with a batting average of .369.) ◗

The work of baseball statisticians is a good example of **descriptive statistics**.

DEFINITION   **Descriptive statistics** consists of methods for organizing and summarizing information.

The purpose of this organizing and summarizing is to help you make some sense out of a mass of information. Descriptive methods include the construction of graphs, charts, and tables, and the calculation of various kinds of averages and indices.

Example 2 ◖ In the fall of 1948, President Truman had good reason to be concerned about statistics. The Gallup poll prior to the election in November predicted that he would win only 44.5 percent of the vote and would lose the presidency. In this case, the statisticians had predicted incorrectly. Mr. Truman won more than 49 % of the vote and the presidency. The Gallup poll modified some of its procedures and has not predicted the wrong winner since. ◗

The work of political polling provides us with an example of **inferential statistics**. It would be tremendously expensive to interview all Americans on their voting preferences. Statisticians who wish to gauge the sentiment of the entire **population** of American voters can afford to interview only a carefully chosen group of a few thousand voters. This collection of a few thousand voters is referred to as a **sample** of the entire population. Statisticians analyze the results obtained from this sample to make an inference (or educated guess) about the voting preferences of the entire voting population. Inferential statistics provides methods for making such educated guesses. The terminology used above is always used in inferential statistics in the following way:

Population: The set of all individuals or items under consideration.
Sample: That part of the population from which information is collected.

DEFINITIONS

Using this terminology, we can define inferential statistics as follows:

Inferential statistics consists of methods for making inferences about a population based on information obtained from a sample of the population.

DEFINITION

Classification of statistical problems   1.2

Both descriptive and inferential methods will be studied in more detail in the following chapters. *At this point you need only be able to classify problems as descriptive or inferential, and identify the population and sample in inferential problems.* The examples below are intended to give you some practice with this. In each example the results of a statistical study are given and the study is classified as descriptive or inferential. You should attempt to classify each study yourself before reading our explanation of it.

◀ *The study* In 1948 the total final popular votes for the major presidential tickets were as follows:

Example **3**

Ticket	Votes	Percent
Truman–Barkley (Dem.)	24,179,345	49.7
Dewey–Warren (Rep.)	21,991,291	45.2
Thurmond–Wright (States Rights)	1,176,125	2.4
Wallace–Taylor (Progressive)	1,157,326	2.4
Thomas–Smith (Socialist)	139,572	0.3

Classification: This study is *descriptive*. It summarizes the results of all votes cast by the entire population of voters. ▶

◀ *The study.* For the 101 years preceding 1977, the baseballs used by the major leagues were purchased from the Spalding Company. In 1977 that company stopped manufacturing major-league baseballs, and the major leagues arranged to buy their baseballs from the Rawlings Company. Early in the 1977 season, pitchers began to complain that the Rawlings ball was "livelier" than previous balls; it was harder, bounced farther and faster, and gave an unfair advantage to hitters. There was some evidence for this. In the first 616 games of 1977, 1033 home runs were hit. (Only 762 home runs were hit in the first 616 games of the previous year.) *Sports Illustrated* magazine sponsored a careful study of this question and the results appeared in the June 13, 1977 issue. In this study, an independent testing company randomly selected 85 baseballs from the supplies of various major league teams, carefully measured the bounce, weight, and hardness of these balls, and compared these results with figures

Example **4**

obtained from similar tests on baseballs used in the years 1952, 1953, 1961, 1963, 1970, and 1973. The conclusion given in the *Sports Illustrated* article (page 24) was that "the 1977 Rawlings ball is livelier than the 1976 Spalding, but not as lively as it could be under big league rules, or as the ball has been in the past."

Classification. This is an *inferential* study. The population consists of all major-league baseballs manufactured for 1977 and the other years for which comparison figures were available. (The number of balls used by the major leagues in 1977 was estimated to be approximately 360,000.) The sample consists of the 85 balls selected for testing in 1977 and the balls previously selected for testing in the comparison years.

This study provides an excellent example of a situation in which the entire population cannot be sampled. After the bounce and hardness tests, all sampled balls were taken to a butcher in Plainfield, New Jersey, to be sliced in half so that the researchers could look inside them. It would not be very practical to test every new baseball in this way. Many other statistical studies are of this kind. Medical studies of the effects of various drugs on laboratory animals often end with the dissection of the animals. ◗

Example **5**   ◖ In the late 1940's and early 1950's there was great public concern over epidemics of polio. A vaccine for polio was developed by Jonas Salk of the University of Pittsburgh. Various tests indicated that this vaccine was safe and potentially effective, but it was necessary to have a large-scale test to determine whether this vaccine would be truly effective in preventing polio. A test was devised involving nearly 2,000,000 grade-school children. All of them were inoculated, but only half received the Salk vaccine. The remaining children were inoculated with a harmless solution (a *placebo*) which would have no known effects. The children's parents and doctors did not know who had received the Salk vaccine and who had not. (This was done to prevent any possible bias in later analysis of the results.) An evaluation center kept records of those who had actually received the Salk vaccine, and determined that the incidence of polio among the children actually receiving Salk vaccine was significantly lower than among those receiving the placebo. The Salk vaccine was then made available for general use.

Classification. This is an *inferential* study. The group of children inoculated may have been quite large, but it was still just a sample of the much larger group of *all* American schoolchildren. This larger group is the population of interest to American parents and doctors. ◗

Example 5 illustrates that there may be some disagreement over definition of the population. Many readers here are probably asking themselves why we should be interested only in American schoolchildren. It would certainly seem better to protect all children in the world against polio. Unfortunately, the test above involved only schoolchildren from the United States and was possibly not representative of what might occur under very different conditions in other parts of the world. In Chapter 14 we will discuss the problem of choosing a sample so that it represents the population of interest.

The science of statistics includes both descriptive and inferential statistics. *Descriptive statistics* appeared first; censuses were taken as long ago as Roman times. Over the years, recordings of births and deaths and tax listings have led naturally to descriptive statistics. *Inferential statistics* is a newer arrival. It is based on the theory of probability, which was not established until the middle of the seventeenth century. Modern inferential statistics did not begin to develop seriously until Karl Pearson and Ronald Fisher published their research in the early years of this century. Since the work of Pearson and Fisher, inferential statistics has grown quite rapidly, and has come to be applied in a wide range of subject areas such as history, biology, psychology, and physics. An understanding of the basic concepts of statistics should help you with work in almost any professional area. In addition, knowing about statistics should help you to make more sense out of many things you read in newspapers and magazines. For example, when you read our description of the *Sports Illustrated* baseball test, it may have struck you as unreasonable to make a conclusion about a group of 360,000 baseballs after sampling only 85 of them. By the time that you have finished Chapter 8 of this book you should see that this is not unreasonable, and understand why people are able to make conclusions from tests of this kind.

The primary objective of this text is to present the fundamentals of *inferential statistics*. However, almost any inferential study involves aspects of descriptive statistics. Therefore, in Chapters 2 and 3, we shall first consider the basic principles and methods of descriptive statistics.

Development of inferential statistics   1.3

EXERCISES

Classify each of the studies in Exercises 1 through 5 as descriptive or inferential, and give a reason for your answer. For each inferential study, describe the sample and the population under consideration.

1 Data collected from a sample of American television viewers, male and female, yielded the following estimates of average TV viewing time per week for all Americans.

Sex	Age group	Time (in hours and minutes per week)
Females	12–17	21 : 04
	18–24	30 : 05
	25–54	32 : 25
	55 and over	35 : 23
Males	12–17	22 : 29
	18–24	21 : 21
	25–54	27 : 38
	55 and over	32 : 40

2 In 1936, the voters of North Carolina cast their presidential votes as follows.

Candidate	Number of votes
Roosevelt, Dem.	616,414
Landon, Rep.	223,283
Thomas, Soc.	21
Browder, Com.	11
Lemke, Union	2

3 The Public Broadcasting System keeps complete records on educational television stations and their broadcast schedules. In 1974, PBS collected the following information.

Number of educational T.V. stations in the United States: 238

Total broadcast hours for 227 stations representing primary broadcasters	18,320 hours
General programs	10,868 hours
Percent of total hours	59.3 %
Instructional programs	7,452 hours
Percent of total hours	40.7 %

4 The Division of Vital Statistics of the Public Health Service made the following estimates of the leading causes of death in the United States for 1976, by sampling 10% of all death certificates.

Cause	Number of deaths for year (rounded to nearest 10)
1. Major cardiovascular diseases	977,410
2. Malignant neoplasms (cancers)	374,780
3. Cerebrovascular diseases	189,000
4. Accidents	100,430
5. Influenza and pneumonia	62,980

5 The New York Stock Exchange keeps records of the selling prices for seats on the Exchange. Below are the high and low prices for some years in this century.

Year	High price	Low price
1900	$ 47,500	$ 37,500
1920	115,000	85,000
1929	625,000	550,000
1935	140,000	65,000
1940	60,000	33,000
1970	320,000	130,000
1976	104,000	40,000

6 Decide whether descriptive or inferential statistics would be used in each of the following situations, and give a reason for your answer.

a) A tire dealer wishes to estimate the average life of a tire.

b) A sports writer wishes to list the winning times in all swimming events in the 1976 Olympics.

c) A politician wishes to know the exact number of votes cast for his opponent in the election of 1976.

d) A medical researcher wishes to test an anticancer drug that may have harmful side effects.

e) A candidate for President wishes to know what percent of American voters prefer him to all other candidates.

f) A banker wishes to estimate the average income of all residents of California in 1978.

g) A small business owner has twenty employees and would like to find their average salary.

7 a) The chairman of a mathematics department wants to know the average final exam score for the 2000 students in her department's basic algebra course. She randomly selects 50 exams from the 2000 and finds that their average was 78.3%. She estimates that the average for all 2000 students was about 78.3%. What kind of study has she done?

b) What kind of study would result if the chairman averaged all 2000 exams?

c) Would you want to average 2000 exams?

CHAPTER REVIEW

Key Terms **descriptive statistics** **sample**
population **inferential statistics**

You should be able to

1 Classify statistical problems and studies as either *descriptive* or *inferential*.

2 Identify the *sample* and *population* in inferential problems and studies.

REVIEW TEST

Classify each of the following studies as descriptive or inferential.

1 During important elections, the major television networks predict final election results for a city, county, or state early in the evening before all returns are in. These predictions are based on the partial results obtained from the first precincts to report.

2 A newspaper reporter who was doing library research for an article on civil aviation found that the total numbers of deaths in civil aviation accidents for 1960, 1965, 1970, and 1975 were:

Year	Deaths
1960	1286
1965	1279
1970	1454
1975	1448

3 A sociologist studying "marriage and the family" looked up the total numbers of marriages and divorces in the state of California in 1976. They were:

Marriages	Divorces
150,654	133,672

4 A cafeteria chain from France decided to see if there was a market in the United States for cafeterias specializing in French cuisine. In 1977, three French cafeterias were opened in the metropolitan area of Phoenix, Arizona, to test this idea.

FURTHER READINGS

FREEDMAN, D., R. PISANI, and R. PURVIS, *Statistics*. New York: W. W. Norton & Company, 1978. In Chapters 1 and 2 these authors describe the work of statisticians by giving a number of interesting, real-life examples.

TANUR, J., *et al.* (Eds.), *Statistics: A Guide to the Unknown*. San Francisco: Holden-Day, Inc., 1972. More than 400 pages of brief (4–10 page) articles on the use of statistics in many areas of interest. The articles do not require any special mathematical background. Taken together, they give an excellent picture of the wide range of applicability of statistics.

In our library

Organizing data

2.1

Data

The information collected and analyzed by statisticians is called *data*. There are different kinds of data, and the statistician's choice of methodology is partly determined by the kind of data he has. In this section we will look at some examples of the most important kinds of data.

Example 1

◖ At noon on April 18, 1977, nearly 3,000 men and women set out to run from Hopkinton Center to the Prudential Plaza in Boston. Their run would cover 26 miles and 385 yards, and would be watched by thousands of people lining Boston streets and millions more on television news reports. It was the eighty-first running of the Boston Marathon.

A great deal of information was accumulated that afternoon. The men's competition was won by Jerome Drayton of Canada with a time of two hours, fourteen minutes, and forty-six seconds. The first woman to finish was Miki Gorman of California; her time was two hours, forty-eight minutes, and thirty-three seconds. 101 women and 2215 men finished before the official cutoff time of four hours. (After four hours no official times are recorded, although some runners may still be on the course.) An elite group of 846 men and 60 women had their names, times, and places printed in *Runner's World* magazine, because they recorded high-quality times—better than three hours for men and three and one-half hours for women. ◗

In the example above, we can find each of the major kinds of data. The simplest kind is the information that puts each entrant into one of the two categories "male" or "female." Such data, which give qualitative information about an individual or item, are called **qualitative data**. The information that Miki Gorman is a female is *qualitative data.*

Most racing fans are more interested in the *place* of their favorite than in the kinds of data given above. Information on place is called **ordinal data**—data that put individuals or objects in order. The information that Jerome Drayton and Veli Bally finished first and second and the information that Thomas Berg and James Williams finished in places 500 and 501 among the men are examples of *ordinal data.*

Ordinal data give information about place, but do not measure the difference in performance between places. Consider the example above. Jerome Drayton actually finished 58 seconds ahead of Veli Bally, while Thomas Berg beat James Williams by one second. More can be learned about what happened in a race by looking at the times of runners. Differences between times indicate exactly how far apart two finishers are, while differences in places don't. Data such as time data, which provide more exact and meaningful measurement of differences between individuals, are referred to as **metric data**.* The information that Jerome Drayton ran his race in 2:14:46 is an example of *metric data.*

---

* The term *metric data* refers to *measurement* data, and should not be confused with data given in metric units. Metric data can be given in other units, such as feet.

Below is a summary of the types of data discussed in the previous paragraphs.

Qualitative data: data that refers to nonnumerical qualities or attributes, such as gender, eye color, and blood type.
Ordinal data: data about order or rank on a scale such as 1, 2, 3, . . . or A, B, C, ...
Metric data: data obtained from measurement of such quantities as time, height, and weight.

DEFINITIONS

Another important type of data is called **count data**. Counting the number of individuals or items falling into various categories, such as "male" and "female," gives count data. For instance, the information given in Example 1, that 101 women and 2215 men finished the Boston Marathon in under four hours, is *count data*.

Count data: data on the number of individuals or items falling into certain classes or categories.

DEFINITION

As you will see in the examples that follow, count data can be obtained from any of the types of data previously discussed—i.e., from qualitative data, ordinal data, or metric data:

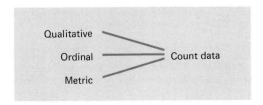

◀ Human beings are classified as having one of the four blood types A, B, AB, and O.

Example **2**

a) What kind of data do you receive when you are told your blood type?
b) Geneticists and anthropologists record the numbers of individuals having each blood type in various populations. What kind of data are they collecting?

a) Your blood type is a piece of *qualitative data*. It places you in one of four categories, A, B, AB, or O. (It is important to know your blood type for purposes of blood transfusions; reactions may occur if donor and recipient are not properly matched.)
b) This is *count data*, obtained by counting the number of individuals in each blood-group category. So this illustrates that *count data can be obtained from qualitative data*. ◗

Solution

◀ At most universities, students completing a course receive a grade of A, B, C, D, or E (failure). What kind of data is:
a) the information that Jane Doe received a grade of A?

Example **3**

b) the information that the final grades in Professor H's statistics class were 17 A's, 16 B's, and 8 C's, with no D's or E's?

Solution a) The information that Jane Doe received a grade of A is *ordinal data*—the grades A, B, C, D, and E put student performances in order from A to E, and are replaced at many schools with the numbers 4, 3, 2, 1, 0.

b) The information on numbers of A's, B's and C's is *count data*. There are five categories here—the category of A students, the category of B students, and so on. We have simply counted the number of students in each category. (Professor H is real, and we actually did this count using his final grades.) This example shows that ordinal ranking of large groups of individuals leads naturally to the establishment of categories based on rank, and also leads to count data based on these categories. That is, this illustrates that *count data can be obtained from ordinal data.* ▶

Note that Professor H's grade distribution is a bit higher than what is usually seen in math classes. This illustrates that not all ranking schemes are the same. Professor H's grade of A may have been easier to get than Professor W's. *Ordinal data such as grade data may mean different things when the ranking is done by different people.*

Example **4** ◀ Angel Falls in Venezuela is 3,281 feet high, more than twice as high as Ribbon Falls at Yosemite, California, which is 1,612 feet high. What kind of data are the heights given here?

Solution These heights are *metric data*, obtained by taking precise measurements. These data were obtained from a list of the world's highest waterfalls, in the *Information Please Almanac.* From this list we can also show that of the world's forty highest waterfalls, four are over 1,700 feet high, five are between 1,000 and 1,700 feet high, and 31 are less than 1,000 feet high. This is *count data obtained from metric data.* ▶

Exercise **A**[*] Classify each of the following types of data as qualitative, ordinal, metric, or count data.

a) In the U.S. in 1975, there were (about) 14,098,000 single males, 49,409,000 married males, 1,817,000 widowed males, and 2,545,000 divorced males.

b) In 1978, the final standings in the National League Western Division were:

1. Los Angeles Dodgers
2. Cincinnati Reds
3. San Francisco Giants
4. San Diego Padres
5. Houston Astros
6. Atlanta Braves

c) The nicotine content of a certain brand of cigarettes is 0.8 mg.

[*] The answers to the reinforcement exercises can be found at the end of the chapter.

Before leaving this section, there is one final comment we wish to make concerning types of data.

Qualitative and *ordinal data* are referred to by statisticians as **discrete data**, because they sort things into separate, discrete classes. Track and field authorities classify entrants as male or female (qualitative data) for competition, and assign the first two runners either place 1 or place 2 (ordinal data). There is no way of being classified *between* male and female, or being assigned place 1.357. On the other hand, most *metric data* is called **continuous** because it involves measurement on a continuous scale: If the measuring device is precise enough, it can conceivably record measurements of any size between any two given values. For instance, electronic timers are now available that record times to a hundreth of a second. Thus a timer could conceivably record a marathon time of 2 : 38 : 44.53, between the times of 2 : 38 : 44 and 2 : 38 : 45.

Statisticians make other distinctions in classifying data, but the distinctions in this section are sufficient for this book.

The problem of data classification is sometimes a difficult one. There are cases where statisticians will disagree over the classification of a data set. However, in most cases the classification is fairly clear and serves as an aid in the choice of the correct statistical method.

EXERCISES Section **2.1**

Classify the data given in Exercises 1 through 8 as qualitative, ordinal, metric, or count data.

1 The principal languages of the world in 1977 were as follows:

Rank	Language	Millions of speakers
1	Mandarin (China)	670
2	English	369
3	Russian	246
4	Spanish	225
5	Hindustani	218
6	Arabic	134
7	Portuguese	133

a) What kind of data are the ranks 1, 2, . . . , 7?
b) What kind of data is contained in the information that Ronald Reagan speaks English?
c) What kind of data is given in the last column above (under "Millions of speakers")?

2 Tobacco production in the United States from 1972–1976 was given as follows by the U.S. Department of Agriculture.

Year	Thousands of bushels
1972	1,749,085
1973	1,742,105
1974	1,989,728
1975	2,181,775
1976	2,118,560

What kind of data is given in the second column of this table?

3 On May 4, 1961, Commander Malcolm Ross, USNR, ascended 113,739.9 feet in a free balloon. What kind of data is the height given here?

4 In 1975, there were 12,272 forest fires on federal land. These fires burned a total of 408,000 acres. What kind of data are contained in the numbers:
a) 12,272?   b) 408,000?

5 In 1974, some of the leading recreational expenditures for all Americans were:

Type of recreation	Money spent
Books and maps	$3,049,000,000
Radios, TV's, records and musical instruments	13,270,000,000
Motion pictures	2,034,000,000
Spectator sports	1,265,000,000

What kind of data are the dollar amounts given?

6 Below are figures on U.S. industrial employment for 1975.

Industry	Employees (in thousands)
Agriculture, forestry, fisheries	3,476
Mining	732
Construction	5,015
Manufacturing	19,275
Transportation	5,623
Trade	17,470
Finance	4,665
Services	30,132

What kind of data are given in the employee numbers?

7 In 1977 there were 3,920,000 males and 3,371,000 females between the ages of 16 and 24 in the civilian labor force.
a) What kind of data is contained in the classification of an individual as male or female?
b) What kind of data are the numbers 3,920,000 and 3,371,000?

8 What kinds of data would be collected in the following situations?
a) A quality-control engineer measures the lifetime of electric lightbulbs.
b) A businessman wishes to know the number of families with preteen children in Pueblo, Colorado.

c) A sporting goods manufacturer is going to classify each major league baseball player as righthanded or lefthanded, and count the number in each category.
d) A sociologist wants to estimate the average annual income of residents of Ossining, New York.
e) A political pollster wishes to classify each individual in a sample of voters as Democrat or Republican, and count the total numbers in each group.
f) An administrator at a community college needs to know how many men and women participated in varsity sports during the 1980 spring semester and how much money was spent on men's sports and on women's sports.

9 There are two different kinds of metric data—data measured on an *interval scale* and data measured on a *ratio scale*.

An example of an interval scale is the Fahrenheit system of temperature measurement. On an interval scale, you can measure differences such as the difference between today's high temperature and yesterday's. But an interval scale has an arbitrary zero point: You can also measure temperatures in Celsius, which has a different zero temperature point.

An example of a ratio scale is the measurement of weight in pounds. This measurement has a zero point that cannot be arbitrarily changed.

The situation can be pictured as follows:

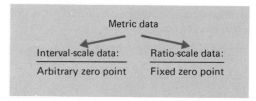

Identify each of the following as interval or ratio data.
a) Measurements of heights of American males
b) Ages of American females
c) Calendar dates of important historical events

10 Can you identify the reasons that the names "interval" and "ratio" are used for the two types of metric data? (See Exercise 9.)

2.2 Grouped data

The amount of data that can be collected in a "real-world" situation is sometimes overwhelming. For example, the results on Boston Marathon finishers in *Runner's World* magazine occupy four pages—with four columns of small type to a page! Results are listed for 906 individuals and ten teams. Making some sense of this data is the work of *descriptive statistics*: It organizes and simplifies data. In this section one such method of organization is discussed: *grouping* of data. We will not use the marathon data in our examples, since a few of the calculations necessary here are a bit messy for the marathon times. Instead, we will use some grade data, which can be organized more simply.

◖ Table 2.1 lists the grades for twenty statistics students on a 100-point test. Group this data so as to obtain some idea of the overall performance of the 20 students.

Example **5**

Student	Grade	Student	Grade	**Table 2.1**
1	88	11	89	
2	85	12	100	
3	90	13	76	
4	81	14	75	
5	67	15	89	
6	82	16	70	
7	63	17	86	
8	96	18	34	
9	64	19	84	
10	39	20	96	

People usually think of grades as 90's, 80's, 70's, etc. We will set up grade categories that correspond to this: 90–99, 80–89, etc. These grade categories are listed in the left part of Table 2.2. To see how many grades there are in each category, go through the data list in Table 2.1 and make a tally mark for each grade on the appropriate line in Table 2.2. (For example, the Student 1 grade of 88 calls for a tally mark on the line for 80–89.) When the entire data list has been tallied, count marks to find the number of student grades at each level. The result is Table 2.2. ◗

Solution

Grade category	Tally	Number of student grades	**Table 2.2** Grouped grade data
0–9		0	
10–19		0	
20–29		0	
30–39	||	2	
40–49		0	
50–59		0	
60–69	|||	3	
70–79	|||	3	
80–89	|||| |||	8	
90–99	|||	3	
100			1	
	Total	20	

The number of individuals in each category of grouped data is called the **frequency** of that category. Thus, tables like Table 2.2 are called **frequency distribution tables**. The categories used for grouping are called **classes**. Example 5 uses a common-sense approach to grouping data into classes. Some of this common sense can be written down as rules and guidelines for setting up frequency distributions. For example:

1. *The number of classes should not be too small or too large.* We suggest that the number of classes be between 5 and 12. This is a rule of thumb only. In Example 5, eleven categories are used.
2. *Each class should be of the same width.* To satisfy this rule we need to change one class. The class of 100's includes only one possible grade, while all other classes include ten possible grades. To keep classes of the same width, we change the 100's class into 100–109. This keeps things orderly, even if it is a bit unnatural.
3. *Each piece of data must belong to one class,* and only one. Careless planning could lead to classes like 90–100, 80–90, 70–80, etc. A student with a grade of 90 would belong in two classes and would be counted twice. The classes in Table 2.2 do not cause such confusion. They cover all possible grades, and they do not overlap.

The list of such "rules" could go on, but this would be artificial. The purpose of data grouping is to break data up into a reasonable number of classes of equal length in order to learn something about the data. If possible, intervals that have a natural interpretation should be used (as illustrated above in grouping grades). In general, a statistician uses common sense to group data as neatly and meaningfully as possible.

Table 2.2 does not tell us what percentage of the twenty students fall in each class. The percentage in a class can be found by dividing the number of students in that class by the total number of students. For example, the percentage of students in the class 80–89 is

$$\text{Number in class} \longrightarrow \frac{8}{20} = \frac{4}{10} = 0.40 \quad (\text{or } 40\%).$$
$$\text{Total number} \longrightarrow$$

The percentage in a class, expressed as a decimal, is called the **relative frequency** of that class. In Table 2.3 a list of the information from Table 2.2 is presented, along with the relative frequency of each class. In Table 2.3 the 100's class is changed to 100–109.

Data interpretation

Many observations can be made from such grouped data. In the case of this particular data set, the grouped data were presented to the students in the course. Each student then had some idea of where he stood in relation to the others. This was especially important for the two students with grades in the thirties; they might otherwise have continued their work at the same level in the mistaken belief that other students had low grades and curving would carry them through. After seeing this data, one of the students in the thirties dropped the course. The other began to study harder, and finished the course with a grade of B.

Class	Frequency	Relative frequency
0–9	0	0.00
10–19	0	0.00
20–29	0	0.00
30–39	2	0.10 ← 2/20
40–49	0	0.00
50–59	0	0.00
60–69	3	0.15 ← 3/20
70–79	3	0.15 ← 3/20
80–89	8	0.40 ← 8/20
90–99	3	0.15 ← 3/20
100–109	1	0.05 ← 1/20

Table 2.3
Frequencies and relative frequencies for grade data

Although the basic ideas of data grouping are common sense, this topic has some technical terminology associated with it. Below is a list of all the terminology introduced so far in this section, and a few additional terms. We will illustrate most of these terms by interpreting them for the class 80–89 in the grade data.

Terminology

Class: A category for grouping data.

Frequency: The number of individuals or objects in a class. The frequency of the class 80–89 is 8.

Relative frequency: The percentage, expressed as a decimal, of the total number of individuals in a given class. The relative frequency of the class 80–89 is 0.40.

Frequency distribution: A listing of classes with their frequencies. The first and second columns of Table 2.3 make up a frequency distribution.

Relative-frequency distribution: A listing of classes with their relative frequencies. The first and third columns of Table 2.3 make up a relative frequency distribution.

Lower class limit: The smallest value that can go into a class. For the class 80–89 the lower class limit is 80.

Upper class limit: The largest value that can go into a class. For the class 80–89 this is 89.

Class width (or *Class interval*): The difference between the lower class limit of the given class and the lower limit of the next higher class. For the class 80–89 this is 10:

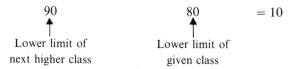

$$90 \qquad\qquad 80 \qquad = 10$$

Lower limit of next higher class   Lower limit of given class

The class width tells how wide the range of data values is for the given class.

Class mark: The value exactly in the middle of a class, found by adding the class limits and dividing the sum by 2. For the class 80–89, it is $(80 + 89)/2 = 84.5$.

Table 2.4 summarizes the above discussion for the grade data.

Table 2.4
Grade data

Class	Frequency	Relative frequency		Class mark
0–9	0	0.00		4.5
10–19	0	0.00		14.5
20–29	0	0.00		24.5
30–39	2	0.10	(10%)	34.5
40–49	0	0.00		44.5
50–59	0	0.00		54.5
60–69	3	0.15	(15%)	64.5
70–79	3	0.15	(15%)	74.5
80–89	8	0.40	(40%)	84.5
90–99	3	0.15	(15%)	94.5
100–109	1	0.05	(5%)	104.5
	20	1.00		

Lower class limits Upper class limits

Example **6** ◖ A pediatrician began testing the cholesterol levels of young patients, and was alarmed to find that a large number had cholesterol levels over 200. A list of the readings of 40 high-level patients is given below.

210	209	212	208	205	219	206	215
217	207	210	203	208	211	208	213
208	210	210	199	213	219	212	219
215	221	213	218	217	221	221	221
202	218	200	214	206	204	208	198

Group these data using a class width of 5, starting at 195.

Solution   We first construct a table similar to Table 2.2 (page 15).

Table 2.5

Cholesterol level	Tally	Frequency
195–199	‖	2
200–204	‖‖‖	4
205–209	ᚼᚼ ᚼᚼ	10
210–214	ᚼᚼ ᚼᚼ ‖	11
215–219	ᚼᚼ ‖‖‖	9
220–224	‖‖‖	4
	Total	40

In Table 2.6 we show the grouped data table for the cholesterol-level data.

Class	Frequency	Relative frequency		Class mark
195–199	2	0.050	(5%)	197
200–204	4	0.100	(10%)	202
205–209	10	0.250	(25%)	207
210–214	11	0.275	(27.5%)	212
215–219	9	0.225	(22.5%)	217
220–224	4	0.100	(10%)	222
	40	1.000		

Lower limits Upper limits

Table 2.6
Cholesterol-level data

The next illustration involves grouping data that are not whole numbers.

◀ A statistician interested in obtaining information about the weights of newborn babies gathered the following data. The data were obtained from hospital records listing neonatal weights to the nearest tenth of a pound.

Example **7**

9.8	6.5	9.5	5.1	4.8	8.8	6.5	9.5
7.7	6.9	6.6	6.0	7.9	7.7	6.9	6.6
5.8	7.1	6.8	8.4	6.9	5.8	7.1	6.8
8.6	9.8	3.8	7.4	7.2	8.6	9.8	3.8
10.3	7.4	5.7	4.5	7.7	10.3	7.4	5.7
9.4	7.8	8.9	5.8	8.6	9.4	7.8	8.9
5.6	7.4	6.7	4.5	7.8	6.6	7.4	5.7
8.8	9.4	6.0	5.9	7.4	8.8	9.4	6.0
7.2	10.5	9.4	7.4	8.9	7.2	10.5	8.4
10.4	7.8	5.0	4.6	8.0	10.4	7.8	5.0

Group this data using a class width of 1 pound (10 tenths), starting at 3.0 pounds.

The tallying procedure results in the frequencies shown in the second column of Table 2.7.

Solution

Table 2.7
Birth-weight data
(lbs)

Class	Frequency	Relative frequency	Class mark
3.0–3.9	2	0.0250	3.45
4.0–4.9	4	0.0500	4.45
5.0–5.9	11	0.1375	5.45
6.0–6.9	14	0.1750	6.45
7.0–7.9	21	0.2625	7.45
8.0–8.9	12	0.1500	8.45
9.0–9.9	10	0.1250	9.45
10.0–10.9	6	0.0750	10.45
	80	1.0000	

Lower
limits

Upper
limits

To illustrate some typical computations for the third and fourth columns, we use the *third* class (5.0–5.9).

$$\text{Relative frequency} = \frac{11}{80} = 0.1375$$

$$\text{Class mark} = \frac{5.0 + 5.9}{2} = 5.45$$

Exercise **B** A group of 50 elementary-school students was given an IQ test. The results of the test are given below.

112	109	102	111	112	109	102	111	89	93
102	91	100	96	102	91	100	96	93	86
92	107	110	97	92	107	109	97	105	93
106	99	116	82	106	99	116	82	100	101
115	115	102	92	115	116	103	92	86	87

Group this data using a class width of 5, starting at 80. Construct a table similar to Table 2.7.

In all of the examples we have considered, each class contained several possible data values. For instance, in Example 6, each class had *five* possible cholesterol levels. The third class, for example, contained 205, 206, 207, 208, and 209. In some cases, however, data are grouped in classes based on a *single* value. The example below is of this kind.

Example **8** ◖ A planner is collecting data on the number of school-age children in a small town, selecting 40 families considered to be representative of the town as a whole. (This selection is actually a sampling problem of the kind to be discussed in Chapter 14.) The planner then finds the number of school-age children in each family. Results are given in Table 2.8. For example, the table shows that seven (7) of the 40 families selected have two (2) school-age children.

Number of children	Frequency	Relative frequency
0	18	0.450
1	8	0.200
2	7	0.175
3	4	0.100
4	3	0.075
Totals	40	1.000

Table 2.8
Frequency and relative-frequency distributions for family data

What are the classes, class limits, and class marks for this grouped data?

The classes are based on number of children; they are "0 children," "1 child," etc. Thus, each class contains a *single* numerical value instead of a range of values. We will compute the class limits and class mark for the class "1 child."

Solution

Lower class limit: 1, the smallest value in the class
Upper class limit: 1, the largest value in the class
Class mark: $\dfrac{1+1}{2} = 1$

This shows that, *when a class for data grouping is based on a single value*, such as 1, *this value will turn out to be the class mark and the lower and upper class limit for that class*. Since the class mark is the same as the number of children for each class, it is unnecessary to add a "class mark" column, and Table 2.8 may also serve as the grouped data table.

Note. Although this data was artificially constructed, it is based on the actual percentage distributions given for all American families in the *Statistical Abstract of the United States* (1976). ▶

The computation of class limits and class marks makes sense for metric data (which, for example, is the kind of data used in Example 7). These computations can also be done for ordinal data. However, they are not possible for qualitative data. For example, with data that classifies runners as "male" or "female," it makes no sense to look for class limits. This is an example of a statistical method (computation of class limits) that cannot be used on certain kinds of data (qualitative data). *The grouping methods based upon class limits cannot be used for qualitative data.*

Grouping and data type

EXERCISES Section **2.2**

In Exercises 1 through 7, construct an appropriate table similar to Table 2.7 or Table 2.8 in this section.

1 Below are the annual incomes of 20 American families taken in a sample during 1975. (The incomes are given in thousands of dollars. They were rounded to the nearest 1000.)

30	23	17	12	7
27	21	16	12	6
27	19	14	10	6
24	19	13	8	4

For your table group the data into six classes beginning with 1–5.

2 A planning director collected age data from a sample of 20 households in a development. She found the ages in the households to be:

43, 40, 17	36, 31, 14	50, 46	49, 43, 17, 16, 14
27, 3, 5	56	47, 42, 16, 15	39, 39
32, 28, 6, 5	27, 24, 1	29, 27	76, 44, 39, 17, 15
66, 62	35, 33, 11, 8	39, 35, 13, 10	43, 38, 18
29, 28, 28	29, 11, 10	32, 30, 5, 3	53, 24

For your table, group the ages into classes beginning with 0–9.

3 A social worker canvassed her new clients in order to determine how many children each had living at home. The results are given below.

2	1	2	1	1
0	3	0	2	3
0	0	4	2	2
1	2	0	1	2

Construct an appropriate table. (Use a single value for each class.)

4 In a fifth-grade class an eight-question quiz on fractions was given. The number of incorrect answers for each student is listed below.

3	4	2	3	1
1	1	6	1	2
2	1	0	3	4
0	3	2	0	3
0	5	1	1	1

Construct an appropriate table. (Use a single value for each class.)

5 Two hikers in the **Rocky Mountains** found a dense stand of lodgepole pine trees. The cones remain on these trees for many years and open only after forest fires, to drop their seeds. The hikers think that they may be able to guess the time since the last fire by estimating the heights of the trees. Their estimates, in years, are listed below:

13	17	22	17	27
11	19	24	15	17
14	5	18	17	19
8	10	16	22	13
25	19	21	19	16

Group the measurements into appropriate classes for your table, beginning with the class 5–9.

6 A prairie dog town in a southwestern zoo had 28 adults of the following lengths (excluding tail). The lengths are given in inches.

11.6	12.0	10.9	13.2	11.6
11.1	12.5	12.4	11.2	11.2
11.7	12.3	11.4	12.2	12.4
12.3	11.2	12.8	11.4	12.9
11.8	11.9	12.1	12.7	
10.9	12.7	11.8	11.9	

Use a class width of 0.5 inches in your table.

7 At the end of the season, a fast-pitch softball coach added up the number of home runs each player on her college softball team hit. The results were as follows:

1	4	8	2	0
3	2	5	2	0
10	0	3	1	3
5	6	0	7	4

Construct an appropriate table.

8 Below are 30 scores from a 100-point test. We have arranged them in order to simplify your tallying work.

51	60	65	74	77	91
51	62	66	74	81	97
52	62	70	74	83	98
53	63	73	75	84	98
56	64	73	75	86	99

a) Find frequencies and relative frequencies using the classes 50–59, 60–69, . . . , 90–99.

b) Find frequencies and relative frequencies using the classes 50–54, 55–59, 60–64, . . . , 90–94, 95–99.

c) Compare these two groupings of the data, and decide which one is more useful. (Answers may vary.)

9 Below is a list of 20 students and their grades in two courses.

Student	Statis-tics	English	Student	Statis-tics	English
1	B	B	11	C	C
2	C	A	12	B	C
3	C	B	13	B	D
4	A	B	14	E	E
5	D	E	15	A	A
6	A	A	16	C	A
7	C	A	17	C	B
8	B	C	18	B	C
9	B	A	19	C	C
10	A	C	20	C	B

This kind of data is called *bivariate* (it measures two things, not one). Such data can be tallied using the following kind of table.

Statistics

English	A	B	C	D	E	Totals
A						
B						
C						
D						
E						
Totals						

To tally a student such as Student 6, place a mark in the box under the "A" for statistics and next to the "A" for English—that is, in the upper lefthand corner box.

a) Using the table above, tally all 20 students and find the frequency of each pair of grades.
b) Add up each row and column, and place the sum in the space marked "Totals" at the end.
c) What do these totals tell you?
d) Make another table and use it to show the relative frequency of each pair of grades (the relative frequency for each square).
e) Total the rows and columns in this table. (What do they tell you?)

10 The following data on 16 community college students tells their year in school and whether or not they are taking mathematics.

Student	Year	Math?	Student	Year	Math?
1	1	Yes	9	1	Yes
2	1	No	10	2	No
3	2	No	11	2	No
4	2	No	12	1	No
5	1	Yes	13	1	No
6	2	Yes	14	2	Yes
7	1	No	15	1	No
8	1	Yes	16	2	Yes

Use the procedure indicated in Exercise 9(a) through (e) above, to construct a relative frequency table.

Another way of making sense out of a data set is to make some kind of picture of it. This section will show various kinds of pictorial representations of grouped data. We will illustrate these representations using the grouped data from Sec. 2.2.

Graphs and charts   2.3

◀ The grade data in Table 2.3 is given in Table 2.9 at the top of page 24.

Example **9**

The simplest picture of this data is a **frequency histogram**, as shown in Fig. 2.1. The height of each bar is equal to the frequency of the class it represents. Each bar extends from its lower class limit on the left to the lower class limit of the next class on the right.

Percentage data is displayed using a bar graph in which the bar heights are relative frequencies of classes. Such a graph is called a **relative-frequency histogram**. The relative-frequency histogram for the grade data is given in Fig. 2.2.

Although it is quite natural to draw this histogram with limits 0, 10, 20, 30, 40, 50, 60, 70, 80, 90, 100, 110 on the bottom, you should be careful in reading it. The rectangle from 80 to 90, for example, includes only grades in the 80's. A grade of 90 would be included in the next rectangle to the right. ◗

Table 2.9

Class	Frequency	Relative frequency
0–9	0	0.00
10–19	0	0.00
20–29	0	0.00
30–39	2	0.10
40–49	0	0.00
50–59	0	0.00
60–69	3	0.15
70–79	3	0.15
80–89	8	0.40
90–99	3	0.15
100–109	1	0.05

Figure 2.1
Frequency histogram for grade data.

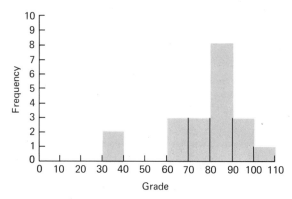

Figure 2.2
Relative-frequency histogram for grade data.

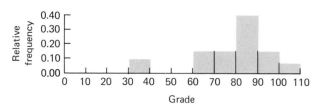

In Example 6, Sec. 2.2, a pediatrician obtained data on the cholesterol levels of some young patients. From Table 2.6, page 19, we repeat the relative-frequency distribution for this data (Table 2.10).

Example **10**

Class	Relative frequency
195–199	0.050
200–204	0.100
205–209	0.250
210–214	0.275
215–219	0.225
220–224	0.100

Table 2.10
Cholesterol-level data

The relative-frequency histogram for this data is shown in Fig. 2.3. ◗

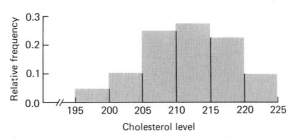

Figure 2.3
Relative-frequency histogram for cholesterol-level data.

In Exercise B, page 20, you were asked to group IQ data. The relative-frequency distribution for this data is given in Table 2.11.

Exercise **C**

Class	Relative frequency	Class	Relative frequency
80–84	0.04	100–104	0.20
85–89	0.08	105–109	0.16
90–94	0.18	110–114	0.10
95–99	0.12	115–119	0.12

Table 2.11

Draw the *relative-frequency histogram* for this data.

In Example 9 there were ten possible grades in each class, and we extended our histogram bars over the ten possible grades (see Figs. 2.1 and 2.2). In some cases, as in the family-size illustration of Example 8, data are grouped in classes based on a *single* value. In such cases, we put the *middle* of each bar directly over the only number in the class. We will illustrate what we are saying in the following example.

Example **11** ◖ In Example 8, we found that the relative-frequency distribution for the family-size data is as shown in Table 2.12. In this case, each class is based on a single value. As we said, in such cases we put the middle of each bar in the histogram directly over the single value in the class. So the relative-frequency histogram for this grouped data is as shown in Fig. 2.4. ◗

Table 2.12

Class	Relative frequency
0	0.450
1	0.200
2	0.175
3	0.100
4	0.075

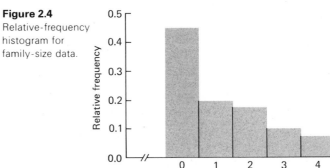

Figure 2.4
Relative-frequency histogram for family-size data.

Although we will usually use histograms to graphically portray data, there are many other ways to display data pictorially. We will illustrate some other common methods for graphing data in the example below.

Example **12** ◖ We return to the grade data of Example 5. In Table 2.13 we have reproduced part of Table 2.4.

Table 2.13
Grade data

Class	Relative frequency	Class mark
0– 9	0.00	4.5
10–19	0.00	14.5
20–29	0.00	24.5
30–39	0.10	34.5
40–49	0.00	44.5
50–59	0.00	54.5
60–69	0.15	64.5
70–79	0.15	74.5
80–89	0.40	84.5
90–99	0.15	94.5
100–109	0.05	104.5

This data is plotted in Fig. 2.5 in a **relative-frequency polygon**.

In this graph, a point is plotted above each *class mark* (class midpoint) at a height equal to the relative frequency of that class. These points are then joined with connecting lines.

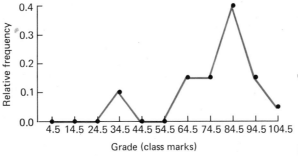

Figure 2.5
Relative-frequency polygon
for grade data.

[handwritten margin notes] More like the histogram because its not cumulative

(Is like the Ogive in that it has lines)

Often we wish the answer to questions like "What percentage of the students' grades were below 80?" Before we draw the graph that displays the answer to the question, we need some more terminology. The frequency of students with grades under 80 is called a **cumulative frequency**. It is found by adding the frequencies of all classes below the 80–89 class (see Table 2.4):

Cumulative frequency—Grades under 80

$$0 + 0 + 0 + 2 + 0 + 0 + 3 + 3 = 8$$

0–9   class   70–79   class

The **cumulative relative frequency** of students with grades under 80 is 8/20, or 0.40. This could also be found by adding the relative frequencies of all classes below the 80–89 class.

Cumulative relative frequency—Grades under 80

$$\frac{0}{20} + \frac{0}{20} + \frac{0}{20} + \frac{2}{20} + \frac{0}{20} + \frac{0}{20} + \frac{3}{20} + \frac{3}{20} = \frac{8}{20}$$

or

$$0.0 + 0.0 + 0.0 + 0.10 + 0.0 + 0.0 + 0.15 + 0.15 = 0.40.$$

Table 2.14 shows all cumulative frequencies and cumulative relative frequencies for the grade data.

Now it is possible to make a graph that answers questions like "What *percentage* of the grades were under 80?" Plot the cumulative relative frequencies on a graph. The **cumulative relative-frequency polygon** for this data is given in Fig. 2.6. Note that the cumulative relative frequencies are plotted at *class limits*. For instance, the cumulative relative frequency of 0.40, for grades less than 80, is plotted above the point 80 on the horizontal scale. ▸

Table 2.14

Less than	Cumulative frequency	Cumulative relative frequency	Cumulative percentage
0	0	0.00	0 %
10	0	0.00	0 %
20	0	0.00	0 %
30	0	0.00	0 %
40	2	0.10	10 %
50	2	0.10	10 %
60	2	0.10	10 %
70	5	0.25	25 %
80	8	0.40	40 %
90	16	0.80	80 %
100	19	0.95	95 %
110	20	1.00	100 %

Figure 2.6
Cumulative relative-frequency polygon for grade data.

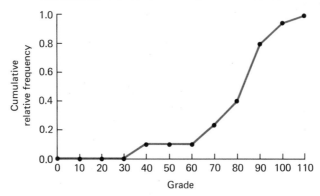

Exercise **D**   For the cholesterol-level data of Example 10 (page 25), draw the *relative-frequency polygon* and the *cumulative relative-frequency polygon*.

Histograms, frequency polygons, and cumulative frequency polygons are very useful when we are discussing such topics as probability theory and the normal distribution. There are countless other ways to make pictorial displays of data that will not be discussed in this text. Fortunately, graphical displays are based mostly on common sense and are usually easy to interpret. Figures 2.7–2.11 on pages 29 and 30 are a few examples of such graphs taken from *Statistical Abstract of the United States, 1976*.

Graphing and data type   Since there are no numerical class limits and marks for qualitative data classes, *it is not possible to use the graphing methods that rely on marks and limits when we are working with qualitative data.* Pie charts are quite suitable for qualitative data; the data in the pie charts in Fig. 2.8 come from the qualitative classes "Military," "Parks," etc. Bar graphs can also be used for qualitative data classes. The information in Fig. 2.8 can be represented by the bar graph given in Fig. 2.12. Note that, on this graph, the bars are separated instead of touching. Pie charts are preferable to bar graphs for qualitative

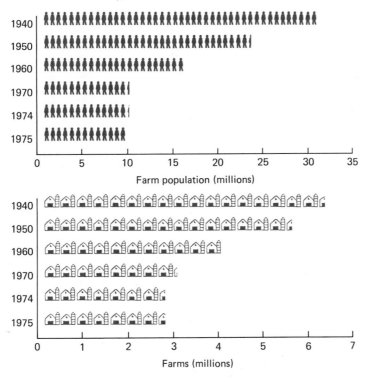

Figure 2.7
Pictograms showing changes in farming, 1940–1975.

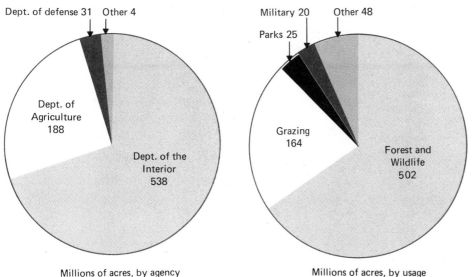

Figure 2.8
Pie charts: land owned by the Federal Government, 1975. (Data from General Services Admin.)

data, since people are accustomed to finding some sort of *order* on the horizontal axis of a bar graph. Someone might read into Fig. 2.12 the implication that park use is "inferior" to grazing use, because park use is shown to the left of grazing use. Pie charts lead to no such inference.

Figure 2.9
Bar charts: foreign travelers, 1960–1975. (Based on data from Bureau of the Census.)

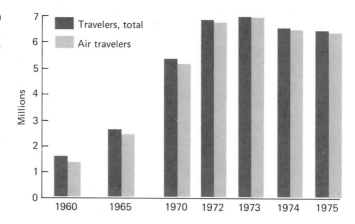

Figure 2.10
Bar charts: U.S. foreign aid since World War II. (Data from U.S. Bureau of Economic Analysis and Board of Governors of the Federal Reserve System.)

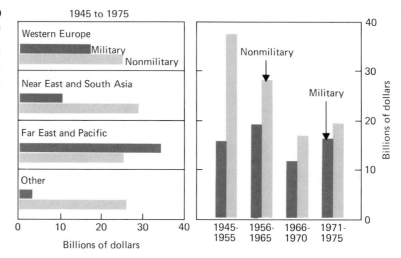

Figure 2.11
Line graphs: exports and imports of merchandise, 1960–1975. (Data from U.S. Bureau of the Census.)

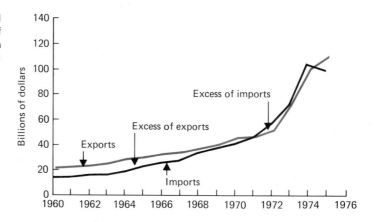

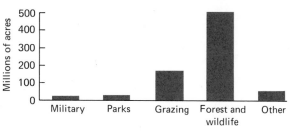

Figure 2.12
Bar graph for pie chart in Fig. 2.8.

EXERCISES Section **2.3**

Exercises 1 through 4: Use the tables given below (from Exercises 1 through 4 in Sec. 2.2) to:

a) Construct a frequency histogram.

b) Construct a relative-frequency histogram.

c) Construct a relative-frequency polygon.

For Exercises 1 and 2, also:

d) Make a table of cumulative percentages.

e) Construct a cumulative relative-frequency polygon similar to the one in Fig. 2.6.

1   Annual incomes in thousands of dollars

Class	Frequency	Relative frequency	Class mark
1– 5	1	0.05	3
6–10	5	0.25	8
11–15	4	0.20	13
16–20	4	0.20	18
21–25	3	0.15	23
26–30	3	0.15	28

2   Ages in 20 households

Class	Frequency	Relative frequency	Class mark
0– 9	8	0.129	4.5
10–19	15	0.242	14.5
20–29	11	0.177	24.5
30–39	13	0.210	34.5
40–49	9	0.145	44.5
50–59	3	0.048	54.5
60–69	2	0.032	64.5
70–79	1	0.016	74.5

3   Number of children living at home

Class	Frequency	Relative frequency
0	5	0.25
1	5	0.25
2	7	0.35
3	2	0.10
4	1	0.05

4   Number of incorrect answers on fraction quiz

Class	Frequency	Relative frequency
0	4	0.16
1	8	0.32
2	4	0.16
3	5	0.20
4	2	0.08
5	1	0.04
6	1	0.04
7	0	0.00
8	0	0.00

5 (Refer to your answers from Exercise 8 of Sec. 2.2. You must have done that exercise before you can begin this one.)

In Sec. 2.2, Exercise 8, you were asked to group a set of 30 grades into classes in two different ways. Using that data,

a) Draw the histogram for the grouping with classes 50–59, 60–69, . . . , 90–99.

b) Draw the histogram for the grouping with classes 50–54, 55–59, . . . , 90–94, 95–99.

c) Compare the histograms. (Which one tells you more about the data?)

d) Can you observe a pattern in the grading of this test? (If you find one, give a reason as to how it might have occurred.)

6 The value halfway between the lower class limit of a given class and the upper class limit of the class below it is the *lower class boundary*. In the grade data from Table 2.4, the lower class boundary for the class 80–89 is $(79 + 80)/2 = 79.5$.

The *upper class boundary* is just like the lower class boundary, but on the upper end of the class. For the class 80–89, the upper class boundary is 89.5.

a) Find the lower and upper class boundaries for the grade data in Exercise 5 of this section. (Use the classes given in Exercise 5(a).)

b) Construct a frequency histogram of the grade data, using class boundaries as the dividing points between the bars

rather than the class limits.

c) What are the advantages of using class boundaries instead of limits?

d) What are the disadvantages?

7 Here is a table of heights for 60 college athletes:

Heights	Frequency
60–64.9	10
65–69.9	20
70–74.9	20
75–84.9	10

Draw a histogram for this data. Pay special attention to the rectangle for the last class. Its height should *not* be 10. Find the correct height and explain your answer.

2.4 How to lie with graphs

Up to this point, we have been concerned with methods for arranging data honestly and accurately. It is possible to abuse some of these methods and to display data in a misleading way. This section has some examples of the misuse of graphs.

Example 13 ◀ Each year the director of the reading program in a certain school district gives a standard test of reading skill and compares the average scores for that school district with the average score for the nation. The district scores are usually higher than the national average. A chart of the scores is made for the school board as shown in Fig. 2.13. This looks good, but could look better if the graph started off at a score of 17.0, as in Fig. 2.14.

This cutoff or *truncated graph* distorts the information displayed; for example, it gives the impression that the district average was twice as good as that of the nation in 1975. The district scores could look even better if the graph started off at 18.0. (To see this, slide a piece of paper over the bottom of Fig. 2.14 so that the bars start at 18.0.) ▶

Figure 2.13
Average reading scores.

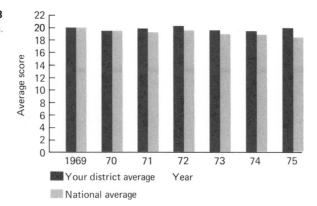

■ Your district average
■ National average

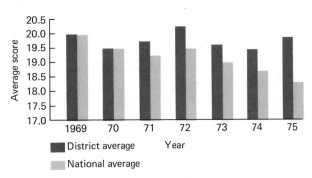

Figure 2.14
Truncation of figure.

Truncated graphs have been a target of statisticians for a long time. Many elementary statistics books discuss them and warn against their use. However, they are still used today, even in reputable publications. The example above was based on a 1978 newspaper article. National magazines and newspapers often use truncation to save space—not to mislead you. However, the results can still be misleading. *Scaling tricks* can also be used to make graphs lie, as indicated below.

◖ A developer is preparing a brochure to attract investors for a new shopping center to be built in a suburban area. This suburb is growing rapidly; twice as many homes were built there in 1976 as in 1975. To depict this, the developer draws a pictogram with a small house showing the number of homes built in 1975 and a house twice as tall showing the number of homes built in 1976. But a house that is twice as tall is also twice as wide, and will end up being *four* times as large. (In Fig. 2.15, each of the four squares made by dotted lines in the larger house is the same size as the entire square part of the smaller house.) So this developer's brochure may mislead the unwary investor. ◗

Example 14

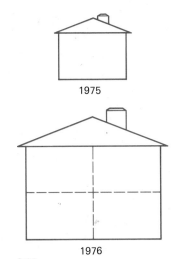

Figure 2.15

There are countless ways to construct misleading figures, besides the ones displayed here. Many more can be found in the entertaining (and classic) book *How to Lie with Statistics* by Darrell Huff (New York: W. W. Norton & Co., 1955). The purpose of this section is not to list all possible forms of misleading graphs, but only to indicate that graphs should be read carefully.

EXERCISES Section **2.4**

1 Find two examples of possibly misleading graphs in a current newspaper or magazine.

The graphical methods covered in Secs. 2.2 and 2.3 are not new. Bar graphs and pie charts date back to William Playfair who used them in the late eighteenth and early nineteenth centuries. We have emphasized these time-honored graphs and charts because they are the most commonly used devices for displaying data. (Histograms are given special emphasis because they are used to explain probability distributions in Chapter 4.) However, there is much more to graphical methods in statistics than bar graphs, pie charts, and histograms. People who deal with data are constantly inventing new ways to

Stem-and-leaf diagrams—a new way of looking at data **2.5**

Skip

display data. In this section we will discuss the stem-and-leaf diagrams invented by Professor John Tukey of Princeton University. These ingenious diagrams are often easier to construct than histograms, and give you much more information.

Example **15**   ◖ The test grade data from Sec. 2.2 is given in Table 2.15.

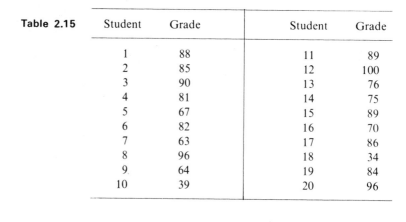

Table 2.15	Student	Grade	Student	Grade
	1	88	11	89
	2	85	12	100
	3	90	13	76
	4	81	14	75
	5	67	15	89
	6	82	16	70
	7	63	17	86
	8	96	18	34
	9	64	19	84
	10	39	20	96

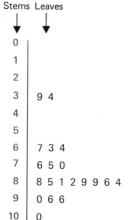

Figure 2.16
Stem-and-leaf diagram for grade data.

In Sec. 2.2, this data was grouped in the categories 100's, 90's, 80's, 70's, etc.; and in Sec. 2.3 the frequency histogram of that grouped data was drawn (see Fig. 2.1, page 24). Using stem-and-leaf diagrams, you can group the data and obtain a display that looks like a histogram at the same time. This is done as follows:

Step 1 Remove the "0" from the category numbers 100, 90, 80, etc., to obtain new numbers 10, 9, 8, 7, 6, 5, 4, 3, 2, 1, 0.

Step 2 List these new numbers in a column on the lefthand side of the page. (The result is the column of numbers displayed in color in Fig. 2.16.)

Step 3 Go through the data set, writing the last digit of each grade to the right of the appropriate first digits. For example, Student 1 got an 88—write an "8" to the right of the "8" in colored type. For student 2 with an 85, write a "5" on the same line, and for Student 3, with a 90, write a "0" to the right of the "9" in colored type. The entire diagram is presented in Fig. 2.16.

The first digits, used to identify categories, are called **stems**. The final digits, used to show the actual grades, are called **leaves**. The resulting **stem-and-leaf diagram** is something like a bar graph or a histogram, since the length of the row of leaves in a class represents the number of grades in that class. In fact, if each row of leaves is shaded over, as in Fig. 2.17, the result is a bar graph for the grade data. The advantage of a stem-and-leaf diagram is that it is not shaded over completely. It displays the value of each individual grade, as well as showing the size of each grade class. This cannot be done with a bar graph. ◗

Figure 2.17
Shaded stem-and-leaf diagram for grade data.

◀ In any marathon race, there is a great deal of excitement around the finish line in the few minutes before the time of three hours is reached. Runners who sense that they have a chance to "break three" push themselves through the final minutes of the race, often accompanied by countdowns from spectators with stopwatches. At Boston in 1977, seventy-seven runners crossed the finish line in the last two minutes before three hours. Their times are listed below:

Example **16**

2 : 58 : 01	2 : 58 : 13	2 : 58 : 27	2 : 58 : 52	2 : 59 : 04	2 : 59 : 31	2 : 59 : 44
2 : 58 : 01	2 : 58 : 13	2 : 58 : 29	2 : 58 : 55	2 : 59 : 12	2 : 59 : 34	2 : 59 : 47
2 : 58 : 02	2 : 58 : 15	2 : 58 : 31	2 : 58 : 56	2 : 59 : 13	2 : 59 : 35	2 : 59 : 49
2 : 58 : 07	2 : 58 : 16	2 : 58 : 36	2 : 58 : 56	2 : 59 : 14	2 : 59 : 36	2 : 59 : 50
2 : 58 : 08	2 : 58 : 17	2 : 58 : 36	2 : 58 : 56	2 : 59 : 17	2 : 59 : 37	2 : 59 : 51
2 : 58 : 08	2 : 58 : 18	2 : 58 : 41	2 : 58 : 57	2 : 59 : 19	2 : 59 : 37	2 : 59 : 51
2 : 58 : 09	2 : 58 : 20	2 : 58 : 42	2 : 58 : 58	2 : 59 : 19	2 : 59 : 38	2 : 59 : 52
2 : 58 : 10	2 : 58 : 21	2 : 58 : 42	2 : 58 : 59	2 : 59 : 23	2 : 59 : 38	2 : 59 : 56
2 : 58 : 10	2 : 58 : 21	2 : 58 : 45	2 : 58 : 59	2 : 59 : 23	2 : 59 : 38	2 : 59 : 57
2 : 58 : 11	2 : 58 : 25	2 : 58 : 50	2 : 59 : 01	2 : 59 : 24	2 : 59 : 41	2 : 59 : 58
2 : 58 : 11	2 : 58 : 27	2 : 58 : 51	2 : 59 : 02	2 : 59 : 30	2 : 59 : 42	2 : 59 : 59

2 : 58 : 0		1 1 2 7 8 8 9
	1	0 0 1 1 3 3 5 6 7 8
	2	0 1 1 5 7 7 9
	3	1 6 6
	4	1 2 2 5
	5	0 1 2 5 6 6 6 7 8 9 9
2 : 59 : 0		1 2 4
	1	2 3 4 7 9 9
	2	3 3 4
	3	0 1 4 5 6 7 7 8 8 8
	4	1 2 4 7 9
	5	0 1 1 2 6 7 8 9

Figure 2.18

Although a reader can get some picture of the events during these two minutes from the list above, a stem-and-leaf diagram based on ten-second intervals is easier to read: see Fig. 2.18.

The stem-and-leaf diagram gives a much clearer picture of the events at the finish line. Runners finish in clusters, with very few finishing in some ten-second intervals and many finishing in others. There are large numbers of ties and near-ties, and even a few threeway ties. Such conclusions can be drawn from a careful study of the original time list, but they are obtained more easily from a simple stem-and-leaf diagram. ▶

Draw a stem-and-leaf diagram for the birth-weight data of Example 7 (page 19). Use one-pound intervals. After you construct the table, shade the leaves to obtain a bar graph.

Exercise **E**

Construct a stem-and-leaf diagram for each data set in Exercises 1 through 5.

1 Below are the average maximum temperatures for 34 selected cities for the period 1941–1970. (They are rounded to the nearest degree.)

Mobile	77	Omaha	63	Washington, D.C.	67	Oklahoma City	71
Juneau	47	Atlantic City	64	Miami	83	Pittsburgh	60
Phoenix	85	Albuquerque	70	Honolulu	83	Providence	59
Los Angeles	69	Buffalo	55	Boise	63	Sioux Falls	57
Denver	64	Charlotte	71	Chicago	59	Nashville	70
Hartford	60	Cleveland	59	Des Moines	58	Houston	80
				Wichita	68	Salt Lake City	64
				Baltimore	65	Burlington	54
				Boston	59	Norfolk	68
				Detroit	58	Seattle	59
				St. Louis	66	Milwaukee	55

Source: *Statistical Abstract of the United States*

2 As of the morning of August 21, 1978, the hitters of the Denver Bears and the Wichita Aeros of the American Association had hit the following numbers of home runs during the 1978 season:

9	16	12	3
1	8	2	7
11	17	8	6
13	1	1	23
4	10	23	7
17	11	13	

3 Thirty 18-year-old males in a jogging club were weighed for a health study. Their weights were:

141	135	131	142	144	129
140	124	126	158	158	155
154	156	143	143	127	140
147	148	142	128	135	140
136	153	120	132	152	136

4 The yield in bushels per acre of corn sold for grain in 1977 for each of the corn-producing states is listed below.

29	35	102	72	36	104	36
116	24	88	55	76	51	59
116	90	96	85	99	105	65
56	105	90	100	80	92	74

5 Electric energy sales (in billions of kilowatthours) for each state in 1976 were as follows:

7.1	45.0	27.9	5.3	74.0	24.5	22.1
5.3	92.1	21.1	36.0	33.0	136.5	8.3
3.2	112.1	34.0	42.7	72.7	9.9	7.8
31.2	56.3	4.0	18.1	44.3	13.2	61.0
4.8	88.1	4.3	55.0	20.0	4.9	35.0
19.5	69.7	12.1	32.3	20.6	16.7	156.0
98.4	32.5	18.1	44.3	39.8	7.2	2.2
						5.6

6 a) Construct both a frequency histogram and a stem-and-leaf diagram for the age data in Exercise 2, Sec. 2.2.

43, 40, 17	36, 31, 14	50, 46	49, 43, 17, 16, 14
27, 3, 5	56	47, 42, 16, 15	39, 39
32, 28, 6, 5	27, 24, 1	29, 27	76, 44, 39, 17, 15
66, 62	35, 33, 11, 8	39, 35, 13, 10	43, 38, 18
29, 28, 28	29, 11, 10	32, 30, 5, 3	53, 24

   b) Which of the two do you find more useful? Give reasons for your answer.

7 a) If you used the methods given in our examples, your stem-and-leaf diagram in Exercise 2 will have only three rows. Can you modify these methods to construct a stem-and-leaf diagram for the same data with more rows?

   b) Which is more informative — the old stem-and-leaf diagram or the new one?

2.6   Computer packages*

Powerful computers are ideal tools for performing tedious statistical calculations and sorting tasks. It is rarely necessary to write your own computer programs for statistical work; there are programs already written to do most routine statistical tasks. These programs are stored in computers, and can be called up for use with a few commands to the computer. The most commonly used programs for statistical work are taken from standard *computer packages*—collections of statistical computer programs that are written by some organization or individual and sold to computer centers. These packages are known by acronyms such as **BMD, MINITAB**, and **SPSS**. In this book we will give examples of printout from one of the most widely used packages, the SPSS (Statistical Package for the Social Sciences) Batch System.†

In order to use SPSS, a person must learn to write the ten or twenty computer commands that get the program ready to run. It is also necessary to learn how to get data into the computer. We will not cover those procedures in this book. (Most

---

* This section is optional.

† SPSS is a registered trademark of SPSS Inc., for its proprietary computer software. No material describing such software may be produced or distributed without the written permission of SPSS Inc.

computer centers have consultants who will show you how to do this if you need to.) Our objective here will be simply to show you the results of SPSS programs, so that you will know that they are available for use, and will know how to read them.

In this chapter we showed you how to group raw data and display the results in a histogram. In Example 9 we did this for 20 grades on a statistics test. Below are computer printouts showing how the same work is done by SPSS.

Printout 2.1

GRADE

CATEGORY LABEL	CODE	ABSOLUTE FREQ	RELATIVE FREQ (PCT)	ADJUSTED FREQ (PCT)	CUM FREQ (PCT)
30–39	35.	2	10.0	10.0	10.0
60–69	65.	3	15.0	15.0	25.0
70–79	75.	3	15.0	15.0	40.0
80–89	85.	8	40.0	40.0	80.0
90–99	95.	3	15.0	15.0	95.0
100–109	105.	1	5.0	5.0	100.0
		------	------	------	
	TOTAL	20	100.0	100.0	

Printout 2.2

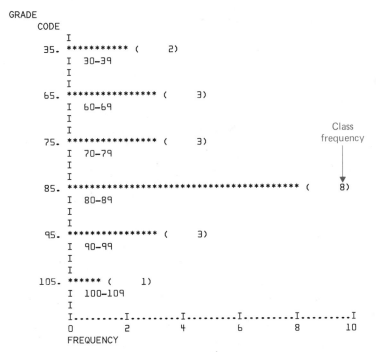

```
   GRADE
     CODE
       I
  35. ********** (        2)
       I  30-39
       I
       I
  65. *************** (        3)
       I  60-69
       I
       I
  75. *************** (        3)
       I  70-79
       I
       I
  85. ****************************************** (        8)
       I  80-89
       I
       I
  95. *************** (        3)
       I  90-99
       I
       I
 105. ****** (        1)
       I  100-109
       I
       I.........I.........I.........I.........I.........I
       0        2        4        6        8        10
       FREQUENCY
```

Class frequency

The circled columns in the table contain the table entries that we calculated in this chapter. There are also columns whose calculations were not covered. The column marked "CODE" gives an identifying code to each class. [The code number here is just the class mark rounded up.] The column marked "ADJUSTED FREQ" is available to adjust relative frequencies in case of missing data values. Here no adjustment is necessary, because none of our 20 grades was missing. The column marked "CUM FREQ" gives the cumulative relative frequency of all grades in a category or below it. For example, 40% of the grades were in the class 70–79 *or* lower classes.

The "HISTOGRAM" is actually a bar graph printed sideways; it is easier to program a computer to do this than to draw the rectangles of a histogram. The bars in the bar graph are made by printing stars across the page; and a scale for reading the bars is printed in dots below the graph. Each category (class) is identified both by printing the code number on the left of its bar, and by printing the class limits below that bar. The number of entries in each class (i.e., the class frequency) is also printed at the end of the bar.

We analyzed the grade data here simply because you are familiar with it. For only 20 pieces of data, sorting by hand is simpler than machine sorting; in this case, it takes less time to sort the data and draw a histogram by hand than it does to walk to the computer center. The real purpose of SPSS and other computer packages is to enable you to handle larger data sets and very time-consuming calculations. If you wished to analyze 2000 grades instead of 20, you would certainly use a.computer instead of hand-sorting.

CHAPTER REVIEW

Key Terms

data
qualitative data
ordinal data
metric data
count data
discrete data
continuous data
grouped data
classes
frequency (*f*)
relative frequency
frequency distribution
relative-frequency distribution

lower class limit
upper class limit
class width
class mark
frequency histogram
relative-frequency histogram
relative-frequency polygon
cumulative frequency
cumulative relative frequency
stems
leaves
stem-and-leaf diagram

You should be able to

1 Classify data as either *qualitative, ordinal, metric,* or *count data.*

2 Group data into a *frequency* and *relative-frequency distribution.*

3 Construct a *frequency* and *relative-frequency histogram.*

4 Construct a *relative-frequency polygon* and a *cumulative relative-frequency polygon.*

5 Draw a *stem-and-leaf diagram.*

REVIEW TEST

1 In 1977, there were 105.7 million males and 111.1 million females living in the United States. What kind of data is given

a) in the classification of an individual as male or female?

b) in the numbers 105.7 million and 111.1 million?

2 The three highest mountains in North America are:

1	Mount McKinley, Alaska	20,320
2	Mount Logan, Canada	19,850
3	Citlaltepec, Mexico	18,700

What kind of data is given in

a) the leftmost column?   b) the rightmost column?

3 Information concerning the number of people in each of 15 families is contained in the table below. Construct a table for the frequencies and relative frequencies for this data. (Use a *single* value for each class.)

1	1	2	2	3
4	2	1	3	2
2	4	5	6	4

4 City government employees in 20 cities in 1977 earned the average monthly salaries given below:

1396	988	1345	1205	1001
1390	1297	906	1146	1522
1464	1463	1150	1116	1024
1049	984	1390	972	1088

Group this data using a class width of $100, starting at $900. Construct a table using the format of the table below.

Class	Tallies	Frequency	Relative frequency	Class mark

5

Class	Frequency	Relative frequency	Class mark
50–99	2	0.1	74.5
100–149	0	0.0	124.5
150–199	6	0.3	174.5
200–249	8	0.4	224.5
250–299	4	0.2	274.5

Use the data from the above table to:

a) construct a frequency histogram,

b) construct a relative-frequency histogram.

6 Use the data from Problem 5 to

a) construct a relative-frequency polygon,

b) construct a cumulative relative-frequency polygon.

7 Make a stem-and-leaf diagram for the following data (ages of employees in a shop). Use ten-year intervals.

27	47	43	52	47
32	35	31	61	36
55	33	42	53	39
28	24	45	52	41

FURTHER READINGS

BENIGER, J. R., and D. L. ROBYN, "Quantitative graphs in statistics: A brief history," *The American Statistician*, February 1978. This is a short and entertaining article on the history of statistics. It contains historical graphs and an outline of developments in graphics from 3800 B.C. to the present.

HABER, A., and R. RUNYAN, *General Statistics*. Reading, Massachusetts: Addison-Wesley, 1977 (third edition). The early chapters of this text contain a great deal of descriptive statistics presented on an elementary level. Several topics we have omitted in descriptive statistics are covered.

HUFF, D., *How to Lie with Statistics*. New York: W. W. Norton & Company, 1955. A classic book on the misuse of descriptive statistics. This book is entertaining and makes enjoyable light reading.

MOSTELLER, F., *et al.* (Eds.), *Statistics by Example: Exploring Data.* Reading, Massachusetts: Addison-Wesley, 1973. This is a collection of fourteen different applications of descriptive statistics to a wide variety of areas, ranging from the stock market to beer-tasting.

TUKEY, J., *Exploratory Data Analysis*. Reading, Massachusetts: Addison-Wesley, 1977. This book is written by the inventor of the stem-and-leaf diagram. It contains a large number of clever methods for looking at data. The presentations are by example; the approach is not theoretical.

Possibly in CCC Library

ANSWERS TO REINFORCEMENT EXERCISES

A a) For individuals, the categories "single," "married," "widowed," and "divorced" are qualitative data. The numbers 14,098,000; 49,409,000; 1,817,000; and 2,545,000 are count data.

b) The places of the teams are ordinal data.

c) Metric

B

Class	Frequency	Relative frequency	Class mark
80–84	2	0.04	82
85–89	4	0.08	87
90–94	9	0.18	92
95–99	6	0.12	97
100–104	10	0.20	102
105–109	8	0.16	107
110–114	5	0.10	112
115–119	6	0.12	117
	50	1.00	

Lower limits   Upper limits

C

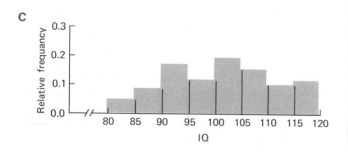

D

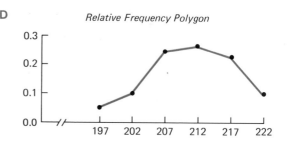

Relative Frequency Polygon

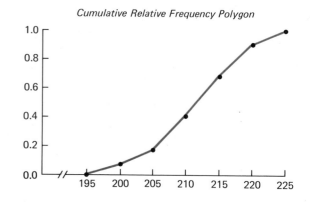

Cumulative Relative Frequency Polygon

E

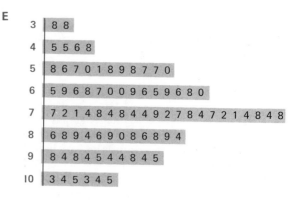

Descriptive measures

Interesting sets of data are often very large; the Boston marathon data mentioned in the last chapter consisted of 906 individual times for 846 men and 60 women. The graphing and grouping methods of the previous chapter are ways of handling large data sets. Another way of summarizing a large data set is to compute a number such as an average; an average replaces a large collection of numbers by a single number. Many college students forget their exact grades in last year's courses, but they do not forget their grade-point averages. In Sec. 3.1 we will use some income data to illustrate three different kinds of "average": the mean, the median, and the mode. We will also show how to obtain an even better description of a data set by computing a number called a "measure of dispersion," in addition to an average.

3.1 Measures of central tendency

The different kinds of averages are often used to describe where the center, or most typical value, of a set of data lies. For this reason. they are called **measures of central tendency**. The three most common measures of central tendency are illustrated using the data in Example 1 below.

Example **1**   One of the authors spent a few summers working for a small mathematical consulting firm. This firm employed a few senior consultants who made from $500 to $700 per week, a few junior consultants who made from $150 to $200 per week, and a number of clerical helpers who earned $100 per week. There was more work in the first half of the summer than in the second, so the staff was larger during the first half.

Typical lists of weekly earnings for the two halves of a summer would look something like Tables 3.1 and 3.2.

Table 3.1 Data set 1	Week ending July 15	Table 3.2 Data set 2	Week ending August 12
	700		700
	600		600
	550		180
	180		160
	180		160
	160		100 } 10 employees
	160 } 13 employees		100
	100		100
	100		100
	100		100
	100		
	100	Total pay	$2300
	100		
Total pay	$3130		

The mean   The most commonly used measure of central tendency is the **mean**. It is the mean that most people refer to when they speak of taking an average. *The mean of each of the data* *sets above is found by simply adding up all the salaries and then dividing this total by the number of employees:*

$$\text{Mean} = \frac{\text{Sum of the data}}{\text{Number of pieces of data}}$$

$$\text{Mean of Data set 1} = \frac{3130}{13} = \$240.77 \quad \text{(to the nearest cent)}$$

$$\text{Mean of Data set 2} = \frac{2300}{10} = \$230.00. \;\blacktriangleright$$

A mathematical consulting firm, similar to the one in Example 1, reported the weekly earnings shown in Tables 3.3 and 3.4. Find the *mean* of each of the data sets given.

Exercise **A**

Data set 3	**Table 3.3**
	Data set 3
Week ending July 15	
750	
620	
525	
175	
175	
175	
150	
100	
100	
100	
100	
100	

Data set 4	**Table 3.4**
	Data set 4
Week ending August 12	
750	
620	
175	
175	
175	
150	
100	
100	
100	

The junior consultants and clerical helpers in Example 1 probably would consider it misleading to say that the average (mean) salary is $240.77 or $230.00. (All but two or three employees were making less than $200 at any time during that summer.) They might prefer to calculate the **median**. The exact determination of the median depends on whether the number of pieces of data is odd or even.

The median

1. *Odd number of pieces of data.* The number of salaries in Data set 1 is 13, an odd number. If the salaries are arranged in order *from lowest to highest*, the seventh salary on the list will be exactly in the middle of the list. This will be the *median*.

100, 100, 100, 100, 100, 100,   160,   160, 180, 180, 550, 600, 700

6 lowest salaries   6 highest salaries

Median = 160

For any data set with an odd number of entries, there will be a number exactly in the middle of its ordered list and this number will be the median.

2. *Even number of pieces of data.* The number of salaries in Data set 2 is 10, an even number. If this data set is listed in order, there is no *single* value that is exactly in the middle of the list. To get a middle value, *average* the highest salary in the bottom 50 percent and the lowest one in the top 50 percent, as below:

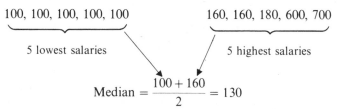

The general idea behind the median is to define a number that is a dividing point between the top 50 percent of the data and the bottom 50 percent of the data.

Exercise B Find the *median* for each of the data sets in Exercise A.

The mode The clerical helpers in Example 1 are probably still upset. The median may be nearer to their salary of $100, but it is still higher than their own $100—and there are more clerical helpers than any other kind of employee. The "average" that the clerical helpers might like to compute is the **mode**. *The mode is the value that appears most often in the data set.* Tables 3.5 and 3.6 show that the mode for each data set is $100.

Table 3.5
Data set 1

Salary	Frequency
700	1
600	1
550	1
180	2
160	2
Mode → 100	6

Table 3.6
Data set 2

Salary	Frequency
700	1
600	1
180	1
160	2
Mode → 100	5

Table 3.7

Salary	Frequency
700	1
600	1
550	1
180	2
Modes → 160	4
→ 100	4

A data set may have more than one mode, if there is a tie for the most frequent value. If two of the clerical workers in Data set 1 were promoted to $160-per-week jobs, the new salary list would be as shown in Table 3.7. This data set has two modes, 160 and 100. More promotions could create even more modes.

Exercise C Find the *mode* for each of the data sets in Exercise A (page 43).

The mean, median, and mode may be different for the same data set. Table 3.8 summarizes the salary "averages" in Example 1.

Comparison

Table 3.8

Term	Explanation	Data set 1	Data set 2
Mean	$\dfrac{\text{Sum of data}}{\text{Number of pieces of data}}$	240.77	230
Median	Dividing line between the top half and bottom half of data	160	130
Mode	Most frequent value	100	100

The mean is much larger than the median in these data sets, because it is affected very strongly by a few large salaries. The mode differs from both the mean and the median. The mode is not really aimed at finding the "middle" of the data set; the most frequent value may not be in the middle at all. A data set can have more than one mode, but it can have only one median and only one mean.

It should be clear that these three different averages generally give different information. There is no simple rule for deciding which "average" to use in any situation. Two different researchers may prefer two different "averages" for the same data set. Example 2 gives a number of different kinds of data, and the kind of average that is commonly used for each.

◖ a) A student takes four exams in a course. His grades are 90, 75, 95, 100. If asked for his average, he would compute the *mean*, which in this case is 90.

b) The number reported in the *Statistical Abstract of the United States* under "average size of families" is a mean. For 1975, this number was 3.42. For 1965 it was 3.70, and for 1955 it was 3.59.

c) Medians are generally used for income reports. The *median* family income for the United States in 1975 was $13,719.

d) The median age is often reported for large groups of people. The *median* age for the United States population in 1975 was 28.8 years.

e) In 1975, the *Current Population Survey* reported that in the United States there were 55,712,000 families. Their sizes were as shown in Table 3.9. The *mode* of these family sizes is two persons.

f) In the 1977 Boston Marathon, there were two categories of official finishers.

Example **2**

Table **3.9**

Family size	Frequency*
2 persons	20,823,000
3 persons	12,137,000
4 persons	11,002,000
5 persons	6,313,000
6 persons	3,005,000
7 or more persons	2,432,000

* Frequencies are given to the nearest thousand.

Sex	Frequency
Male	846
Female	60

The *mode* here is "male." (*Note*: Numbers of female finishers are increasing rapidly. Some physiologists have pointed out that females have special physical characteristics that may make them better suited for endurance events than are males.) ▶

Measuring central tendency for different types of data

One can often decide which average to use for a data set simply by looking at the type of data that is in the set. The marathon data above is count data based on the *qualitative data* "male" and "female." There is no way to compute a mean or a median for these two categories; "(man + woman) ÷ 2" makes no sense at all. The mode is the only "average" that makes sense here. In general, *the mode is the measure of central tendency to use for qualitative data.*

Most statisticians recommend using the median for ordinal data, but some researchers also calculate means for ordinal data sets. The different possibilities for ordinal data are shown in Examples 3 and 4.

Example 3 ◀ A distance runner has entered seven marathons. (Each of these races had at least 300 finishers.) His finishing places in the first six races were 4, 2, 5, 7, 3, and 4. In the seventh race he decided to go all out to win, and ran in first place for 20 miles. This tired him out so badly that he ended up walking parts of the last six miles. He did finish, but only in 72nd place.

This runner's places make up an *ordinal data* set. Arranged in order, the places are:

$$2, 3, 4, 4, 5, 7, 72$$
$$\uparrow$$
$$\text{Median}$$

Although the median is 4, the mean is much larger

$$\text{Mean} = \frac{2+3+4+4+5+7+72}{7} = \frac{97}{7} \doteq 13.9.^*$$

The median gives the best description of this runner's ability. The mean is affected too much by the extreme value of 72. ▶

Example 4 ◀ In the summer of 1977, twenty-one algebra students were asked to rate the change in "test anxiety" produced by their algebra course. Negative ratings meant that they worried more over tests at the end of the course than at the beginning; positive ratings meant that they worried less at the end. Their ratings were:

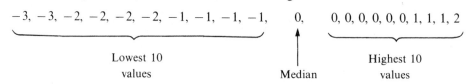

$$-3, -3, -2, -2, -2, -2, -1, -1, -1, -1, \quad 0, \quad 0, 0, 0, 0, 0, 0, 1, 1, 1, 2$$

Lowest 10 values   Median   Highest 10 values

---

* The symbol "$\doteq$" will be used to denote "approximately equals".

The median of these ratings is 0, but it is easy to see that negative results were more common than positive ones. The mean of these ratings is

$$\text{Mean} = \frac{\text{Sum of data}}{21} = -\frac{13}{21} \doteq -0.62.$$

This negative mean describes the overall survey results a bit more accurately than the median does. ▶

One of the main reasons that the median is recommended for ordinal data is that methods of inferential statistics for ordinal data usually rely on the median, rather than the mean. (See Chapter 13.) For this reason, *the median is preferred for ordinal data, although the mean is sometimes used.*

For metric data, either the mean or the median may be used. As with ordinal data, the median is preferred for descriptive purposes if there are a few very large or very small values that might have too much effect on the mean. Much of the statistical theory for metric data uses the mean. For this reason, *the mean is preferred for metric data, although the median is also used for some applications.*

◀ A physical anthropologist studies physical characteristics of different human populations. One such physical characteristic is height. An anthropologist who finds the heights of 100 adult males in a tribe is collecting *metric data*, and in describing the data, could report both the mean and the median for descriptive purposes. However, an anthropologist is more likely to report only the mean, since the mean is the number to be used when making inferences from the data. ▶   Example **5**

Table 3.10 summarizes the relation between data type and average.

Data type	Measure of central tendency		**Table 3.10**
	Preferred	Also used	
Qualitative	Mode		
Ordinal	Median	Mean	
Metric	Mean	Median	

We have defined the mean, median, and mode for a set of data, such as the income data in Example 1. This set of data may represent the entire population of interest, or it may be only a sample of a larger population. For a researcher who is interested in the earnings of the employees in this particular consulting firm, Data set 1 is a *population*, and the mean, median, and mode of Data set 1 are the *population mean*, *population median*, and *population mode*. For another researcher who is interested in the earnings of employees in *all* such consulting firms, Data set 1 is only a *sample*, and the mean,   Population and sample averages median, and mode of Data set 1, are the *sample mean*, *sample median*, and *sample mode*. The diagram below illustrates the two ways in which the mean of a data set may be interpreted.

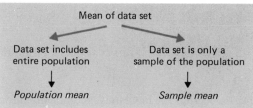

We do not usually have information about an entire population of interest. For example, a lumber company researcher who wants to determine the average height of all trees in a large forest does not have the time or resources to measure all of the trees. It will be necessary to *estimate* the average height by measuring a sample of, say, 200 trees, and calculations will be done with sample data.

Since population data is so rare in practice, *we will concentrate on the descriptive measures for sample data* to prepare you for *inferential statistics*, which uses samples to make inferences about populations.

EXERCISES Section **3.1**

Find the mean, median and mode(s) for each data set given in Exercises 1 through 6. If one of these measures is inappropriate or useless, explain why it is.

1 The hourly temperatures in Colorado Springs, Colorado, for Tuesday, August 22, 1978 are given below. They are given in degrees Fahrenheit, and begin at 1 A.M.

69	61	70	88	85	70
66	60	74	87	73	68
65	61	79	85	74	65
63	64	84	87	71	62

2 In 1974, the percentages of unemployed persons in the labor forces of the Appalachian regions of thirteen states were:

Alabama	5.7	Ohio	5.6
Georgia	5.3	Pennsylvania	5.6
Kentucky	5.4	South Carolina	4.5
Maryland	5.3	Tennessee	5.4
Mississippi	4.7	Virginia	5.3
New York	5.5	West Virginia	5.9
North Carolina	4.7		

3 A middle-class family in the Southwest decided to analyze its check-writing. A sample of 50 checks written in Spring 1978 were for the following amounts (rounded to the nearest dollar).

$12	$88	$9	$105	$11	$19	$29	$475	$139	$42
105	69	27	14	77	78	18	8	86	5
14	100	19	21	10	100	166	120	143	51
120	9	12	121	11	22	55	100	200	59
135	100	175	103	77	100	35	76	5	49

4 The family in Exercise 3 was also interested in its grocery costs. Here are the amounts of the checks written for 26 consecutive weekly supermarket trips (rounded to the nearest dollar).

$ 76	$ 67	$ 74	$ 65	$ 51
71	63	80	81	75
74	49	70	70	75
125	125	86	81	136
78	76	111	101	87
				143

5 The numbers of pages in 20 of the paperback books on my leisure reading shelf are:

294	313	343	238	391
221	191	190	58	255
639	261	191	124	153
576	218	478	245	387

6 Most record albums indicate playing times for songs or sides. The total playing times (for both sides) of 20 of my rock albums are given to the nearest minute at the top of page 49.

42	32	42	30	45
37	34	23	34	34
30	35	34	35	42
49	35	34	53	33

7   For each of the data sets in Exercise 1 through 6, describe

 a) a situation in which the data would be a sample from a population of interest, and

 b) a situation in which the data set would be regarded as *the* population of interest.

8   In some data sets, there are values called *outliers* that probably should not be included in the data set. Suppose, for example, that you are interested in the ability of high-school algebra students to find square roots. You decide to give a test on square roots to ten algebra students to test their abilities. Unfortunately, one of the students had a fight with his girlfriend and can't concentrate. He gets a 0. The ten scores are

0, 58, 61, 63, 67, 69, 70, 71, 78, 80.

↑
Lovesick

The score of 0 is an outlier. Many data sets contain obvious outliers. You may not know why they are there, but you will have a strong feeling that you should throw them out before doing any statistical analysis. Statisticians have a systematic way of avoiding outliers in calculating means. They compute *trimmed means*, in which they "trim off" high and low values for the data given before computing means. For example, to compute the 10% trimmed mean for the data above you would throw out the top 10% and the bottom 10% of the data before computing a mean:

$$10\% \text{ trimmed mean} = \frac{58 + 61 + 63 + 67 + 69 + 70 + 71 + 78}{8}$$

(We have removed 0, which is the bottom 10%, and 80, which is the top 10%.)

Below is a set of algebra final-exam scores for a 40-question test.

2	16	19	21	26
4	16	20	24	27
15	16	21	25	27
15	17	21	25	28

 a) Do any of the scores look like outliers?
 b) Compute the usual mean for the data.
 c) Compute the 5% trimmed mean for the data.
 d) Compute the 10% trimmed mean for the data.
 e) Compare your results. Which mean tells you most about "typical" algebra students?

3.2 Summation notation—the sample mean

In Sec. 3.1, the mean of a set of data was defined by a "word equation":

$$\text{Mean} = \frac{\text{Sum of the data}}{\text{Number of pieces of data}}$$

There are shorthand notations that allow us to write this more simply. We will introduce the mathematical notation for "sum of the data" in Example 6 below.

◀ Example 2(a) discussed the mean of four test grades: 90, 75, 95, and 100. To describe this computation in mathematical shorthand, we can use letters x_i to stand for the numbers in this data set: Example **6**

90	75	95	100
↕	↕	↕	↕
x_1	x_2	x_3	x_4
↓	↓	↓	↓
First number in data set	Second number in data set	Third number in data set	Fourth number in data set

The numbers 1, 2, 3, 4 written below the x's are called *subscripts*. The sum of the data can now be written as:

$$x_1 + x_2 + x_3 + x_4.$$

Summation notation provides a shorthand for $x_1 + x_2 + x_3 + x_4$. It uses the Greek letter $\sum$ (sigma). This letter is a Greek capital "S", and can be thought of as standing for "sum" or "add up the values that follow." One summation notation for $x_1 + x_2 + x_3 + x_4$ is $\sum x$, read "summation x".

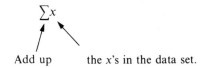

Add up   the x's in the data set.

For our example,

$$\sum x = x_1 + x_2 + x_3 + x_4 = 90 + 75 + 95 + 100 = 360.$$

Since most people add numbers using columns of figures, we often give data in columns with an x above and $\sum x$ below. See the table at the left.

Grade data, x
90
75
95
100
$\sum x = 360$

Example **7**   ◖ In Table 3.11 we give a data set in which the x's are the sales (in thousands of dollars) of the 12 leading retail grocery chains for 1975. Find $\sum x$.

Table 3.11

	Grocery chain sales (\$000's)
Safeway	9,716,889
A & P	6,537,897
Kroger	5,339,225
American	3,207,248
Lucky	3,109,406
Winn Dixie	2,962,165
Jewel	2,817,754
Food Fair	2,482,539
Grand Union	1,611,195
Supermarkets General	1,550,408
National Tea	1,472,341
Fisher Foods	1,379,994

Solution   Adding the numbers in the table we find that $\sum x = 42,187,061$.

(Since the amounts in Table 3.11 are given in thousands of dollars, the actual total dollar sales for these chains is $42,187,061,000—more than 42 billion dollars.) ▶

There is also a notation to save us the trouble of writing the phrase "sample mean." Namely, *the symbol $\bar{x}$ (read "x bar") is used to stand for a sample mean.* If, in addition, we use the letter "*n*" to represent the number of pieces of data, then we can write down the formula for computing the sample mean in a very concise form:

Notation
for the
sample mean

The **sample mean** of *n* pieces of (sample) data is

DEFINITION

$$\bar{x} = \frac{\sum x}{n}$$

◀ A cigarette manufacturer is interested in obtaining information about the tar content of the cigarettes his factory produces, and decides to test a sample of ten cigarettes for tar content. The results are shown in Table 3.12. Find the (sample) mean tar content of these ten cigarettes.

Example **8**

Cigarette	Tar content (mg)
1	10.39
2	10.88
3	10.58
4	10.51
5	10.78
6	11.48
7	10.87
8	10.84
9	10.14
10	12.13
	$\sum x = 108.60$

Table 3.12

Adding the second column of numbers in the above table gives $\sum x = 108.60$. Since there are ten pieces of data ($n = 10$), the sample mean tar content is

Solution

$$\bar{x} = \frac{\sum x}{n} = \frac{108.60}{10} = 10.86 \text{ mg.} \quad ▶$$

◀ A wildlife biologist is interested in the length of red foxes in her area. She has measured the length (to the nearest tenth of an inch) of a sample of twelve red foxes, and compiled the following (sample) data:

Example **9**

37.3	36.8	37.8	37.6
37.4	37.9	36.5	37.7
37.8	37.4	38.3	37.5

Find the sample mean length of these twelve red foxes.

Solution Here $n = 12$. The sum of the data is $\sum x = 450.0$. Consequently, the sample mean length of the 12 foxes is

$$\bar{x} = \frac{\sum x}{n} = \frac{450.0}{12} = 37.5 \text{ in.} \quad \blacktriangleright$$

Exercise **D** A tire manufacturer wants to get some information about the life of a new steel-belted radial he is going to sell. The results of tests on a sample of 16 of these tires are given below (data in miles).

43,725	44,473	39,783	38,686
40,652	43,097	44,652	44,019
37,732	37,396	38,740	40,220
41,868	42,200	39,385	40,742

Find the (sample) mean life of these 16 steel-belted radial tires.

EXERCISES Section **3.2**

1 Suppose $x_1 = 1$, $x_2 = 7$, $x_3 = 4$, $x_4 = 5$, $x_5 = 10$. Compute:

a) $\sum x$ b) $\bar{x}$

2 Suppose $x_1 = 10$, $x_2 = 8$, $x_3 = 9$, $x_4 = 17$, $x_5 = 8$, $x_6 = 9$, $x_7 = 12$. Compute:

a) $\sum x$ b) $\bar{x}$

3 The heights (in inches) of members of two families are given below:

68	62	64	60	40
67	48	60	61	33

a) Compute $\sum x$. b) Find n. c) Compute $\bar{x}$.

For Exercises 4 through 7:

a) Compute $\sum x$. b) Find n. c) Compute $\bar{x}$.
d) Fill in the square of each x value under x^2.
e) Compute $\sum x^2$. (This will be useful later.)

4 The numbers of cars owned by six families on a local street are:

x = number of cars	x^2
1	
2	
3	
2	
1	
4	

5 The numbers of children in the six families in Exercise 4 are:

x = number of children	x^2
2	
3	
4	
4	
0	
3	

6 The amounts of money a salesperson earned in five days were:

x = amount of money	x^2
75	
63	
98	
112	
130	

7 The number of years seven women have been married are:

x = number of years	x^2
11	
3	
8	
4	
6	
14	
1	

Up to this point, we have discussed only measures of central tendency (averages). However, two data sets can have the same mean or the same median and still be quite different. The next example shows two such data sets.

Measures of dispersion—the sample standard deviation

3.3

◖ Below are the heights of the five starting players on two college basketball teams.

Example **10**

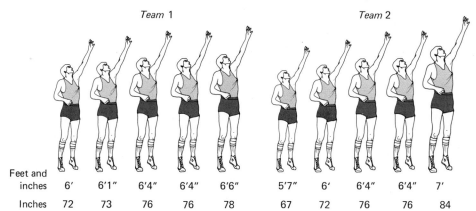

	Team 1					Team 2				
Feet and inches	6′	6′1″	6′4″	6′4″	6′6″	5′7″	6′	6′4″	6′4″	7′
Inches	72	73	76	76	78	67	72	76	76	84

The means for the two groups (calculated using inches) are the same—75″ or 6′3″. The medians for the two groups are also identical—6′4″ for each team. It is clear, however, that the two starting fives are not the same. There is much more variation in the heights of the second group. A **measure of dispersion** is a number used to show how much variation there is in a data set. Just as there are different kinds of averages, there are different measures of dispersion. In this section we will discuss two of the most frequently used measures of dispersion—the *range* and the *standard deviation*. Since we are concerned with samples here, we will think of these teams as samples of all collegiate basketball players.

If we place the smallest player on each team next to the largest, the contrast between the two teams is clear.

The range

Team 1   Team 2

6′   6′6″   5′7″   7′

The range measures variation by looking at the difference between the smallest and largest data value:

DEFINITION | **Range = Largest value − Smallest value**

For the two teams, the ranges are as follows:

Team 1:   Range = 78″ − 72″ = 6″

Team 2:   Range = 84″ − 67″ = 17″ ◗

Exercise **E**   In Example 1 (page 42) we looked at two income data sets (Tables 3.1 and 3.2). We reproduce them here in Table 3.13. Find the *range* of each data set.

Table 3.13	Data set 1	Data set 2
	$700	$700
	600	600
	550	180
	180	160
	180	160
	160	100
	160	100
	100	100
	100	100
	100	100
	100	
	100	
	100	

The range of a data set is very easy to compute. However, the range depends only on the *largest and smallest* values; it says nothing about what happens *between* the largest and smallest values.

The sample standard deviation*

The sample standard deviation, which we shall look at next, uses all of the data values, and fits in very nicely with the theory of statistics. For these reasons, the standard deviation is far more widely used than the range as a measure of dispersion. The standard deviation is much harder to compute than the range, but this is not a serious problem with computers and sophisticated calculators available to do the work of calculation. In what follows we will go through the calculation of the (sample) standard deviation for the height data set of Team 1 in Example 10.

---

* As we said before, most data sets we work with will be *sample* data. Consequently, all data considered in this section will be assumed to be sample data. We will discuss population data and the *population* standard deviation in Sec. 3.7.

◀ The first step in defining the standard deviation is to ask the question "Deviation   Example **11**
from what?" The answer is "Deviation from the mean."

Step 1 Find the mean, $\bar{x}$, and calculate the difference between each data value and the
mean. This difference is called a *deviation from the mean.*

For Team 1, the mean is $\bar{x} = 75''$ (from Example 10). The deviations are shown in
column 2 of Table 3.14.

Height x	Deviation from mean $x - \bar{x}$	Squared deviation $(x - \bar{x})^2$
72	-3	9
73	-2	4
76	1	1
76	1	1
78	3	9
	0	24

Table 3.14

Note that some of the differences are negative, since $\bar{x}$ is larger than the *x*-value in
some cases. Although the numbers $x - \bar{x}$ express deviations from the mean, adding
them up to get a total deviation from the mean is of no value because the numbers $x - \bar{x}$
always add up to zero. In the calculation of the standard deviation, the $x - \bar{x}$ values are
squared, to obtain positive numbers that do not add up to zero.

Step 2 Square the $x - \bar{x}$ values, and add them. The sum is called the *sum of squared
deviations.*

We perform this calculation and get the sum of squared deviations in column 3 of
Table 3.14:

$$\text{Sum of squared deviations} = \sum (x - \bar{x})^2 = 24$$

Step 3 Divide the sum of squared deviations by $n - 1$ (that is, one less than the number
of pieces of data). The number obtained is called the **sample variance**.*

$$\text{Sample variance} = \frac{\sum (x - \bar{x})^2}{n - 1} = \frac{24}{5 - 1} = 6$$

Since the sample variance involves squared unit values (in this case, in²), we finish
the calculation by taking square roots.

---

* If, instead of dividing by $n - 1$, we divided by n, then the sample variance would just be the mean (average)
of the squared deviations. Dividing by n (instead of $n - 1$) seems more natural and, in fact, this used to be the
way the sample variance was defined. However, statisticians now usually divide by $n - 1$. Basically, the reason
for this is as follows: One of the main uses of the sample variance is to estimate the population variance (we
will discuss this later). Division by n tends to underestimate the population variance, whereas dividing by
$n - 1$ yields, on the average, a better estimate of the population variance.

Step 4 Find the square root of the sample variance. This square root is the **sample standard deviation**, and is denoted by the letter "*s*".

$$s = \text{Sample standard deviation} = \sqrt{\frac{\sum(x - \bar{x})^2}{n - 1}} = \sqrt{6} \doteq 2.4'' \blacktriangleright$$

In summary, the sample variance is the sum of squared deviations ($\sum(x - \bar{x})^2$) divided by one less than the number of pieces of data ($n - 1$); and the sample standard deviation is the square root of the sample variance.

DEFINITION

The **sample standard deviation** of n pieces of (sample) data is

$$s = \sqrt{\frac{\sum(x - \bar{x})^2}{n - 1}}$$

Example **12**   ◖ Compute the sample standard deviation for Team 2 of Example 10.

$$\bar{x} = \frac{\sum x}{n} = \frac{67 + 72 + 76 + 76 + 84}{5} = 75$$

x	$x - \bar{x}$	$(x - \bar{x})^2$
67	-8	64
72	-3	9
76	1	1
76	1	1
84	9	81

$$\sum(x - \bar{x})^2 = 156$$

$$s = \sqrt{\frac{\sum(x - \bar{x})^2}{n - 1}} = \sqrt{\frac{156}{5 - 1}} = \sqrt{39} \doteq 6.2''$$

$$s \doteq 6.2'' \blacktriangleright$$

Note. From Examples 11 and 12 we see that Team 1, which has less variation in height than Team 2, also has a smaller standard deviation (2.4″ for Team 1 compared with 6.2″ for Team 2). This is the way a measure of dispersion ought to work: The less variation there is, the smaller the measure of dispersion should be.

◀ In Example 1 we considered two data sets of earnings. The second data set is repeated below.   Example **13**

Data set 2

$700
600
180
160
160
100
100
100
100
100

$n = 10$

$\sum x = \$2300$

Find the sample standard deviation of this data.

Here $n = 10$, and $\sum x = 2300$. So   Solution

$$\bar{x} = \frac{\sum x}{n} = \frac{2300}{10} = 230$$

To find s we proceed as before.

x	$x - \bar{x}$	$(x - \bar{x})^2$
700	470	220,900
600	370	136,900
180	-50	2,500
160	-70	4,900
160	-70	4,900
100	-130	16,900
100	-130	16,900
100	-130	16,900
100	-130	16,900
100	-130	16,900
	$\sum(x - \bar{x})^2 = 454,600$	

$$s = \sqrt{\frac{\sum(x - \bar{x})^2}{n-1}} = \sqrt{\frac{454,600}{10-1}} = \sqrt{50,511.11} \doteq \$224.75 \ \blacktriangleright$$

Exercise **F** A telephone company is interested in obtaining information concerning the duration of telephone conversations. A sample of eight telephone conversations was monitored for duration. The results (to the nearest minute) are given in Table 3.15.

Table 3.15	Phone conversation	Duration (minutes)
	1	1
	2	3
	3	6
	4	15
	5	8
	6	1
	7	4
	8	2

a) Compute $\bar{x}$.

b) Compute s.

The calculation in Example 13 is time-consuming. It would be even worse if $\bar{x}$ ($= 230$) weren't a whole number. There is a shortcut formula for computing the sample standard deviation. It is given below:

FORMULA **Shortcut formula for sample standard deviation:**

$$s = \sqrt{\frac{n(\sum x^2) - (\sum x)^2}{n(n-1)}}$$

We should comment on the similar-looking expressions $\sum x^2$ and $(\sum x)^2$. Here $\sum x^2$ tells you to first square each x, and then add up those squared values, while $(\sum x)^2$ tells you to first add up the x values, and then square the result.

Example **14** ◖ Use the shortcut formula to find the sample standard deviation of the earnings data in Example 13.

Solution Here $n = 10$. To use the shortcut formula, we need to compute $\sum x^2$ and $\sum x$. We do this in Table 3.16. (See the top of page 59.)

Since $\sum x = 2300$, we have $(\sum x)^2 = (2300)^2 = 5{,}290{,}000$. Also, from the table we see that $\sum x^2 = 983{,}600$. So, using the shortcut formula, we get

$$s = \sqrt{\frac{n(\sum x^2) - (\sum x)^2}{n(n-1)}} = \sqrt{\frac{10(983{,}600) - 5{,}290{,}000}{10(9)}}$$

$$= \sqrt{\frac{4{,}546{,}000}{90}} = \sqrt{50{,}511.11} \doteq \$224.75$$

x	x^2	**Table 3.16**
700	490,000	
600	360,000	
180	32,400	
160	25,600	
160	25,600	
100	10,000	
100	10,000	
100	10,000	
100	10,000	
100	10,000	
$\sum x = 2300$	$\sum x^2 = 983,600$	

The answer given by the shortcut formula is, of course, exactly the same as the answer given by the long formula. The shortcut formula is almost always preferred for calculating the sample standard deviation by hand. ▶

Use the shortcut formula to find s for the telephone data of Exercise F, page 58.   Exercise **G**

EXERCISES Section **3.3**

1 Below are eleven IQ scores:

110	132	101
97	91	111
122	107	142
115	125	

a) Get a watch or timer you can use to time your calculations.
b) Calculate the sample standard deviation for the eleven IQ scores using the formula

$$s = \sqrt{\frac{\sum(x - \bar{x})^2}{n - 1}}$$

and record the time your calculation required.
c) Calculate the sample standard deviation for the eleven IQ scores using the formula

$$s = \sqrt{\frac{n(\sum x^2) - (\sum x)^2}{n(n - 1)}}$$

and record the time your calculation required.
d) Was the shortcut formula really a time-saver? (To answer this question, you might want to compare results with others in your class.)

In Exercises 2 through 5, find
a) the range,   b) the sample variance,
c) the sample standard deviation.

2 The hourly temperatures from Exercise 1 of Sec. 3.1.

69	61	70	88	85	70
66	60	74	87	73	68
65	61	79	85	74	65
63	64	84	87	71	62

3 The grocery-check amounts from Exercise 4 of Sec. 3.1.

76	78	125	70	81	51	87
71	67	76	86	70	75	143
74	63	74	111	81	75	
125	49	80	65	101	136	

4 The record-album playing times for 20 rock albums given in Exercise 6 of Sec. 3.1.

42	32	42	30	45
37	34	23	34	34
30	35	34	35	42
49	35	34	53	33

5 Each of the following four data sets.

Data set 1	Data set 2	Data set 3	Data set 4
1	1	5	10
1	1	5	10
2	1	5	4
2	1	5	4
5	1	5	4
5	9	5	4
8	9	5	4
8	9	5	4
9	9	5	4
9	9	5	2

6 Another measure of dispersion that is sometimes given is the *mean absolute deviation*. The formula for this measure of dispersion is

$$\frac{\sum |x - \bar{x}|}{n}$$

Compute the mean absolute deviation for:

a) the IQ scores in Exercise 1 of this section,
b) the data sets in Exercise 5 of this section.

7 a) In Exercise 8 of Sec. 3.1 we discussed extreme values called *outliers*. Investigate the effect of extreme values on the standard deviation by computing the standard deviation for each of the following sets of scores on a 25-point quiz.

Data set 1			Data set 2	
0	14	16	10	15
0	14	17	12	15
10	15	23	14	15
12	15	24	14	16
14	15		14	17

b) Compute the range for each data set in part (a).
c) Compute the mean absolute deviation for each data set in (a).

8 (This requires some algebra.) Derive the identity

$$\frac{1}{n-1} \sum (x - \bar{x})^2 = \frac{n \sum x^2 - (\sum x)^2}{n(n-1)}$$

This will show that the defining and shortcut formulas for the sample variance s^2 are equivalent.

3.4 Interpretation of the standard deviation — Chebychev's theorem

The (sample) standard deviation was introduced to measure how much variation there is in a data set. In Examples 11 and 12 (pages 55 and 56) we *illustrated* that the standard deviation has an important property that any measure of dispersion should have: *The more variation there is in a data set, the bigger the standard deviation.*

Up to this point, we have been concentrating on the calculation of the standard deviation, rather than its meaning. In this section we will present some material that should give you a better feeling as to how the standard deviation measures the variation in a data set. To gain a real understanding of the use of the standard deviation in statistics, it is necessary to read through some of the later chapters. However, we can illustrate the way in which the standard deviation measures variation by looking at two data sets with different levels of variation. This is done in Example 15.

Example **15** ◀ Table 3.17 gives two sets of exam scores. It is clear, by inspection, that there is much more variation in the second data set than in the first.

Table 3.17	Set 1		Set 2	
	65	84	30	93
	78	85	59	95
	81	86	79	97
	82	91	87	100
	84	94	90	100

Using our formulas

$$\bar{x} = \frac{\sum x}{n}, \qquad s = \sqrt{\frac{n(\sum x^2) - (\sum x)^2}{n(n-1)}},$$

we compute the sample mean and sample standard deviation of each data set. We get

Set 1	Set 2
$\bar{x} = 83$	$\bar{x} = 83$
$s \doteq 7.85$	$s \doteq 22.32$

So, as we would expect, the standard deviation of the second data set is much larger than the first.

To enable you to look at these variations in a picture, we have drawn some graphs. On each graph we have marked the data values (with crosses) and the sample mean, and measured off intervals equal in size to the standard deviation. This should help you to "see" the standard deviation.

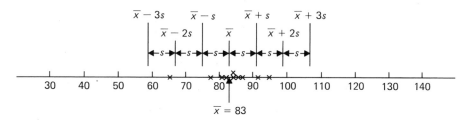

Figure 3.1
Set 1: $\bar{x} = 83$, $s = 7.85$.

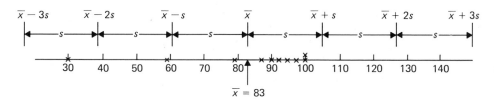

Figure 3.2
Set 2: $\bar{x} = 83$, $s = 22.32$.

The graphs in Figs. 3.1 and 3.2 illustrate vividly that the grades from Set 2 are much more spread out than those from Set 1. In addition, the graphs show that, for each data set, all the data lie within a range of a few standard deviations to either side of the mean. This is no accident: *Almost all of the data in any data set will lie within three standard deviations measured off on either side of the mean.* Data sets with a great deal of variation will have large standard deviations, and their range of a few standard deviations will be quite extensive, as in Fig. 3.2. Data sets with less variation will have smaller standard deviations, and their range of a few standard deviations will be smaller, as in Fig. 3.1. ▶

In Exercise F, a telephone company interested in the lengths of telephone conversations monitored the duration of eight sample telephone conversations. The resulting data (to the nearest minute) are given in Table 3.18. Draw a graph for these data similar to Figs. 3.1 and 3.2.

Exercise **H**

Table 3.18	Phone Conversation	Duration (minutes)	Phone Conversation	Duration (minutes)
	1	1	5	8
	2	3	6	1
	3	6	7	4
	4	15	8	2

Chebychev's theorem*

We mentioned in the previous example that in *any* data set, *almost all of the data will lie within three standard deviations to either side of the mean.* Chebychev's Theorem puts this last statement in a more precise form.

> The portion of data that lies within k standard deviations to either side of the mean is at least $1 - (1/k^2)$.

We emphasize that this is true for *any* data set.

Let's examine some special cases of Chebychev's Theorem. For example, if $k = 2$ then $1 - (1/k^2) = 1 - \frac{1}{4} = \frac{3}{4}$, and so we conclude: The portion of data that lies within two standard deviations to either side of the mean is at least $\frac{3}{4}$. In other words,

For any data set, *at least 75 per cent* of the data lies within two standard deviations to either side of the mean.

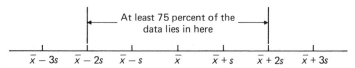

For $k = 3$, $1 - (1/k^2) = 1 - \frac{1}{9} = \frac{8}{9} \doteq 89\%$; and so, in this case, Chebychev's Theorem tells us the following:

For any data set, *at least 89 per cent* of the data lies within three standard deviations to either side of the mean.

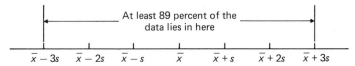

Exercise I

What does Chebychev's Theorem say about the percentage of data that lies within four standard deviations to either side of the mean? Draw a picture similar to the one above.

Example 16

◀ A statistician working for the Bureau of the Census is interested in obtaining information concerning the birth weights of babies in a certain city. A sample of 80

* The remainder of this section is optional.

babies had a mean weight of 7.48 lbs with a standard deviation of 1.70 lbs.

Although we don't have the data here, we can still make some observations about the birth weights of the 80 babies. We have

$$\bar{x} = 7.48 \text{ lbs}, \qquad s = 1.70 \text{ lbs}$$

and so we get Fig. 3.3. (*Note*, for example, that $\bar{x} - 3s = 7.48 - 3 \cdot (1.70) = 7.48 - 5.10 = 2.38$.)

By Chebychev's Theorem (with $k = 2$), we know that *at least* 75 % of the 80 babies (i.e., at least 60 of the babies) had birth weights within two standard deviations of the mean. From the picture we can conclude that *at least* 60 of the babies had birth weights between 4.08 and 10.88 lbs. ▶

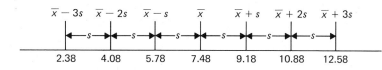

Figure 3.3

---

a) In Example 16, use Chebychev's Theorem with $k = 3$ to make some comments concerning the birth weights of the 80 babies in the statistician's sample.

b) Suppose, in Example 16, that $s = 1.12$ (instead of 1.70). Draw a picture similar to Fig. 3.3, and use Chebychev's Theorem to make some comments on the data, as you did in part (a).

Exercise **J**

EXERCISES Section **3.4**

For each data set in Exercises 1 through 6 you will be given $\bar{x}$ and s. Draw a graph for each data set similar to the graphs in Figs. 3.1 and 3.2 of this section.

1 The record-album playing times

42	32	42	30	45
37	34	23	34	34
30	35	34	35	42
49	35	34	53	33

$\bar{x} = 36.7; \qquad s = 7$

2 The numbers

1	2	8	9
1	5	8	
2	5	9	

$\bar{x} = 5; \qquad s = 3.3$

3 The numbers

1	1	9	9
1	1	9	
1	9	9	

$\bar{x} = 5; \qquad s = 4.2$

4 The hourly temperatures in Colorado Springs on August 22, 1978, from Exercise 1, Sec. 3.1.

69	61	70	88	85	70
66	60	74	87	73	68
65	61	79	85	74	65
63	64	84	87	71	62

$\bar{x} = 72.1; \qquad s = 9.4$

5 The I.Q. scores from Exercise 1, Sec. 3.3.

110	122	132	107	101	142
97	115	91	125	111	

$\bar{x} = 113.9; \qquad s = 15.3$

6 The numbers of pages in paperbacks, from Exercise 5, Sec. 3.1.

294	313	343	238	391
221	191	190	58	255
639	261	191	124	153
576	218	478	245	387

$\bar{x} = 288.3; \qquad s = 146.6$

<u>7</u> a) Use Chebychev's theorem with $k = 3$ to make some comments about the record-playing times in Exercise 1.

b) Use your picture from Exercise 1 to interpret your comments in (a). (Do they make sense?)

<u>8</u> a) Use Chebychev's theorem with $k = 2$ to make some comments about the numbers in Exercise 3.

b) What percentage of your data actually lies within two standard deviations on either side of the mean?

9 Using the results of Exercise 8, interpret the statement "Chebychev's theorem gives a *conservative* estimate of the percentage of data that lies within k standard deviations of its mean."

3.5 Computing $\bar{x}$ and s for grouped data

In Secs. 3.2 and 3.3 we showed how to find the sample mean, $\bar{x}$, and sample standard deviation, s, of a data set. Frequently the data we encounter will be grouped in a frequency distribution table, especially if there is a large amount of data. In this section we shall show how to compute $\bar{x}$ and s when the data is given in a frequency distribution table.

Example **17** ◖ Consider again the salary data from Example 1 (page 42). In Table 3.19 we have repeated the data from Data set 2. Its frequency distribution appears in Table 3.20.

Table 3.19
Data set 2

Data set 2

$700
600
180
160
160
100
100
100
100
100

$2300

Table 3.20
Frequency distribution table

Salary	Frequency
700	1
600	1
180	1
160	2
100	5
	10

The (sample) mean of this data set is

$$\bar{x} = \frac{\text{Sum of salaries}}{\text{Number of salaries}} = \frac{2300}{10} = \$230$$

In finding the sample mean we had to find the sum of all the salaries in Data set 2. We will now show how we can use the frequency distribution table to find this sum. Writing out the sum, we get

$$\overset{1}{\overbrace{700}} + \overset{1}{\overbrace{600}} + \overset{1}{\overbrace{180}} + \overset{2}{\overbrace{160 + 160}} + \overset{5}{\overbrace{100 + 100 + 100 + 100 + 100}}$$

Sum of salaries =

The frequency of each of the *different* salaries is written above each group.

Consequently, we can rewrite the sum as follows:

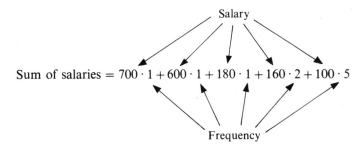

Sum of salaries $= 700 \cdot 1 + 600 \cdot 1 + 180 \cdot 1 + 160 \cdot 2 + 100 \cdot 5$

In other words, we can find the sum of the data by multiplying each of the *different* salaries by its corresponding *frequency*, and then adding the results. Using the frequency distribution table, we can do this quickly and easily, as illustrated in Table 3.21.

Table 3.21

Salary x	Frequency f	Salary $\cdot$ Frequency xf
700	1	700
600	1	600
180	1	180
160	2	320
100	5	500
$n = \sum f = 10$		$\sum xf = 2300$

In this table we have introduced some mathematical notation: x stands for the data value "salary" that defines the class; f stands for the class frequency; and xf stands for the salary times its frequency. Note that xf is just the total of the salaries in each class.

Using this notation we can write a simple mathematical formula for finding the mean of the data when it is in grouped form. To see how to do this first recall that

$$\bar{x} = \frac{\text{Sum of salaries}}{\text{Number of salaries}}$$

As we saw above, we can find the "sum of salaries" by adding the third column of Table 3.21:

$$\text{Sum of salaries} = \sum xf$$

The number of salaries, which is 10, can be found by adding the frequencies, i.e., by adding the second column of Table 3.21:

$$\text{Number of salaries} = \sum f$$

Consequently,

$$\bar{x} = \frac{\text{Sum of salaries}}{\text{Number of salaries}} = \frac{\sum xf}{\sum f}$$

Previously we used the letter n to stand for the number of pieces of data (in this case, the number of salaries). To minimize the amount of notation, we will usually use n instead of $\sum f$. However, you should remember that, if data is given in a frequency distribution table, then to find n, add up the frequencies (that is, $n = \sum f$). ◗

In summary, we compute the mean of grouped data by using the following formula.

FORMULA

The **sample mean of grouped data** is

$$\bar{x} = \frac{\sum xf}{n},$$

where f = frequency, n = number of pieces of data.

Example **18** ◖ The frequency distribution table for the salary Data set 1 from Example 1 is given in Table 3.22. Compute the mean $\bar{x}$, of this data.

Table 3.22 Frequency distribution table for Data set 1	Salary, x	Frequency, f
	700	1
	600	1
	550	1
	180	2
	160	2
	100	6

Solution We append an "xf" column to the table above, and then proceed as in Example 17 (see Table 3.23).

Table 3.23	x	f	xf
	700	1	700
	600	1	600
	550	1	550
	180	2	360
	160	2	320
	100	6	600
		13	3130

$$\bar{x} = \frac{\sum xf}{n} = \frac{3130}{13} \doteq \$240.77 \text{ ◗}$$

In Example 8 of Chapter 2 we looked at some family size data. The frequency distribution table for this data is given as Table 3.24. Find the (sample) mean of this data.

Exercise **K**

Number of children, x	Number of families, f	Table 3.24
0	18	
1	8	
2	7	
3	4	
4	3	

We have just seen how to compute the (sample) mean of grouped data. By using similar reasoning, we can also get a formula for finding the sample standard deviation of grouped data.

The **sample standard deviation of grouped data** is

FORMULAS

$$s = \sqrt{\frac{\sum (x - \bar{x})^2 f}{n - 1}},$$

and the **shortcut formula** is

$$s = \sqrt{\frac{n(\sum x^2 f) - (\sum xf)^2}{n(n - 1)}},$$

where f = frequency, n = number of pieces of data.

◖ A team of medical researchers has developed a new exercise program that they feel will be helpful in reducing hypertension (high blood pressure). The team selected a sample of 35 hypertensive individuals, and subjected them to the exercise program. Table 3.25 shows the (diastolic) blood pressures in grouped form, of the 35 individuals *before* the exercise program. Find the sample standard deviation of this grouped data.

Example **19**

Blood pressure (diastolic)	Number of individuals	Table 3.25
91	1	
92	1	
93	6	
94	5	
95	9	
96	3	
97	6	
98	4	

Solution We use the formula

$$s = \sqrt{\frac{n(\sum x^2 f) - (\sum xf)^2}{n(n-1)}}$$

As you can see from this formula, we will need a table (Table 3.26) with columns for x, f, xf, x^2, and $x^2 f$. We will also need to add the second, third, and fifth columns of this table to find n, $\sum xf$, and $\sum x^2 f$.

Table 3.26

x	f	xf	x^2	$x^2 f$
91	1	91	8281	8281
92	1	92	8464	8464
93	6	558	8649	51894
94	5	470	8836	44180
95	9	855	9025	81225
96	3	288	9216	27648
97	6	582	9409	56454
98	4	392	9604	38416
	35	3328		316562
	↑	↑		↑
	n	$\sum xf$		$\sum x^2 f$

Then

$$s = \sqrt{\frac{n(\sum x^2 f) - (\sum xf)^2}{n(n-1)}} = \sqrt{\frac{35 \cdot (316{,}562) - (3{,}328)^2}{35 \cdot 34}}$$

$$= \sqrt{\frac{11{,}079{,}670 - 11{,}075{,}584}{1190}} = \sqrt{\frac{4086}{1190}}$$

$$\doteq \sqrt{3.43} \doteq 1.85 \quad \blacktriangleright$$

Table 3.27

Number of accidents	Frequency
0	1
1	2
2	9
3	1
4	6
5	3
6	2

Exercise **L** A highway department official is interested in obtaining information about the number of accidents per month on a portion of a busy street. Records from the previous two years on the number of accidents per month yielded the data shown in Table 3.27. Find the sample mean and sample standard deviation of this data.

You might have noticed that all the grouped data we considered in this section was the type in which each class is based on a *single* value. That is, each class represented only one data value. If we are dealing with grouped data, in which each class represents a number of different values (such as the grouped grade data on page 18), there is usually no way to recover the exact data. Consequently, in such cases, it is not possible to calculate the mean and standard deviation exactly. However, there is a way to calculate them *approximately*.

In the formulas

$$\bar{x} = \frac{\sum xf}{n}, \qquad s = \sqrt{\frac{n(\sum x^2 f) - (\sum xf)^2}{n(n-1)}}$$

for the mean and standard deviation, we use the *class marks* for the x-values. We will explore this in the exercises.

EXERCISES Section **3.5**

In Exercises 1 through 4:

  a) Find the sample mean $\bar{x}$ for the given grouped data.
  b) Find the sample standard deviation s.

1 (Course ratings by 100 students in an algebra course)

	Rating	Frequency
Highest ⟶	4	32
	3	44
	2	14
	1	6
Lowest ⟶	0	4

2 (Course ratings by 100 students who had the same course as the students in Exercise 1, but another teacher)

Rating	Frequency
4	14
3	42
2	27
1	10
0	7

3 (Heights of 25 students at a military school)

Height	Frequency	Height	Frequency
65	2	70	6
66	0	71	4
67	2	72	3
68	0	73	4
69	3	74	1

4 (Numbers of cars at 50 homes)

Number of cars	Frequency	Number of cars	Frequency
1	10	4	3
2	25	5	2
3	10		

For grouped data, in which each class represents a number of values, we use the same formulas:

$$\bar{x} = \frac{\sum xf}{n} \quad \text{and} \quad s = \sqrt{\frac{n(\sum x^2 f) - (\sum xf)^2}{n(n-1)}}$$

However, we use as x, the class mark for each class. (This assumes that the class mark is very close to the average of all entries in its class.) These formulas do not give exact values for $\bar{x}$ and s, but usually give close approximations.

5 a) Complete Table 3.28 for the grade data that we looked at in Chapter 2.
  b) Use the value of $\sum xf$ to find $\bar{x}$.
  c) Use the values of $\sum xf$ and $\sum x^2 f$ to find s.

Table 3.28

Grade class	Class mark, x	Frequency, f	xf	$x^2 f$
0– 9	4.5	0	0	0
10–19	14.5	0	0	0
20–29	24.5	0	0	0
30–39	34.5	2	(34.5)2 = 69	(34.5)²2 = 2380.5
40–49	44.5	0	0	0
50–59	54.5	0	0	
60–69		3	193.5	
70–79		3		16650.75
80–89		8	676	
90–99		3		26790.75
100–109		1		
		20	1550	126345
		↑ n	↑ $\sum xf$	↑ $\sum x^2 f$

<u>6</u> Find approximate values for $\bar{x}$ and s for the temperature data:

Temperature class	Frequency
60–69	11
70–79	7
80–89	6

<u>7</u> The formula for the mean of grouped data is given as

$$\bar{x} = \frac{\sum xf}{n}$$

This formula can be rewritten as

$$\bar{x} = \sum x\frac{f}{n}.$$

Since f/n represents the relative frequency of a class, we have

$$\bar{x} = \sum x\frac{f}{n} = \sum (\text{class mark}) \cdot (\text{relative frequency of class}).$$

a) Use this alternative version of the formula for $\bar{x}$ to compute the mean for our family-size data given in the form:

Number of children, x	Relative frequency of families, f/n
0	0.450
1	0.200
2	0.175
3	0.100
4	0.075

b) Compute the approximate mean grade for the grade data set given in the form:

Grade class	Relative frequency, f/n	Grade class	Relative frequency, f/n
30–39	0.10	70–79	0.15
40–49	0.00	80–89	0.40
50–59	0.00	90–99	0.15
60–69	0.15	100–109	0.05

<u>8</u> Below is a set of 20 grades on a 100-point psychology quiz:

70, 70, 70, 70, 70, 70, 70, 70,
80, 80, 80, 80, 80, 80, 80, 80, 80, 80
90, 90

The mean of these grades is $\bar{x} = 77$. If we group them in the traditional fashion, we get

Grade	Frequency
70–79	8
80–89	10
90–99	2

a) If we use the class-mark method explained above to find $\bar{x}$ from this grouped-data table, we get $\bar{x} = 81.5$. Why is this approximation so bad?

b) What would you suggest doing about this data grouping, to make the estimate of $\bar{x}$ more accurate?

3.6   Percentiles; Box-and-whisker diagrams

We have already studied one example of a percentile. The *median* divides a set of data into a top 50 percent and a bottom 50 percent. For this reason, it is referred to as the *fiftieth percentile* for a data set. In addition to percentiles, which can divide the data according to any percentage value, researchers commonly find *quartiles*, which divide data into quarters, and *deciles*, which divide them into tenths. In the next two examples we find quartiles, deciles, and some other percentiles for a few data sets.

Example **20**   To find quartiles, we split our data into quarters, and then look for dividing lines between the quarters. This is done below for the supermarket sales data from Example 7.

First
quarter $\left\{\begin{array}{ll}\text{Fisher Foods} & 1,379,994 \\ \text{National Tea} & 1,472,341 \\ \text{Supermarkets} \\ \quad\text{General} & 1,550,408\end{array}\right.$

First quartile

$$\frac{1,550,408 + 1,611,195}{2} = 1,580,801.5$$

Second
quarter $\left\{\begin{array}{ll}\text{Grand Union} & 1,611,195 \\ \text{Food Fair} & 2,482,539 \\ \text{Jewel} & 2,817,754\end{array}\right.$

Second quartile

$$\frac{2,817,754 + 2,962,165}{2} = 2,889,959.5$$

Third
quarter $\left\{\begin{array}{ll}\text{Winn Dixie} & 2,962,165 \\ \text{Lucky} & 3,109,406 \\ \text{American} & 3,207,248\end{array}\right.$

Third quartile

$$\frac{3,207,248 + 5,339,225}{2} = 4,273,236.5$$

Fourth
quarter $\left\{\begin{array}{ll}\text{Kroger} & 5,339,225 \\ \text{A\&P} & 6,537,897 \\ \text{Safeway} & 9,716,889\end{array}\right.$

In summary then, for the supermarket sales data we have:

First quartile   = 1,580,801.5
Second quartile = 2,889,959.5 = Median
Third quartile   = 4,273,236.5 ▮

◀ Below we have reproduced the grade data from Example 5 of Sec. 2.2.   Example **21**

88	67	64	76	86
85	82	39	75	34
90	63	89	89	84
81	96	100	70	96

To find the quartiles for this data we:

Step 1 Arrange the data in increasing order,

Step 2 Divide the data into quarters, and

Step 3 Find the dividing lines between quarters.

The result of applying steps 1 through 3 is given below.

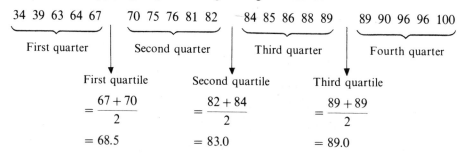

Here the arrangement of grades in order puts one 89 in the third quarter, and one in the fourth quarter. This is artificial. You would actually tell either student with a grade of 89 that he was *tied* for the fifth highest grade, and probably not talk to him in terms of quarters at all. This indicates a problem that commonly comes up with discrete data: It doesn't divide naturally into separate quarters because of ties, or because the number of pieces of data isn't divisible by 4. Many statistics books spend some time giving rules for the definition of quartiles in this situation. These complex rules obscure a simple, common-sense concept: *Quartiles divide the data into four quarters.* If you do not have exact quarters, use common sense, get as close to quarters as possible, and furnish any user of your data with an explanation of what you did.

To illustrate deciles, we group the grade data in tenths, and compute deciles as below.

First
tenth $\begin{cases} 34 \\ 39 \end{cases}$

$\longrightarrow \dfrac{39 + 63}{2} = 51.0 = $ 1st decile, or 10th percentile

Second
tenth $\begin{cases} 63 \\ 64 \end{cases}$

$\longrightarrow \dfrac{64 + 67}{2} = 65.5 = $ 2nd decile, or 20th percentile

Third
tenth $\begin{cases} 67 \\ 70 \end{cases}$

$\longrightarrow \dfrac{70 + 75}{2} = 72.5 = $ 3rd decile, or 30th percentile

Fourth
tenth $\begin{cases} 75 \\ 76 \end{cases}$

$\longrightarrow \dfrac{76 + 81}{2} = 78.5 = $ 4th decile, or 40th percentile

Fifth $\begin{cases} 81 \\ 82 \end{cases}$
tenth

$\longrightarrow \dfrac{82 + 84}{2} = 83.0 = $ 5th decile, or 50th percentile

Sixth $\begin{cases} 84 \\ 85 \end{cases}$
tenth

$\longrightarrow \dfrac{85 + 86}{2} = 85.5 = $ 6th decile, or 60th percentile

Seventh $\begin{cases} 86 \\ 88 \end{cases}$
tenth

$\longrightarrow \dfrac{88 + 89}{2} = 88.5 = $ 7th decile, or 70th percentile

Eighth $\begin{cases} 89 \\ 89 \end{cases}$
tenth

$\longrightarrow \dfrac{89 + 90}{2} = 89.5 = $ 8th decile, or 80th percentile

Ninth $\begin{cases} 90 \\ 96 \end{cases}$
tenth

$\longrightarrow \dfrac{96 + 96}{2} = 96.0 = $ 9th decile or 90th percentile

Tenth $\begin{cases} 96 \\ 100 \end{cases}$
tenth

In practice, quartiles, deciles and percentiles are most useful with *large* data sets. We have used the grade data for illustration only. The grade and class standing or rank would convey just as much information, if not more, to a student in a class of 20. ◗

In Exercise F, page 58, we considered some data on the duration of telephone conversations. The data is repeated as Table 3.29. Find the quartiles for this data.

Exercise **M**

Phone conversation	Duration (minutes)	Phone conversation	Duration (minutes)
1	1	5	8
2	3	6	1
3	6	7	4
4	15	8	2

Table 3.29

Box-and-whisker
diagrams

These diagrams, like stem-and-leaf diagrams, were invented by John Tukey. They provide a nice picture of data variation. Below we list the steps in construction of these diagrams. We will perform each step for the grade data.

Step 1 Find the highest and lowest data values:

High = 100;   Low = 34.

Step 2 Find the median of the data set:

Median = 83.

Step 3 Find the median of the top half of the data set and the median of the bottom half. These medians are called *hinges*.

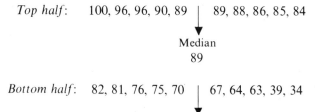

Top half: 100, 96, 96, 90, 89 | 89, 88, 86, 85, 84

Median
89

Bottom half: 82, 81, 76, 75, 70 | 67, 64, 63, 39, 34

Median
68.5

Step 4 Draw a vertical axis on which these values can be marked. Next to this axis, mark the high, low, median and hinges as below. Then connect the hinges to each other to make a box and then to the high or low by lines (called *whiskers*):

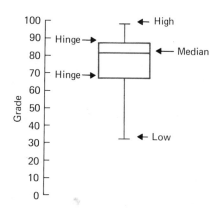

By taking the median of the top and bottom half, we are really coming as close as we can to dividing the data into quarters. Thus the two boxes in a box-and-whisker diagram show us how "spread out" the two middle quarters of the data are. The whiskers show us the spread of the top quarter and bottom quarter. From this box-and-whisker diagram we can see that the two higher quarters of the data are less spread out than the two lower quarters, and that the lowest quarter has the greatest spread of all.

Find the box-and-whisker diagram for the telephone conversation data in Exercise M, page 73.

Exercise **N**

Box-and-whisker diagrams give information that cannot be seen directly from histograms or grouped-data tables. Although they seem quite simple, they are recent inventions. The reader should keep in mind that the various techniques given in books such as this are not the only possible ones. It is quite possible to invent new statistical tools which, like box-and-whisker diagrams, are both simple and clever.

EXERCISES Section **3.6**

1 Divide the Colorado Springs temperature data into quarters and find the quartiles.

69	63	61	74	88	87	74	68
66	61	64	79	87	85	71	65
65	60	70	84	85	73	70	62

2 For the record-playing times below, find the
   a) quartiles   b) deciles

42	32	42	30	45
37	34	23	34	34
30	35	34	35	42
49	35	34	53	33

3 For the number of pages from the paperbacks, find the
   a) quartiles   b) deciles

294	313	343	238	391
221	191	190	58	255
639	261	191	124	153
576	218	478	245	387

4 Divide the following 24 numbers into quarters and find the quartiles.

76	78	125	86	70	75
71	67	76	111	81	75
74	63	74	65	101	136
125	49	80	81	51	87

5 Construct a box-and-whisker diagram for the temperature data in Exercise 1.

6 Construct a box-and-whisker diagram for the record-playing time data in Exercise 2.

7 Construct a box-and-whisker diagram for the number of pages from the paperbacks in Exercise 3.

8 Construct a box-and-whisker diagram for the numbers in Exercise 4.

9 For the record-playing time data of Exercise 2, write down on one page:
   a) The mean $\bar{x}$ and standard deviation s.
   b) The stem-and-leaf *and* box-and-whisker diagrams.
   c) It is usual to describe a data set by giving $\bar{x}$ and s. How would you prefer to be told about this data? Would you prefer
      i) being given $\bar{x}$ and s, or
      ii) being given the stem-and-leaf and box-and-whisker diagrams?

Up to this point, we have concentrated on descriptive measures for sample data (i.e., data obtained from a sample of the population). This is because the data we actually work with in practice is almost always sample data. However, what we really want to do is describe the entire population, and the reason we resort to a sample is because it is usually more practical. The following example illustrates this point.

Use of samples; **3.7**
the population mean
and population
standard deviation

◀ A statistician is interested in obtaining information about the mean birth weight of U.S. babies born in 1979. To obtain the complete population data of all birth weights for that year would be extremely expensive and time-consuming. As we shall see in Chapters 7 and 8, it is also unnecessary, because the statistician can obtain sufficiently

Example **22**

accurate information about the mean birth weight of U.S. babies born in 1979 from the mean birth weight of a *sample* of such babies.

The statistician decides to take a sample of 250 birth weights, and use the data he obtains from the sample to make an inference about the mean birth weight of *all* U.S. babies born in 1979.

The population of interest here is the birth weights of all babies born in the U.S. in 1979. The mean birth weight of all such babies is the *population mean*. The sample, in this case, consists of the birth weights of the 250 babies the statistician selects, and the mean birth weight of these babies is the *sample mean*. The following picture summarizes the situation.

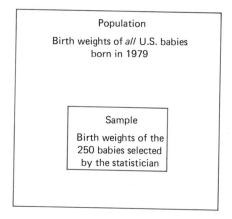

Once the statistician takes his sample, he can compute the sample mean, $\bar{x}$, of the 250 birth weights. Based on this information, he can then make an inference (educated guess) about the population mean of all such birth weights (we will learn how to make these kinds of inferences in Chapter 7).

Just as the symbol $\bar{x}$ is used to stand for the mean of a *sample*, there is also a symbol for the mean of a *population*—the Greek letter μ (pronounced "mew").

$$\mu = \text{population mean}, \qquad \bar{x} = \text{sample mean}$$

In this example, μ is the mean birth weight of all U.S. babies born in 1979, and $\bar{x}$ is the mean birth weight of the statistician's sample of 250 such babies. The statistician wants to use $\bar{x}$ to make an inference about μ. ◗

Recall that for a *sample* of size n (n pieces of data), the sample mean is given by

$$\bar{x} = \frac{\sum x}{n}$$

In some cases, when we know an entire population, we can compute the population mean instead of a sample mean. The mean, μ, of a (finite) *population* is defined in the same way, but in order to emphasize that it is a *population* mean, we will use N (instead of n) to denote the size of the population.

The **population mean** of a (finite) population of size N is $$\mu = \frac{\sum x}{N}$$	DEFINITION

◀ An instructor is teaching a night-school class of 20 students, and is curious about their mean age. These ages are given below.

Example **23**

29	30	25	30	28
27	28	30	26	30
24	22	24	27	26
28	27	26	25	28

The (population) mean age of this night school class is

$$\mu = \frac{\sum x}{N} = \frac{540}{20} = 27 \text{ years.}$$

If the instructor were interested in the mean age of *all* students in similar night-school courses, then the 20 students in this particular class would be considered a sample, instead of a population. The mean age of the 20 students would then be a sample mean, instead of a population mean, and we would write $\bar{x} = 27$, instead of $\mu = 27$. This last point illustrates that whether a data set is considered to be population data or sample data often depends on the *use* to be made of the data. In practice, it will be clear from the context of the problem whether the data is population data or sample data. ▶

In statistical inference problems, the population size (N) is ordinarily quite large, and the population mean (μ) is usually unknown. However, *for the sake of illustration, we will sometimes consider small populations for which we can easily calculate μ.*

The areas of the ten largest states (largest by area) are given in Table 3.30 (to the nearest 1000 square miles). Find the (population) mean area of these ten largest states.

Exercise **O**

Table **3.30**

State	Area (in 1000 sq mi)
Alaska	586
Texas	267
California	159
Montana	147
New Mexico	122
Arizona	114
Nevada	110
Colorado	104
Wyoming	98
Oregon	97

In the next examples we shall illustrate how the sample standard deviation is used.

Example **24**   ◖ A typical problem in *quality control* is the following. A manufacturer needs to produce bolts approximately 10 mm in diameter to fit into a circular hole that will be 10.4 mm in diameter. A salesperson is trying to sell him a machine that makes bolts with a mean diameter of $\mu = 10$ mm. Although the machine produces bolts that, *on the average*, are 10 mm in diameter, the manufacturer still needs to be concerned about the *variability* in diameter of the bolts produced. If too many bolts produced by the machine have diameters much larger or smaller than 10 mm, then too many bolts won't fit correctly into the 10.4 mm hole.

In other words, the manufacturer needs information about the standard deviation of all bolts produced by the machine (the *population* standard deviation), so he must try to estimate it by looking at the *sample* standard deviation from a *sample* of bolts produced by the machine. The manufacturer decides to take a sample of twenty bolts, and use the diameters of these bolts to make an inference about the standard deviation of all such bolts.

We introduced the letter *s* to stand for the standard deviation of a *sample*. The symbol we will use to denote the standard deviation of a *population* is the Greek letter σ (pronounced "sigma").

$$s = \text{sample standard deviation}, \qquad \sigma = \text{population standard deviation}.$$

Just as s^2 is the sample variance, σ^2 *is the population variance.*

In this example, σ is the standard deviation of the diameters of all bolts that will ever be produced by this machine, and *s* is the standard deviation of the manufacturer's sample of 20 bolt diameters. The manufacturer wants to use *s* to make an inference about σ. (We will learn how to make these kinds of inferences in Chapter 9.) ◗

The standard deviation of a *sample* of size *n* was defined to be

$$s = \sqrt{\frac{\sum(x - \bar{x})^2}{n - 1}}.$$

In some cases we will know an entire population of interest, and can compute its standard deviation σ. The formula for this is given below.

DEFINITION   The **population standard deviation** of a (finite) population of size *N* is

$$\sigma = \sqrt{\frac{\sum(x - \mu)^2}{N}}. \, ^*$$

---

* You have probably noted that for the population standard deviation (σ) we divide by the population size (N), whereas for the sample standard deviation (*s*) we divide by *one less* than the sample size ($n - 1$). The reason for this was explained in the footnote on page 55. Namely, in statistical inference, the main purpose of *s* is to estimate σ. Dividing by *n*, instead of $n - 1$, in the formula for *s* tends to underestimate σ, whereas dividing by $n - 1$, results, on the average, in a better estimate.

◀ Find the population standard deviation, σ, of the ages of the twenty night-school students in Example 23.

Example **25**

We will use Table 3.31.

Solution

	$\mu = 27$ (from Example 23)		Table **3.31**
x	$x - \mu$	$(x - \mu)^2$	
29	2	4	
27	0	0	
24	-3	9	
28	1	1	
30	3	9	
28	1	1	
22	-5	25	
27	0	0	
25	-2	4	
30	3	9	
24	-3	9	
26	-1	1	
30	3	9	
26	-1	1	
27	0	0	
25	-2	4	
28	1	1	
30	3	9	
26	-1	1	
28	1	1	
		$\sum (x - \mu)^2 = 98$	

Now we compute σ:

$$\sigma = \sqrt{\frac{\sum (x - \mu)^2}{N}} = \sqrt{\frac{98}{20}} = \sqrt{4.90} \doteq 2.21 \quad ▶$$

Find the population standard deviation, σ, of the areas of the ten largest states given in Exercise O (page 77).

Exercise **P**

There are also formulas for finding the mean and standard deviation of a population when the data is given in *grouped* form. These formulas will be used, especially in Chapter 4, when we discuss the mean and standard deviation of a random variable. The formulas are given at the top of page 80.

FORMULAS

The **population mean** and **standard deviation** of a (finite) population whose data values are given in **grouped form** are

$$\mu = \frac{\sum xf}{N},$$

$$\sigma = \sqrt{\frac{\sum (x - \mu)^2 f}{N}},$$

where f = frequency, N = population size.

Example 26 ◖ The heights of the players on the 1978 Arizona State University basketball team are given in Table 3.32 in *grouped* form. Find the (population) mean and standard deviation of the heights of these players.

Table 3.32

Height (in.), x	Number of players, f	Height (in.), x	Number of players, f
70	1	78	4
73	2	80	1
74	2	82	2
76	2	83	1
77	1		

Solution We use the formulas

$$\mu = \frac{\sum xf}{N}, \qquad \sigma = \sqrt{\frac{\sum (x - \mu)^2 f}{N}}.$$

First compute μ using the first three columns of Table 3.33.

Table 3.33

x	f	xf	$x - \mu$	$(x - \mu)^2$	$(x - \mu)^2 f$
70	1	70	-7	49	49
73	2	146	-4	16	32
74	2	148	-3	9	18
76	2	152	-1	1	2
77	1	77	0	0	0
78	4	312	1	1	4
80	1	80	3	9	9
82	2	164	5	25	50
83	1	83	6	36	36
	16	1232			200
	$\uparrow$	$\uparrow$			$\uparrow$
	N	$\sum xf$			$\sum (x - \mu)^2 f$

To compute μ:

$$\mu = \frac{\sum xf}{N} = \frac{1232}{16} = 77$$

Next we compute σ using the last three columns of Table 3.33.

$$\sigma = \sqrt{\frac{\sum (x - \mu)^2 f}{N}} = \sqrt{\frac{200}{16}} = \sqrt{12.5} \doteq 3.54 \ \blacktriangleright$$

The age data for the twenty night-school students in **Example 23** is given in *grouped* form in Table 3.34. Compute the (population) mean and standard deviation of the ages of these twenty students.

We should like to emphasize again the following fact. In inferential statistics the population mean, μ, and population standard deviation, σ, are rarely known. In fact, two major problems in inferential statistics are:

1. Using the mean, $\bar{x}$, of a sample of a population to make inferences about the population mean μ, and
2. Using the standard deviation, s, of a sample of a population to make inferences about the population standard deviation σ.

Problem 1 will be covered in Chapters 7 and 8, and problem 2 in Chapter 9. In Chapters 4 through 6 we will discuss the mathematical prerequisites for solving problems 1 and 2.

Exercise Q

Table 3.34

Age, x	Number of students, f
22	1
24	2
25	2
26	3
27	3
28	4
29	1
30	4

EXERCISES Section **3.7**

1 In Example 11 of this chapter we looked at the heights of five basketball players. The heights were 72, 73, 76, 76, 78. For these heights we found the sample mean $\bar{x}$ and sample standard deviation s.
   a) Find the population mean μ for these five heights.
   b) Find the population standard deviation σ.
   c) Give examples of situations where these five heights would be
      i) a sample,   ii) a population.

2 In Example 13 of this chapter we looked at a set of earnings:

700	180	160	100	100
600	160	100	100	100

We found the sample mean and sample standard deviation to be $\bar{x} = 230$ and $s = 224.75$.
   a) Find the population mean μ.
   b) Find the population standard deviation σ.

c) Give examples of situations in which these salaries would constitute:
   i) a sample,   ii) a population.

3 In Exercise K, we looked at grouped data on 40 families. The data are given in Table 3.35. Suppose these 40 families are a population of interest to you.

Table 3.35

Number of children (class)	Number of families (frequency)
0	18
1	8
2	7
3	4
4	3

   a) find μ,   b) find σ.

4 Suppose you are given the data of Exercise 2 in grouped form:

Earnings	Frequency
700	1
600	1
180	1
160	2
100	5

Use the grouped-data formulas to find μ and σ.

5 For each of the following data sets, describe situations in which the data could be looked at as:

i) a sample, ii) a population.

a) The heights of twenty children in Evansville, Indiana.

b) The numbers of cars at each of fifteen different homes in Chicago, Illinois.

c) The price of gasoline at fifty gas stations in Denver, Colorado on March 7, 1979.

6 a) Calculate *both* the sample standard deviation s and the population standard deviation σ for each of the following sets of numbers:

Data set 1	Data set 2		Data set 3	
2	7	5	4	3
7	9	6	4	9
4	5	3	7	4
3	8		5	7
			8	5

b) For any finite data set, you can compute a value for s or σ, as you did above. Will these values be closer together for large or for small data sets?

7 Derive a formula that gives the exact relationship between the values of s and σ calculated from the same data.

━━

3.8 Computer packages*

Packages such as SPSS will also calculate means, standard deviations, and many other descriptive measures for you. Below are sections of a printout from an SPSS program entitled CONDESCRIPTIVE. The data that was processed by the computer is once again the grade data given in Example 5 of Sec. 2.2 (page 15).

```
VARIABLE GRADE

MEAN         77.700
VARIANCE    308.747
RANGE        66.000
SUM        1554.000

VALID OBSERVATIONS -    20




STD ERROR     3.929     STD DEV      17.571
KURTOSIS      1.414     SKEWNESS     -1.275
MINIMUM      34.000     MAXIMUM     100.000

        MISSING OBSERVATIONS - 0
```

You should observe a characteristic of computer packages here: They give you many different statistics. You need to pick out the ones you understand: MEAN, VARIANCE, RANGE, SUM, MAXIMUM, MINIMUM, and STANDARD DEVIATION. (The computer can also be programmed in advance to print out *only* these statistics.)

---

* This section is optional.

Many grouped-data methods for samples have been made obsolete by the computer. There was a time when people wanted to group sample data into classes and use grouped-data formulas; this simplified the work of hand computation. With computers, we no longer have to worry about hand computations; and grouped-data calculations become a bother, because their formulas are more complex and their results less accurate than those used with ungrouped data. We have introduced some such methods solely to prepare you for work with random variables in Chapter 4.

CHAPTER REVIEW

Key Terms

measures of central tendency	s, s^2
mean	Chebychev's theorem
median	percentiles
mode	quartiles
population mean	deciles
sample mean	box-and-whisker diagrams
$\sum x$	μ
$\bar{x}$	population standard deviation
range	population variance
sample standard deviation	σ, σ^2
sample variance	

[handwritten annotations: s^2 = variance / stand. dev. squared; variance / stan. dev., squared (for population)]

Formulas

sample mean:

$$\bar{x} = \frac{\sum x}{n} \quad (n = \text{sample size})$$

sample standard deviation:

$$s = \sqrt{\frac{\sum (x - \bar{x})^2}{n-1}} \quad \text{or} \quad s = \sqrt{\frac{n(\sum x^2) - (\sum x)^2}{n(n-1)}}$$

[handwritten: original formula; shortcut formula; These 2 are algebraically equivalent; Use it in freq. data not in dist table.]

sample mean (grouped form):

$$\bar{x} = \frac{\sum xf}{n} \quad (f = \text{frequency})$$

sample standard deviation (grouped form):

$$s = \sqrt{\frac{\sum (x - \bar{x})^2 f}{n-1}} \quad \text{or} \quad s = \sqrt{\frac{n(\sum x^2 f) - (\sum xf)^2}{n(n-1)}}$$

[handwritten: original; shortcut; use if data in freq. dist. table.]

if data not in freq dist table.

population mean:

$$\mu = \frac{\Sigma x}{N} \quad (N = \text{population size})$$

population standard deviation:

$$\sigma = \sqrt{\frac{\Sigma (x - \mu)^2}{N}}$$

If data in freq. dist. in freq. dist table.

population mean (grouped form):

$$\mu = \frac{\Sigma xf}{N} \quad (f = \text{frequency})$$

population standard deviation (grouped form):

$$\sigma = \sqrt{\frac{\Sigma (x - \mu)^2 f}{N}}$$

You should be able to

1   Use each of the preceding formulas. *Know which formula to use - Very important*

2   Find the *mean, median, and mode*(s) for a data set.

3   Choose an appropriate measure of central tendency to use for a data set. *Biggest thing is outliers.*

usually $ money use median, metric use mean?

4   Distinguish between a *sample mean* ($\bar{x}$) and a *population mean* (μ).

5   Use and understand *summation notation*.

6   *Interpret* the standard deviation as a measure of dispersion (Sec. 3.4). *Chebychev's Theory.*

7   Find *quartiles* and *deciles* for a set of data. *Will be in % percentiles - will not do.*

8   Construct a *box-and-whisker diagram* for a data set. *will not do*

9   Distinguish between a *sample standard deviation* (s) and a *population standard deviation* (σ).

Be able to interpret a percentile.

REVIEW TEST

Find (a) the mean, (b) the median, and (c) the mode(s) for the data sets in problems 1 and 2.

1 | 1500 | 50 | 1500 | 1000 | 1000 |
    |------|-----|------|------|------|
    | 500 | 500 | 1000 | 50 | |

2 | 8 | 6 | 5 | 7 | 3 |
    |---|---|---|---|---|
    | 3 | 5 | 9 | 10| 4 |

3   The weights of ten coeds are given below:

125	116	122	146	113
110	131	105	128	137

Compute $\bar{x}$.

4   Compute $\bar{x}$ for the 100-meter dash finishing times listed below:

10.2	10.5	11.1	10.6	11.0	10.3

5   Calculate the *sample* standard deviation for the following data:

6	3	5	5	9	4
1	6	7	2	4	8

6   The mean of the data in Problem 5 is $\bar{x} = 5$.
   a) Locate $\bar{x} = 5$ on the graph on page 85, and mark off three standard deviations to either side of $\bar{x}$.

b) Graph the data from Problem 5 on the graph below.

c) Fill in the blank: By Chebychev's Theorem, at least ____ percent of the data lies within two standard deviations of the mean.

d) What percent of the data actually lies within two standard deviations of the mean?

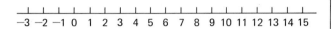

7 Find the *sample* mean $\bar{x}$ for the grouped data below:

x	f
26	10
27	19
28	23
29	17
30	8

8 Find the *sample* standard deviation for the grouped data below:

x	f
25	2
26	0
27	10
28	20
29	10
30	8

9 a) Divide the following data into quarters and find the quartiles.

b) Construct a box-and-whisker diagram for the data below.

16	31
8	16
26	11
6	7
13	19
11	20
21	13
14	17
8	19
30	27
27	4
15	20

10 The weights (to the nearest pound) of the starting players on a baseball team are as follows:

204	200
188	172
174	196
194	130
207	

Find:
a) the *population* mean weight of these players;
b) the *population* standard deviation of these weights.

FURTHER READINGS

Note. The references below were also given at the end of Chapter 2. Chapters 2 and 3 together cover the fundamentals of *descriptive statistics*, and these references apply to both chapters.

HABER, A., and R. RUNYAN, *General Statistics* (Third edition). Reading, Massachusetts: Addison-Wesley, 1977.

HUFF, D. *How to Lie with Statistics*. New York: W. W. Norton and Company, 1955.

MOSTELLER, F., *et al.* (Eds.), *Statistics by Example: Exploring Data.* Reading, Massachusetts: Addison-Wesley, 1973.

TUKEY, J. *Exploratory Data Analysis.* Reading, Massachusetts: Addison-Wesley, 1977.

ANSWERS TO REINFORCEMENT EXERCISES

A   Data set 3

$$\frac{3070}{12} = 255.83$$

Data set 4

$$\frac{2345}{9} = 260.56$$

B   Data set 3

$$\frac{150 + 175}{2} = 162.50$$

Data set 4

175

C   Data set 3

100

Data set 4

175, 100

D $\bar{x} = \dfrac{\sum x}{n} = \dfrac{657370}{16} = 41085.6$

E   Data set 1

Range $= 700 - 100 = 600$

Data set 2

Range $= 700 - 100 = 600$

F $\bar{x} = 5,$ $s = 4.7$

G $\sum x = 40$ $\sum x^2 = 356$

$$s = \sqrt{\frac{8(356) - (40)^2}{8\cdot 7}} = \sqrt{22.286} = 4.7$$

H $\bar{x} = 5,$ $s = 4.7$

$\bar{x} + s = 5 + 4.7 = 9.7$ $\bar{x} - s = 5 - 4.7 = 0.3$

$\bar{x} + 2s = 5 + 9.4 = 14.4$ $\bar{x} - 2s = 5 - 9.4 = -4.4$

$\bar{x} + 3s = 5 + 14.1 = 19.1$ $\bar{x} - 3s = 5 - 14.1 = -9.1$

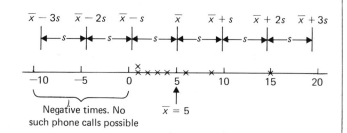

Negative times. No
such phone calls possible

$\bar{x} = 5$

I   If $k = 4,$ $1 - \dfrac{1}{4^2} = \dfrac{15}{16} = 0.94$

At least 94 percent lies within four standard deviations to
either side of the mean.

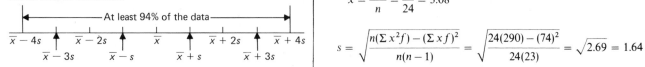

J   a) $1 - \dfrac{1}{k^2} = 1 - \dfrac{1}{3^2} = 0.89$

At least 89 percent of the data lies within three standard
deviations of the mean. From Fig. 3.3, three standard
deviations to either side of the mean extend from 2.38 to
12.58. Since 89% of 80 = 71.2, at least 72 of the 80 babies
will have weights between 2.38 and 12.58.

b) If $s = 1.12,$

$\bar{x} + s = 8.60$ $\bar{x} - s = 6.36$

$\bar{x} + 2s = 9.72$ $\bar{x} - 2s = 5.24$

$\bar{x} + 3s = 10.84$ $\bar{x} - 3s = 4.12$

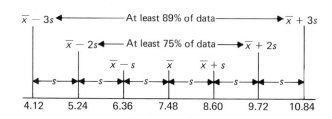

At least 60 babies will have weights between 5.24 and 9.72.
At least 72 babies will have weights between 4.12 and 10.84.

K $\sum xf = 46,$ $\bar{x} = \dfrac{\sum xf}{n} = \dfrac{46}{40} = 1.15.$

L

x	f	xf	x^2	$x^2 f$
0	1	0	0	0
1	2	2	1	2
2	9	18	4	36
3	1	3	9	9
4	6	24	16	96
5	3	15	25	75
6	2	12	36	72
24		74		290
n		$\sum xf$		$\sum x^2 f$

$\bar{x} = \dfrac{\sum xf}{n} = \dfrac{74}{24} = 3.08$

$s = \sqrt{\dfrac{n(\sum x^2 f) - (\sum xf)^2}{n(n-1)}} = \sqrt{\dfrac{24(290) - (74)^2}{24(23)}} = \sqrt{2.69} = 1.64$

M Data in order: 1, 1, 2, 3, 4, 6, 8, 15

Dividing points:

First quartile: $\dfrac{1+2}{2} = 1.5$

Second quartile (Median): $\dfrac{3+4}{2} = 3.5$

Third quartile: $\dfrac{6+8}{2} = 7$

N

O $\mu = 180.4$

P $\sigma = 143.5$

Q

x	f	xf	$x - \mu$	$(x - \mu)^2$	$(x - \mu)^2 f$
22	1	22	-5	25	25
24	2	48	-3	9	18
25	2	50	-2	4	8
26	3	78	-1	1	3
27	3	81	0	0	0
28	4	112	1	1	4
29	1	29	2	4	4
30	4	120	3	9	36
	20	540			98

$$\mu = \frac{\sum xf}{N} = \frac{540}{20} = 27$$

$$\sigma = \sqrt{\frac{\sum (x - \mu)^2 f}{N}} = \sqrt{\frac{98}{20}} = \sqrt{4.9} \doteq 2.21$$

CHAPTER

Probability

In Chapter 1 we distinguished between two types of statistics: *descriptive statistics* and *inferential statistics*. Descriptive statistics involves organizing and summarizing data. Inferential statistics deals with making inferences (educated guesses) about a population based on a sample from the population. For example, a Gallup poll uses voting information collected from a sample of the voting population to make *inferences* concerning the preferences of the entire voting population.

Since inferential statistics involves (educated) guessing, it is always possible to make an incorrect inference. We need to know how likely it is that our inference is correct. In order to measure our chance of being right in a statistical inference, we need to understand the theory of *probability*, which is the mathematical foundation for inferential statistics.

4.1 Introduction

We introduce the concept of probability by using relative frequency distributions.

Example 1 ◖ A school planner in a small town of forty families classified each family according to the number of children (under 18 years old) in the family. The planner's findings are summarized in Table 4.1.* The table shows, for example, that 17.5 percent (0.175) of the forty families have two children (under 18 years of age.)

Table 4.1

Children under 18	Number of families (frequency)	Percentage	Relative frequency
0	18	45.0 %	0.450
1	8	20.0 %	0.200
2	7	17.5 %	0.175
3	4	10.0 %	0.100
4	3	7.5 %	0.075
	40	100.0 %	1.000

Now suppose one of the forty families is selected **at random**—meaning that each family is *equally likely* to be the one selected. What is the probability that the family selected has three children? The answer, as you can probably guess, is 4/40. We have simply divided the total number of families with three children (4) by the total number of families (40). Note that the probability, 4/40, of selecting a family with three children is exactly the same as the relative frequency, 0.100, of the number of families with three children. ◗

---

* This data is based on information found in the *Statistical Abstract of the United States*, 1976. The data is, in fact, very close to the actual figures for the entire United States.

The above example illustrates how to compute probabilities when dealing with experiments in which each outcome is equally likely to occur.

Suppose there are N *equally likely* outcomes in an experiment. The **probability** that an event occurs is the number of ways, f, that the event can occur, divided by the total number, N, of possible outcomes. That is, the probability is

$$\frac{f}{N} \begin{array}{l} \leftarrow \text{Total number of ways event can occur} \\ \leftarrow \text{Total number of possible outcomes} \end{array}$$

Computing probabilities for experiments with equally likely outcomes

In Example 1, $N = 40$ since there are forty families. If we are looking at the event that the family selected has three children, then $f = 4$, because four of the families have three children. So

$$\text{Probability of three children} = \frac{f}{N} = \frac{4}{40} = \frac{1}{10}$$

Here are some additional examples:

◖ A card is selected from a well-shuffled deck of 52 cards. What is the probability that a queen is selected?

Example **2**

$N = 52$ since there are 52 cards. There are four queens in the deck, so $f = 4$. The probability of selecting a queen is

Solution

$$\frac{f}{N} = \frac{4}{52} = \frac{1}{13} \ ◗$$

In Example 2, what is the probability that a club is selected?

Exercise **A**

◖ A senator is selected at random from the U.S. Senate to serve as the chairman of a subcommittee. What is the probability that the senator selected is from Arizona?

Example **3**

$N = 100$ since there are 100 U.S. senators; $f = 2$ because there are two senators from Arizona. Consequently, the probability is

Solution

$$\frac{f}{N} = \frac{2}{100} = \frac{1}{50} \ ◗$$

◖ A pair of fair dice is rolled. What is the probability that the sum of the two dice is 7?

Example **4**

Solution   There are 36 equally likely outcomes when two dice are rolled. (See Fig. 4.1.) So $N = 36$. The sum of the dice can be 7 in $f = 6$ ways. The probability that the sum is 7 is thus

$$\frac{f}{N} = \frac{6}{36} = \frac{1}{6} \blacktriangleright$$

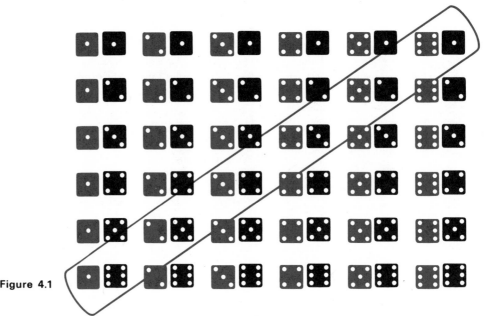

Figure 4.1

Exercise **B**   In Example 4, find the probability that

a) the sum of the dice is 11;

b) doubles are rolled (i.e., both dice come up the same number).

It is important to note some simple but useful facts about probabilities.

1. The probability of an event is always between 0 and 1; that is, $0 \leqslant f/N \leqslant 1$. (This serves as a useful check on calculations. If you calculate a probability and get an answer like 5 or -0.23, you made an error.)
2. The probability of an event that cannot occur is 0.
3. The probability of an event that is certain to occur is 1.

In this section we have seen how to compute probabilities when dealing with experiments in which each outcome is equally likely. In situations where this is not the case, it is necessary to resort to other methods for computing probabilities. We shall discuss some of these methods in Secs. 4.8 and 4.10.

1   In the dice illustration of Example 4, calculate the probabilities of the following events.
a) The sum of the dice is 6.
b) The sum of the dice is even.
c) The sum of the dice is 7 or 11.
d) The sum of the dice is either 2 or 7.

2   For the card illustration of Example 2, find the probabilities for each of the events below.
a) The card selected is a heart.
b) The card selected is a face card (king, queen, or jack).

3   The number of births per month in 1970, in thousands of people, is given below.

Month	Number (in thousands)	Month	Number (in thousands)	Month	Number (in thousands)
Jan	299	May	292	Sept	317
Feb	284	June	285	Oct	305
Mar	306	July	295	Nov	294
Apr	290	Aug	306	Dec	287

If a person born in 1970 is selected at random, what is the probability the person was born in:
a) March?   b) June or July?
c) the first half of the year?

4   The numbers of housing units (in thousands) in the U.S. in 1976 is given below.

Owner-occupied	Renter-occupied	Vacant year round
47,904	26,101	5,311

What is the probability that a housing unit selected at random is:
a) owner-occupied?   b) renter-occupied?
c) vacant year round?

5   In 1972 there were 141,500 inmates in U.S. prisons in the categories listed below. (These numbers are rounded to the nearest 100.)

Serving sentence	Awaiting trial	Other adjudication status
60,200	50,800	30,500

What is the probability that a prisoner selected at random
a) was awaiting trial?
b) was serving a sentence?

6   In 1977, Americans living on farms were divided into age groups as follows:

Age group	Number (in thousands)	Age group	Number (in thousands)	Age group	Number (in thousands)
Under 14	1555	25–34	730	65 and over	950
14–19	1076	35–44	891		
20–24	517	45–64	2086		

What is the probability that a person selected at random was
a) under 14?   b) 65 or over?
c) between 20 and 24?   d) between 14 and 19?
e) 35 or over?   f) under 25?

7   In 1975, the numbers (in thousands) of U.S. businesses in various categories were as follows

Proprietorships	Partnerships	Corporations
10,882	1,073	2,024

What is the probability that a business is
a) a proprietorship?   b) a partnership?
c) a corporation?

8   At a large university the IQ's of the students have been grouped as follows:

IQ	Number of students
90–99	3,600
100–110	11,900
111–120	12,000
121–130	4,800
Over 130	2,700

If a student is selected at random, what is the probability that the student's IQ
a) exceeds 110?   b) is at most 120?
c) is between 100 and 130 (inclusive)?
d) is not between 100 and 120 (inclusive)?

<u>9</u> A study of the methods of transportation people use to get to work in the U.S. revealed the following data (in hundreds of thousands of workers).

Method of transportation	Type of worker	
	Urban	Rural
Automobile	450	150
Public transportation	65	5

If a worker is selected at random, what is the probability that he or she
a) is an urban worker who drives an automobile to work?
b) drives an automobile to work?
c) is a rural worker who uses public transportation?
d) is an urban worker?

<u>10</u> In a city of 120,000, the party affiliations by religion are given below. (For the sake of this problem assume that all voters are either Catholic, Jewish, or Protestant, and that all voters are either Democrats or Republicans.)

Religion	Democratic	Republican
Catholic	13,200	10,800
Jewish	3,600	2,400
Protestant	36,000	54,000

What is the probability that a voter selected at random from this city is
a) a Republican?
b) a Protestant and a Republican?
c) a Catholic?
d) a Democrat and Jewish?

<u>11</u> An automobile manufacturer has three production plants. Each plant produces both regular cars and station wagons. For every 1000 automobiles manufactured, the plants produce the following quantities:

Type	Plant		
	I	II	III
Regular cars	405	200	285
Station wagons	45	50	15

If an automobile produced by this manufacturer is selected at random, what is the probability that
a) it is a station wagon produced by Plant I?
b) it is a regular car from Plant III?
c) it is a station wagon?

<u>**12**</u> The N outcomes in an experiment are called *basic outcomes*, and the collection of all basic outcomes is called the *sample space*. For example, in Exercise 2, each of the 52 cards in the deck is a basic outcome, and the sample space has 52 elements—one for each card.
a) What is the sample space in Exercise 3?
b) Find two different ways of looking at the dice experiment in Exercise 1 that lead to two different sample spaces for the same experiment. (The basic outcomes in one of these sample spaces will *not* be equally likely.)

<u>**13**</u> Explain in your own words what is wrong with the following argument:

When you roll two dice, the totals that can come up are 2, 3, 4, 5, 6, 7, 8, 9, 10, 11, 12. This gives 11 possibilities, so the probability of rolling a 2 is

$$\frac{1}{11} \doteq 0.0909.$$

4.2 Random variables; probability distributions

In Example 1 we looked at a survey that classified each of forty families according to the number of children in the family. Since the "number of children" *varies* from family to family, it is called a *variable*. When we select a family at random the "number of children" is called a **random variable** since its value depends on chance (namely, upon which family is selected).

DEFINITION **Random variable:** A variable quantity whose value depends on chance.

Example **5** ◀ For our forty families, we have noted that the number of children in a family selected at random is a random variable. The possible values for this random variable are 0, 1, 2, 3, and 4. ▶

◖ If a pair of fair dice is rolled, then the sum of the two dice is a random variable,   Example **6**
because the sum of the two dice depends on chance. The possible values for this random
variable are 2, 3, 4, 5, 6, 7, 8, 9, 10, 11, and 12. ◗

◖ The 1975 production of U.S. coins (by denomination) is given in Table 4.2. Values   Example **7**
given are in millions of pieces.

Half-dollars	Quarters	Dimes	Nickels	Pennies	Table **4.2**
476	1394	902	587	9960	

Suppose a 1975 coin is selected at random, and the "value of the coin" (in cents) is
noted. The "value of the coin" is a random variable, with possible values 50, 25, 10, 5,
and 1. ◗

It is customary to denote random variables by letters such as x, y, z. By doing this
we can develop a useful shorthand for probability and random variables. We introduce
these concepts by appealing to our example involving the number of children in each of
40 families.

◖ The relative-frequency distribution for the number of children under 18 in each of   Example **8**
forty families is repeated in Table 4.3. Suppose we select a family at random, and
let x denote the "number of children" in this randomly selected family. Then, x is a
random variable. We can conveniently describe the event that "the family selected
has three children" simply by

$$(x = 3).$$

Children under 18	Number of families (frequency)	Relative frequency	Table **4.3**
0	18	0.450	
1	8	0.200	
2	7	0.175	
3	4	0.100	
4	3	0.075	
	40	1.000	

Children under 18, x	Probability
0	0.450
1	0.200
2	0.175
3	0.100
4	0.075

Table **4.4**

Since probabilities in this case are just relative frequencies, we can list the
probabilities in Table 4.4.

Just as a table of relative frequencies is called a relative-frequency distribution
table, a table of probabilities is called a *probability distribution table*. So Table 4.4
gives us the **probability distribution of the random variable x**. (It tells us the
distribution of probabilities for the various values of the random variable x.)

The statement "The probability that the family selected at random has three children is 0.100" can be written compactly as:

"$P(x = 3) = 0.100$"

This is read "The probability x equals 3 is 0.100." When there is no question about what random variable is under discussion, we will often write $P(3)$ instead of $P(x = 3)$. ◗

Example **9** ◖ Suppose a pair of fair dice is rolled. Let x denote the sum of the dice. From Fig. 4.1 on page 92 we can compute the probabilities for the various values of x. For example, the probability that the sum is 4 is

$$P(x = 4) = \frac{f}{N} = \frac{3}{36} = \frac{1}{12}$$

This is because there are three ways to get a sum of 4, and 36 possible outcomes altogether. The probability distribution of the random variable x is given in Table 4.5. We see, for example, that $P(x = 8) = 5/36$. ◗

Table 4.5

Sum of dice, x	Probability, $P(x)$	Sum of dice, x	Probability, $P(x)$
2	1/36	8	5/36
3	2/36	9	4/36
4	3/36	10	3/36
5	4/36	11	2/36
6	5/36	12	1/36
7	6/36		

Exercise **C** The heights of the players on the 1978 Arizona State University basketball team are given in Table 4.6.

Table 4.6

Height (in.)	Number of players (frequency)	Height (in.)	Number of players (frequency)
70	1	78	4
73	2	80	1
74	2	82	2
76	2	83	1
77	1		

Let x be the height of a player selected at random.
a) Describe, in words, the event $(x = 78)$.
b) $P(x = 78) = ?$ c) $P(82) = ?$
d) Find the probability distribution of the random variable x. (Use the same format as in Table 4.5.)

1 Table 4.7 gives the frequency distribution for the record-playing times given in the exercises in Sec. 3.1. If x is the playing time of one of these albums selected at random, find the probability distribution of x.

Table 4.7

Time, x	Frequency, f	Time, x	Frequency, f
23	1	37	1
30	2	42	3
32	1	45	1
33	1	49	1
34	5	53	1
35	3		

2 Below is a set of final exam scores for a 40-question test:

2	16	19	21	26
4	16	20	24	27
15	16	21	25	27
15	17	21	25	28

Let x be the score of an exam selected at random. Find the probability distribution of x.

3 100 students in an experimental class were asked to fill out an evaluation form. One item on the form requested that the student evaluate the overall course experience on a scale of 4 = very good, 3 = good, 2 = fair, 1 = poor, 0 = very poor. The responses were as given in Table 4.8.

Table 4.8

Response	Frequency
4	32
3	44
2	14
1	6
0	4

Let x be the response (to the overall course-experience question) of an evaluation form selected at random.
a) Find the probability distribution for x.
b) What is the probability that x is 3 $(P(x = 3))$?
c) What is the probability that x is 4 $(P(x = 4))$?
d) What is the probability that x is 2 $(P(x = 2))$?

4 A pair of fair dice is rolled (see Fig. 4.1, page 92). Let x be the largest number showing on the two dice. For example, if the outcome is

then $x = 5$.
a) Find the probability distribution for x.
b) Find $P(x = 4)$.   c) Find $P(x = 2)$.

5 A card is selected at random from a deck of 52 cards. Let x denote the value of the card selected (ace = 11; and all face cards equal 10).
a) Find the probability distribution of x.
b) What is $P(x = 10)$?   c) What is $P(x = 6)$?

6 A fair coin is tossed three times. The possible outcomes are HHH, HHT, HTH, HTT, THH, THT, TTH, TTT. Since the coin is fair, each of these eight outcomes is equally likely. So the probability is $\frac{1}{8}$ for each one. Let x denote the number of heads obtained.
a) Find the probability distribution of x.
b) What is $P(x = 2)$?
c) What is the probability that x is even? *an even # is divisible*
d) Find $P(x = 0)$. *by 2 exactly & to remainder - so it include 0.*

7 According to the *Statistical Abstract of the United States*, the 1976 income-level distribution of families was as given in Table 4.9. Let x denote the income class of a family selected at random in 1976.

Table 4.9

Income class	Yearly income	Number of families (in thousands)
1	Under $2,000	1,181
2	$2,000–$2,999	1,350
3	$3,000–$3,999	1,912
4	$4,000–$4,999	2,306
5	$5,000–$6,999	4,668
6	$7,000–$9,999	7,199
7	$10,000–$11,999	5,006
8	$12,000–$14,999	7,537
9	$15,000–$24,999	17,042
10	$25,000 and over	7,930

a) Find the probability distribution of x.
b) What is $P(x = 3)$?   c) Find $P(x = 9)$.
d) What is $P(x = 8)$?

8 The data on the number of births per month in 1970 from Exercise 3, Sec. 4.1, is given below:

Month	Number (in thousands)	Month	Number (in thousands)
1	299	7	295
2	284	8	306
3	306	9	317
4	290	10	305
5	292	11	294
6	285	12	287

Let x be the birth month of a person selected at random who was born in 1970.

a) Find the probability distribution for x.
b) Find $P(x = 6)$.   c) Find $P(x = 2)$.   d) Find $P(x = 12)$.

9 Tests on the gas mileage for fifteen cars produced by a manufacturer are given below.

Gas mileage	32	33	34	35	36	37
Number of cars	1	1	4	4	1	4

Let x be the gas mileage of one of these cars selected at random.
a) Find the probability distribution for x.
b) Find $P(x = 33)$.   c) Find $P(x = 35)$.
d) Find $P(x = 32)$.   e) Find $P(x = 37)$.

10 Tests on 30 acres of farmland produced the following yields (in bushels) per acre of wheat.

Bushels	30	40	41	42	43	44	45
Number of acres	3	6	7	6	5	2	1

Let x be the yield of one of these acres selected at random.
a) Find the probability distribution for x.
b) Find $P(x = 39)$.   c) Find $P(x = 45)$.
d) Find $P(x = 42)$.

4.3   Compound events

In this section we will discuss some of the fundamental rules for dealing with probabilities.

Example 10   ❮ Let us return to our survey of the number of children under 18 years of age in each of forty families.

We have used the word "event" intuitively in the previous sections. To be a little more precise, an event consists of an outcome or collection of outcomes. For example, "the family has three children," and "the family has at least two children [i.e., has 2, 3, or 4 children]" are examples of events.

Just as we used letters x, y, z for random variables, it will be convenient for us to use letters such as $A, B, C, D, \ldots$, to represent events. For example, when a family is selected at random, we might let

A = event the family has three children,
B = event the family has at least two children (2, 3, or 4),
C = event the family has at most two children (0, 1, or 2),
D = event the family has more than two children (3 or 4).

We have already learned how to find probabilities of events such as A:

$$P(A) = \frac{f}{N} = \frac{4}{40} = 0.100 \quad \text{(see Table 4.10)}$$

We have also considered probabilities of events such as B, C, and D. However, these probabilities cannot be found *directly* in Table 4.10, because they are each a *collection* of outcomes listed in Table 4.10. For example, the event B consists of the three

Children under 18, x	Number of families (frequency), f	Probability, $P(x)$	**Table 4.10**
0	18	0.450	
1	8	0.200	
2	7	0.175	
3	4	0.100	
4	3	0.075	
	40	1.000	

outcomes "two children", "three children", and "four children." Events consisting of more than one outcome are called **compound events**. So B, C, and D are compound events.

But it is not at all difficult to compute probabilities of compound events such as B, C, and D. One way is to simply use the definition of probability given in Sec. 4.1 (page 91). Here, we have $N = 40$, and for the event B (2, 3, or 4 children),

$$
\begin{array}{ccccccc}
& \text{2 children} & & \text{3 children} & & \text{4 children} & \\
& \downarrow & & \downarrow & & \downarrow & \\
f = & 7 & + & 4 & + & 3 & = 14
\end{array}
$$

(See Table 4.10.) So,

$$P(B) = \frac{f}{N} = \frac{14}{40} = 0.350$$

Similarly,

$$P(C) = \frac{f}{N} = \frac{18 + 8 + 7}{40} = \frac{33}{40} = 0.825 \; \blacktriangleright$$

In Example 10, find $P(D)$; that is, find the probability that the family selected at random has *more than two* children.   Exercise **D**

There is another way to express the probabilities of compound events. Consider, for example, the event B:

$$
P(B) = \frac{f}{N} = \frac{7 + 4 + 3}{40} = \overset{P(2)}{\underset{}{\frac{7}{40}}} + \overset{P(3)}{\underset{}{\frac{4}{40}}} + \overset{P(4)}{\underset{}{\frac{3}{40}}}.
$$

The three numbers in the sum on the right are just the probabilities of the outcomes "two children," "three children," and "four children" that make up the compound event B.

The principle illustrated above can be stated as follows:

> **Probability of a compound event:**   The probability of a compound event is the sum of the probabilities of the individual outcomes that make up the event.

Exercise **E**   In Example 10, find $P(C)$ and $P(D)$ using the above principle, along with Table 4.10.

Example **11**   ◖ Suppose a pair of fair dice is rolled. Let

$$A = \text{event the sum is 7 or 11,}$$
$$B = \text{event the sum is 2, 3, or 12.}$$

Find $P(A)$ and $P(B)$.

Solution   The probability distribution for the sum of the dice is given in Table 4.11.

Table 4.11

Sum of dice, x	Probability, $P(x)$	Sum of dice, x	Probability, $P(x)$
2	1/36	8	5/36
3	2/36	9	4/36
4	3/36	10	3/36
5	4/36	11	2/36
6	5/36	12	1/36
7	6/36		

From this table we can compute:

$$P(A) = \underset{P(7)}{\frac{6}{36}} + \underset{P(11)}{\frac{2}{36}} = \frac{8}{36} = \frac{2}{9} \qquad P(B) = \underset{P(2)}{\frac{1}{36}} + \underset{P(3)}{\frac{2}{36}} + \underset{P(12)}{\frac{1}{36}} = \frac{4}{36} = \frac{1}{9} \quad ◗$$

It is often convenient to use random-variable terminology to describe compound events. This is illustrated in the following examples.

Example **12**   ◖ In Example 10, we considered four events. If we let x be the number of children in a randomly selected family, then we can describe the events in a very compact form in terms of the random variable x as follows:

$$A = \text{event the family has three children} = (x = 3),$$
$$B = \text{event the family has at least two children} = (x \geqslant 2),$$
$$C = \text{event the family has at most two children} = (x \leqslant 2),$$
$$D = \text{event the family has more than two children} = (x > 2).$$

Then we can also write $P(B)$ as $P(x \geqslant 2)$.   ◗

Example **13**   ◖ The height data for the 1978 Arizona State University basketball team is given in Table 4.12.

Height (in.), x	Number of players (frequency), f	Probability $P(x)$	**Table 4.12**
70	1	1/16	
73	2	2/16	
74	2	2/16	
76	2	2/16	
77	1	1/16	
78	4	4/16	
80	1	1/16	
82	2	2/16	
83	1	1/16	

Here x is the height of a player selected at random. The (compound) event that the player selected at random is at least 6′6″ (78″) tall can be described as

$$(x \geqslant 78).$$

So

$$P(x \geqslant 78) = \frac{4}{16} + \frac{1}{16} + \frac{2}{16} + \frac{1}{16} = \frac{8}{16} = 0.5$$

The event that the player selected is between 6′4″ (76″) and 6′6″ (78″) tall, inclusive, can be written as

$$(76 \leqslant x \leqslant 78)$$

and

$$P(76 \leqslant x \leqslant 78) = \frac{2}{16} + \frac{1}{16} + \frac{4}{16} = \frac{7}{16} = 0.4375 \blacktriangleright$$

For the dice experiment of Example 11 (page 100), let x be the sum of the dice.   Exercise **F**

a) Let

> A = event the sum of the dice is at least 7,
> B = event the sum is less than 4,
> C = event the sum is between 4 and 10, inclusive.

Describe the events A, B, and C in terms of the random variable x.

b) Describe, in words, each of the following events:

   i) $(x \leqslant 5)$,
   ii) $(x > 7)$,
   iii) $(3 < x < 7)$.

c) Find the probabilities of the events in part (b).

EXERCISES Section **4.3**

In Exercises 1 through 5 calculate the probability of the given compound events, by using the fact that *the probability of a compound event is the sum of the probabilities of the individual outcomes that make up the event.*

1 Suppose a fair coin is tossed three times. Let x be the number of heads obtained. The probability distribution for x is (see Exercise 6, Sec. 4.2):

x	0	1	2	3
$P(x)$	1/8	3/8	3/8	1/8

Find:

a) $P(0 < x < 3)$ b) $P(x \geqslant 2)$
c) $P(x \leqslant 2)$ d) $P(x < 3)$

2 According to the *Statistical Abstract of the United States*, the 1976 income-level distribution of the families was as given in Table 4.13.

Table 4.13

Yearly income	Number of families (in thousands)	Percentage
Under $ 2,000	1,181	2.1 %
$ 2,000–$ 2,999	1,350	2.4 %
$ 3,000–$ 3,999	1,912	3.4 %
$ 4,000–$ 4,999	2,306	4.1 %
$ 5,000–$ 6,999	4,668	8.3 %
$ 7,000–$ 9,999	7,199	12.8 %
$10,000–$11,999	5,006	8.9 %
$12,000–$14,999	7,537	13.4 %
$15,000–$24,999	17,042	30.4 %
$25,000 and over	7,930	14.1 %

Suppose that in 1976 a family was selected at random. Let

A = event the family made at least $3,000,
B = event the family made at least $5,000 but less than $15,000,
C = event the family made under $2,000 or at least $25,000.

Find

a) $P(A)$ b) $P(B)$ c) $P(C)$

Interpret your results in parts (a) through (c) in terms of percentages.

3 In a horse race, the odds indicate that the probabilities for the various horses winning are as follows:

Horse x	Probability $P(x)$	Horse x	Probability $P(x)$
#1	1/8	#5	1/32
#2	1/16	#6	3/32
#3	1/4	#7	5/32
#4	3/16	#8	3/32

Let

A = event one of the two "favorites" wins.
    (The favorites are the two horses with the two highest win probabilities.)
B = the winning horse has a number above 4.
C = the winning horse's number is under 5.

a) Rewrite events A, B, and C in terms of the random variable x.

Find:

b) $P(A)$ c) $P(B)$ d) $P(C)$

4 An article by R. D. Clarke in the *Journal of the Institute of Actuaries* (1946), gave statistics for flying-bomb hits in the south of London during World War II. The area was divided into 576 equal areas, and the following statistics were compiled.

No. of hits	No. of areas
0	229
1	211
2	93
3	35
4	7
5 or more	1

Suppose one of the 576 areas is selected at random. Let

A = event the area received at least three hits,
B = event the area received fewer than four hits,
C = event the area received at most one hit.

a) Complete the table:

No. of hits x	$P(x)$	No. of hits x	$P(x)$
0		3	
1		4	
2		5 or more	

b) Rewrite A, B, and C in terms of the random variable x.

Find:

c) $P(A)$ d) $P(B)$ e) $P(C)$

Interpret your results in terms of percentages.

5 The age distribution of the employees in a large company are shown in Table 4.14.

Age	Percentage	
Under 18	5%	**Table 4.14**
18–24	17%	
25–34	36%	
35–49	25%	
50–64	15%	
65 and over	2%	

Suppose an employee is selected at random. Let:

A = event the employee is 50 or over;

B = event the employee is between 25 and 49, inclusive;

C = event the employee is either under 18 or at least 65.

Find

a) $P(A)$ b) $P(B)$ c) $P(C)$

6 Below is the course evaluation data from Exercise 3, Sec. 4.2.

Response, x	Probability
0	0.04
1	0.06
2	0.14
3	0.44
4	0.32

Find:

a) $P(x \geqslant 2)$ b) $P(x \leqslant 3)$

c) $P(0 < x < 4)$ d) $P(1 \leqslant x \leqslant 3)$

7 Below is the probability distribution table on card selection from Exercise 5, Sec. 4.2.

x	$P(x)$	x	$P(x)$
2	1/13	7	1/13
3	1/13	8	1/13
4	1/13	9	1/13
5	1/13	10	4/13
6	1/13	11	1/13

Find:

a) $P(3 < x < 8)$ b) $P(x \geqslant 10)$ c) $P(x \leqslant 7)$

8 Below is the probability distribution table on income-level distribution from Exercise 7, Sec. 4.2.

x	$P(x)$	x	$P(x)$
1	0.021	6	0.128
2	0.024	7	0.089
3	0.034	8	0.134
4	0.041	9	0.304
5	0.083	10	0.141

Find:

a) $P(x \geqslant 5)$ b) $P(4 \leqslant x \leqslant 9)$ c) $P(x < 4)$

9 Below is the probability distribution table for the gas mileage of the cars in Exercise 9, Sec. 4.2.

x	$P(x)$	x	$P(x)$
32	0.067	35	0.267
33	0.067	36	0.067
34	0.267	37	0.267

Find:

a) $P(33 \leqslant x \leqslant 36)$ b) $P(x > 35)$ c) $P(x \leqslant 34)$

10 The following table is based on the data in Exercise 10 in Sec. 4.1. It gives the probabilities of combinations of party affiliation and religion in a large city.

	Democrat	Republican
Catholic	0.11	0.09
Jewish	0.03	0.02
Protestant	0.30	0.45

Find the probability that a voter is:

a) Jewish,

b) A Democrat who is either Catholic or Jewish,

c) A Catholic Republican,

d) A Republican and not a Protestant.

11 Mathematicians have discovered axioms that probability measures must satisfy. (These axioms use the concepts of *basic outcome* and *sample space* covered in Exercise 12 of Sec. 4.1. The single outcomes of an experiment are called basic outcomes,

and the sample space is the collection of all basic outcomes.) The axioms for probability are:

Axiom 1. For any basic outcome e,

$$0 \leqslant P(e) \leqslant 1.$$

Axiom 2. If an event E consists of basic outcomes e_i,

$$P(E) = \sum P(e_i).$$

Axiom 3. Let S be the event consisting of all basic outcomes in the sample space. Then

$$P(S) = 1.$$

a) Identify the basic outcomes and sample space in Exercise 1 of this section.

b) Discuss the meaning of each of the three axioms, using examples from the sample space of Exercise 1.

c) Suppose the sample space consists of basic outcomes $e_1, e_2, \ldots, e_N$. Find

$$\sum_{i=1}^{N} P(e_i)$$

and prove that your answer is correct, using Axioms 2 and 3.

4.4 Rules of probability

As we have seen, compound events arise by considering collections of individual outcomes. However, they can also arise by looking at two or more compound events.

Example 14 ◖ In the game of "craps" (rolling a pair of fair dice), the game ends *after the first roll* if either one or the other of these events occur:

A = the event the sum of the dice is 7 or 11 (the player wins),
B = the event the sum of the dice is 2, 3, or 12 (the player loses).

Events A and B are both compound events. The event that the game ends after the first roll occurs if *either* the event A *or* the event B occurs. We write this simply as the event

$$(A \text{ or } B).$$

The event $(A \text{ or } B)$ consists of the five individual outcomes

$$\underbrace{7, 11,}_{A} \underbrace{2, \ 3, \ 12}_{B} \ ◗$$

DEFINITION **(A or B):** The event $(A \text{ or } B)$ occurs if *either* event A occurs *or* event B occurs, or *both* occur. The event $(A \text{ or } B)$ consists of all the individual outcomes in either event A or event B.

Example 15 ◖ In Example 10 let

A = event the family has an even number of children (0, 2, or 4),

B = event the family has at least two children (2, 3, or 4).

The event $(A \text{ or } B)$ occurs if either the family has an even number of children *or* the family has at least two children. In terms of individual outcomes, we can describe

(A or B) as

      (A or B) = event the family has either 0, 2, 3, or 4 children. ▶

◀ In rolling a pair of dice, let   Example **16**

        A = event doubles are rolled (both dice
                come up the same number),

        B = event the sum of the dice is 8.

The event A can occur in six ways:

The event B can occur in five ways:

The event (A or B) occurs if either doubles are rolled *or* the sum of the dice is 8. Note that there are only ten ways that (A or B) can occur since the outcome ⚃⚃ occurs in both A and B, and should be counted only once. ▶

Consider the experiment of selecting a card at random from an ordinary deck of 52   Exercise **G**
cards. Let

        A = event a king is selected,

        B = event a heart is selected.

a) Describe these events in terms of individual outcomes. Use the notation ah for ace of hearts, 2h for two of hearts, ks for king of spades, and so on.
b) Describe in words, the event (A or B).
c) List the individual outcomes in (A or B).

    We will develop a formula for computing $P(A \text{ or } B)$ in terms of $P(A)$ and $P(B)$. But before we can do this we need to introduce the idea of two events being **mutually exclusive**.

Mutually exclusive events: Events A and B are called *mutually exclusive* if they   DEFINITION
have no common individual outcomes and therefore cannot both occur at the same time.

Example **17** ◖ In Example 14 the events A (sum of dice is 7 or 11), and B (sum of dice is 2, 3, or 12) are mutually exclusive since they have no common individual outcomes and thus cannot both occur at the same time. ◗

Example **18** ◖ In Example 15 the events A (0, 2, or 4 children) and B (2, 3 or 4 children) are *not* mutually exclusive since they can both occur at the same time if the family selected has either two or four children. ◗

Example **19** ◖ In Example 16 the events A (doubles are rolled) and B (the sum is 8) are *not* mutually exclusive since they can occur simultaneously. They have the common individual outcome

 ◗

Exercise **H** Consider again the experiment of selecting a card at random from an ordinary deck of 52 cards. Let

A = event a king is selected,
B = event a heart is selected,
C = event a 10 is selected,
D = event a face card (king, queen, or jack) is selected.

Determine which of the following pairs of events are mutually exclusive. If a given pair is *not* mutually exclusive, list their common individual outcomes.

a) A, B b) A, C c) A, D
d) B, C e) B, D f) C, D

We will now give a formula for computing $P(A \text{ or } B)$ in terms of $P(A)$ and $P(B)$, when A and B are mutually exclusive.

Example **20** ◖ In the game of craps, the game ends after the first roll if either one or the other of the following events occurs:

A = the sum of the dice is 7 or 11 (the player wins),
B = the sum of the dice is 2, 3, or 12 (the player loses).

What is the probability that the game ends after one roll?

Solution We need to compute $P(A \text{ or } B)$. We can, of course, resort to writing the event $(A \text{ or } B)$ in terms of individual outcomes. Then using Table 4.11 (page 100), we get

$$P(A \text{ or } B) = P(7, 11, 2, 3, \text{ or } 12) = \frac{6}{36} + \frac{2}{36} + \frac{1}{36} + \frac{2}{36} + \frac{1}{36} = \frac{12}{36} = \frac{1}{3}.$$

However, we have, in Example 11, already computed

$$P(A) = \frac{6}{36} + \frac{2}{36} = \frac{2}{9}, \qquad P(B) = \frac{1}{36} + \frac{2}{36} + \frac{1}{36} = \frac{1}{9}.$$

Hopefully, we can compute $P(A \text{ or } B)$ in terms of $P(A)$ and $P(B)$. To see how to do this, let's look more carefully at what we did in terms of individual outcomes.

$$P(A \text{ or } B) = P(7, 11, 2, 3, \text{ or } 12)$$

$$= \underbrace{\frac{6}{36} + \frac{2}{36}}_{P(A)} + \underbrace{\frac{1}{36} + \frac{2}{36} + \frac{1}{36}}_{P(B)} = P(A) + P(B).$$

So, in this case,

$$P(A \text{ or } B) = P(A) + P(B).$$

This formula is not always true, but it *is* when the events A and B are mutually exclusive (as they are in this example). ◗

The addition principle of probability for mutually exclusive events: FORMULA

If A and B are *mutually exclusive* events, then

$$P(A \text{ or } B) = P(A) + P(B)$$

◖ In the survey of forty families of Example 10, find the probability that the family selected at random has either (exactly) three children or at most two children.   Example **21**

Let   Solution

A = event the family has three children,
C = event the family has at most two children
    (that is, 0, 1, or 2 children).

Then A and C are mutually exclusive, and since $P(A) = 0.100$ and $P(C) = 0.825$ (from Example 10), we have

$$P(A \text{ or } C) = P(A) + P(C) = 0.100 + 0.825 = 0.925 ◗$$

◖ Consider once more the height data for the 1978 Arizona State University basketball team. Let x be the height of a player selected at random. The probability distribution of x is given in Table 4.15. Find $P(x < 74 \text{ or } x > 80)$. That is, find the probability that the player selected at random is less than 74″ tall *or* greater than 80″ tall.   Example **22**

x	$P(x)$	x	$P(x)$	Table **4.15**
70	1/16	78	1/4	
73	1/8	80	1/16	
74	1/8	82	1/8	
76	1/8	83	1/16	
77	1/16			

Solution   The events $(x < 74)$ and $(x > 80)$ are mutually exclusive, and

$$P(x < 74) = \tfrac{1}{16} + \tfrac{1}{8} = \tfrac{3}{16},$$
$$P(x > 80) = \tfrac{1}{8} + \tfrac{1}{16} = \tfrac{3}{16}.$$

Consequently, by the addition principle,

$$P(x < 74 \quad \text{or} \quad x > 80) = P(x < 74) + P(x > 80)$$
$$= \tfrac{3}{16} + \tfrac{3}{16} = \tfrac{3}{8} \; \blacktriangleright$$

Exercise **I**   Refer to Exercise H (page 106). Let

A = event a king is selected,
B = event a heart is selected,
C = event a 10 is selected,
D = event a face card is selected.

Use the addition principle to find

a) $P(A \text{ or } C)$ b) $P(C \text{ or } D)$

There is a special case of the addition principle of probability for mutually exclusive events that is extremely useful. We illustrate the concept in the following example.

Example **23**   ◖ 100 students in an experimental class were asked to fill out an evaluation form. One item on the form requested that the student evaluate the *overall course experience* on a scale of 4 = very good, 3 = good, 2 = fair, 1 = poor, and 0 = very poor. The responses were as given in Table 4.16. If an evaluation form is selected at random, what is the probability that the response (to the overall course experience) is at least 1?

	Response	Frequency	Relative frequency
Table 4.16	4	32	0.32
	3	44	0.44
	2	14	0.14
	1	6	0.06
	0	4	0.04

Solution   Let

A = event the response is at least 1 (that is, 1, 2, 3, or 4)

We want to find $P(A)$. Using Table 4.16, we get

$$P(A) = P(1) + P(2) + P(3) + P(4)$$
$$= 0.06 + 0.14 + 0.44 + 0.32 = 0.96$$

So $P(A) = 0.96$. There is, however, an easier way to find this probability. Consider the event that "*A does not occur*," which we write as (not A). Since the event A is that the response is at least 1, the event (not A) is the event that the response is *not* at least 1— that is, that the response is 0.

By the table above

$$P(\text{not } A) = P(0) = 0.04$$

Now here is the important point. Either the event A occurs or it doesn't occur. That is, either the event A occurs, or the event (not A) occurs. Therefore,

$$P(A \text{ or } (\text{not } A)) = 1$$

Moreover, the events A and (not A) are *mutually exclusive*. Consequently, by the addition principle,

$$P(A \text{ or } (\text{not } A)) = P(A) + P(\text{not } A)$$

Putting these last two equations together we get

$$P(A) + P(\text{not } A) = 1,$$

or rearranging terms,

$$P(A) = 1 - P(\text{not } A)$$

Since $P(\text{not } A) = P(0) = 0.04$, we can quickly find $P(A)$:

$$P(A) = 1 - P(\text{not } A) = 1 - 0.04 = 0.96 \blacktriangleright$$

Below we summarize what was done in the above example and illustrate further.

If E is an event, then the event that "E does *not* occur" is written as **(not E)**. We have   FORMULA

$$P(E) = 1 - P(\text{not } E)$$

◀ The age distribution of the employees in a large company is given in Table 4.17. If an employee is selected at random, what is the probability he or she is under 65?   Example **24**

Let   Solution

$$A = \text{event the employee is under 65.}$$

The problem is to find $P(A)$. Note that

$$(\text{not } A) = \text{event the employee is 65 or over.}$$

From Table 4.17 we see immediately that

$$P(\text{not } A) = 0.02 \quad (2\%).$$

Consequently, using the above principle we find that

$$P(A) = 1 - P(\text{not } A) = 1 - 0.02 = 0.98.$$

Table 4.17

Age	Percentage
Under 18	5%
18–24	17%
25–34	36%
35–49	25%
50–64	15%
65 and over	2%

[*Note.* We could also find $P(A)$ by adding the first five percentages in the table, but the solution given is quicker.] ▶

Exercise **J**   In Example 24, find the probability that an employee selected at random is over 24.

Example **25**   ◀ The probability distribution for the basketball player height data of Example 22 is given in Table 4.18. Here x is the height of a player selected at random.

Table **4.18**

x	70	73	74	76	77	78	80	82	83
$P(x)$	1/16	1/8	1/8	1/8	1/16	1/4	1/16	1/8	1/16

Find the probability that the player selected is *less than* 82″ tall; that is, find $P(x < 82)$.

Solution   It is easier to find the probability of the *nonoccurrence* of the event $(x < 82)$, which is the event $(x \geqslant 82)$. From Table 4.18 we see that

$$P(x \geqslant 82) = P(82) + P(83) = \frac{1}{8} + \frac{1}{16} = \frac{3}{16}$$

Therefore

$$P(x < 82) = 1 - P(x \geqslant 82) = 1 - \frac{3}{16} = \frac{13}{16}. \ ▶$$

Exercise **K**   In Example 25, find the probability that a player selected at random is over 70″ tall.

EXERCISES Section **4.4**

In Exercises 1 through 5 you will be given pairs of events. For each pair, determine whether or not the given events are mutually exclusive. If the events are mutually exclusive, use *the addition principle of probability for mutually exclusive events* to find the probability that either one event or the other event occurs.

1   Suppose a fair coin is tossed three times. Let x be the number of heads obtained. The probability distribution for x is (see Exercise 6, Sec. 4.2).

x	0	1	2	3
$P(x)$	1/8	3/8	3/8	1/8

Let

A = event 1 or 2 heads,
B = event at least 2 heads,
C = event at most 2 heads.

Pairs of events:
a) A, B b) A, C c) B, C

2   According to the *Statistical Abstract of the United States*, the 1976 income-level distribution of families was as given in Table 4.19. Suppose that in 1976 a family was selected at random.

Table **4.19**

Yearly income	Number of families (in thousands)	Percentage
Under $2,000	1,181	2.1%
$2,000–$2,999	1,350	2.4%
$3,000–$3,999	1,912	3.4%
$4,000–$4,999	2,306	4.1%
$5,000–$6,999	4,668	8.3%
$7,000–$9,999	7,199	12.8%
$10,000–$11,999	5,006	8.9%
$12,000–$14,999	7,537	13.4%
$15,000–$24,999	17,042	30.4%
$25,000 and over	7,930	14.1%

Let

A = event the family made at least \$3,000,
B = event the family made at least \$5,000 but less than \$15,000,
C = event the family made under \$2,000 or at least \$25,000.

Pairs of events:
a) A, B b) A, C c) B, C

3. In a horse race, the odds indicate that the probabilities for the various horses winning are as follows:

Horse x	Probability $P(x)$	Horse x	Probability $P(x)$
#1	1/8	#5	1/32
#2	1/16	#6	3/32
#3	1/4	#7	5/32
#4	3/16	#8	3/32

Let

A = event one of the two "favorites" wins.
    (The two favorites are horses #3 and #4.)
B = the winning horse has a number above 4.
C = the winning horse's number is even.
D = one of the two "longshots" wins. (The longshots are horses #2 and #5.)

Pairs of events:
a) A, B b) A, C c) A, D
d) B, C e) B, D f) C, D

4. An article by R. D. Clarke in the *Journal of the Institute of Actuaries* (1946) gave statistics for flying-bomb hits in the south of London during World War II. The area was divided into 576 equal areas, and the following statistics were compiled.

No. of hits	0	1	2	3	4	5 or more
No. of areas	229	211	93	35	7	1

Suppose one of the 576 areas is selected at random. Let

A = event the area received at least 3 hits,
B = event the area received fewer than 4 hits,
C = event the area received at most 1 hit.

Pairs of events:
a) A, B b) A, C c) B, C

5. The age distribution of the employees in a large company is given in Table 4.20. Suppose an employee is selected at random. Let

A = event the employee is 50 or over;
B = event the employee is between 25 and 49, inclusive;

Age	Percentage	**Table 4.20**
Under 18	5%	
18–24	17%	
25–34	36%	
35–49	25%	
50–64	15%	
65 and over	2%	

C = event the employee is either under 18 or in the 65 and over group.

Pairs of events:
a) A, B b) A, C c) B, C

6. Suppose that A and B are mutually exclusive events.
a) If $P(A) = 0.2$ and $P(B) = 0.7$, what is $P(A \text{ or } B)$?
b) If $P(A) = 0.3$ and $P(A \text{ or } B) = 0.62$, what is $P(B)$?

7. Use the rule, $P(A) = 1 - P(\text{not } A)$, to find the probabilities of the following events.
a) A fair coin is tossed three times. What is the probability of getting at least one head?
b) In Exercise 5, what is the probability that an employee selected at random is at least 18?

8. For the events in Exercise 1, find:
a) $P(\text{not } A)$ b) $P(\text{not } B)$ c) $P(\text{not } C)$

9. For the events in Exercise 3, find:
a) $P(\text{not } A)$ b) $P(\text{not } B)$ c) $P(\text{not } C)$

10. Suppose z is a random variable, and $P(z > 1.96) = 0.025$. What is $P(z \leqslant 1.96)$?

11. Suppose t is a random variable with $P(t > 2.02) = 0.05$. Also suppose $P(t < -2.02) = 0.05$. What is $P(-2.02 \leqslant t \leqslant 2.02)$?

12. Suppose z is a random variable with $P(-1.64 \leqslant z \leqslant 1.64) = 0.90$. Also assume $P(z > 1.64) = P(z < -1.64)$. Find $P(z > 1.64)$.

13. Suppose A, B, and C are three mutually exclusive events; that is, any two are mutually exclusive. Use the addition principle of probability to show

$$P(A \text{ or } B \text{ or } C) = P(A) + P(B) + P(C).$$

14. Suppose x is a random variable with $P(x > c) = \alpha$, where c and α are numbers, and $0 \leqslant \alpha \leqslant 1$. What is $P(x \leqslant c)$?

15. Assume y is a random variable with $P(y > c) = \alpha/2$, where c and α are numbers, and $0 \leqslant \alpha \leqslant 1$. Also suppose $P(y < -c) = P(y > c)$. Find $P(-c \leqslant y \leqslant c)$.

16. Suppose t is a random variable with $P(-c \leqslant t \leqslant c) = 1 - \alpha$. Assume also that $P(t < -c) = P(t > c)$. Find $P(t > c)$ in terms of α.

4.5   Further rules of probability

We have seen that if two events *A* and *B* are *mutually exclusive*, then the probability of (*A* or *B*) is $P(A) + P(B)$. This is the addition principle of probability for mutually exclusive events. What happens if *A* and *B* are *not* mutually exclusive? To answer this question, we first look at another way of forming a compound event from *A* and *B*.

Example **26**   ◖ In Example 15 we looked at the two events

A = event the family has an even number of children (0, 2, or 4).

B = event the family has at least two children (2, 3, or 4).

We noted that *A* and *B* both occur if the family selected has either two or four children. The event that *both A and B occur* is designated by

(*A* & *B*)

and consists of the individual outcomes "two children" and "four children." ◗

DEFINITION   **(A & B):** The event (*A* & *B*) occurs if *both* event *A* *and* event *B* occur and consists of all individual outcomes common to both event *A* and event *B*.

Example **27**   ◖ In rolling a pair of fair dice, let

A = event doubles are rolled,

B = event the sum is 8.

We saw in Example 19 (see also page 105) that the events *A* and *B* both occur if a

is rolled. This is the only outcome common to both *A* and *B*. So (*A* & *B*), in this case, consists of the individual outcome

 ◗

Example **28**   ◖ In Example 17 (rolling dice), we noted that the events

A = event sum is 7 or 11,

and

B = event sum is 2, 3, or 12,

are mutually exclusive. So *A* and *B* cannot both occur at the same time—they have no common outcomes. In this case, (*A* & *B*) is impossible. ◗

Suppose a card is selected at random from an ordinary deck of 52 cards. Let

$$A = \text{event a king is selected,}$$

$$B = \text{event a heart is selected,}$$

$$C = \text{event a 10 is selected,}$$

$$D = \text{event a face card is selected.}$$

a) Describe in words the events
   i) (A & B)   ii) (A & C)   iii) (A & D)
   iv) (B & C)   v) (B & D)   vi) (C & D)

b) For the events in (a) that are *not* impossible, list the individual outcomes.

We can now develop our last general rule for computing probabilities of compound events.

◀ In rolling a pair of fair dice, let   Example **29**

A = event doubles are rolled (both dice show the same number),
B = event the sum is 8.

The event A can occur in six out of the 36 possible rolls:

So,

$$P(A) = \frac{f}{N} = \frac{6}{36} = \frac{1}{6}.$$

The event B can occur in five out of the 36 possible rolls:

So,

$$P(B) = \frac{f}{N} = \frac{5}{36}$$

The event $(A$ or $B)$ can occur in ten out of 36 possible rolls:

◀ In the family survey of Example 10, page 93, what is the probability that the family   Example **30**
selected at random has either an even number of children, or at least two children?

So,

$$P(A \text{ or } B) = \frac{f}{N} = \frac{10}{36} = \frac{5}{18}.$$

Note that

$$P(A \text{ or } B) \neq P(A) + P(B)$$

$$\frac{10}{36} \neq \frac{6}{36} + \frac{5}{36}$$

The reason is that A and B are *not* mutually exclusive. They have the common outcome

So when we add $P(A)$ to $P(B)$ we "count"

twice, instead of once. If we subtract the probability of the common outcome

or $(A \& B)$ we do get the correct answer:

$$P(A \text{ or } B) = P(A) + P(B) - P(A \& B)$$

$$\frac{10}{36} = \frac{6}{36} + \frac{5}{36} - \frac{1}{36}$$

We can now state the general formula for computing $P(A \text{ or } B)$.

FORMULA **The addition principle of probability:**

If A and B are any two events, then

$$P(A \text{ or } B) = P(A) + P(B) - P(A \& B)$$

Note that this formula is consistent with the addition principle for mutually exclusive events since, if the events A and B are mutually exclusive, then $(A \& B)$ is impossible and $P(A \& B) = 0$.

Example **30** ❰ In the family survey of Example 10, page 98, what is the probability that the family selected at random has either an even number of children, or at least two children?

Let

A = event the family has an even number of children (0, 2, or 4),

B = event the family has at least 2 children (2, 3, or 4).

Using Table 4.10 (page 99), we find that $P(A) = 0.700$ and $P(B) = 0.350$. The outcomes common to A and B are 2 and 4:

$(A \ \& \ B)$ = event the family has an even number of children, *and* has at least two children = event 2 or 4 children,

$P(A \ \& \ B) = 0.175 + 0.075 = 0.250$

So,

$$P(A \text{ or } B) = P(A) + P(B) - P(A \ \& \ B)$$
$$= 0.700 + 0.350 - 0.250 = 0.800 \ \blacktriangleright$$

As in Exercise L, suppose a card is selected at random from an ordinary deck of 52 cards. Let

A = event a king is selected,

B = event a heart is selected,

C = event a 10 is selected,

D = event a face card is selected.

Exercise **M**

Using the *addition principle of probability*, determine

a) $P(A \text{ or } B)$ b) $P(A \text{ or } C)$ c) $P(A \text{ or } D)$

d) $P(B \text{ or } C)$ e) $P(B \text{ or } D)$ f) $P(C \text{ or } D)$

Remark. You may have noted that, in Example 30, it is easier to compute $P(A \text{ or } B)$ directly in terms of individual outcomes; $(A \text{ or } B)$ consists of the individual outcomes 0, 2, 3, 4; and so

$$P(A \text{ or } B) = 0.450 + 0.175 + 0.100 + 0.075 = 0.800$$

We have derived the addition principle because it is quite important in developing the *theory* of probability and, more practically, there are many cases where using the formula is the easier or only way to compute $P(A \text{ or } B)$. The following example illustrates the latter point, as well as demonstrating our earlier rules.

◀ In 1975, according to the FBI[†], a total of 1,513,196 people were arrested for serious crimes. Of these, 1,206,596 were male and 306,600 were female. Of the males arrested, 539,169 were under 18, and 124,271 of the females arrested were under 18.

Example **31***

1. What percentage of those arrested were female?
2. What percentage of those arrested were under 18?
3. What percentage of those arrested were either female or under 18?

---

* This example is optional.

† Uniform Crime Reports for the United States.

Solution   Since relative frequencies are the same as probabilities for population data, we can phrase the problem in terms of probability. Let

$$M = \text{event person arrested is male,}$$
$$F = \text{event person arrested is female,}$$
$$E = \text{event person arrested is under 18.}$$

1. $P(F) = \dfrac{f}{N} = \dfrac{306,600}{1,513,196} = 0.203$. So 20.3 *percent of those arrested were female.*

2. $P(E)$ cannot be calculated directly from the data, but the data does tell us that the probability that the person arrested is male *and* under 18 is

Males under 18 arrested

$$P(M \ \& \ E) = \frac{539,169}{1,513,196} = 0.356$$

Total arrests

Similarly the probability that the person arrested is female *and* under 18 is

$$P(F \ \& \ E) = \frac{124,271}{1,513,196} = 0.082$$

Now the event E (under 18) occurs when either $(M \ \& \ E)$—(male and under 18), *or* $(F \ \& \ E)$—(female and under 18)—occurs. In other words,

$$E = ((M \ \& \ E) \ \text{ or } \ (F \ \& \ E))$$

Since the events $(M \ \& \ E)$ and $(F \ \& \ E)$ are mutually exclusive (why?), we have, by the addition principle,

$$P(E) = P(M \ \& \ E) + P(F \ \& \ E)$$
$$= 0.356 + 0.082 = 0.438$$

Consequently, 43.8 *percent of those arrested were under* 18.

3. $P(F \text{ or } E) = P(F) + P(E) - P(F \ \& \ E) = 0.203 + 0.438 - 0.082 = 0.559$. So 55.9 *percent of those arrested were either female or under* 18. ▶

EXERCISES Section **4.5**

In each of Exercises 1 through 8, two events, A and B, are given. For each exercise

a) Find $P(A)$.　　　　　　b) Find $P(B)$.
c) Describe in words (or using individual outcomes) the event $(A \ \& \ B)$.
d) Find $P(A \ \& \ B)$.
e) Find $P(A \text{ or } B)$ using the formula

$$P(A \text{ or } B) = P(A) + P(B) - P(A \ \& \ B).$$

1　Suppose a fair coin is tossed four times. Let x be the number of heads obtained. The probability distribution for x is:

x	0	1	2	3	4
$P(x)$	1/16	1/4	3/8	1/4	1/16

A = event of at least two heads,

B = event of an even number of heads.

2 In Exercise 4 of Sec. 4.3, we gave the statistics for flying-bomb hits in the south of London during World War II. The entire area was divided into 576 equal areas. The statistics are given below:

No. of hits	0	1	2	3	4	5 or more
No. of areas	229	211	93	35	7	1

If an area is selected at random, let

   A = event that the area received at least two hits,

   B = event that the area received fewer than five hits.

3 The total (estimated) school expenditures for the 20 most populous states in 1975 are given in Table 4.21 in grouped form (in millions of dollars).

Table 4.21

Expenditure (in millions)	Number of states	Relative frequency
0– 999	3	0.15
1000–1999	9	0.45
2000–2999	3	0.15
3000–3999	3	0.15
4000–4999	0	0.00
5000–5999	0	0.00
6000–6999	1	0.05
7000–7999	1	0.05

Suppose a state is chosen at random from the 20 most populous states (in 1975). Let

   A = event that the total school expenditures for the state in 1975 were at least 1000 but less than 5000.

   B = event that the total school expenditures for the state in 1975 were at least 3000.

4 A card is selected at random from an ordinary deck of 52 cards. Let

   A = event that the card is a heart

   B = event that the card is a face card

5 Table 4.22 shows the probability distribution for the ages of travelers in the United States in 1977.

Table 4.22

Age class	Probability
Under 18	0.220
18–24	0.150
25–44	0.331
45–64	0.232
65 and over	0.067

Let

   A = event that a traveler is 18 or older,

   B = event that a traveler is 64 or younger.

6 In 1977 there were 99.9 million cars in use in the United States. The distribution of these cars by age was:

Years	Under 3	3–5	6–8	9–11	12 and over
Proportion	0.242	0.301	0.223	0.142	0.092

Suppose a car (in 1977) is selected at random. Let

   A = event that the car is five years old or less,

   B = event that the car is nine or more years old.

7 In a city of 120,000, the party affiliations by religion are given below.

	Democrat	Republican
Catholic	13,200	10,800
Jewish	3,600	2,400
Protestant	36,000	54,000

Suppose a voter is selected at random from this city. Let

   A = event that the person is Catholic,

   B = event that the person is a Democrat.

8 An automobile manufacturer has three production plants. Each plant produces both regular cars and station wagons. For every 1000 automobiles manufactured, the plants produce the following quantities.

Type	Plant		
	I	II	III
Regular cars	405	200	285
Station wagons	45	50	15

Suppose an automobile produced by this manufacturer is selected at random. Let

A = event it is a regular car,

B = event it was produced by plant II.

9 Suppose A and B are events, and $P(A) = \frac{1}{4}$, $P(B) = \frac{1}{3}$, and $P(A \text{ or } B) = \frac{1}{2}$.
a) Are A and B mutually exclusive?
b) What is $P(A \& B)$?

10 If $P(A) = \frac{1}{3}$, $P(A \text{ or } B) = \frac{1}{2}$, and $P(A \& B) = \frac{1}{10}$, what is $P(B)$?

11 The following is a generalization of the rule

$$P(A \text{ or } B) = P(A) + P(B) - P(A \& B).$$

Suppose A, B, and C are events. Show that

$$P(A \text{ or } B \text{ or } C) = P(A) + P(B) + P(C)$$
$$- P(A \& B) - P(A \& C)$$
$$- P(B \& C)$$
$$+ P(A \& B \& C).$$

[*Hint*: Let $D = (B \text{ or } C)$, and apply the rule for finding $P(A \text{ or } D)$. Also notice that $(A \& D) = ((A \& B) \text{ or } (A \& C))$.]

12 Suppose A, B, and C are the events given in Exercise 1 of Sec. 4.4.
a) Calculate the probabilities of the events A, B, C, $(A \& B)$, $(A \& C)$, $(B \& C)$, $(A \& B \& C)$, and $(A \text{ or } B \text{ or } C)$, by using the fact that the probability of an event is the sum of the probabilities of the individual outcomes that make up the event.
b) Show that the probabilities you got in (a) satisfy the equation given in Exercise 11.

4.6 Conditional probability

In this section we will discuss the concept of conditional probability. This concept is of both theoretical and practical value in the study of probability and statistics.

Example **32** ◀ A study* of the method of transportation used to get to work in the U.S. revealed the data (in hundreds of thousands of workers) given in Table 4.23. Suppose a worker is chosen at random from the (working) population of 67,000,000 workers in Table 4.23.

Table **4.23**

Mode of transportation to work	Living area		
	Urban	Rural	Total
Automobile	450	150	600
Public transportation	65	5	70
Total	515	155	670

---

*· This data is based on a 1970 report of the U.S. Bureau of the Census. The data has been altered to make the numbers easier to work with.

Let

U = event the person lives in an urban area,
R = event the person lives in a rural area,
A = event the person uses an automobile to travel to work,
T = event the person uses public transportation to travel to work.

In this case, probabilities are just relative frequencies, and so

$$P(U) = \frac{f}{N} = \frac{515}{670} = 0.769$$

$$P(R) = \frac{f}{N} = \frac{155}{670} = 0.231$$

$$P(A) = \frac{f}{N} = \frac{600}{670} = 0.896$$

$$P(T) = \frac{f}{N} = \frac{70}{670} = 0.104$$

But if we are given additional information concerning the experiment, these probabilities may change. For example, suppose we are told that the worker is being selected from an urban area (that is, we are given that the event U occurs). Now what is the probability that the person uses an automobile to travel to work (that is, what is the probability A occurs)? In short, what is the probability A occurs, *given* that U has occurred? We write this probability as

$$P(A|U),$$

which is read "the probability of A given U".

To find this probability, we simply restrict our attention to the "Urban" column of Table 4.23. The answer is

$$P(A|U) = \frac{\text{Number of urban workers using automobiles}}{\text{Total number of urban workers}} = \frac{450}{515} = 0.874$$

Note that $P(A|U) = 0.874$ while $P(A) = 0.896$, so that $P(A|U) \neq P(A)$. The additional (given) information that we are selecting an urban worker (that is, that U occurs) has affected the probability of A occurring. ▶

If E and F are events, then the **conditional probability**, $P(F|E)$, is the probability that the event F occurs, *given* that the event E occurs.   DEFINITION

In Example 32, find the (conditional) probability that a worker selected at random uses an automobile to travel to work, given that the worker is being selected from a rural area. That is, find $P(A|R)$.   Exercise **N**

As illustrated in Example 32, it is often the case that $P(F|E) \neq P(F)$. That is, the knowledge that event E has occurred will often affect the probability of event F occurring.

If we examine carefully what we did in computing $P(A|U)$ in Example 32, we can derive a formula for conditional probabilities in terms of ordinary probabilities. We found that

$$P(A|U) = \frac{\text{Number of urban workers using automobiles}}{\text{Total number of urban workers}} = \frac{450}{515}$$

The event corresponding to selecting "an urban worker who uses an automobile to travel to work" is just $(U \ \& \ A)$—"is an urban worker *and* uses an automobile." From Table 4.23 we see that

$$P(U \ \& \ A) = \frac{450}{670} \quad \text{and} \quad P(U) = \frac{515}{670}$$

So, by arithmetic, we see that

$$P(A|U) = \frac{450}{515} = \frac{450/670}{515/670} = \frac{P(U \ \& \ A)}{P(U)}$$

Thus $P(A|U)$ can be found if we know $P(U \ \& \ A)$ and $P(U)$. This formula holds in general.

FORMULA

Conditional probabilities:

If E and F are events, then

$$P(F|E) = \frac{P(E \ \& \ F)}{P(E)}.$$

We shall further illustrate this formula by presenting several more examples. The following step-by-step method will be helpful in solving conditional probability problems using our formula above. These steps will be illustrated in the examples that follow.

Procedure for solving conditional probability problems

Step 1. Write down the events of interest in the problem, and choose a letter to represent each event.

Step 2. Determine which conditional probability you must find, and use the above formula to express this conditional probability in terms of ordinary probabilities.

Step 3. Find the (ordinary) probabilities of the events from Step 2.

Step 4. Use the formula to find the desired conditional probability.

Example **33**   ❲ In Example 32, what is the (conditional) probability that a person selected at random is a rural worker, given that the person drives an automobile to work?

Step 1. The events of interest in this problem are "rural worker" and "drives an automobile to work." So, let's set

$$R = \text{event that the person is a rural worker}$$

and

$$A = \text{event that the person drives an automobile to work.}$$

Step 2. We want to find the *conditional* probability that a person selected at random is a rural worker (R), given that the person drives an automobile to work (A). That is, find $P(R|A)$. We can compute this using

$$P(R|A) = \frac{P(A \ \& \ R)}{P(A)}.$$

Step 3. We need to compute the (ordinary) probabilities $P(A \ \& \ R)$ and $P(A)$. But, from Table 4.23 we see that

$$P(A \ \& \ R) = \frac{150}{670} \quad \text{and} \quad P(A) = \frac{600}{670}$$

Step 4. From Steps 2 and 3 we have

$$P(R|A) = \frac{P(A \ \& \ R)}{P(A)} = \frac{150/670}{600/670} = \frac{150}{600} = 0.25$$

This can be interpreted to mean that 25 percent of the people using automobiles to travel to work are rural workers. ▶

Our formula for computing conditional probabilities in terms of ordinary probabilities works equally well when we are dealing with problems where probabilities are not relative frequencies.

◀ Suppose we are playing craps with one red die and one black die. Suppose we roll the dice and the black die rolls out of view, but we see that the red die comes up 6. What is the probability that the sum of the dice is either 7 or 11?   Example **34**

We are *given* that the red die comes up 6, and we want to compute the (conditional) probability that the sum of the dice is 7 or 11.

Step 1. The events of interest are "the red die comes up 6" and "the sum of the dice is 7 or 11". So, let

$$E = \text{event the red die comes up 6}$$

and

$$F = \text{event the sum of the dice is 7 or 11.}$$

Step 2. The problem is to find $P(F|E)$.

$$P(F|E) = \frac{P(E \ \& \ F)}{P(E)}$$

Step 3. We need to compute the probabilities $P(E \ \& \ F)$ and $P(E)$. From Fig. 4.1 we see that the outcomes making up the events E and F are as shown below.

$F =$

$E =$

and so,

$(E \ \& \ F) =$

Remembering that there are 36 equally likely outcomes in this experiment, we conclude that

$$P(E \ \& \ F) = \frac{2}{36} = \frac{1}{18}$$

and

$$P(E) = \frac{6}{36} = \frac{1}{6}.$$

Step 4. $P(F|E) = \dfrac{P(E \ \& \ F)}{P(E)} = \dfrac{1/18}{1/6} = \dfrac{1}{3}.$

Note, by the way, that $P(F) = 8/36 = 2/9$ and so $P(F|E) \neq P(F)$. That is, knowing the red die comes up 6 *does* affect the probability that the sum of the dice is 7 or 11. ▶

Exercise **O**   In a city of 120,000 the party affiliations by religion are as shown in Table 4.24.

Table 4.24

	Democrat	Republican	Total
Catholic	13,200	10,800	24,000
Jewish	3,600	2,400	6,000
Protestant	36,000	54,000	90,000
Total	52,800	67,200	120,000

If a voter is selected at random, let

$$C = \text{event that the voter is Catholic,}$$

$$R = \text{event that the voter is Republican.}$$

Determine the following probabilities:
a) $P(C)$ b) $P(R)$ c) $P(R \& C)$
d) Use (b) and (c) to find $P(C|R)$.
e) Interpret your results in (d) in terms of percentages.

We have been computing conditional probabilities using the formula

$$P(F|E) = \frac{P(E \& F)}{P(E)}$$

For the problems so far considered, it has been relatively easy to find $P(E \& F)$ and $P(E)$. However, it is often the case that it is the conditional probability $P(F|E)$ and the probability $P(E)$ that are easy to find, or are clear from the context of the problem. In such cases we can then use the equation above to find $P(E \& F)$. We obtain the formula for doing this by multiplying both sides of the above equation by $P(E)$. The result is

$$P(E \& F) = P(E)P(F|E)$$

This is called the **multiplication principle of probability**.
   We illustrate this principle in the following examples.

◖ Suppose that two cards are selected at random from an ordinary deck of 52 cards. Assume that the first card selected is *not* replaced before the second card is drawn. Find the probability that: Example **35**
a) Both cards selected are hearts.
b) The first card selected is a heart, and the second a spade.

Let   Solution

   H1 = event the first card selected is a heart,
   H2 = event the second card selected is a heart,
   S2 = event the second card selected is a spade.

a) In this part we want to find the probability that both cards selected are hearts—that is, $P(\text{H1 \& H2})$. First of all,

$$P(\text{H1}) = \frac{13}{52} = \frac{1}{4}.$$

Now, given that the first card selected is a heart (i.e., given that H1 occurs), there are then 51 cards left in the deck, 12 of which are hearts. So the (conditional) probability that the second card selected is a heart, *given* that the first card selected is a heart, is

$$P(\text{H2}|\text{H1}) = \frac{12}{51}$$

Note that we did *not* use the formula for finding $P(H2|H1)$. This conditional probability was found directly.

Using the *multiplication principle of probability*, we can now find $P(H1 \& H2)$:

$$P(H1 \& H2) = P(H1)\,P(H2|H1) = \frac{1}{4} \cdot \frac{12}{51} = \frac{3}{51} \doteq 0.059$$

That is, the probability that both cards selected are hearts is $3/51$.

b) Here we want to find $P(H1 \& S2)$. Now,

$$P(H1) = \frac{13}{52} = \frac{1}{4},$$

$$P(S2|H1) = \frac{13}{51}. \quad \longleftarrow \text{No. of spades left}$$
$$\longleftarrow \text{No. of cards left}$$

Consequently, by the *multiplication principle*,

$$P(H1 \& S2) = P(H1) \cdot P(S2|H1) = \frac{1}{4} \cdot \frac{13}{51} = \frac{13}{204} \doteq 0.064 \quad \blacktriangleright$$

Example 36 ◖ The frequency distribution for the heights of the sixteen players on the 1978 Arizona State University basketball team is given in Table 4.25.

Table 4.25

Height (in.)	70	73	74	76	77	78	80	82	83
Frequency	1	2	2	2	1	4	1	2	1

Suppose two players are selected at random *without replacement*—i.e., the first player selected is excluded from the second selection. What is the probability that both players selected are at least 78″ tall?

Solution Let

$$E = \text{event the first player selected is at least 78″ tall,}$$
$$F = \text{event the second player selected is at least 78″ tall.}$$

We need to find $P(E \& F)$. Now,

$$P(E) = \frac{\text{No. of players at least 78″ tall}}{\text{No. of players}} = \frac{8}{16}$$

Also,

$$P(F|E) = \frac{\text{No. of players left at least 78″ tall}}{\text{No. of players left}} = \frac{7}{15}.$$

Consequently, by the multiplication principle,

$$P(E \ \& \ F) = P(E) P(F|E) = \frac{8}{16} \cdot \frac{7}{15} = \frac{7}{30} \ \blacktriangleright$$

In Example 36 find the probability that:

a) both players are under 76″ tall;
b) the first player is under 76″ tall, and the second at least 76″ tall;
c) both players are 78″ tall;
d) both players are 80″ tall.

Exercise **P**

EXERCISES Section **4.6**

1 Suppose $P(E) = \frac{1}{3}$, $P(F) = \frac{1}{4}$, $P(E \ \& \ F) = \frac{1}{6}$.
a) Find $P(E|F)$. b) Find $P(F|E)$.

2 Suppose a fair coin is tossed twice. Outcomes: HH, HT, TH, TT.
a) Find the (conditional) probability that both coins come up heads, given that the first one comes up heads.
b) Find the (conditional) probability that both coins come up heads, given that at least one of the coins comes up heads.

3 In a city of 120,000 the party affiliations by religion are as given below:

	Democrat	Republican
Catholic	13,200	10,800
Jewish	3,600	2,400
Protestant	36,000	54,000

a) What is the probability that a voter selected at random is a Democrat, given that the voter is Catholic?
b) What is the probability that a voter selected at random is Jewish, given that the voter is a Republican?
c) Interpret your results in (a) and (b) in terms of percentages.

4 An automobile manufacturer has three production plants. Each plant produces both regular cars and station wagons. For every 1000 automobiles manufactured, the plants produce the following quantities:

	Plant		
Type	I	II	III
Regular cars	405	200	285
Station wagons	45	50	15

If an automobile produced by this manufacturer is selected at random, what is the probability that:
a) It is a station wagon, given that it was produced by Plant I?
b) It was produced by Plant II, given that it is a station wagon?
c) It is a regular car, given that it was produced by either Plant I or Plant II?
d) Interpret your results in (a)–(c) in terms of percentages.

5 According to the *Statistical Abstract of the United States*, the composition of the 93rd Congress by political party was as follows:

	House	Senate
Democrat	239	56
Republican	193	42
Other	1	2

If a Congressman was selected at random from the 93rd Congress, what is the probability that:
a) He was a Democrat, given that he was a Senator?
b) He was in the House, given that he was a Democrat?
c) Interpret your results in (a) and (b) in terms of percentages.

6 In 1977, the ages of cars and trucks in use in a western city were as follows. (Numbers of vehicles are in thousands.)

Age	Under 3 years	3–5	6–8	9–11	12 years and over	Total
Cars	26.2	30.6	21.3	15.1	9.3	102.5
Trucks	6.9	7.6	4.8	3.7	4.8	27.8

a) What is the probability that a vehicle selected at random is a truck, given that the vehicle is 6–8 years old?
b) What is the probability that a vehicle selected at random is 6–8 years old, given that the vehicle is a truck?

c) What is the probability that a vehicle selected at random is under three years old, given that the vehicle is a car?

d) What is the probability that a vehicle selected at random is twelve years old or over, given that the vehicle is a car?

7 A high school had the following numbers of students participating in varsity sports for each class:

	Male	Female
Freshman	50	39
Sophomore	71	56
Junior	73	45
Senior	61	32

Find the probability that a student participating in varsity sports
a) is a junior, given that the student is a male,
b) is a female, given that the student is a junior,
c) is a junior, given that the student is a female.

8 Cards numbered 1, 2, 3, . . . , 10 are placed in a box. The box is shaken and a blindfolded person selects two successive cards. (He does not replace the first card selected.)
a) What is the probability that the first card selected is numbered 6?
b) Given that the first card was a 6, what is the probability that the second is numbered 9?
c) What is the probability of selecting first a 6 and then a 9?

9 A person has agreed to participate in an ESP experiment. He is asked to pick two random numbers between 1 and 6. (The second number must be different from the first.) Let

H = the event that the first number picked is a 3,
K = the event that the second number picked is greater than 4.

Find:
a) $P(H)$ b) $P(K \mid H)$ c) $P(H \& K)$
Find the probability that:
d) Both numbers picked are less than 3.
e) Both numbers picked are greater than 3.

10 Use the table on party affiliations by religion in Exercise 3 above to:
a) Find the probability distribution of religion for Democrats (i.e., make a table showing the probability that a Democrat is Catholic, Protestant or Jewish).
b) Find the probability distribution of religion for Republicans.
c) The above probability distributions are called *conditional distributions*. Find three other conditional distributions in Exercise 3 that would interest a political analyst.

11 In Exercise 3 above, let

D = event that a randomly selected voter is a Democrat,
R = event that a randomly selected voter is a Republican,
C = event that a randomly selected voter is a Catholic.

a) Find $P(C)$.   b) Find $P(D \& C) + P(R \& C)$.
c) What is the relation between $P(C)$ and $[P(D \& C) + P(R \& C)]$?
d) Without calculating any further numerical values, prove that $P(C) = P(D)P(C \mid D) + P(R)P(C \mid R)$.

4.7   Independence

In this section we consider the important concept in probability of two events being *independent* of each other. Intuitively this means that the occurrence (or nonoccurrence) of one of the events in no way affects the probability of the occurrence of the other event. We will illustrate this concept in the following example.

Example 37   ◀ As in Example 34, suppose we are playing craps with one red die and one black die. Again, suppose we roll the dice, and the black die rolls out of view, but we see that the red die comes up 6. What is the probability that the black die shows a 5?

Solution   We are *given* that the red die comes up 6, and we want to compute the (conditional) probability that the black die shows a 5. We use the procedure given on page 120.

Step 1. Let the event "the red die comes up 6" be denoted by E, and the event "the black die shows a 5" by H.

Step 2. The problem is then to find $P(H|E)$. By our formula,

$$P(H|E) = \frac{P(E \& H)}{P(E)}$$

Step 3. We need to compute the probabilities $P(E \& H)$ and $P(E)$. Referring to Fig. 4.1 on page 92, we see that E consists of six outcomes:

and $(E \& H)$ consists of the single outcome

Therefore, $P(E) = \dfrac{6}{36} = \dfrac{1}{6}$ and $P(E \& H) = \dfrac{1}{36}$.

Step 4. $P(H|E) = \dfrac{P(E \& H)}{P(E)} = \dfrac{1/36}{1/6} = \dfrac{1}{6}$.

So, $$P(H|E) = \frac{1}{6}.$$

Now, note from Fig. 4.1 that H consists of 6 out of 36 possible outcomes, and so we also have

$$P(H) = \frac{6}{36} = \frac{1}{6}.$$

So, in contrast to our previous examples, we have, in this case,

$$P(H|E) = P(H)$$

(both probabilities equal $\frac{1}{6}$). So, knowing that the red die comes up 6 *does not* affect the probability that the black die comes up 5. This, of course, makes sense, because what comes up on the red die should have nothing to do with what comes up on the black die. We say that H is independent of E. ▶

Independence: Suppose E and F are events, and that

$$P(F|E) = P(F).$$

Then F is said to be *independent* of E, because the knowledge that E has occurred does not affect the probability of F occurring.

DEFINITION

◀ In Example 32 (page 118), the event A, that the person uses an automobile, is *not* independent of the event U, that the person is an urban worker. This is because

Example **38**

$$P(A|U) = 0.874 \quad \text{while} \quad P(A) = 0.896,$$

so that

$$P(A|U) \neq P(A). \blacktriangleright$$

Example **39**   ◖ In Exercise O, we considered the party affiliations by religion in a city of 120,000 voters. The table is repeated below as Table 4.26.

Table 4.26	Democrat	Republican	Total
Catholic	13,200	10,800	24,000
Jewish	3,600	2,400	6,000
Protestant	36,000	54,000	90,000
Total	52,800	67,200	120,000

If a voter is selected at random, let

J = event the voter is Jewish,
D = event the voter is a Democrat.

Is the event D independent of the event J?

Solution   We need to check whether or not $P(D|J) = P(D)$. Using our formula for computing conditional probabilities, along with Table 4.26, we find

$$P(D|J) = \frac{P(J \& D)}{P(J)} = \frac{3,600/120,000}{6,000/120,000} = \frac{36}{60} = 0.60$$

and

$$P(D) = \frac{52,800}{120,000} = 0.44.$$

Consequently, $P(D|J) \neq P(D)$. In other words the event D is *not* independent of the event J. This means that the probability of a Democratic voter being selected is affected by the knowledge that the voter is being selected from among Jewish voters. ▶

Exercise **Q**   In Example 39, let

P = event the voter selected is Protestant,
R = event the voter selected is a Republican.

Is the event P independent of the event R?

The multiplication principle of probability is

$$P(E \& F) = P(E) P(F|E).$$

However, if F is *independent* of E, then $P(F|E) = P(F)$ and $P(F|E)$ can be replaced by $P(F)$ in the multiplication principle, resulting in the following principle:

Multiplication principle of probability for independent events:

FORMULA

If F is independent of E, then

$$P(E \ \& \ F) = P(E)P(F).$$

The multiplication principle of probability for independent events is extremely important both in theory and in practice.

◖ Suppose that 52 percent of the population is female, and that 0.05 percent of the population is blind. Assume that sex and blindness are independent. What is the probability that a person selected at random is a blind female?

Example **40***

Let F be the event that the person selected is female, and B the event that the person selected is blind. Then we want to find $P(B \ \& \ F)$. Now,

Solution

$$P(F) = 0.52,$$
$$P(B) = 0.0005.$$

By assumption, B is independent of F. So, by the multiplication principle,

$$P(B \ \& \ F) = P(B)P(F) = 0.52(0.0005) = 0.00026$$

In other words, 0.026 percent of the population consists of blind females. ◗

In many cases in probability we are *told* (as in Example 40) or can *reasonably assume* that certain events, say E and F, are independent. In such cases, the multiplication principle can then be used to calculate $P(E \ \& \ F)$, when we know $P(E)$ and $P(F)$. This is what we did in Example 40, and we will use this idea again in the next section.

Finally, we point out that the concept of independence for two events can be extended to three or more events. The definition of independence in the case of more than two events is more complicated than that for two events. The only thing of importance for us is that the *multiplication principle* still holds for more than two independent events. For example, if E, F, and G are independent events, then

$$P(E \ \& \ F \ \& \ G) = P(E)P(F)P(G).$$

◖ A roulette wheel contains 38 numbers, of which 18 are red, 18 are black, and 2 are green. In three plays of the wheel, what is the probability that the ball lands on red the first time, black the second time, and green the third time?

Example **41**

Let

Solution

R = event that the ball lands on red the first time,
B = event that the ball lands on black the second time,
G = event that the ball lands on green the third time.

---

* This data is based on information obtained from the *Statistical Abstract of the United States*, 1976.

Then

$$P(R) = \frac{18}{38} = \frac{9}{19}, \qquad P(B) = \frac{18}{38} = \frac{9}{19}, \qquad P(G) = \frac{2}{38} = \frac{1}{19}.$$

If A is the event that the ball lands on red the first time, on black the second time, and on green the third time, then $A = (R \; \& \; B \; \& \; G)$. Also it is reasonable to assume that R, B, and G are *independent*. By the multiplication principle for independent events,

$$P(A) = P(R \; \& \; B \; \& \; G) = P(R)\,P(B)\,P(G) = \frac{9}{19} \cdot \frac{9}{19} \cdot \frac{1}{19} = \frac{81}{6859} \doteq 0.0118 \;\blacktriangleright$$

Exercise **R**   Suppose a fair die is rolled three times. Use the method illustrated in Example 41 to find the probability that the die comes up "5" all three times. (Assume that the rolls of the die are independent of each other.)

EXERCISES Section **4.7**

1 (From Exercise 5, Sec. 4.6) The 93rd Congress was composed as follows:

	House	Senate
Democrat	239	56
Republican	193	42
Other	1	2

Suppose a Congressman was selected at random. Let

E = event the Congressman selected was a Republican,
F = event the Congressman selected was a Senator.

Are the events E and F independent?

2 Suppose a pair of fair dice is rolled (one red, one black). Let

E = event the red die comes up even,
F = event the sum of the dice is even.

a) Find $P(E|F)$.   b) Find $P(F|E)$.
c) Are E and F independent?

3 Two cards are drawn from an ordinary deck of 52 cards. What is the probability that both cards are aces, if:
a) The first card is replaced before the second card is drawn?
b) The first card is not replaced before the second card is drawn?

4 In the game of Yahtzee, five fair dice are rolled.
a) What is the probability of rolling all 6's?
b) What is the probability that all the dice come up the same number?

[Use the fact that the outcomes on different dice are independent, and apply the multiplication principle (extended to five events).]

5 Suppose E and F are independent events, and $P(E) = \frac{1}{3}$, $P(F) = \frac{1}{4}$. What is
a) $P(E \; \& \; F)$?   b) $P(E \text{ or } F)$?

6 Suppose $P(E) = \frac{1}{3}$, $P(F) = \frac{1}{4}$, and $P(E|F) = \frac{1}{3}$.
a) Find $P(E \; \& \; F)$.
b) Are E and F independent?

7 Table 4.27 gives the numbers of persons injured in the United States in 1974, broken down by sex of the person injured and the circumstances in which the injury occurred (numbers in millions).

	Male	Female	Total	**Table 4.27**
Work	8.0	1.3	9.3	
Home	9.8	11.6	21.4	
Other	17.8	12.9	30.7	
Total	35.6	25.8	61.4	

Let

M = event that an injured person is a male,
F = event that an injured person is a female,
W = event that a person is injured at work,
H = event that a person is injured at home.

a) Find $P(M|W)$.
b) Are M and W independent?
c) Are M and H independent?

8 Suppose E and F are independent events with $P(E) > 0$ and $P(F) > 0$. Can E and F be mutually exclusive?

9 Three events A, B, and C are said to be *independent* if:

$$P(A \& B) = P(A) P(B),$$
$$P(A \& C) = P(A) P(C),$$
$$P(B \& C) = P(B) P(C),$$

and

$$P(A \& B \& C) = P(A) P(B) P(C).$$

What do you think is required for *four* events to be independent?

10 Consider the experiment of rolling a pair of fair dice. Let

A = event first die comes up even,
B = event second die comes up even,
C = event sum of dice is even,
D = event first die comes up 1, 2, or 3,
E = event first die comes up 3, 4, or 5,
F = event sum of dice is 5.

a) Are A, B, and C independent?

b) Show that $P(D \& E \& F) = P(D) P(E) P(F)$, but that D, E, and F are *not* independent?

[Use the definition of three events being independent given in Exercise 9.]

11 Suppose a fair coin is tossed four times. There are sixteen equally likely outcomes:

HHHH	THHH	THHT	THTT
HHHT	HHTT	THTH	TTHT
HHTH	HTHT	TTHH	TTTH
HTHH	HTTH	HTTT	TTTT

Let

A = event first toss is heads,
B = event second toss is tails,
C = event last two tosses are heads.

Use the definition of three independent events given in Exercise 9 to show that A, B, and C are independent.

One of the most important probability distributions, because it arises so frequently in so many different types of practical problems, is called the **binomial distribution**. We will introduce the binomial distribution in some examples.

The binomial distribution   4.8

◀ In an ESP experiment, a person in one room selects at random one of ten cards numbered 1 through 10, and a person in another room tries to guess the number. This experiment is repeated three times (with the card chosen being replaced before the next selection). Assuming that the person guessing lacks ESP, what is the probability that he guesses correctly exactly twice?

Example **42**

We let s denote a "success" on a given trial, and f represent a "failure." There are eight possible outcomes:

Solution

sss	sfs	fss	ffs
ssf	sff	fsf	fff

For example, ssf represents success on the first two trials and failure on the third.

These eight possible outcomes, however, are *not* equally likely. To figure out the probability of a typical outcome, we use the fact that the various trials are *independent* of one another, and apply the *multiplication principle of probability for independent events* (see Sec. 4.7, page 129).

Since there are ten cards, the probability of a success (correct guess) on any given trial is $\frac{1}{10}$. The probability of failure on any given trial is $1 - \frac{1}{10} = \frac{9}{10}$. So, for example, the probability of the outcome ssf is

$$P(\text{ssf}) = P(\text{s})\,P(\text{s})\,P(\text{f}) = \frac{1}{10} \cdot \frac{1}{10} \cdot \frac{9}{10} = \left(\frac{1}{10}\right)^2 \cdot \frac{9}{10} = \frac{9}{1000}$$

while the probability of the outcome fsf is

$$P(\text{fsf}) = P(\text{f})\,P(\text{s})\,P(\text{f}) = \frac{9}{10} \cdot \frac{1}{10} \cdot \frac{9}{10} = \frac{1}{10} \cdot \left(\frac{9}{10}\right)^2 = \frac{81}{1000}$$

The probabilities of the various outcomes of the experiment are given in Table 4.28. A **tree diagram** (see Fig. 4.2) is also useful for displaying the probabilities of the various outcomes.

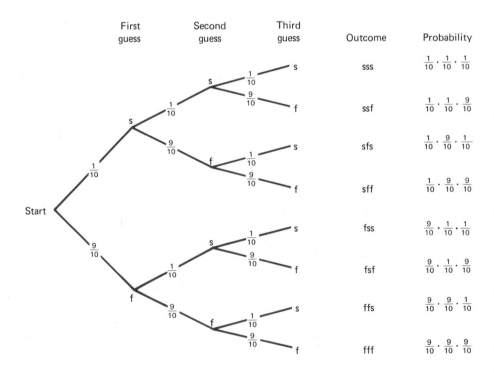

Figure 4.2

Outcome	Probability	Table 4.28
sss	$\frac{1}{10} \cdot \frac{1}{10} \cdot \frac{1}{10} = \left(\frac{1}{10}\right)^3 = \frac{1}{1000}$	
ssf	$\frac{1}{10} \cdot \frac{1}{10} \cdot \frac{9}{10} = \left(\frac{1}{10}\right)^2 \cdot \frac{9}{10} = \frac{9}{1000}$	
sfs	$\frac{1}{10} \cdot \frac{9}{10} \cdot \frac{1}{10} = \left(\frac{1}{10}\right)^2 \cdot \frac{9}{10} = \frac{9}{1000}$	
sff	$\frac{1}{10} \cdot \frac{9}{10} \cdot \frac{9}{10} = \frac{1}{10} \cdot \left(\frac{9}{10}\right)^2 = \frac{81}{1000}$	
fss	$\frac{9}{10} \cdot \frac{1}{10} \cdot \frac{1}{10} = \left(\frac{1}{10}\right)^2 \cdot \frac{9}{10} = \frac{9}{1000}$	
fsf	$\frac{9}{10} \cdot \frac{1}{10} \cdot \frac{9}{10} = \frac{1}{10} \cdot \left(\frac{9}{10}\right)^2 = \frac{81}{1000}$	
ffs	$\frac{9}{10} \cdot \frac{9}{10} \cdot \frac{1}{10} = \frac{1}{10} \cdot \left(\frac{9}{10}\right)^2 = \frac{81}{1000}$	
fff	$\frac{9}{10} \cdot \frac{9}{10} \cdot \frac{9}{10} = \left(\frac{9}{10}\right)^3 = \frac{729}{1000}$	

Now that we have the probabilities of the various outcomes we can answer the question "What is the probability of exactly two correct guesses?"

This event consists of the three outcomes ssf, sfs, and fss. From Table 4.28 we see that *each of these three outcomes has the same probability*—9/1000. So,

$$P(\text{exactly two correct guesses}) = P(\text{ssf}) + P(\text{sfs}) + P(\text{fss})$$

$$= \frac{9}{1000} + \frac{9}{1000} + \frac{9}{1000} = 3 \cdot \frac{9}{1000}$$

$$= \frac{27}{1000}.$$

If we let x be the random variable representing the number of successes (correct guesses), then we have just shown that

$$P(x = 2) = \frac{27}{1000}$$

Similarly, we can determine the entire probability distribution of x from Table 4.28. This is shown in Table 4.29.

Table 4.29	x	$P(x)$
	0	$\dfrac{729}{1000}$
	1	$\dfrac{243}{1000}$
	2	$\dfrac{27}{1000}$
	3	$\dfrac{1}{1000}$

A little later on in this section we will develop a general formula for the binomial distribution. To help motivate this formula it will be instructive to rewrite Table 4.29 as Table 4.30.*

Table 4.30	x	$P(x)$	Description
	0	$1 \cdot \left(\dfrac{1}{10}\right)^0 \cdot \left(\dfrac{9}{10}\right)^3$	0 successes, 3 failures
	1	$3 \cdot \left(\dfrac{1}{10}\right)^1 \cdot \left(\dfrac{9}{10}\right)^2$	1 success, 2 failures
	2	$3 \cdot \left(\dfrac{1}{10}\right)^2 \cdot \left(\dfrac{9}{10}\right)^1$	2 successes, 1 failure
	3	$1 \cdot \left(\dfrac{1}{10}\right)^3 \cdot \left(\dfrac{9}{10}\right)^0$	3 successes, 0 failures

So, for example, the probability of exactly two successes is

$$P(x = 2) = \underset{\substack{\text{Number of} \\ \text{ways to get} \\ \text{two successes}}}{3} \cdot \underset{\substack{\text{Probability} \\ \text{of success}}}{\overset{\text{2 successes}}{\left(\frac{1}{10}\right)^2}} \cdot \underset{\substack{\text{Probability} \\ \text{of failure}}}{\overset{\text{1 failure}}{\left(\frac{9}{10}\right)^1}}$$

Before developing the general formula, let's look at one more example.

---

* In Table 4.30, we make use of the fact that if b is any nonzero number, then $b^0 = 1$.

◖ A certain drug is known to be 80 percent effective in curing a certain disease. If four people with the disease are given the drug, what is the probability that exactly three are cured?

Example **43**

Here we have basically the same situation as in Example 42. Success in this case is "a cure," instead of "a correct guess." The probability of a success on any given trial is 0.8, instead of $\frac{1}{10}$. There are sixteen outcomes instead of eight. These outcomes, along with their probabilities, are listed in Table 4.31. These are obtained in the same way as those in Example 42.

Solution

Outcome	Probability	Description	
ssss	$(0.8)^4(0.2)^0$	4 successes, 0 failures	**Table 4.31**
sssf	$(0.8)^3(0.2)^1$		
ssfs	$(0.8)^3(0.2)^1$	3 successes, 1 failure	
sfss	$(0.8)^3(0.2)^1$		
fsss	$(0.8)^3(0.2)^1$		
ssff	$(0.8)^2(0.2)^2$		
sfsf	$(0.8)^2(0.2)^2$		
sffs	$(0.8)^2(0.2)^2$	2 successes, 2 failures	
fssf	$(0.8)^2(0.2)^2$		
fsfs	$(0.8)^2(0.2)^2$		
ffss	$(0.8)^2(0.2)^2$		
sfff	$(0.8)^1(0.2)^3$		
fsff	$(0.8)^1(0.2)^3$	1 success, 3 failures	
ffsf	$(0.8)^1(0.2)^3$		
fffs	$(0.8)^1(0.2)^3$		
ffff	$(0.8)^0(0.2)^4$	0 successes, 4 failures	

If we let x represent the number of successes (cures), then using Table 4.31 we get the probability distribution of x given in Table 4.32. So the probability the drug cures exactly three out of the four people is:

$$P(x = 3) = \underset{\substack{\uparrow \\ \text{Number of} \\ \text{ways to get} \\ \text{three successes}}}{4} \cdot \underset{\substack{\uparrow \\ \text{Probability} \\ \text{of success}}}{\overset{\overset{\text{3 successes}}{\downarrow}}{(0.8)^3}} \cdot \underset{\substack{\uparrow \\ \text{Probability} \\ \text{of failure}}}{\overset{\overset{\text{1 failure}}{\downarrow}}{(0.2)^1}} = 0.4096.$$

Table 4.32

x	$P(x)$
0	$1 \cdot (0.8)^0(0.2)^4$
1	$4 \cdot (0.8)^1(0.2)^3$
2	$6 \cdot (0.8)^2(0.2)^2$
3	$4 \cdot (0.8)^3(0.2)^1$
4	$1 \cdot (0.8)^4(0.2)^0$

Exercise **S**   A basketball player has a 75 percent success percentage at the free-throw line (i.e., the probability that he makes a free throw is $\frac{3}{4}$). Let s denote a success on a given free throw and f a failure. If the player is shooting two free throws,

a) Construct a table similar to Table 4.28 (page 133) for the various outcomes and their probabilities.
b) Draw a tree diagram for this problem similar to the one given in Fig. 4.2 (page 132).
c) Find the probability distribution of the random variable x, if x is the number of successful free throws in two attempts.

To write down a general formula for the binomial distribution we first need to discuss factorial notation and binomial coefficients.

DEFINITION   **Factorial notation:**   If n is a positive integer, we define $n!$ (read "n factorial") to be

$$n(n-1) \cdots 2 \cdot 1$$

That is, to obtain $n!$, we start with n and continually multiply by the *next smallest integer* until we get to 1.

Example **44**   ◀ Find 3!, 5!, and 7!

Solution

$$3! = 3 \cdot 2 \cdot 1 = 6$$
$$5! = 5 \cdot 4 \cdot 3 \cdot 2 \cdot 1 = 120$$
$$7! = 7 \cdot 6 \cdot 5 \cdot 4 \cdot 3 \cdot 2 \cdot 1 = 5040 \ ▶$$

It will be necessary to define 0! to be equal to 1 (not 0).

$$0! = 1$$

DEFINITION   **Binomial coefficients:**   If n is a positive integer and k is a nonnegative integer less than or equal to n, then we define the *binomial coefficient* as

$$\binom{n}{k} = \frac{n!}{k!\,(n-k)!}$$

Example **45**   ◀ Find the values of the following binomial coefficients:

i) $\binom{4}{1}$ ii) $\binom{5}{3}$ iii) $\binom{6}{2}$ iv) $\binom{8}{0}$

Solution   i) $\binom{4}{1} = \frac{4!}{1!\,(4-1)!} = \frac{4!}{1! \cdot 3!} = \frac{4 \cdot 3 \cdot 2 \cdot 1}{(1) \cdot (3 \cdot 2 \cdot 1)} = 4.$

ii) $\binom{5}{3} = \frac{5!}{3!\,(5-3)!} = \frac{5!}{3!\,2!} = \frac{5 \cdot 4 \cdot 3 \cdot 2 \cdot 1}{(3 \cdot 2 \cdot 1) \cdot (2 \cdot 1)} = 10.$

iii) $\binom{6}{2} = \dfrac{6!}{2!\,(6-2)!} = \dfrac{6!}{2!\,4!} = \dfrac{6\cdot5\cdot4\cdot3\cdot2\cdot1}{(2\cdot1)(4\cdot3\cdot2\cdot1)} = 15.$

iv) $\binom{8}{0} = \dfrac{8!}{0!\,(8-0)!} = \dfrac{8!}{0!\,8!} = \dfrac{8\cdot7\cdot6\cdot5\cdot4\cdot3\cdot2\cdot1}{(1)\cdot(8\cdot7\cdot6\cdot5\cdot4\cdot3\cdot2\cdot1)} = 1.\ \blacktriangleright$

Find the values of the following binomial coefficients:

Exercise **T**

a) $\binom{3}{3}$ b) $\binom{4}{2}$ c) $\binom{5}{0}$ d) $\binom{7}{3}$

The significance of the binomial coefficients for what we are doing is this: *If n success–failure experiments are performed, then the number of ways to get exactly k successes is* $\binom{n}{k}$. Let us check this for what we did in Examples 42 and 43. In Example 42, we wanted exactly two ($k = 2$) successes in three trials ($n = 3$). From Table 4.28 we see that there are three ways this can happen; namely, ssf, sfs, and fss. Thus it should be the case that $\binom{3}{2} = 3$. This is so because

$$\binom{3}{2} = \frac{3!}{2!\,(3-2)!} = \frac{3!}{(2!)\,(1!)} = \frac{3\cdot2\cdot1}{(2\cdot1)\cdot(1)} = 3.$$

In Example 43, we wanted exactly three ($k = 3$) successes in four trials ($n = 4$). From Table 4.31, this can happen in the four ways sssf, ssfs, sfss, and fsss. So we should have $\binom{4}{3} = 4$, which it does:

$$\binom{4}{3} = \frac{4!}{3!\,(4-3)!} = \frac{4!}{(3!)\,(1!)} = \frac{4\cdot3\cdot2\cdot1}{(3\cdot2\cdot1)\cdot(1)} = 4.$$

We now state formally the assumptions for the binomial distribution, along with a general formula.

Assumptions for the binomial distribution

1. n success–failure experiments are performed.
2. The probability of a success on any given trial is p.
3. The trials are independent of each other.

If x represents the number of successes, then the probability formula for x is

k successes $(n-k)$ failures

$$P(x = k) = \binom{n}{k} \cdot p^k \cdot (1-p)^{n-k}$$

Number of ways Probability Probability
to get k successes of success of failure
in n trials

The random variable x is said to have the **binomial distribution** *with parameters n and p.*

The following step-by-step procedure will be useful for solving binomial distribution problems.

Procedure for solving binomial distribution problems

1. Identify a success.
2. Determine p (the success probability).
3. Determine n (the number of trials).
4. If x denotes the number of successes, use the formula

$$P(x = k) = \binom{n}{k} p^k (1-p)^{n-k}$$

to find the desired probability.

Example **46**   ◖ Use the procedure above to solve the problem in the ESP experiment of Example 42 (see page 131).

Solution   1. A success here is guessing the correct number on the card.
2. The success probability is $p = \frac{1}{10}$.
3. The number of trials is $n = 3$.
4. From the formula we have that the probability of exactly two correct guesses $(k = 2)$ is

$$P(x = 2) = \binom{3}{2}\left(\frac{1}{10}\right)^2\left(1 - \frac{1}{10}\right)^{3-2}$$

$$= \frac{3!}{2!\,1!} \cdot \left(\frac{1}{10}\right)^2\left(\frac{9}{10}\right)^1$$

$$= 3 \cdot \left(\frac{1}{10}\right)^2\left(\frac{9}{10}\right)^1 = \frac{27}{1000},$$

which agrees with what we found in Example 42. ◗

Example **47**   ◖ Apply the procedure given above to solve the problem in Example 43.

Solution   The problem is to find the probability that a drug with 80 percent effectiveness cures exactly three out of four people.

1. A success here is counted when the drug cures a person with the disease.
2. The success probability is $p = 0.8$.
3. The number of trials is $n = 4$ (four people are given the drug).
4. Consequently, the desired probability is

$$P(x = 3) = \binom{4}{3}(0.8)^3(1 - 0.8)^{4-3}$$

$$= \frac{4!}{3!\,1!}(0.8)^3(0.2)^1$$

$$= 4 \cdot (0.8)^3(0.2)^1 = 0.4096.$$

This agrees with the answer we obtained in Example 43. ◗

◖ A student takes a multiple-choice test with ten questions. Each question has four   Example **48**
possible answers. If the student guesses randomly on each question, what is the
probability that the student gets

a) exactly 5 correct?
b) exactly 9 correct?
c) exactly 10 correct?
d) at least 90% on the exam?

There are ten questions ($n = 10$). Each question has four possible answers. Since the   Solution
student guesses at each question, the probability of success on any given question is
$\frac{1}{4}$ ($p = \frac{1}{4}$). Let x be the number of successes (that is, the number of questions
answered correctly). Then, by our formula on page 138,

$$P(x = k) = \binom{10}{k} \left(\frac{1}{4}\right)^k \left(1 - \frac{1}{4}\right)^{10-k}$$

a) $P(x = 5) = \binom{10}{5} \left(\frac{1}{4}\right)^5 \left(1 - \frac{1}{4}\right)^{10-5}$

$$= 252 \cdot \left(\frac{1}{4}\right)^5 \left(\frac{3}{4}\right)^5 \doteq 0.058399 \qquad \textit{use table.}$$

b) $P(x = 9) = \binom{10}{9} \left(\frac{1}{4}\right)^9 \left(1 - \frac{1}{4}\right)^{10-9}$

$$= 10 \cdot \left(\frac{1}{4}\right)^9 \left(\frac{3}{4}\right)^1 \doteq 0.000029$$

c) $P(x = 10) = \binom{10}{10} \left(\frac{1}{4}\right)^{10} \left(1 - \frac{1}{4}\right)^{10-10}$

$$= 1 \cdot \left(\frac{1}{4}\right)^{10} \left(\frac{3}{4}\right)^0 \doteq 0.000001$$

d)   The student gets at least 90 percent on the exam if $x = 9$ or $x = 10$. The probability
of this happening is

$$P(x = 9) + P(x = 10) \doteq 0.000029 + 0.000001 = 0.000030 \ \blacktriangleright$$

Refer to Exercise S where a basketball player, with a 75 percent success percentage at   Exercise **U**
the free-throw line, is shooting two free throws. Use the procedure on page 138 to find
the probability that he makes

a) exactly 1,
b) exactly 2,
c) at least 1.

EXERCISES Section **4.8**

1   A fair die is rolled three times. Suppose we consider a success (s) on a roll to be a "5" coming up. So, sfs means the first and third roll yielded a "5", and the second roll didn't.
a) Construct a table for this experiment similar to Table 4.28.
b) Use this table to find the probability of exactly one success (i.e., exactly one "5" in the three rolls).
c) Use the table to find the probability distribution of the random variable x, where x denotes the number of successes in three rolls.

In Exercises 2 through 7, use the formula

$$P(x = k) = \binom{n}{k} p^k (1 - p)^{n-k}$$

to find the probabilities in question.

2   From past experience, a salesman knows that, on the average, 10 percent of the people he contacts will buy his product. If on a given day he will contact four people, what is the probability he will sell his product to
a) exactly two people?   b) exactly three people?
c) at least one person?

3   Based on data obtained in the *Statistical Abstract of the United States*, the probability of a girl being born is about 0.49. If a family has three children, what is the probability they have:
a) exactly one girl?   b) at least one girl?
c) all girls?

4   An automobile manufacturer claims that out of 100 cars, on the average, five will have some sort of defect. If ten cars produced

by this manufacturer are selected at random, what is the probability that
a) at most one will be defective?
b) at least one will be defective?

5   A basketball player has an 80 percent success ratio at the free-throw line. If the player shoots two throws, what is the probability he makes:
a) at least 1?   b) both of them?

6   From past experience, the owner of a restaurant knows that, on the average, 4 percent of the parties making reservations never show up. If on a certain night, 15 parties make reservations, what is the probability that:
a) they all show up?
b) at most two parties don't show up?
c) at least one party doesn't show up?

7   A sales representative for a tire manufacturer claims that his steel-belted radials get at least 35,000 miles. A tire dealer decides to check this claim by testing eight of the tires. If 75 per cent or more of the eight tires he tests get at least 35,000 miles, he will purchase tires from the sales representative. If, in fact, 90 percent of the steel-belted radials produced by the manufacturer get at least 35,000 miles, what is the probability the tire dealer purchases tires from the sales representative?

8   From past experience, the owner of a restaurant knows that, on the average, 4 percent of the parties making reservations never show up. How many reservations can the owner take and still be at least 80 percent sure that all parties making a reservation will show up?

4.9   The mean and standard deviation of a random variable

The mean and standard deviation of a random variable are analogous to the mean and standard deviation of a population. In fact, we will see that if the probability distribution of a random variable arises from a relative-frequency distribution for a population, then these concepts are exactly the same.

Example **49**   ◖ We return to our example of the number of children under 18 in each of a *population* of forty families. As before, x denotes the number of children, and N denotes the population size (in our case $N = 40$). The relative-frequency distribution (see Table 4.3) is given here in Table 4.33. Using the first two columns of the table, the *mean* number of children under 18 in the forty families is (see Sec. 3.7, page 80)

$$\mu = \frac{\sum xf}{N} = \frac{0 \cdot 18 + 1 \cdot 8 + 2 \cdot 7 + 3 \cdot 4 + 4 \cdot 3}{40} = 1.15$$

Children under 18 (x)	Number of families (frequency, f)	Relative frequency, f/N, or Probability, P(x)	
0	18	18/40	**Table 4.33**
1	8	8/40	
2	7	7/40	
3	4	4/40	
4	3	3/40	
	40		

Suppose, as before, we select a family at random, and let x be the random variable denoting the number of children under 18 in the randomly selected family. Now, in this case, probabilities are just relative frequencies, listed in the third column of Table 4.33. So, for each value of x, f/N is just $P(x)$. For example, for $x = 3$, f/N and $P(x)$ are both 4/40. Now, we find

$$\mu = \frac{\sum xf}{N} = \frac{0 \cdot 18 + 1 \cdot 8 + 2 \cdot 7 + 3 \cdot 4 + 4 \cdot 3}{40}$$

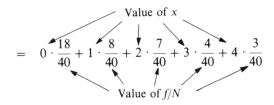

Value of x

$$= \quad 0 \cdot \frac{18}{40} + 1 \cdot \frac{8}{40} + 2 \cdot \frac{7}{40} + 3 \cdot \frac{4}{40} + 4 \cdot \frac{3}{40}$$

Value of f/N

$$= \sum x \frac{f}{N} = \sum x P(x)$$

Since, in this form, the calculation of the mean μ involves the probability distribution of the random variable x, we make the following definition. ▶

Mean of a random variable: The mean, μ, of a random variable x is defined by   DEFINITION

$$\mu = \sum x P(x)$$

It is easier to compute μ by using a probability distribution table. For Example 49 the probability distribution table is given in Table 4.4 (page 95). We repeat this as Table 4.34, along with the values $xP(x)$.

Table 4.34	Children under 18, x	Probability, $P(x)$	$xP(x)$
	0	0.450	0.000
	1	0.200	0.200
	2	0.175	0.350
	3	0.100	0.300
	4	0.075	0.300
			$\sum xP(x) = 1.150$

To compute the mean,

$$\mu = \sum xP(x),$$

we simply add up the last column of Table 4.34. So, in this case, the mean of the random variable x is $\mu = 1.15$.

We should emphasize that the formula $\mu = \sum xP(x)$ applies regardless of whether or not the probability distribution of x arises from a relative frequency distribution for a population.

Example **50** ◖ Suppose a pair of fair dice is rolled. Let x denote the sum of the dice. Find the mean of the random variable x.

Solution The probability distribution of x is given in Table 4.35 along with the values $xP(x)$.

Table 4.35	Sum of dice, x	Probability, $P(x)$	$xP(x)$
	2	1/36	2/36
	3	2/36	6/36
	4	3/36	12/36
	5	4/36	20/36
	6	5/36	30/36
	7	6/36	42/36
	8	5/36	40/36
	9	4/36	36/36
	10	3/36	30/36
	11	2/36	22/36
	12	1/36	12/36
			252/36

So the mean of the random variable x is $\mu = \sum xP(x) = \dfrac{252}{36} = 7$ ▶

Suppose a card is selected at random from an ordinary deck of 52 playing cards. Let x be the value of the card selected [$x = 11$ if an Ace is selected, and $x = 10$ if a face card is selected].

a) Find the probability distribution of the random variable x.
b) Compute the mean μ of the random variable x.

Exercise **V**

The mean of a set of data is just the average value of the data. A similar interpretation can be given to the mean μ of a random variable x. *It is the value of x that one would expect to get, on the average, if the experiment were repeated many, many times.*

Interpretation of the mean μ of a random variable x

◀ In Example 49 (children under 18) we found that $\mu = 1.15$. This means that, if we repeated our experiment of selecting a family at random many, many times then, *on the average*, we would expect to observe 1.15 children. Of course, we can never observe a family with exactly 1.15 children. This is only "on the average." ▶

Example **51**

◀ In Example 50 (sum of two dice) we obtained $\mu = 7$. So this means that if we rolled the dice over and over again, then we would expect that, *on the average*, the sum of the two dice would be 7. ▶

Example **52**

We next examine the concept of the standard deviation of a random variable. We begin with an example.

Standard deviation of a random variable

◀ In Example 49 we considered the number of children in each of forty families. The (population) *standard deviation* of the number of children under 18 is (see Sec. 3.7, page 80):

Example **53**

$$\sigma = \sqrt{\frac{\sum(x - \mu)^2 f}{N}}$$

$$= \sqrt{\frac{(0 - 1.15)^2 \cdot 18 + (1 - 1.15)^2 \cdot 8 + (2 - 1.15)^2 \cdot 7 + (3 - 1.15)^2 \cdot 4 + (4 - 1.15)^2 \cdot 3}{40}}$$

$$= 1.30.$$

If we let x denote the number of children in a family selected at random, then, since probabilities are just relative frequencies in this case, we can rewrite the formula for standard deviation in terms of probabilities. Because this involves the probability distribution of the random variable x, we have the following definition. ▶

Standard deviation of a random variable: The standard deviation, σ, of a random variable x is

$$\sigma = \sqrt{\sum(x - \mu)^2 P(x)}.$$

DEFINITION

Again, we should emphasize that the formula, $\sigma = \sqrt{\sum (x - \mu)^2 P(x)}$ for the standard deviation of a random variable x, applies regardless of whether or not the probability distribution of x arises from a relative-frequency distribution for a population.

Computational form

We learned in Chapter 3 that there are shortcut methods for computing the standard deviation. We won't do the derivation, but the shortcut formula in terms of probability is:

$$\sigma = \sqrt{\sum x^2 P(x) - [\sum x P(x)]^2}$$

Some examples are now presented to illustrate the computations involved in determining the standard deviation of a random variable.

Example **54**

◖ Suppose a pair of fair dice is rolled. Let x denote the sum of the dice. Find the standard deviation of the random variable x.

Solution

In Table 4.36 you will find the probability distribution table for x with the values for $x^2 P(x)$ and $x P(x)$. These are used in the shortcut formula for σ:

$$\sigma = \sqrt{\sum x^2 P(x) - [\sum x P(x)]^2}$$

Table 4.36	Sum of dice, x	Probability, $P(x)$	$x^2 P(x)$	$x P(x)$
	2	1/36	4/36	2/36
	3	2/36	18/36	6/36
	4	3/36	48/36	12/36
	5	4/36	100/36	20/36
	6	5/36	180/36	30/36
	7	6/36	294/36	42/36
	8	5/36	320/36	40/36
	9	4/36	324/36	36/36
	10	3/36	300/36	30/36
	11	2/36	242/36	22/36
	12	1/36	144/36	12/36
			1974/36	252/36
			$\uparrow$ $\sum x^2 P(x)$	$\uparrow$ $\sum x P(x)$

So,

$$\sum x^2 P(x) = \frac{1974}{36} = 54.833 \quad \text{and} \quad \sum x P(x) = \frac{252}{36} = 7.$$

Consequently,

$$\sigma = \sqrt{\sum x^2 P(x) - [\sum x P(x)]^2} = \sqrt{54.833 - 7^2} \doteq 2.42 \ \blacktriangleright$$

Let x be the value of a card selected at random from an ordinary deck of 52 cards (see   Exercise **W**
Exercise V, page 143). Find the standard deviation, σ, of the random variable x.

We conclude this section by giving formulas for the mean and standard deviation of a random variable x having the *binomial distribution*. To motivate these formulas we will first consider an example.

◀ Consider again the ESP experiment of Example 42 (page 131). If x is the number of   Example **55**
correct guesses made by the person, find the mean and standard deviation of the random variable x.

The random variable x, in this case, has the binomial distribution with $n = 3$ (3 trials),   Solution
and $p = \frac{1}{10}$. Using the formula

$$P(x = k) = \binom{3}{k}\left(\frac{1}{10}\right)^k \left(1 - \frac{1}{10}\right)^{3-k}$$

we get the probability distribution of x, which is shown along with the values $x^2 P(x)$ and $x P(x)$ in Table 4.37.

Table 4.37	x	$P(x)$	$x^2 P(x)$	$x P(x)$
	0	0.729	0.000	0.000
	1	0.243	0.243	0.243
	2	0.027	0.108	0.054
	3	0.001	0.009	0.003
			0.360	0.300
			$\uparrow$	$\uparrow$
			$\sum x^2 P(x)$	$\sum x P(x)$

Consequently, we have

$$\mu = \sum x P(x) = 0.3$$

and

$$\sigma = \sqrt{\sum x^2 P(x) - [\sum x P(x)]^2}$$

$$= \sqrt{0.36 - (0.3)^2} = \sqrt{0.36 - 0.09} = \sqrt{0.27} \doteq 0.52$$

On the other hand, since $p = \frac{1}{10}$, we would expect that 10% of all trials would result in success. Since $\frac{1}{10}$ of 3 is 0.3, the mean number of successes is $0.3 = 3 \cdot \frac{1}{10} = np$. This is the same as we found for μ, so that in this case

$$\mu = np.$$

Note also that

$$\sqrt{np(1-p)} = \sqrt{3 \cdot \frac{1}{10} \cdot \left(1 - \frac{1}{10}\right)} = \sqrt{3 \cdot \frac{1}{10} \cdot \frac{9}{10}} = \sqrt{0.27}.$$

This is the same as we found for σ, and so, in this case,

$$\sigma = \sqrt{np(1-p)}. \blacktriangleright$$

The results of the previous example are no accident. Although we won't prove them, the following formulas hold for binomially distributed random variables.

FORMULAS   Suppose x has the *binomial distribution* with parameters n and p. Then the *mean* and *standard deviation* of the random variable x are

$$\mu = np,$$

and

$$\sigma = \sqrt{np(1-p)}.$$

Example **56**   ◖ In Example 48 a student takes a multiple-choice test with ten questions. Each question has four possible answers. If the student guesses at each question, find the mean and standard deviation of the number of questions answered correctly by the student.

Solution   Let x be the number of questions answered correctly by the student. Then x has the binomial distribution with $n = 10$ and $p = \frac{1}{4}$. Consequently,

$$\mu = np = 10 \cdot \frac{1}{4} = 2.50$$

and

$$\sigma = \sqrt{np(1-p)} = \sqrt{10 \cdot \frac{1}{4} \cdot \frac{3}{4}} = \sqrt{\frac{30}{16}} = \sqrt{1.875} \doteq 1.37 \blacktriangleright$$

Exercise **X**   In Example 43, a drug has an effectiveness of 80 percent (0.80) in curing a certain disease. Let x be the number of people cured by the drug, if four people with the disease are given the drug. Find the mean and standard deviation of the random variable x.

EXERCISES Section **4.9**

In each of Exercises 1 through 7, find the *mean* and *standard deviation* of the given random variable.

1 Suppose a fair die is rolled. Let x be the number that comes up.

2 The playing times of 20 albums are (in minutes)

42	32	42	30	45
37	34	23	34	34
30	35	34	35	42
49	35	34	53	33

Let x be the playing time of an album selected at random (see Exercise 1 of Sec. 4.2).

3 A fair coin is tossed three times. The eight *equally likely* outcomes are

HHH	HTH	THH	TTH
HHT	HTT	THT	TTT

Let x be the number of heads obtained (see Exercise 6 of Sec. 4.2).

4 A pair of fair dice is rolled (see Fig. 4.1, page 92). Let x be the smallest number showing on the two dice. For example, if the outcome is

then $x = 3$.

5 A card is selected at random from a deck of 52 cards. Let x be the value of the card selected (Ace = 11, and all face cards equal 10).

6 The number of births per month in 1970, in thousands of people, is given below.

Month	No. of births	Month	No. of births
Jan	299	July	295
Feb	284	Aug	306
Mar	306	Sep	317
Apr	290	Oct	305
May	292	Nov	294
June	285	Dec	287

Suppose a person born in 1970 is selected at random. Let $x = 1$, if the person selected was born in January, $x = 2$, if the person selected was born in February; and so on.

7 A total of 8750 tickets were sold for a rock concert. Tickets were $6, $7.50, and $10. The number of each sold were as follows:

$6.00	$7.50	$10.00
3500	3625	1625

Let x be the price paid for a ticket by a person selected at random from the 8750 people purchasing tickets.

8 Find the mean and standard deviation of the random variables given below:

a)

x	0	1	2	3
$P(x)$	0.1	0.7	0.15	0.05

b)

y	0	10	20	30
$P(y)$	0.1	0.7	0.15	0.05

In Exercises 9 and 10 compute the mean and standard deviation of the given binomial random variable x the long way. That is, first find the probability distribution of x using the formula

$$P(x) = \binom{n}{x} p^x (1-p)^{n-x},$$

Then compute the mean and standard deviation of x by the formulas

$$\mu = \sum xP(x) \quad \text{and} \quad \sigma = \sqrt{\sum x^2 P(x) - \left[\sum xP(x)\right]^2}.$$

Finally, verify your results by computing the mean and standard deviation using the shortcut formulas

$$\mu = np \quad \text{and} \quad \sigma = \sqrt{np(1-p)}.$$

9 Let x be the number of "5's" obtained in rolling a fair die three times (see Exercise 1 of Section 4.8).

10 Let x be the number of sales made by the salesman in Exercise 2, Sec. 4.8.

11 In craps (rolling a pair of fair dice), the probability that the sum is 7 is $\frac{1}{6}$. Therefore, if we roll the dice six times, *on the average*, we will get one (sum of) 7 ($\mu = n \cdot p = 6 \cdot \frac{1}{6} = 1$). But how many times must we roll the dice to be at least 95 percent sure of getting at least one (sum of) 7?

12 Suppose that the probability distribution for a random variable x arises from a relative-frequency distribution for a population. Then the formula for the standard deviation of the population is (see Sec. 3.7, page 80)

$$\sigma = \sqrt{\frac{\Sigma(x-\mu)^2 f}{N}}.$$

On the other hand, the formula given on page 143 for the standard deviation of the random variable x is

$$\sigma = \sqrt{\Sigma(x-\mu)^2 P(x)}.$$

Show that these two formulas are the same. That is, show that

$$\frac{\Sigma(x-\mu)^2 f}{N} = \Sigma(x-\mu)^2 P(x).$$

13 A factory manager collected data on the number of work stoppages per day due to equipment breakdowns. From this data he derived the following probability distribution.

Number of breakdowns	x	0	1	2
Probability	$P(x)$	0.80	0.15	0.05

a) Compute the mean, μ, and variance, σ^2, of this distribution.

b) *Assume that, for any two consecutive days, the number of breakdowns on Day 1 is independent of the number of breakdowns on Day 2.* Fill in the probability chart for breakdowns on two consecutive days given below.

Day 2 breakdowns

Day 1 breakdowns

[*Hint.* For the upper lefthand corner, the answer is P(0 breakdowns on Day 1 & 0 breakdowns on Day 2) = P(0 breakdowns on Day 1)·P(0 breakdowns on Day 2) = (0.8)·(0.8) = 0.64.]

c) Use your answer from part (b) to find the probability distribution for the total number of breakdowns in two days, and complete Table 4.38.

Table 4.38	Total number of breakdowns	Probability
	0	
	1	
	2	
	3	
	4	

d) Find the mean and variance of the random variable in (c), and relate it to your answer in part (a).

14 In Exercise 13, we are really looking at the sum of two *independent* random variables. The two independent random variables are

$$x = \text{number of breakdowns on Day 1,}$$
$$y = \text{number of breakdowns on Day 2.}$$

Their sum is

$$x + y = \text{total number of breakdowns in two days.}$$

Each of these three random variables (x, y, and ($x + y$)) has a mean and a variance.

There are many occasions (as in Exercise 13) when we need to discuss several random variables at once. In such cases we need to make clear to which random variable we are referring, when we talk about things such as mean and variance. To accomplish this we write μ_x for the mean of x, σ_x^2 for the variance of x, μ_y for the mean of y, and so on.

Now, suppose you are given any two random variables x and y. As in Exercise 13, it will be true that $\mu_{x+y} = \mu_x + \mu_y$. If, in addition, x and y are *independent*, then $\sigma_{x+y}^2 = \sigma_x^2 + \sigma_y^2$. Apply these facts to check your results in Exercise 13.

15 The factory manager in Exercise 13 estimates that each work stoppage costs him $100. If x is the number of breakdowns, then $100x$ is his cost due to breakdowns.

a) Fill in the chart below to find the probability distribution of breakdown cost (compare Exercise 13).

Breakdowns, x	0	1	2
Cost, $100x$	0	100	200
Probability			

b) Find the mean breakdown cost μ_{100x}.

c) What is the relation between μ_x and μ_{100x}?

16 In Exercise 15, we were really looking at a random variable obtained by multiplying another random variable by a constant:

$$x = \text{number of breakdowns in a day},$$
$$100x = \text{cost per day for breakdowns}.$$

Now suppose you are given a random variable x (not necessarily the one in Exercise 15) with probability distribution:

x	x_1	x_2	x_3
$P(x)$	p_1	p_2	p_3

a) Find a summation formula for μ_{kx}, where k is a constant.
b) Show that $\mu_{kx} = k\mu_x$.
c) Find a summation formula for σ_{kx}^2.
d) Show that $\sigma_{kx}^2 = k^2 \sigma_x^2$.

The relative-frequency interpretation of probability. Empirical probability* 4.10

In Sec. 4.1 (page 91) we learned how to find probabilities for experiments in which each individual outcome is equally likely to occur. However, there are many instances in which it is unreasonable to assume equally likely occurrence of the individual outcomes of an experiment. For example, for a typical baseball player a hit and a strikeout are not equally likely. It is not equally likely for a person to own or not own a color TV. In such cases then, how can we find probabilities? To answer this question we first need to discuss the *meaning* of probability. This has been the crux of many philosophical discussions, and it is indeed a complicated question. The answer that appears to satisfy many situations is usually called the **relative-frequency interpretation of probability**.

If we say that a coin has probability $\frac{1}{2}$ of coming up heads (i.e., the coin is fair), the relative-frequency interpretation of probability offers the following explanation:

If the coin is tossed over and over again a large number of times, then the proportion of times that the coin comes up heads will be approximately $\frac{1}{2}$.

In general we have the following statement regarding the meaning of probability.

The relative frequency interpretation of probability gives the following meaning to the statement that *an event E has probability p of occurring* (that is, $P(E) = p$): *If the experiment is repeated over and over again a large number of times, then the proportion of times the event E occurs will be approximately equal to p.*

Relative frequency interpretation of probability

◀ A telephone company official in a large city has stated that the probability that a phone call lasts longer than three minutes is about 0.37. Interpret this statement, using the relative-frequency interpretation of probability.

Example 57

The experiment here consists of observing the length (duration) of a telephone conversation. The event E is that the phone call lasts longer than three minutes, and $p = P(E) = 0.37$. According to the relative-frequency interpretation of probability, we can make the following statement: "If a large number of phone conversations are observed, then the proportion of those conversations that last longer than three minutes will be about 0.37 or 37 percent." ▶

Solution

---

* This section is optional.

Exercise **Y**   A teller at a bank claims that the probability that he will serve a customer within three minutes of the time such a customer arrives at his "window" is (about) 0.78.

a) What is the experiment in this case?
b) What is the event E?
c) What is p (that is, $P(E)$)?
d) Interpret the teller's statement using the relative-frequency interpretation of probability.
e) Interpret the teller's statement in terms of percentages.
f) If the teller serves 250 customers in a week, about how many of these customers would you expect to have been served within three minutes?

We can now return to the problem of finding probabilities of events that cannot be found by using the methods of the earlier sections of this chapter. The general procedure for finding such probabilities is based on the relative-frequency interpretation of probability. The procedure is illustrated in the following example.

Example **58**   ◖ A pharmaceutical company is interested in testing the effectiveness of a new drug they have developed to cure a certain disease. That is, the company wants to find the probability that the drug will cure a person with the disease. How can the company find this probability?

The answer to this question is based on the relative-frequency interpretation of probability. Suppose a person with the disease is given the drug. Let E = event that the drug cures the person. The company wants to find $p = P(E)$. If the experiment is repeated a large number of times (i.e., many people with the disease are given the drug), then the proportion of times that the event E (the patient is cured) occurs will be *approximately equal* to p. Consequently, the company can find (approximately) the drug's effectiveness by administering the drug to a large number of people with the disease, and then determining the proportion of these people who are cured.

For instance, suppose the company tests the drug on 800 people with the disease, and 656 are cured. Then the proportion of times the event E occurs is $\frac{656}{800} = 0.82$. According to the relative-frequency interpretation of probability, this proportion is *approximately* equal to p (the probability that the drug cures a person with the disease). That is,

$$p \doteq 0.82$$

It is natural to ask at this point "How close is the proportion 0.82 to the *true* value of p?" Although we know that the proportion 0.82 will be close to p, we don't know *how* close. Such questions will be answered in Chapter 7. (In fact, using the procedure on page 233 we find that we can be 90 percent confident that the proportion 0.82 is within 0.03 of the actual value of p. In other words, we can be 90 percent confident that p is somewhere between 0.79 and 0.85.) ◗

DEFINITION   Probabilities obtained by using the relative-frequency interpretation of probability (as in the previous example) are called **empirical** or **statistical probabilities**.

According to the *Statistical Abstract of the United States* (1976), out of 3,160,000 births in 1974, 1,538,000 were female.   Exercise **Z**

a) If you had no data at all, what guess would you make for the probability, *p*, that a baby born (in the U.S.) is a female?

b) Using the data given above, what guess would you make for *p*?

[*Note.* It has been verified, without a doubt, that male and female births are *not* equally likely—the probability of a female birth is less than 0.5, as indicated by your answer to part (b).].

EXERCISES Section **4.10**

In Exercises 1 through 3:

a) Identify the experiment.   b) Give the event *E*.

c) Find $P(E)$.

d) Interpret the given statement in terms of relative frequencies and percentages.

1. A police dispatcher says that the probability of receiving a report of a burglary between 10:00 and 11:00 P.M. on a weekday night is 0.34.

2   A department store clerk says that the probability of making a sale to a customer stopping at his counter is 0.21.

3   A quality-control engineer for an electronics firm estimates that the probability that his plant will produce a defective calculator is 0.003.

4   Suppose a fair die is rolled 360 times.

   a) About how many 5's would you expect to come up in these 360 rolls?

   b) About how many times, out of these 360 rolls, would you expect the die to come up even?

5   The probability is $\frac{1}{2}$ that a person selected at random has an IQ over 100. If 5000 people are selected at random, about how many would you expect to have IQ's over 100?

6   According to the *Statistical Abstract of the United States*, the probability of a girl being born is about 0.487. If 1500 newborn babies are selected at random, about how many of them would you expect to be girls?

7   A pharmaceutical company is interested in testing the effectiveness of a new drug they have developed to cure a certain disease. The company decides to administer the drug to a random sample of 900 people with the disease. The results show that the drug was effective on 753 of the 900 people. Based on this data, what would you estimate the drug's effectiveness to be (i.e., what estimate would you give for the probability that the drug cures a person with the disease)?

8   A group of businessmen decide to do a market study by estimating the percentages (probabilities) of the various age groups in a large city. To do this, they take a random sample of 250 residents. Their results are as follows:

Age	Number of people	Age	Number of people
0–5	37	22–40	58
6–12	51	41–65	47
13–21	42	Over 65	15

Based on this data, what estimates would you give for the age distribution of the city? That is, what percentage (or probability) estimates would you assign to the various age groups for the entire city?

9   A gambler has been observing a roulette wheel, and wonders if he can find an imperfection in the wheel so that he can improve his odds of winning. The gambler observes 200 spins of the wheel, and finds that out of these 200 trials, the ball lands on red 97 times. Based on this data, what estimate should the gambler give for the probability of the ball landing on red?

According to the relative-frequency interpretation of probability, if an experiment is performed *n* times, and the event *E* occurs *f* times in these *n* trials, then

$$P(E) \doteq \frac{f}{n}.$$

That is, the proportion of times *E* occurs in *n* trials should be approximately equal to $P(E)$. There is a law (theorem) in probability that makes this statement more precise. It is called the *law of large numbers*. Verbally, this law states that:

As *n* (*the number of trials*) *gets very large, the probability is high* (*close to* 1) *that the proportion of times E occurs is close to* $P(E)$.

To illustrate this, consider the experiment of tossing a *fair* coin. Let E be the event that the coin comes up heads. Then

$$P(E) = \tfrac{1}{2}.$$

Let $\hat{p}$ be the proportion of times heads comes up in n tosses of the coin ($\hat{p} = f/n$). Then $\hat{p}$ is a random variable. For example, if $n = 2$, the probability distribution for $\hat{p}$ is:

$\hat{p}$	$P(\hat{p})$
0	$\tfrac{1}{4}$
$\tfrac{1}{2}$	$\tfrac{1}{2}$
1	$\tfrac{1}{4}$

If $n = 3$, the probability distribution for $\hat{p}$ is:

$\hat{p}$	$P(\hat{p})$
0	$\tfrac{1}{8}$
$\tfrac{1}{3}$	$\tfrac{3}{8}$
$\tfrac{2}{3}$	$\tfrac{3}{8}$
1	$\tfrac{1}{8}$

<u>10</u> Find the probability distribution of $\hat{p}$ for $n = 4$.

<u>11</u> Find the probability that $\hat{p}$ is within $\tfrac{1}{4}$ of $P(E)$ for $n = 2, 3, 4$. That is, find

$$P(|\hat{p} - \tfrac{1}{2}| \leqslant \tfrac{1}{4}) \qquad \text{for } n = 2, 3, 4.$$

CHAPTER REVIEW

Key Terms

at random
random variable
probability distribution
compound events
(*A* or *B*)
mutually exclusive
addition principle for mutually
   exclusive events
(not *E*)
(*A* & *B*)
addition principle
conditional probability
$P(F \mid E)$
multiplication principle
independent events

multiplication principle for independent events
binomial distribution
tree diagram
factorial notation
$n!$
binomial coefficients
$\binom{n}{k}$
mean of a random variable
standard deviation of a random
   variable
relative frequency interpretation
   of probability*
empirical probabilities*

Formulas   **probabilities for experiments with equally likely outcomes:**

$$P(E) = \frac{f}{N}$$

(f = number of ways E can occur, N = total number of possible outcomes).

---

* An asterisk refers to material covered in an optional section.

addition principle of probability for mutually exclusive events:

$$P(A \text{ or } B) = P(A) + P(B)$$

(A and B mutually exclusive).

computing the probability of an event in terms of the probability of its nonoccurrence:

$$P(E) = 1 - P(\text{not } E).$$

addition principle of probability:

$$P(A \text{ or } B) = P(A) + P(B) - P(A \text{ \& } B)$$

(A and B are *any* two events).

conditional probability:

$$P(F \mid E) = \frac{P(E \text{ \& } F)}{P(E)}$$

multiplication principle of probability:

$$P(E \text{ \& } F) = P(E) \cdot P(F \mid E).$$

multiplication principle of probability for independent events:

$$P(E \text{ \& } F) = P(E) \cdot P(F)$$

(F independent of E).

factorial notation:

$$n! = n(n-1) \cdots 2 \cdot 1 \qquad (0! = 1).$$

binomial coefficients:

$$\binom{n}{k} = \frac{n!}{k!(n-k)!}.$$

probabilities for a random variable x having the binomial distribution:

$$P(x = k) = \binom{n}{k} p^k (1-p)^{n-k}$$

($p = $ success probability, $n = $ number of trials).

mean of a random variable x:

$$\mu = \sum x P(x).$$

standard deviation of a random variable x:

$$\sigma = \sqrt{\sum (x - \mu)^2 P(x)} \qquad \text{or} \qquad \sigma = \sqrt{\sum x^2 P(x) - \left[\sum x P(x)\right]^2}.$$

mean and standard deviation for a binomially distributed random variable:

$$\mu = np, \qquad \sigma = \sqrt{np(1 - p)}$$

(p = success probability, n = number of trials).

You should be able to

1 Use each of the preceding formulas.

2 Find *probability distributions* for random variables.

3 Calculate probabilities for *compound events*.

4 Describe events using *random-variable terminology*, when appropriate.

5 Find (A or B) and (A & B).

6 Determine whether or not two events are *independent*.

7 State and understand the *relative-frequency interpretation of probability.**

8 Find *empirical probabilities.**

REVIEW TEST

1 The ages of students at a military school are:

Age	18	19	20	21	22	23
Frequency	1200	1000	800	800	100	100

A student is selected at random. Find the probability that
a) The student is 20.
b) The student is younger than 21.
c) The student is 22 or older.

2 a) Use the numbers given in Problem 1 to fill in the probability distribution given below.

Age, x	18	19	20	21	22	23
Probability, $P(x)$						

b) Find $P(x \geqslant 19)$.   c) Find $P(x < 22)$.
d) Find $P(19 \leqslant x \leqslant 21)$.

3 Below is a probability distribution for the numbers of automobiles owned per household in the United States in 1974.

Number of cars, x	0	1	2	3
Probability, $P(x)$	0.185	0.488	0.268	0.059

Suppose that in 1974 a household was selected at random. Consider the following events.

A: The household owns at least one car.
B: The household owns two or more cars.
C: The household has no car.

a) Are A and B mutually exclusive?
b) Are A and C mutually exclusive?

---

* An asterisk refers to material covered in an optional section.

c) Find $P(A \text{ or } C)$.

d) Find the values of x that make up the event $(A \text{ or } B)$.

e) Find $P(A \text{ or } B)$.

f) Find the values of x that make up the event $(A \& B)$.

g) Find $P(A \& B)$.

h) Complete the formula

$$P(A \text{ or } B) = P(A) + P(B) - \boxed{}.$$

i) Calculate $P(A \text{ or } B)$, using the formula from part (h).

4 The physical science degrees conferred by a school between 1972 and 1975 were broken down as follows:

Major	Male	Female	Totals
Physics	25	25	50
Chemistry	60	40	100
Geology	30	20	50
Totals	115	85	200

Suppose a person is selected at random from these graduates. Consider the following events:

M: The person is a male.
F: The person is a female.
P: The person graduated in physics.
C: The person graduated in chemistry.
G: The person graduated in geology.

Find:

a) $P(M)$ b) $P(M \mid P)$

c) Complete the formula

$$P(M \mid P) = P(P \& M) \Big/ \boxed{}.$$

d) Find $P(M \& P)$.   e) Find $P(C \mid F)$.

f) Are the events C and F independent? (Yes or no)

g) Give a reason for your answer in part (f).

5 The probability that a person who enters a certain bookstore in a shopping mall will buy a book is $p = 0.4$. If $n = 10$ customers enter this store, what is the probability that:

a) Exactly one will buy a book?

b) One or two will buy a book?

6 Find the mean and standard deviation of the random variable given below:

x	1	2	3	4
$P(x)$	0.1	0.4	0.3	0.2

7* For each of the following situations,

i) identify the event of interest, and

ii) interpret the statement given in terms of relative frequencies and percentages.

a) A resident of Colorado says that the probability of a snowstorm in his town on March 15 is 0.20.

b) A salesman says that the probability of making a sale to any customer he visits is 0.36.

FURTHER READINGS

INGRAM, J., *Introductory Statistics*. Menlo Park, California: Cummings Publishing Company, 1977. Chapter 3 and the earlier sections of Chapter 4 cover basic probability from a slightly more abstract point of view. The text is still quite readable.

WEISS, N., and M. YOSELOFF, *Finite Mathematics*. New York: Worth Publishers, Inc., 1975. Chapters 2 through 4 give a highly detailed introduction to probability, using set theory and combinatorial analysis as a basis.

KRUSKAL, W., *et al.* (Eds.), *Statistics by Example: Weighing Chances*. Reading, Massachusetts: Addison-Wesley, 1973. Selections 1 and 5 in this book deal with finding probabilities, and selection 6 deals with the binomial distribution. The examples are real and interesting.

ANSWERS TO REINFORCEMENT EXERCISES

A $\dfrac{f}{N} = \dfrac{13}{52} = \dfrac{1}{4}$.

B a) $\dfrac{2}{36} = \dfrac{1}{18}$ b) $\dfrac{6}{36} = \dfrac{1}{6}$

C a) The player selected is 78″ tall.
b) 0.2500 c) 0.1250
d)

Height, x	Probability, $P(x)$	Height, x	Probability, $P(x)$
70	0.0625	78	0.2500
73	0.1250	80	0.0625
74	0.1250	82	0.1250
76	0.1250	83	0.0625
77	0.0625		

D

$$f = \underset{\substack{\uparrow \\ 3\text{ children}}}{4} + \underset{\substack{\uparrow \\ 4\text{ children}}}{3} = 7$$

$$P(D) = \frac{f}{N} = \frac{7}{40} = 0.175.$$

E

$$P(C) = \underset{\substack{\uparrow \\ P(0)}}{0.450} + \underset{\substack{\uparrow \\ P(1)}}{0.200} + \underset{\substack{\uparrow \\ P(2)}}{0.175} = 0.825$$

$$P(D) = \underset{\substack{\uparrow \\ P(3)}}{0.100} + \underset{\substack{\uparrow \\ P(4)}}{0.075} = 0.175$$

F a) A: $(x \geqslant 7)$
 B: $(x < 4)$
 C: $(4 \leqslant x \leqslant 10)$
b) i) $(x \leqslant 5)$: event the sum of the dice is at most 5.
 ii) $(x > 7)$: event the sum is greater than 7.
 iii) $(3 < x < 7)$: event the sum is greater than 3 and less than 7.
c) $P(x \leqslant 5) = P(2) + P(3) + P(4) + P(5)$

$$= \frac{1}{36} + \frac{2}{36} + \frac{3}{36} + \frac{4}{36} = \frac{10}{36} = \frac{5}{18}.$$

$P(x > 7) = P(8) + P(9) + P(10) + P(11) + P(12)$

$$= \frac{5}{36} + \frac{4}{36} + \frac{3}{36} + \frac{2}{36} + \frac{1}{36} = \frac{15}{36} = \frac{5}{12}.$$

$P(3 < x < 7) = P(4) + P(5) + P(6)$

$$= \frac{3}{36} + \frac{4}{36} + \frac{5}{36} = \frac{12}{36} = \frac{1}{3}.$$

G a) A: kh, ks, kc, kd
 B: ah, 2h, 3h, 4h, 5h, 6h, 7h, 8h, 9h, 10h, jh, qh, kh
b) $(A$ or $B)$: We select a card that is a king or a heart.
c) $(A$ or $B)$: ks, kc, kd, ah, 2h, 3h, 4h, 5h, 6h, 7h, 8h, 9h, 10h, jh, qh, kh

H a) Not mutually exclusive (kh in common).
b) Mutually exclusive.
c) Not mutually exclusive: Every king is a face card—kh, ks, kc, kd in common.
d) Not mutually exclusive (10h in common).
e) Not mutually exclusive (jh, qh, kh in common).
f) Mutually exclusive.

I a) $P(A$ or $C) = P(A) + P(C) = \dfrac{4}{52} + \dfrac{4}{52} = \dfrac{8}{52} = \dfrac{2}{13}$.

b) $P(C$ or $D) = P(C) + P(D) = \dfrac{4}{52} + \dfrac{12}{52} = \dfrac{16}{52} = \dfrac{4}{13}$.

J $P(\text{age} > 24) = 1 - P(\text{age} \leqslant 24) = 1 - (0.05 + 0.17) = 1 - 0.22 = 0.78$.

K $P(x > 70) = 1 - P(x \leqslant 70) = 1 - \dfrac{1}{16} = \dfrac{15}{16}$.

L a) i) King of hearts is selected
 ii) The card selected is both a king and a 10 (impossible—mutually exclusive).
 iii) The card selected is both a king and a face card (that is, a king).
 iv) 10 of hearts is selected.
 v) The card selected is a heart and a face card.
 vi) The card selected is a 10 and a face card (impossible—mutually exclusive).
b) i) kh iii) kh, ks, kc, kd
 iv) 10h v) kh, qh, jh

M a) $P(A$ or $B) = P(A) + P(B) - P(A \& B)$

$$= \frac{4}{52} + \frac{13}{52} - \frac{1}{52} = \frac{16}{52} = \frac{4}{13}.$$

b) $P(A$ or $C) = P(A) + P(C) - P(A \& C)$

$$= \frac{4}{52} + \frac{4}{52} - 0 = \frac{8}{52} = \frac{2}{13}.$$

c) $P(A \text{ or } D) = P(A) + P(D) - P(A \& D)$

$$= \frac{4}{52} + \frac{12}{52} - \frac{4}{52} = \frac{12}{52} = \frac{3}{13}.$$

d) $P(B \text{ or } C) = P(B) + P(C) - P(B \& C)$

$$= \frac{13}{52} + \frac{4}{52} - \frac{1}{52} = \frac{16}{52} = \frac{4}{13}.$$

e) $P(B \text{ or } D) = P(B) + P(D) - P(B \& D)$

$$= \frac{13}{52} + \frac{12}{52} - \frac{3}{52} = \frac{22}{52} = \frac{11}{26}.$$

f) $P(C \text{ or } D) = P(C) + P(D) - P(C \& D)$

$$= \frac{4}{52} + \frac{12}{52} - 0 = \frac{16}{52} = \frac{4}{13}.$$

N $P(A \mid R) = \dfrac{150}{155} \doteq 0.968$

O a) $P(C) = \dfrac{24000}{120000} = 0.20.$

b) $P(R) = \dfrac{67200}{120000} = 0.56.$

c) $P(R \& C) = \dfrac{10800}{120000} = 0.09.$

d) $P(C \mid R) = \dfrac{P(R \& C)}{P(R)} = \dfrac{0.09}{0.56} \doteq 0.16.$

e) 16 percent of the Republican voters are Catholic.

P a) Let H = first player under 76″, K = second player under 76″. Then

$$P(H \& K) = P(H) \cdot P(K \mid H) = \frac{5}{16} \cdot \frac{4}{15} = \frac{4}{48} = \frac{1}{12}.$$

b) $P(H \& \text{not } K) = P(H) \cdot P(\text{not } K \mid H) = \dfrac{5}{16} \cdot \dfrac{11}{15} = \dfrac{11}{48}.$

c) S = first player is 78″ tall; T = second player is 78″ tall. Then

$$P(S \& T) = P(S) \cdot P(T \mid S) = \frac{4}{16} \cdot \frac{3}{15} = \frac{1}{20}.$$

d) V = first player is 80″ tall, W = second player is 80″ tall. Then

$$P(V \& W) = P(V) \cdot P(W \mid V) = \frac{1}{16} \cdot \frac{0}{15} = 0.$$

Q $P(P) = 0.75$, $P(R) = 0.56$; $P(R \& P) = 0.45.$

$$P(P \mid R) = \frac{P(R \& P)}{P(R)} = \frac{0.45}{0.56} \doteq 0.804.$$

Not independent since $P(P \mid R) \neq P(P).$

R E = first roll is "5", F = second roll is "5", G = third roll is "5". Then

$$P(E \& F \& G) = \frac{1}{6} \cdot \frac{1}{6} \cdot \frac{1}{6} = \frac{1}{216} \doteq 0.0046.$$

S a)

Outcome	Probability
ss	$(0.75)(0.75) = 0.5625$
sf	$(0.75)(0.25) = 0.1875$
fs	$(0.25)(0.75) = 0.1875$
ff	$(0.25)(0.25) = 0.0625$

b)

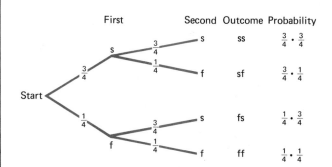

c)

Probability distribution

x	$P(x)$
0	0.0625
1	0.3750
2	0.5625

T a) $\dbinom{3}{3} = 1$ b) $\dbinom{4}{2} = 6$

c) $\dbinom{5}{0} = 1$ d) $\dbinom{7}{3} = 35$

U a) $P(x = 1) = \binom{2}{1}(0.75)^1 \cdot (0.25)^1 = 0.375.$

b) $P(x = 2) = \binom{2}{2}(0.75)^2 (0.25)^0 = 0.5625.$

c) $P(x = 1 \text{ or } 2) = 0.375 + 0.5625 = 0.9375.$

V

Card	x	$P(x)$	$xP(x)$
2	2	1/13	2/13
3	3	1/13	3/13
4	4	1/13	4/13
5	5	1/13	5/13
6	6	1/13	6 13
7	7	1/13	7/13
8	8	1/13	8/13
9	9	1/13	9/13
Face or 10	10	4/13	40/13
Ace	11	1/13	11/13

$$\mu = 95/13$$

W $\sum xP(x) \doteq 7.31$ from Exercise V; $\sum x^2 P(x) = 61.92$. So

$$\sigma = \sqrt{61.92 - (7.31)^2} = 2.91.$$

X $\mu = 4(0.8) = 3.2; \quad \sigma = \sqrt{4(0.8)(0.2)} = 0.8.$

Y a) Serving a customer and observing the time required.
b) E: The customer is served within three minutes.
c) $p = P(E) = 0.78.$
d) If a large number of customers is served, the proportion taken care of in three minutes or less is about 0.78.
e) About 78 percent of customers require three minutes or less.
f) $(0.78)(250) = 195.$

Z a) 0.5   b) $\dfrac{1,538,000}{3,160,000} \doteq 0.487.$

The normal distribution

5.1 Estimating a population mean

A first problem in statistical inference is to obtain information concerning the mean, μ, of a population (or random variable). For instance we might be interested in estimating

1. The mean tar content, μ, of a certain brand of cigarette.
2. The mean life, μ, of a newly developed steel-belted radial tire.
3. The mean gas mileage, μ, of a new model car.
4. The mean income, μ, of the residents of California.

Of course, if the population is "small," then we can usually compute μ exactly. However, in most cases the population will be large, and it will be either impractical, impossible, or extremely expensive to compute μ exactly. Moreover, in most cases, we can obtain sufficiently accurate information about μ by just taking a *sample* of the population, and computing the sample mean, $\bar{x}$.

Example 1 A manufacturer of a certain brand of cigarettes is interested in obtaining information regarding the mean tar content, μ, of all such cigarettes. He obviously cannot test all the cigarettes, so he decides to take a sample of, say, $n = 10$ such cigarettes, and test the ten sampled cigarettes. His results are given below.

Cigarette	1	2	3	4	5	6	7	8	9	10
Tar content (mg)	10.34	10.88	10.58	10.51	10.78	11.49	10.87	10.84	10.14	12.13

The *sample mean* tar content of these ten cigarettes is

$$\bar{x} = \frac{1}{10}\sum x = 10.86 \text{ mg}$$

If you had to guess at the (population) mean tar content, μ, of all cigarettes of this brand, based on this sample of ten cigarettes, you would probably guess that μ is "about" 10.86 mg.

If another sample of ten cigarettes were tested, you probably wouldn't expect to get the same sample mean. Let us assume, for example, that a second sample results in the following data:

Cigarette	1	2	3	4	5	6	7	8	9	10
Tar content (mg)	11.31	11.32	11.01	11.49	10.97	11.08	10.68	10.54	11.23	10.58

The sample mean tar content of these ten cigarettes is

$$\bar{x} = \frac{1}{10}\sum x = 11.02 \text{ mg}$$

So, again, if you had to guess at the (population) mean tar content, μ, of all cigarettes of this brand, based on *this* sample of ten cigarettes, you would probably guess that μ is "about" 11.02 mg. ▶

After considering Example 1, there are several questions that immediately arise. For instance,

1. What justification is there for using a sample mean, $\bar{x}$, to estimate a population mean, μ?
2. How confident can we be about the accuracy involved in using $\bar{x}$ to estimate μ? (For instance, in Example 1, if a sample of ten cigarettes is tested, what is the probability that their (sample) mean tar content, $\bar{x}$, will be within, say, 0.1 mg of the true (population) mean tar content, μ, of all such cigarettes?)
3. How large a sample do we need to take in order to ensure a certain accuracy with high confidence? (For instance, in Example 1, how large a sample would we have to take in order to ensure that 95 percent of all the possible $\bar{x}$ values we could get would be within 0.1 mg of the true mean tar content, μ?)

We will answer the first question in this section. Answers to questions 2 and 3 are more complicated, and will be the topic of discussion later in this chapter and in Chapters 6 and 7.

It is clear from Example 1 that the value of $\bar{x}$ depends on the sample selected. The first sample of ten cigarettes yielded $\bar{x} = 10.86$ mg, while the second gave $\bar{x} = 11.02$ mg.

In sampling from a population, **the sample mean, $\bar{x}$, is a random variable**. Its value depends on chance; namely, upon which sample is selected.

Although the value of $\bar{x}$ will vary from sample to sample, *if the sample size (n) is "large", it is "extremely likely" that $\bar{x}$ will be "close" to the population mean, μ.* This is a rather loose statement of a theorem in mathematics called **the law of large numbers**. The law of large numbers is the mathematical justification for using $\bar{x}$ to estimate μ (that is, it answers question 1).

Below we give an example that illustrates the law of large numbers. It is extremely oversimplified, but it does serve as a nice illustration.

◖ Assume that the *population* under consideration consists of the heights of the starting   Example **2** five players on a basketball team:

Player	1	2	3	4	5
Height (inches)	67	73	75	76	84

The mean of this population is

$$\mu = \frac{67 + 73 + 75 + 76 + 84}{5} = 75 \text{ in.}$$

We are interested in finding how close the sample mean, $\bar{x}$, will be to the population mean μ. This population is so small that we can look at all possible values of $\bar{x}$ and compare them to μ. We will do this for samples of sizes 2, 3, and 4.

There are ten possible samples of size 2. Table 5.1 lists these samples, along with their $\bar{x}$'s, and is followed by a picture that clearly illustrates the situation. In the picture, the location of each $\bar{x}$-value is shown by a cross.

Table 5.1   Sample size $n = 2$

Sample	1	2	3	4	5	6	7	8	9	10
Players selected	1 & 2	1 & 3	1 & 4	1 & 5	2 & 3	2 & 4	2 & 5	3 & 4	3 & 5	4 & 5
Heights (in.)	67 & 73	67 & 75	67 & 76	67 & 84	73 & 75	73 & 76	73 & 84	75 & 76	75 & 84	76 & 84
$\bar{x}$	70.0	71.0	71.5	75.5	74.0	74.5	78.5	75.5	79.5	80.0

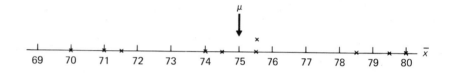

For example, if our sample of size 2 yielded players 1 and 5 (sample 4), then $\bar{x} = (67 + 84)/2 = 75.5$ inches.

If we consider samples of size 3, there are again ten possible samples we could get. Table 5.2 lists them, and is followed by an illustration.

Table 5.2   Sample size $n = 3$

Sample	1	2	3	4	5	6	7	8	9	10
Players selected	1, 2, 3	1, 2, 4	1, 2, 5	1, 3, 4	1, 3, 5	1, 4, 5	2, 3, 4	2, 3, 5	2, 4, 5	3, 4, 5
$\bar{x}$	71.7	72.0	74.7	72.7	75.3	75.7	74.7	77.3	77.7	78.3

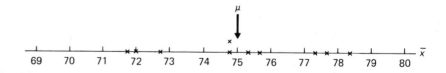

For samples of size 4 there are five possible samples listed in Table 5.3 and the situation is illustrated below it.

Sample size $n = 4$ Table 5.3

Sample	1	2	3	4	5
Players selected	1, 2, 3, 4	1, 2, 3, 5	1, 2, 4, 5	1, 3, 4, 5	2, 3, 4, 5
$\bar{x}$	72.75	74.75	75.00	75.50	77.00

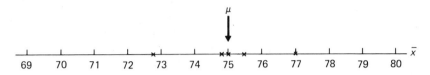

As you can see from the three graphs above, the possible $\bar{x}$ values tend to cluster closer around the true mean, μ, as the sample size increases. For example, when $n = 2$, forty percent of the possible $\bar{x}$ values lie within $3''$ of μ; when $n = 3$, eighty percent of the possible $\bar{x}$ values lie within $3''$ of μ; and when $n = 4$, *all* of the possible $\bar{x}$ values lie within $3''$ of μ. ▶

The law of large numbers is the mathematical justification for using $\bar{x}$ to estimate μ, but it is not particularly useful in answering important practical questions involving the accuracy of such estimates. For instance, the law of large numbers tells us that if we test a "large" sample of cigarettes, then it is "likely" that their sample mean tar content, $\bar{x}$, will be "close" to the population mean tar content, μ, of all such cigarettes. But if we tested a sample of 100 cigarettes, the law of large numbers would *not* tell us the probability that $\bar{x}$ will be within 0.1 mg of μ. In order to find such probabilities, we really need to know how to calculate probabilities for $\bar{x}$. In practice, we usually won't know how to calculate probabilities for $\bar{x}$ *exactly*. However, there is a rather simple way that we can calculate probabilities for $\bar{x}$ *approximately* using areas under bell-shaped curves. We will discuss these curves in the next section.

EXERCISES Section **5.1**

In Exercises 1 through 5:
a) State the population of interest, and
b) Find the sample mean, $\bar{x}$, from the given data.

1   A business analyst wants to estimate the mean number of days, μ, that strikes in the U.S. lasted in 1974. She surveys 15 unions that struck. The responses show that their strikes lasted the following numbers of days.

16	24	50	11	19
39	46	7	31	35
18	29	22	1	6

2   To estimate the mean, μ, of annual food expenditures for U.S. middle class families of four in 1973, a news reporter obtained the expenses for ten such families. The expenses are listed below.

$4,947	3,686	4,491	4,331	3,049
3,727	3,977	4,376	3,913	2,213

3 A sample of high-school seniors who took the College Boards in 1979 received the following scores:

573	427	326	531	383	658	485
565	432	500	337	283	443	252
637	558	522	716	395	751	374
482	449	545	602	618	421	402

4 The following list is a sample of 1975 annual wages in the U.S. for non-farm working males aged 25–34 with exactly four years of high-school education.

$13,379	11,353	17,358	9,540
17,011	14,865	14,573	17,649

5 On five days in October 1978, the following numbers of people applied for unemployment insurance at State Division of Employment and Training Offices in Colorado:

$$3,761 \quad 2,584 \quad 3,296 \quad 3,527 \quad 3,707$$

The following data is used in Exercises 6 through 9 in order to illustrate the law of large numbers (as was done in Example 2).

Below is a list of the lengths in centimeters of six bullfrogs found in a small pond.

Bullfrog	1	2	3	4	5	6
Length (cm)	19	14	15	9	16	17

The six bullfrogs were the only bullfrogs in the pond, and could be considered a population.

6 Calculate the mean length, μ, of the bullfrog population.

7 a) List all possible samples of size 2 that can be selected from the population of six bullfrogs. (There are 15 possible samples of size 2.)
   b) Compute $\bar{x}$ for each sample in (a).
   c) Draw a graph similar to the one in Example 2.

8 Repeat Exercise 7 using samples of size 3. (There are 20 possible samples of size 3.)

9 Repeat Exercise 7 using samples of size 4. (There are 15 possible samples of size 4.)

10 What do the results in Exercises 7 through 9 illustrate?

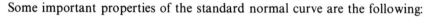

5.2   The standard normal or z-curve

As we saw in Section 5.1, it is important to be able to find probabilities for $\bar{x}$, which we indicated can be done *approximately* using areas under certain bell-shaped curves. In this section and the next, we will explain how to find areas under these curves, which are called **normal curves**.

There are many (in fact, infinitely many) normal curves. But fortunately there is a way to find areas under any normal curve by looking at areas under one specific normal curve. This one specific normal curve is called the **standard normal curve**, or *z*-**curve**. It is pictured at the left.*

Some important properties of the standard normal curve are the following:

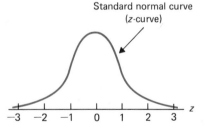

Standard normal curve
(z-curve)

1. The area under the standard normal curve is 1.
2. The standard normal curve extends indefinitely in both directions, approaching the horizontal axis as it does so.
3. The standard normal curve is *symmetric* about 0. That is, the part of the curve to the left of 0 is the *mirror image* of the part of the curve to the right of 0.
4. Most of the area (we will shortly see exactly how much area) under the standard normal curve lies between −3 and 3.

---

* The equation of the standard normal curve is

$$y = \frac{1}{\sqrt{2\pi}} e^{-z^2/2},$$

where $e \doteq 2.718$, and π is the familiar constant from geometry—$\pi \doteq 3.142$.

Because of the importance of finding areas under the standard normal curve, tables of such areas have been constructed. Such a table is found as Table I of the appendix. For ease of reference we have reproduced part of Table I in Table 5.4 below.

				Second decimal place in z						
z	0	1	2	3	4	5	6	7	8	9
$\vdots$				$\vdots$						
−2.9	.0019	.0018	.0017	.0017	.0016	.0016	.0015	.0015	.0014	.0014
−2.8	.0026	.0025	.0024	.0023	.0023	.0022	.0021	.0020	.0020	.0019
$\vdots$				$\vdots$						
−1.0	.1587	.1562	.1539	.1515	.1492	.1469	.1446	.1423	.1401	.1379
−0.9	.1841	.1814	.1788	.1762	.1736	.1711	.1685	.1660	.1635	.1611
−0.8	.2119	.2090	.2061	.2033	.2005	.1977	.1949	.1922	.1894	.1867
$\vdots$				$\vdots$						
1.8	.9641	.9649	.9656	.9664	.9671	.9678	.9686	.9693	.9699	.9706
1.9	.9713	.9719	.9726	.9732	.9738	.9744	.9750	.9756	.9761	.9767
2.0	.9772	.9778	.9783	.9788	.9793	.9798	.9803	.9808	.9812	.9817
$\vdots$				$\vdots$						
2.8	.9974	.9975	.9976	.9977	.9977	.9978	.9979	.9979	.9980	.9981
2.9	.9981	.9982	.9982	.9983	.9984	.9984	.9985	.9985	.9986	.9986
$\vdots$				$\vdots$						

Table 5.4
Areas under the standard normal curve (to four decimal places)

◀ A typical four-decimal-place number in the body of Table 5.4 gives the area under the standard normal curve to the *left* of a specified value. For example, to find the area under the standard normal curve to the left of $z = 1.83$ we proceed as follows. Go down the lefthand column (labeled z) to "1.8." Then go across this row until you are under the "3" in the top row. The number in the body of the table here, which is .9664, is the area under the standard normal curve to the left of $z = 1.83$. ▶

Example **3**

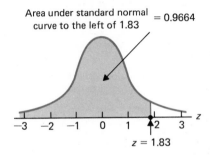

Area under standard normal curve to the left of 1.83 $= 0.9664$

$z = 1.83$

The number in the body of the table corresponding to a specific z-value is the area under the standard normal curve to the *left* of that z-value.

Example **4** ◀ Find the area under the standard normal curve to the left of $z = -0.97$. To look up this area, go down the lefthand column to -0.9. Then go across this row until you are under the 7 in the top row. The area under the standard normal curve to the left of $z = -0.97$ is the number .1660. ▶

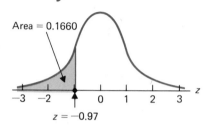

Exercise **A** Find the area under the standard normal curve
a) To the left of $z = -0.84$;
b) To the left of $z = 2.87$.

Next we will see how to compute areas under the standard normal curve *between* two values.

Example **5** ◀ Find the area under the standard normal curve between $z = -1.04$ and $z = 2.06$. From Table 5.4, we see that the area under the standard normal curve to the left of $z = -1.04$ is .1492, and the area to the left of $z = 2.06$ is .9803.

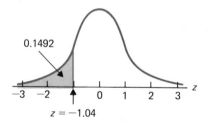
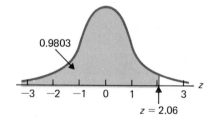

The area under the standard normal curve *between* $z = -1.04$ and $z = 2.06$ is the area to the left of $z = 2.06$ *minus* the area to the left of $z = -1.04$. So we subtract:

$$\text{Area} = 0.9803 - 0.1492 = 0.8311. ▶$$

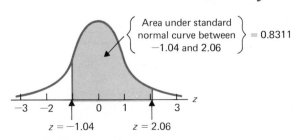

◀ Find the area under the standard normal curve between $z = -1$ and $z = 1$. Use Table I in the appendix. We give two solutions.　　Example **6**

First solution. Proceeding as before, we get the following pictures.　　Solutions

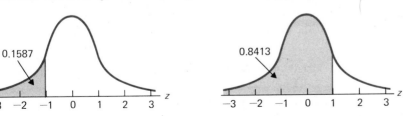

So the area under the standard normal curve between $z = -1$ and $z = 1$ is

$$0.8413 - 0.1587 = 0.6826$$

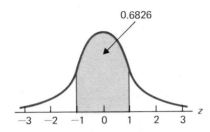

Second solution. In this solution we shall use the facts that the standard normal curve is symmetric about 0 and that the area under the standard normal curve is 1. We know from the first picture above that the area under the curve to the left of $z = -1$ is 0.1587. Since the standard normal curve is symmetric about 0, this means that the area to the *right* of $z = 1$ is also 0.1587:

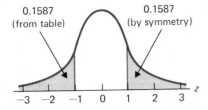

Finally, because the total area under the curve is 1, this means that the area under the curve between $z = -1$ and $z = 1$ (the unshaded area in the above picture) is $1 - (0.1587 + 0.1587) = 0.6826$. This, of course, agrees with our first solution, but requires one less table look-up. ▶

Find the area under the standard normal curve　　Exercise **B**
a) Between $z = -2$ and $z = 2$ (use first solution method given in Example 6);
b) Between $z = -3$ and $z = 3$ (use second solution method given in Example 6).

From Example 6 and Exercise B we get the following useful facts:

1. *About* 68% of the area under the standard normal curve lies between $z = -1$ and $z = 1$.
2. *About* 95% of the area under the standard normal curve lies between $z = -2$ and $z = 2$.
3. *About* 99.7% of the area under the standard normal curve lies between $z = -3$ and $z = 3$.

Example 7 ◖ Find the area under the standard normal curve
a) to the right of $z = 1.64$, and
b) either to the left of $z = -2.34$ *or* to the right of $z = 1.64$.

Solutions a) We give two solutions to part (a).

First solution. The area under the standard normal curve to the left of $z = 1.64$ is 0.9495.

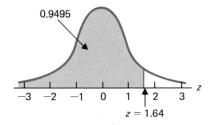

Since the total area under the curve is 1, the area to the right of $z = 1.64$ is $1 - 0.9495 = 0.0505$.

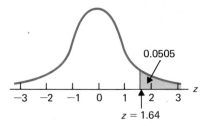

Second solution. By symmetry, the area under the standard normal curve to the right of $z = 1.64$ is the same as the area under the curve to the left of $z = -1.64$, which, from Table I, is 0.0505.

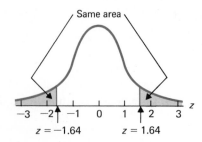

b) The area to the left of $z = -2.34$ is found as usual—it is .0096. Also, we have just seen in (a) that the area to the right of $z = 1.64$ is .0505. Consequently, the area under the standard normal curve *either* to the left of $z = -2.34$ *or* to the right of $z = 1.64$ is

$$0.0096 + 0.0505 = 0.0601 \blacktriangleright$$

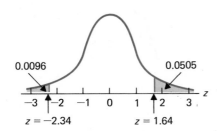

Using only *one* table look-up find the area under the standard normal curve
a) To the left of $z = -1.96$;
b) To the right of $z = 1.96$;
c) *Either* to the left of $z = -1.96$ *or* to the right of $z = 1.96$;
d) Between $z = -1.96$ and $z = 1.96$.

Exercise **C**

EXERCISES Section **5.2**

Use Table I to find the area under the standard normal curve specified in each of the problems below. Sketch a standard normal curve and shade the area of interest.

1 Find the area under the standard normal curve to the *left* of
   a) $z = -2.33$ b) $z = 1.75$ c) $z = 0.23$
   d) $z = -1.64$ e) $z = 3.04$ f) $z = 0$

2 Find the area under the standard normal curve *between*
   a) $z = -2.33$ and $z = 2.33$
   b) $z = -0.47$ and $z = 1.75$
   c) $z = 0.23$ and $z = 2.98$
   d) $z = -1.64$ and $z = 1.64$

3 Find the area under the standard normal curve to the *right* of
   a) $z = 2.98$ b) $z = -1.64$
   c) $z = 0$ d) $z = 0.23$

4 Find the area under the standard normal curve
   a) Either to the left of $z = -2.33$ or to the right of $z = 2.33$
   b) Either to the left of $z = -0.47$ or to the right of $z = 0.23$
   c) Either to the left of $z = -1.5$ or to the right of $z = 2.5$
   d) Either to the left of $z = -2.30$ or to the right of $z = 1.85$

5 Find the area under the standard normal curve
   a) To the left of $z = 1.05$
   b) To the right of $z = 2.57$
   c) Between $z = 0$ and $z = 2.25$
   d) Either to the left of $z = -1.28$ or to the right of $z = 2.63$

6 Find the following areas under the standard normal curve using only one table look-up.
   a) To the left of $z = 1.25$
   b) To the right of $z = -1.25$
   c) Either to the left of $z = -1.25$ or to the right of $z = 1.25$
   d) Between $z = -1.25$ and $z = 1.25$.

In the previous section we learned how to find areas under the *standard* normal curve. We also pointed out that there were infinitely many normal curves. To calculate probabilities for $\bar{x}$ (our main goal here), we need to be able to find areas under *any* normal curve. In this section we will show how to solve such area problems using areas under the *standard* normal curve.

Areas under normal curves   **5.3**

Each normal curve can be identified by two numbers, called **parameters**. These two parameters are usually denoted by μ and σ. The parameter μ tells us where the normal curve is "centered", and the parameter σ indicates the "spread" of the normal curve.* The normal curve with parameters μ and σ looks like this:

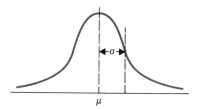

Example **8**   ◖Below are three different normal curves. The normal curve on the left has parameters $\mu = -2$ and $\sigma = 1$; the one in the center has parameters $\mu = 3$ and $\sigma = 2$; and the one on the right has parameters $\mu = 6$ and $\sigma = 3$.

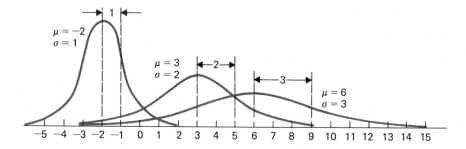

The important things for you to notice are that the normal curves are centered at μ and, the larger the value of σ, the more the curve is "spread out." ◗

We also point out that the **normal curve with parameters $\mu = 0$ and $\sigma = 1$ is the standard normal curve.**

The most important properties of normal curves are the following:

1. The area under any normal curve is 1.
2. Any normal curve extends indefinitely in both directions, approaching the horizontal axis as it does so.
3. The normal curve with parameters μ and σ is symmetric about μ. That is, the part of the curve to the left of μ is the mirror image of the part of the curve to the right of μ.
4. Most (about 99.7 %) of the area under the normal curve with parameters μ and σ lies between $\mu - 3\sigma$ and $\mu + 3\sigma$. (You are asked to show this in Exercise 7.)

---

* Although we will not need it, the equation of the normal curve with parameters μ and σ is $y = (1/\sqrt{2\pi}\sigma)e^{-(x-\mu)^2/2\sigma^2}$.

It will not be necessary to draw normal curves exactly, but it will be useful to sketch normal curves. Using the above facts (especially 3 and 4), we can easily sketch normal curves.

◖ Sketch the normal curve with   Example **9**
a) parameters $\mu = 5$ and $\sigma = 2$,
b) parameters $\mu = 8$ and $\sigma = 3$

a) By property 3, this normal curve is symmetric about $\mu = 5$. Also, by property 4,   Solutions
most of the area under this normal curve lies between

$$\mu - 3\sigma = 5 - 3 \cdot 2 = -1$$

and

$$\mu + 3\sigma = 5 + 3 \cdot 2 = 11$$

So, we sketch this curve as follows:

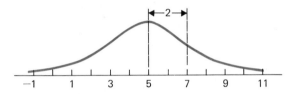

b) This normal curve is symmetric about $\mu = 8$, and most of the area is between $\mu - 3\sigma = 8 - 3 \cdot 3 = -1$ and $\mu + 3\sigma = 8 + 3 \cdot 3 = 17$. So we sketch this curve as follows:

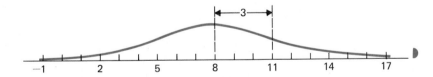

Sketch the normal curve with parameters   Exercise **D**
a) $\mu = 2$ and $\sigma = 1$,
b) $\mu = -3$ and $\sigma = 2$.

We will now see how to find areas under any normal curve. As we said before, we can do this by just looking at areas under the *standard* normal curve. The basic idea is to relate any normal curve to the standard normal curve. We will state the general rule presently, but we first illustrate it with an example.

◖ Find the area under the normal curve with $\mu = 5$ and $\sigma = 2$,   Example **10**
a) to the left of 4,
b) between 3 and 6.

Solutions

a) In Fig. 5.1 we have sketched the normal curve with parameters $\mu = 5$ and $\sigma = 2$, along with the desired area.

Note that we have labeled the horizontal axis with an "x". This will help us to keep things straight, as you will see.

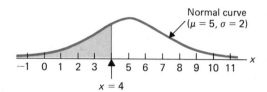

Normal curve
$(\mu = 5, \sigma = 2)$

$x = 4$

Figure 5.1

It can be shown mathematically (see Ex. 8) that the area under the normal curve ($\mu = 5, \sigma = 2$) to the left of $x = 4$ is equal to the area under the *standard* normal curve to the left of

$$z = \frac{x - \mu}{\sigma} = \frac{4 - 5}{2} = -0.5$$

This area is pictured in Fig. 5.2.

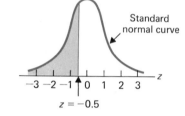

Standard
normal curve

$z = -0.5$

Figure 5.2

In other words, we are saying that the areas shaded in Figs. 5.1 and 5.2 are equal. But by Table I, the area shaded in Fig. 5.2 equals .3085. Thus, the area shaded in Fig. 5.1 is also .3085. That is, the area under the normal curve ($\mu = 5$, $\sigma = 2$) to the left of 4 equals .3085.

b) The area under the normal curve ($\mu = 5$, $\sigma = 2$) between 3 and 6 is pictured in Fig. 5.3.

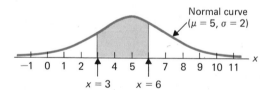

Normal curve
$(\mu = 5, \sigma = 2)$

$x = 3$ $x = 6$

Figure 5.3

As in part (a), the area under the normal curve ($\mu = 5$, $\sigma = 2$) between $x = 3$ and $x = 6$ is equal to the area under the *standard* normal curve between

$$z = \frac{x - \mu}{\sigma} = \frac{3 - 5}{2} = -1$$

and

$$z = \frac{x - \mu}{\sigma} = \frac{6 - 5}{2} = 0.5$$

This area is shown in Fig. 5.4.

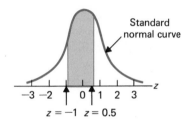

Figure 5.4

By Table I, the shaded area in Fig. 5.4 equals $.6915 - .1587 = .5328$. Consequently, so does the shaded area in Fig. 5.3. ◗

Find the area under the normal curve ($\mu = 5$, $\sigma = 3$) to the right of $x = 8$.   Exercise **E**

The process of getting a z-value from an x-value by subtracting μ and then dividing by σ,

$$z = \frac{x - \mu}{\sigma}$$

is often referred to as **standardizing**. The value z is called the **standard score** or **z-score** for the value x.

We will now summarize the general procedure that was illustrated in Example 10(b):

> The area under the normal curve (μ, σ) between $x = a$ and $x = b$ is equal to the area under the *standard* normal curve between
>
> $$z = \frac{a - \mu}{\sigma} \qquad \text{and} \qquad z = \frac{b - \mu}{\sigma}$$

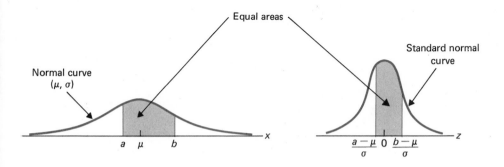

Areas under a normal curve to the right (or left) of a given x-value are found similarly, by converting to z-scores, and then finding the area under the *standard* normal curve to the right (or left) of the z-score.

Example **11**   ◖ Find the area under the normal curve ($\mu = 1$, $\sigma = 1.5$),
a) between 2 and 3,
b) to the right of 4.

Solutions   The solutions can be written compactly as follows:

a)
$$z = \frac{x - \mu}{\sigma} = \frac{x - 1}{1.5}$$

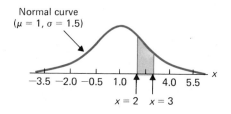

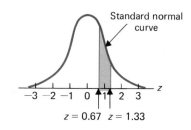

Change to z

$$x = 2 \rightarrow z = \frac{2 - 1}{1.5} = 0.67$$

$$x = 3 \rightarrow z = \frac{3 - 1}{1.5} = 1.33$$

Area $= 0.9082 - 0.7486 = 0.1596$

b)
$$z = \frac{x - \mu}{\sigma} = \frac{x - 1}{1.5}$$

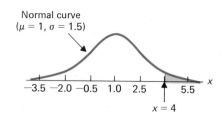

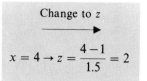

Change to z

$$x = 4 \rightarrow z = \frac{4 - 1}{1.5} = 2$$

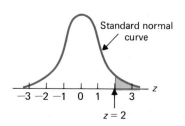

Area $= 1 - 0.9772 = 0.0228$ ◗

Exercise **F**   Find the area under the normal curve ($\mu = 1$, $\sigma = 1.5$) between -3 and 4.

1   Sketch the normal curve with
   a) parameters $\mu = 3$, $\sigma = 3$,
   b) parameters $\mu = 0$, $\sigma = 2$,
   c) parameters $\mu = -2$, $\sigma = 1$.

2   Find the area under the normal curve with $\mu = 3$ and $\sigma = 3$,
   a) to the left of -1,
   b) between 0 and 4,
   c) to the right of 1.5.

3   Find the area under the normal curve with $\mu = 1$ and $\sigma = 2.5$,
   a) to the right of 0,
   b) to the left of -1.5,
   c) between -2 and 2.

4   Find the area under the normal curve with $\mu = -1.5$ and $\sigma = 1$,
   a) between 0 and 1.4,
   b) to the left of -1.5,
   c) to the right of 1.

5   For the normal curve with parameters $\mu = 4$ and $\sigma = 2$, find the area that lies between -2 and 10.

6   Show that about 99.7% of the area under the normal curve with parameters $\mu = 0$ and $\sigma = 1$ (the standard normal curve) lies between $\mu - 3\sigma$ and $\mu + 3\sigma$.

7   Show that about 99.7% of the area under the normal curve with parameters μ and σ lies between $\mu - 3\sigma$ and $\mu + 3\sigma$.

8   This exercise requires elementary calculus and justifies the procedure described on page 173.

   a) Show that the area under the normal curve (μ, σ) between $x = a$ and $x = b$ is

   $$\int_a^b \frac{1}{\sqrt{2\pi}\sigma} e^{-(x-\mu)^2/2\sigma^2} \, dx$$

   (see footnote on page 170).

   b) Making the substitution $z = (x - \mu)/\sigma$, show that the integral in (a) equals

   $$\int_{(a-\mu)/\sigma}^{(b-\mu)/\sigma} \frac{1}{\sqrt{2\pi}} e^{-z^2/2} \, dz$$

   c) What area does this last integral equal? (See the footnote on page 164).

As we have mentioned several times, our chief goal right now is to see how to use areas under normal curves to find probabilities for $\bar{x}$ (approximately). But before we see how to do this, it will be useful for us to consider what are called **normally distributed populations** and **normally distributed random variables**.

Many (but not all) populations and random variables representing quantities such as height, weight, and IQ have probability distributions that can be represented, *at least approximately*, by normal curves. That is, *probabilities for such quantities can be found by looking at areas under normal curves*.

The following example will give you a good frame of reference for the general discussion that follows.

◀ Suppose the *population* under consideration consists of the heights of a group of 100 football players, measured to the *nearest inch*. The population data is summarized in Table 5.5.

Normally distributed populations and random variables **5.4**

Example **12**

Height (in.)	69	70	71	72	73	74	75	76	77	78	79
Frequency	1	2	6	13	17	20	18	12	7	3	1
Relative frequency	0.01	0.02	0.06	0.13	0.17	0.20	0.18	0.12	0.07	0.03	0.01

Table 5.5
Heights of football players

The relative frequency histogram for this data is given in Fig. 5.5.

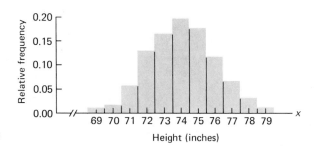

Figure 5.5

As with any population, we can associate with this population of heights a random variable x, the height of a football player selected at random. Of course, probabilities for x are, in this case, just relative frequencies. For example, $P(x = 72) = 0.13$.

The important point in this example is that the histogram in Fig. 5.5 is quite bell-shaped. Because of this, we will be able to approximate probabilities for x using areas under a suitable normal curve. As you might guess, the appropriate normal curve is just the normal curve with parameters μ and σ, where μ is the mean of the population (or of the random variable x) and σ is its standard deviation.

Using formulas from Sec. 4.9, we found the mean and standard deviation of x to be $\mu = 74.06$ and $\sigma = 1.95$.

In Fig. 5.6 we have superimposed the normal curve with parameters $\mu = 74.06$ and $\sigma = 1.95$, upon the histogram given in Fig. 5.5.

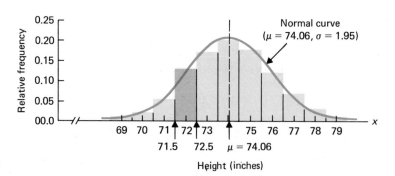

Figure 5.6

Now let us see just how we can approximate probabilities for x by using areas under the normal curve ($\mu = 74.06$, $\sigma = 1.95$). To be specific, let's consider the probability that a football player selected at random is 72 inches tall. Since heights were measured to the nearest inch, we are really asking for the probability that a football player selected at random is between 71.5 and 72.5 inches tall. By Table 5.5, we know the *exact* answer: $P(71.5 < x < 72.5) = 0.13$.

Note that this probability also equals the area of the darkly shaded rectangle between 71.5 and 72.5. This is because the height of the rectangle is 0.13 and the width of the rectangle is 1, as indicated in the figure at the right.

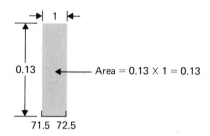

Area = 0.13 × 1 = 0.13

Now let's look at the parts of the histogram and normal curve in Fig. 5.6 between 71.5 and 72.5 (see the second figure on the right).

We have shaded the area under the normal curve between 71.5 and 72.5. As you can see from the picture, *the area under the normal curve is very close to the area of the rectangle.* But we have also seen that the area of the rectangle is just $P(71.5 < x < 72.5)$. Consequently, we see that the probability that x is between 71.5 and 72.5, is approximately equal to the area under the normal curve between 71.5 and 72.5:

$$P(71.5 < x < 72.5) \doteq \frac{\text{Area under normal curve}}{\text{between 71.5 and 72.5}}$$

Although we have already made the point we want to in this example, it is interesting to actually compare the values of the left- and righthand sides of the above equation. We already know that the value of the left side is 0.13. To find the value of the right side, we use the technique of Section 5.3:

$$z = \frac{x - \mu}{\sigma} = \frac{x - 74.06}{1.95}$$

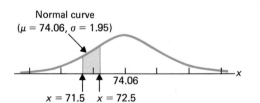

Normal curve
($\mu = 74.06$, $\sigma = 1.95$)

74.06

$x = 71.5$ $x = 72.5$

Change to z

$x = 71.5 \rightarrow z = -1.31$

$x = 72.5 \rightarrow z = -0.80$

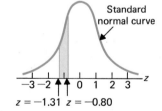

Standard normal curve

$z = -1.31$ $z = -0.80$

Area = $0.2119 - 0.0951 = 0.1168$

So the area under the normal curve between 71.5 and 72.5 is 0.1168, whereas the *exact* value of $P(71.5 < x < 72.5)$ is 0.13. ▷

Exercise **G** In Example 12, compare the values of
a) $P(74.5 < x < 75.5)$ and the area under the normal curve between 74.5 and 75.5.
b) $P(72.5 < x < 76.5)$ and the area under the normal curve between 72.5 and 76.5.

The important point of Example 12 is that, *for certain populations and random variables, we can use areas under normal curves to determine probabilities.* Not all populations (or random variables) have the property that areas under a normal curve can be used to determine probabilities. But, if a population or random variable *does* have that property, then we say it is **normally distributed.**

DEFINITION

> When we say a population or random variable is (approximately) **normally distributed**, we mean that probabilities for the population or random variable are (approximately) equal to areas under a normal curve. Of course, if the population or random variable has mean μ and standard deviation σ, then the normal curve that is used is just the one with parameters μ and σ.

We can summarize the above discussion "visually" in the following way:

If x is a random variable that is *normally distributed* with mean μ and standard deviation σ, then:

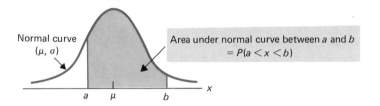

Normal curve
(μ, σ)

Area under normal curve between a and b
$= P(a < x < b)$

If a population or random variable is normally distributed with mean $\mu = 0$ and standard deviation $\sigma = 1$, then we say that it has the **standard normal distribution** (because we find probabilities for such a population or random variable by looking at areas under the *standard* normal curve). We now give an example that illustrates how to compute probabilities for normally distributed populations or random variables.

Example **13** ◀ Suppose the weights at birth of newborn babies are normally distributed with mean 7.3 pounds and standard deviation 2.0 pounds. Find the probability that a newborn baby weighs

a) less than 10 pounds,
b) between 6 and 8 pounds.

Solutions Let x be the weight of a newborn baby selected at random. Then, by assumption, x is a normally distributed random variable with mean $\mu = 7.3$ and standard deviation $\sigma = 2.0$. So we can find probabilities for x using areas under the normal curve with parameters $\mu = 7.3$ and $\sigma = 2.0$.

a) We want $P(x < 10)$ and, consequently, we need to find the area under the normal curve ($\mu = 7.3$, $\sigma = 2.0$) to the left of 10.

$$z = \frac{x - \mu}{\sigma} = \frac{x - 7.3}{2.0}$$

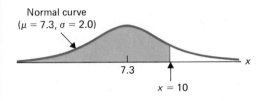

Normal curve
($\mu = 7.3$, $\sigma = 2.0$)

7.3

$x = 10$

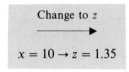

Change to z

$x = 10 \rightarrow z = 1.35$

Area = 0.9115

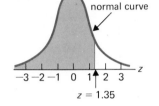

Standard normal curve

$-3\ -2\ -1\quad 0\quad 1\ 2\quad 3$

$z = 1.35$

So $P(x < 10) = 0.9115$. In other words, the probability that a newborn baby weighs less than 10 pounds is 0.9115.

b) To obtain $P(6 < x < 8)$, we need to find the area under the normal curve ($\mu = 7.3$, $\sigma = 2.0$) between 6 and 8.

$$z = \frac{x - \mu}{\sigma} = \frac{x - 7.3}{2.0}$$

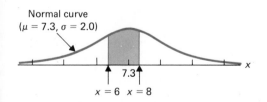

Normal curve
($\mu = 7.3$, $\sigma = 2.0$)

7.3

$x = 6\quad x = 8$

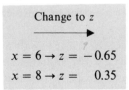

Change to z

$x = 6 \rightarrow z = -0.65$
$x = 8 \rightarrow z = \quad 0.35$

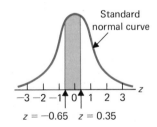

Standard normal curve

$-3\ -2\ -1\quad 0\quad 1\quad 2\quad 3$

$z = -0.65\quad z = 0.35$

Area = $0.6368 - 0.2578 = 0.3790$

Thus, $P(6 < x < 8) = 0.3790$. ▶

In Example 13, find the probability that a newborn baby weighs between 5 and 10 pounds.

Exercise **H**

EXERCISES Section **5.4**

1 Using the data from Example 12, compare the values of
   a) $P(72.5 < x < 75.5)$ and the area under the normal curve between 72.5 and 75.5.
   b) $P(x > 75.5)$ and the area under the normal curve to the right of 75.5.

2 In February 1979, the mean monthly rent for a one-room (studio) apartment in the U.S. was $\mu = \$179$ with $\sigma = \$30$. Find the probability that an apartment rented for between $140 and $200 if the prices were normally distributed.

3 At the 1978 Denver Stock Show, purebred steers sold for a mean price $\mu = \$1,750$, with $\sigma = \$185$. Find the probability that a buyer paid between $1,500 and $2,000 for a steer. (Assume that the prices were normally distributed.)

4 In 1916, ranchers sold their cattle for a mean price of $\mu = \$5.85$ per 100 pounds with $\sigma = \$.30$. Find the probability that a steer was sold for between $5 and $6 per 100 pounds if the prices were normally distributed.

5 Find the probability that the height of a mature saguaro cactus in an area of Arizona is between 25 and 35 feet if the heights are normally distributed with $\mu = 30$ feet and $\sigma = 3.3$ feet.

6 If the mean annual wage (excluding board) of farm laborers in 1926 in the U.S. was normally distributed with $\mu = \$586$ and $\sigma = \$97$, find the probability that a laborer earned between $500 and $600.

7 Find the probability that the food expenditure for a U.S. middle-class family of four in 1973 was between $3,500 and $4,000, if the expenditures were normally distributed with $\mu = \$3,875$ and $\sigma = \$291$.

8 In the U.S. in 1929, males who were 22 years old and 5′ 10″ tall had a mean weight $\mu = 154$ pounds. Find the probability that such a man weighed less than 150 pounds. (Assume that these weights were normally distributed with $\sigma = 11$.)

9 In the U.S. in 1974, males who were 22 years old and 5′ 10″ tall had a mean weight $\mu = 168$ pounds. Find the probability that such a male weighed less than 150 pounds. (Assume that these weights were normally distributed with $\sigma = 15$.)

10 If the lengths of adult yellow-bellied sapsuckers are normally distributed, with $\mu = 8.5$ inches and $\sigma = 0.17$ inch, find the probability that an adult yellow-bellied sapsucker is at least 8 inches long.

5.5 Standardizing a random variable

We learned, in Section 5.3, that we find areas under the normal curve with parameters μ and σ by *standardizing*. That is, we convert x-values to z-values by the formula

$$z = \frac{x - \mu}{\sigma},$$

and then use the *standard* normal curve.

It is useful to consider the process of **standardizing a random variable**.

Example **14**   ◖ Let x be the IQ of a person selected at random. Then x is a random variable with mean $\mu = 100$ and standard deviation $\sigma = 10$.* We can get a new random variable (which we call z) by *standardizing* the random variable x; namely,

$$z = \frac{x - 100}{10}$$

If we select a person at random, then the random variable x "tells us" the IQ of the person. What does the *standardized* random variable "tell us"? It tells us *how many standard deviations the person's IQ is away from the mean*. For example, if $x = 120$

* The value of σ actually depends on the particular IQ test used. In this text we will assume $\sigma = 10$. However you may see different values in other texts (for example, $\sigma = 16$).

(i.e., the IQ of the person selected is 120), then

$$z = \frac{x - 100}{10} = \frac{120 - 100}{10} = 2.$$

So the person's IQ is (+)2 standard deviations away from the mean. We can illustrate this pictorially as follows:

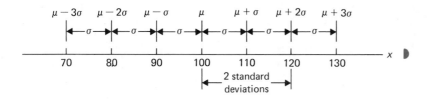

Suppose x is a random variable with mean μ and standard deviation σ. Then the **standardized random variable**

$$z = \frac{x - \mu}{\sigma}$$

tells us how many standard deviations x is away from the mean.

In Example 14, how many standard deviations away from the mean is the IQ of   Exercise **I**
a) 110?   b) 115?   c) 70?

Of course, we can use the process of standardizing "in reverse."

◀ A person selected at random has an IQ that is 1.5 standard deviations away from the   Example **15**
mean (that is, $z = 1.5$). What is this person's IQ?

Here we know z ($z = 1.5$) and we want to find x. All we have to do is solve for x in the   Solution
equation

$$z = \frac{x - 100}{10}$$

Doing the necessary algebra we get

$$x = 100 + 10z = 100 + 10 \cdot (1.5) = 115$$

So, the person's I.Q. is 115. ▶

In Example 14, find the IQ if   Exercise **J**

a) $z = 3$ b) $z = 0$ c) $z = -1.5$

Now to the main point of this section.

Suppose x is a *normally distributed* random variable with mean μ and standard deviation σ. Then the standardized random variable

$$z = \frac{x - \mu}{\sigma}$$

has the *standard normal distribution*. Thus we can find probabilities for the *standardized* random variable by looking at areas under the *standard* normal curve.

We will use this fact (that a standardized normally distributed random variable has the *standard* normal distribution) throughout the remainder of the book.

As a first application of this fact, we consider the following example.

Example **16**   ◖ In a large city, the heights of adult males are normally distributed with mean μ = 68″ and standard deviation $\sigma = 2.5″$. Let x be the height of an adult male (in this city) selected at random. Then x is normally distributed ($\mu = 68$, $\sigma = 2.5$).

Now, we know that the *standardized* random variable

$$z = \frac{x - \mu}{\sigma} = \frac{x - 68}{2.5}$$

has the *standard* normal distribution. Then, for example, using Table I, we find that

$$P(-2 < z < 2) = 0.9544$$

(The area under the standard normal curve between -2 and 2 is 0.9544.)

Since z is just the number of standard deviations that x is away from its mean, we can say that the probability that x is within ($\pm$)2 standard deviations of its mean is 0.9544. But -2 (2 in the negative direction) standard deviations away from the mean is

$$\underset{\underset{\mu}{\uparrow}}{68} - \underset{\underset{\sigma}{\uparrow}}{2 \cdot (2.5)} = 63,$$

and $+2$ standard deviations away from the mean is

$$68 + 2 \cdot (2.5) = 73.$$

So $P(63 < x < 73) = 0.9544$. In other words, about 95 percent of the adult males in the city are between 63 and 73 inches tall (5′ 3″ to 6′ 1″).

With a little practice you'll be able to do computations like this very quickly. ◗

If you use Table I (or look back to Section 5.2, page 168) you will find that, if z has the *standard* normal distribution, then

1. $P(-1 < z < 1) = 0.6826$—About 68 % of the area under the standard normal curve lies between -1 and 1.
2. $P(-2 < z < 2) = 0.9544$—About 95 % of the area under the standard normal curve lies between -2 and 2.

3. $P(-3 < z < 3) = 0.9974$—About 99.7% (almost all) of the area under the standard normal curve lies between -3 and 3.

If we interpret these facts in terms of standardized variables we get the following information (which is mainly useful for "visualizing" a random variable or population).

> Suppose a population is *normally distributed*. Then,
>
> 1.   About 68% of the population lies within one standard deviation of its mean.
> 2.   About 95% of the population lies within two standard deviations of its mean.
> 3.   About 99.7% (almost all) of the population lies within three standard deviations of its mean.

◀ IQ's are normally distributed with $\mu = 100, \sigma = 10$. Using the above rule, we see that   Example **17**
1.   About 68% of the population have IQ's between 90 ($= 100 - 1 \cdot 10$) and 110 ($= 100 + 1 \cdot 10$).
2.   About 95% of the population have IQ's between 80 and 120.
3.   About 99.7% of the population have IQ's between 70 and 130. ▶

(Refer to Example 16.) Fill in the blanks: Exercise **K**
a)   About 68% of the adult males in the city are between _____ and _____ inches tall.
b)   About 99.7% of the adult males in the city are between _____ and _____ inches tall.

EXERCISES Section **5.5**

1   If studio apartments in February 1979 rented for a mean $\mu = \$179$ with $\sigma = \$30$, how many standard deviations from the mean was a rental charge of
a) \$209?   b) \$119?

2   If the mean annual wage of a farm worker in 1926 in the U.S. was $\mu = \$596$ with $\sigma = \$97$, how many standard deviations from the mean was a wage of
a) \$877?   b) \$828.50?

3   If the mean food expenditure for a middle-class family of four in 1973 was $\mu = \$3875$ with $\sigma = \$291$, how many standard deviations from the mean was an expenditure of
a) \$3002?   b) \$3729.50?

4   At the 1978 Denver Stock Show purebred steers sold for a mean $\mu = \$1,750$ with $\sigma = \$185$. How much did a steer cost if the price was
a) 1 standard deviation from the mean?
b) -2 standard deviations from the mean?
c) 2.5 standard deviations from the mean?
d) -0.5 standard deviation from the mean?

5   In 1929 males who were 22 years old and 5' 10" tall had a mean weight of $\mu = 154$ pounds. Using a value of $\sigma = 11$, find a person's weight if
a) $z = 1$ b) $z = -2.5$ c) $z = 3$ d) $z = 0$

In Exercises 6 and 7, assume that the population given is (approximately) normally distributed.

6   Refer to Exercise 5 above.
a) In 1929 about 68% of males who were 22 years old and 5' 10" inches tall weighed between _____ and _____ pounds.
b) In 1929 about 95% of males who were 22 years old and 5' 10" inches tall weighed between _____ and _____ pounds.

7   In 1916 ranchers sold their cattle for a mean price of $\mu = \$5.85$ per 100 pounds. Fill in the blanks, using $\sigma = \$0.30$.
a) About 68% of the cattle sold for a price of between _____ and _____ per 100 pounds.
b) About 95% of the cattle sold for a price of between _____ and _____ per 100 pounds.
c) About 99.7% of the cattle sold for a price of between _____ and _____ per 100 pounds.

5.6

The normal approximation to the binomial distribution *

One of the earliest uses of the normal distribution was in approximating the binomial distribution.[†] Specifically, we shall see, that under certain conditions on n and p, we can use areas under normal curves to find (approximately) binomial probabilities.

Before beginning, we will briefly review the binomial distribution (see Section 4.8 for details).

1. n success–failure experiments are performed.
2. The probability of success on any given trial is p.
3. The trials are independent of each other.
4. If x represents the number of successes, then

$$P(x = k) = \binom{n}{k} p^k (1 - p)^{n-k}$$

You might be wondering why we need to use areas under normal curves to find probabilities for x *approximately*, when we have an *exact* formula for finding such probabilities. The following example should help explain why we need to do this.

Example 18

◖ A new drug is claimed to be 67 percent effective in curing a certain disease. Assume that this claim is correct. Suppose 500 people with this disease are given the drug. Find the probability that the drug cures
a) exactly 335 of them;
b) between 320 and 350 of them, inclusive.

Solution

Let x be the number of people cured by the drug. There are $n = 500$ trials. Also, since there is a 67% chance of cure, $p = 0.67$. So x has the binomial distribution with $n = 500$ and $p = 0.67$. Consequently, we "can" find probabilities for x *exactly*, by using the binomial probability formula

$$P(x = k) = \binom{500}{k} (0.67)^k (0.33)^{500-k}$$

Let's apply this formula to (a) and (b).

a) Here we want exactly 335 successes. So, the "answer" is

$$P(x = 335) = \binom{500}{335} (0.67)^{335} (0.33)^{165}$$

Try to get the actual answer on your calculator. That is, try to compute

$$\binom{500}{335} (0.67)^{335} (0.33)^{165}$$

---

* This section is optional.

† The mathematical theory for doing this is credited to Abraham DeMoivre (1667–1754), and Pierre Laplace (1749–1827).

on your calculator. (Don't really do it—it's a mess!) Even if you do figure out how to do it, you'll have to admit it's rather difficult, and extremely time-consuming. As you will see, we'll be able to find the answer (approximately) by using areas under a normal curve, and *it will be easy.*

b) Here we want between 320 and 350 successes, inclusive. So, the "answer" is

$$P(x = 320, 321, \ldots, \text{or } 350)$$
$$= P(x = 320) + P(x = 321) + \cdots + P(x = 350)$$
$$= \binom{500}{320}(0.67)^{320}(0.33)^{180} + \binom{500}{321}(0.67)^{321}(0.33)^{179}$$
$$+ \cdots + \binom{500}{350}(0.67)^{350}(0.33)^{150}$$

If you thought it would be tough to get the answer in (a), look at this one! Here we would have to do 31 computations like the one in (a). [There are 31 terms in the above sum.] So it seems rather hopeless to actually figure out the answer by using the formula. However, you will soon see that, by using areas under a normal curve, you'll be able to get an (approximate) answer quickly and easily. ◗

The previous example should make it clear to you why we want to approximate binomial probabilities. For even though we have an exact formula for computing binomial probabilities, the formula is not of much practical value when we are dealing with large values of n. Under certain conditions on n and p (the conditions will be stated shortly), the histogram of the binomial distribution is "bell-shaped" and we can use areas under normal curves to approximate binomial probabilities.

The following example illustrates a "bell-shaped" binomial distribution. For this example, it would actually be simple to calculate probabilities exactly by using the binomial probability formula. But it is a good illustration of using normal curves to find binomial probabilities approximately.

◖ A student is taking a true–false exam with ten questions. If the student *guesses* at all ten questions, what is the probability she gets either seven or eight correct?   Example **19**

Let x be the number of correct guesses by the student. There are ten questions, so here   Solution
$n = 10$. Since the student guesses at each question, the success probability is $p = 0.5$. So x has the binomial distribution with $n = 10$ and $p = 0.5$. That is,

$$P(x = k) = \binom{10}{k}(0.5)^k(1 - 0.5)^{10 - k}$$

Using this formula we can get the exact probability distribution of x.

x	0	1	2	3	4	5	6	7	8	9	10
$P(x)$	0.0010	0.0098	0.0439	0.1172	0.2051	0.2461	0.2051	0.1172	0.0439	0.0098	0.0010

Table 5.6
Probability distribution for x

The problem is to find $P(x = 7$ or 8). In this case the exact answer is (see Table 5.6):

$$P(x = 7 \text{ or } 8) = P(7) + P(8) = 0.1172 + 0.0439 = 0.1611$$

Now let's see how we can use areas under a normal curve to find (approximately) $P(x = 7$ or 8). The normal curve we will use, of course, is the one with parameters μ and σ, where μ is the mean of x and σ is its standard deviation.

Now (see Section 4.9, page 146), in this case,

$$\mu = np = 10 \cdot (0.5) = 5,$$

and

$$\sigma = \sqrt{np(1-p)} = \sqrt{10 \ (0.5)(1-0.5)} \doteq 1.58$$

In Fig. 5.7 we have drawn the histogram for x, along with the normal curve ($\mu = 5$, $\sigma = 1.58$).

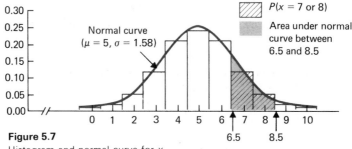

Figure 5.7
Histogram and normal curve for x.

As in Example 12, $P(x = 7$ or 8) equals the area of the corresponding rectangles in the histogram. Also, you should note that this area, which is shown in Fig. 5.7, is *about* the same as the area under the normal curve between 6.5 and 8.5. It should be clear to you from the picture why we consider the area under the normal curve between 6.5 and 8.5. This is referred to as the **correction for continuity**. (Keep this in mind when we set forth the general procedure after this example.)

In summary then, we see from the picture that

$$P(x = 7 \text{ or } 8) \doteq \frac{\text{Area under normal curve}}{\text{between 6.5 and 8.5}} \blacktriangleright$$

Exercise **L**   Find the value of the righthand side of the above equation, and compare it to the value of the lefthand side (which is 0.1611).

From Example 19 we see that, at least under certain circumstances, it is reasonable to use areas under normal curves to approximate binomial probabilities. The rule of thumb is given as follows.

It is reasonable to use areas under a normal curve to approximate binomial probabilities when *both np and n(1 − p) are at least 5.*

By examining carefully what we did in Example 19, we can write down the general procedure for using areas under normal curves to approximate binomial probabilities.

1. Determine *n* (the number of trials), and *p* (the success probability).
2. Check that both *np* and *n(1 − p)* are at least 5. If they are not, you should not use the normal-curve approximation.
3. Find μ and σ using the formulas

$$\mu = np \quad \text{and} \quad \sigma = \sqrt{np(1-p)}$$

4. Determine the *k*-values in question. (In Example 19, the *k*-values are 7 and 8.)
5. Make the *correction for continuity* (a quick sketch is helpful here), and find the appropriate area under the normal curve with parameters μ and σ.

Procedure for using areas under a normal curve to approximate binomial probabilities

We will now use the procedure explained above to solve Example 18, part (b).

Solution to Example 18(b)

1. From page 184, $n = 500$ and $p = 0.67$.
2. Check that both np and $n(1 − p)$ are at least 5:

$$np = 500(0.67) = 335;$$

$$n(1 − p) = 500(0.33) = 165.$$

3. Find μ and σ:

$$\mu = np = 500(0.67) = 335;$$

$$\sigma = \sqrt{np(1-p)} = \sqrt{500(0.67)(0.33)} \doteq 10.51$$

4. The *k*-values in question here are $k = 320, 321, \ldots, 350$.
5. Make the correction for continuity, and find the appropriate area under the normal curve ($\mu = 335$, $\sigma = 10.51$):

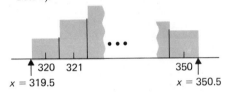

So we need to find the area under the normal curve ($\mu = 335$, $\sigma = 10.51$) between $x = 319.5$ and $x = 350.5$. We do this as usual.

$$z = \frac{x - 335}{10.51}$$

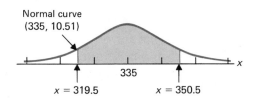

Change to z

$x = 319.5 \rightarrow z = -1.47$

$x = 350.5 \rightarrow z = 1.47$

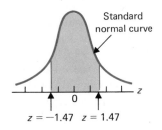

Standard
normal curve

Area $= 0.9292 - 0.0708 = 0.8584$

Thus the probability is (about) 0.8584 that between 320 and 350 of the people given the drug will be cured by it.

Exercise **M** a) Solve part (a) of Example 18 (page 184).

b) In Example 18, find the probability that *at least* 350 of the 500 people are cured by the drug.

EXERCISES Section **5.6**

1 Use Table 5.6 of this section to find the probability that the student in Example 19 guesses
a) 4 or 5 questions correctly, $P(x = 4 \text{ or } 5)$.
b) Between 3 and 7 questions, inclusive, correctly, $P(x = 3, 4, 5, 6, \text{ or } 7)$.
c) At most 5 questions correctly, $P(x \leqslant 5)$.
d) At least 6 questions correctly, $P(x \geqslant 6)$.

2 Find the area under the normal curve in Fig. 5.7 between
a) 3.5 and 5.5   b) 2.5 and 7.5
c) -0.5 and 5.5   d) 5.5 and 10.5
e) compare these results with the results from Exercise 1.

3 If, in Example 19, the true–false exam had 30 questions instead of 10, what normal curve would you use to approximate probabilities for the number of correct guesses?

In Exercises 4 through 10, follow the procedure given on page 187, for using areas under a normal curve to approximate the binomial probabilities.

4 A student is taking a test with 25 multiple-choice questions. Each multiple-choice question has four choices. If the student guesses at all 25 questions, what is the probability that he gets between 8 and 12 questions, inclusive, correct?

5 On the average, a salesperson makes a sale in one out of every ten presentations. Using the normal approximation to the binomial distribution, find the probability that between 5 and 10 sales, inclusive, are made in the next 50 presentations.

6 A basketball player has a probability equal to 0.5 of making a free throw. Find the probability of between 5 and 7 successes, inclusive, in the next ten free throws.

7 An absent-minded professor remembers, on the average, to come home on time for dinner 7 out of 10 times. Using the normal approximation to the binomial distribution, find the probability that the professor will be home on time between 10 and 15 times, inclusive, on the next 20 teaching days.

8 In 1941, Ted Williams of the Boston Red Sox was the last major league player to bat over .400. His batting average that year

was .406. Using the normal approximation to the binomial distribution, find the probability that Ted would get 10 or less hits in 20 times at bat during that season. (Use $p = 0.4$.)

9 A soccer team has a probability equal to 0.6 of winning. If the team wins 16 or more of their remaining 20 games they will win their league championship. What is the probability that they become the league champions?

10 The probability that a certain commuter will miss the 8 : 02 train any *weekday* morning is 0.25. Find the probability of missing the train at least two times during four weeks of commuting.

CHAPTER REVIEW

Key Terms

law of large numbers
normal curves
standard normal curve
parameters (μ and σ)
standardizing
standard score (z-score)
normally distributed random
   variables and populations

standard normal distribution
standardizing a random variable
normal approximation to the
   binomial*
correction for continuity*

Formulas and Key Facts

sample mean (ungrouped data):†

$$\bar{x} = \frac{\Sigma x}{n} \qquad (n = \text{sample size})$$

finding areas under a normal curve with parameters μ and σ:

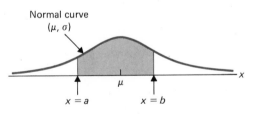

Change to z

$$x = a \rightarrow z = \frac{a - \mu}{\sigma}$$

$$x = b \rightarrow z = \frac{b - \mu}{\sigma}$$

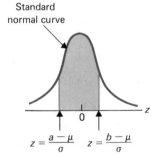

Standard normal curve

$$z = \frac{a - \mu}{\sigma} \qquad z = \frac{b - \mu}{\sigma}$$

(Similar diagrams are used for finding areas to the left or right of a given x-value.)

finding probabilities for normally distributed populations and random variables:

If the population (or random variable) is normally distributed with mean μ and standard deviation σ, then probabilities can be found by looking at areas under the normal curve with parameters μ and σ.

---

* An asterisk refers to material covered in an optional section.

† This is from Chapter 3, but is repeated here for emphasis.

standardizing a random variable:

$$z = \frac{x - \mu}{\sigma}$$

(μ = mean of x, σ = standard deviation of x.)

standardized normally distributed random variable:

If x is a normally distributed random variable with mean μ and standard deviation σ, then the standardized random variable

$$z = \frac{x - \mu}{\sigma}$$

has the *standard normal distribution*.

normal approximation to the binomial:*

If x has the binomial distribution with parameters n and p, and if both np and $n(1 - p)$ are at least 5, then probabilities for x can be found (approximately) by finding areas under the normal curve with parameters $\mu = np$ and $\sigma = \sqrt{np(1 - p)}$.

You should be able to

1 Use all the formulas and key facts.

2 Give an intuitive statement of the *law of large numbers*.

3 Sketch a *normal curve*, given the *parameters* μ and σ.

4 Make statements concerning the *percentage* of a *normally distributed population that lies within 1, 2, or 3 standard deviations of its mean*.

REVIEW TEST

1 Use Table I to find the area under the standard normal curve to the left of $z = -1.27$.

2 Use Table I to find the area under the standard normal curve to the right of $z = 2.64$.

3 Use Table I to find the area under the standard normal curve between $z = -1.85$ and $z = 2.03$.

4 Use Table I to find the area to the left of $x = 6$ under the normal curve with $\mu = 2$ and $\sigma = 3$.

5 For the normal curve with parameters $\mu = 3.5$ and $\sigma = 5$, use Table I to find the area that lies between $x = -1$ and $x = 7.6$.

6 The mean cost of putting on a new roof is $\mu = \$2000$ with $\sigma = \$350$. If the costs are normally distributed, find the probability that a new roof costs between \$1800 and \$2500.

---

* An asterisk refers to material covered in an optional section.

7 If the mean height of women in 1976 was $\mu = 66$ inches with $\sigma = 2.2$, how many standard deviations from the mean is a woman whose height is 60.5 inches?

8 The mean summer earnings of a class of community college students is $\mu = \$2700$ with $\sigma = \$400$. About 95 % of the students earned between ____ and ____. (Assume such earnings are normally distributed.)

9 The probability of getting a head in a single coin toss is $p = 0.5$. Use the normal approximation to the binomial to find the probability of tossing between 60 and 64 heads (inclusive) in 100 tosses.

FURTHER READINGS

DIXON, W., and MASSEY, F., *Introduction to Statistical Analysis.* New York: McGraw-Hill, 1969. Chapter 5 of this book gives a more detailed treatment of the normal distribution (including such topics as normal-probability graph paper, and a discussion of the normal curve equation).

KRUSKAL, W., *et al.* (Eds.), *Statistics by Example: Detecting Patterns.* Reading, Massachusetts: Addison-Wesley, 1973. Of special interest here are the articles: "Normal Probability Distributions" (by Roger Carlson) and "How Much Does a 40-Pound Box of Bananas Weigh?" (by Ralph D'Agostino).

ANSWERS TO REINFORCEMENT EXERCISES

A a) .2005   b) .9979

B a)

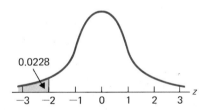

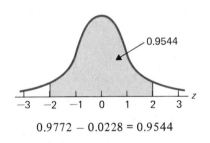

$$0.9772 - 0.0228 = 0.9544$$

b)

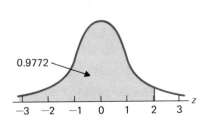

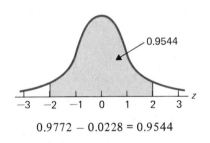

$$1 - (0.0013 + 0.0013) = 0.9974$$

B b)

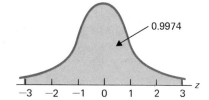

0.9974

C a)

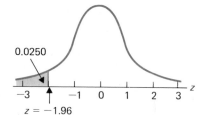

0.0250

$z = -1.96$

b) 0.0250 by symmetry
c) $0.0250 + 0.0250 = 0.0500$
d) $1 - 0.0500 = 0.9500$

D a)

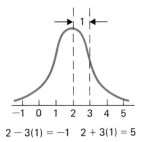

$2 - 3(1) = -1 \quad 2 + 3(1) = 5$

b)

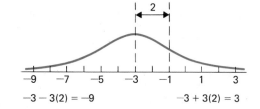

$-3 - 3(2) = -9 \qquad -3 + 3(2) = 3$

E

Normal curve
($\mu = 5, \sigma = 3$)

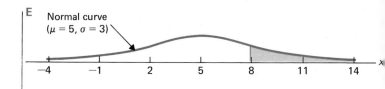

$$z = \frac{8 - 5}{3} = 1$$

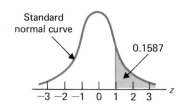

Standard
normal curve

0.1587

$1 - 0.8413 = 0.1587$

F

Normal curve
($\mu = 1, \sigma = 1.5$)

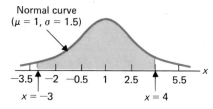

$x = -3 \qquad x = 4$

Change to z
$\longrightarrow$

$$x = -3 \rightarrow z = \frac{-3 - 1}{1.5} = -2.67$$

$$x = 4 \rightarrow z = \frac{4 - 1}{1.5} = 2$$

Standard
normal curve

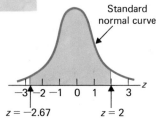

$z = -2.67 \qquad z = 2$

$0.9772 - 0.0038 = 0.9734$

G a) i) Area under normal curve:

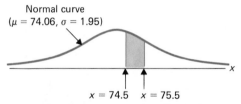

Normal curve
($\mu = 74.06$, $\sigma = 1.95$)

$x = 74.5$ $x = 75.5$

Change to z

$x = 74.5 \to z = \dfrac{74.5 - 74.06}{1.95} = 0.23$

$x = 75.5 \to z = \dfrac{75.5 - 74.06}{1.95} = 0.74$

Standard normal curve

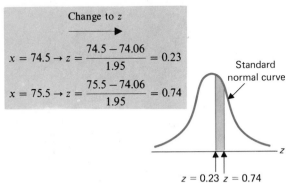

$z = 0.23$ $z = 0.74$

Normal curve approximation: $0.7704 - 0.5910 = 0.1794$

ii) From Table 5.5, $P(74.5 < x < 75.5) = P(x = 75) = 0.18$

b) i) Area under normal curve:

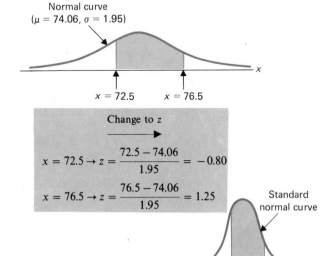

Normal curve
($\mu = 74.06$, $\sigma = 1.95$)

$x = 72.5$ $x = 76.5$

Change to z

$x = 72.5 \to z = \dfrac{72.5 - 74.06}{1.95} = -0.80$

$x = 76.5 \to z = \dfrac{76.5 - 74.06}{1.95} = 1.25$

Standard normal curve

$z = -0.80$ $z = 1.25$

Normal curve approximation: $0.8944 - 0.2119 = 0.6825$

ii) From Table 5.5,

$$P(72.5 < x < 76.5) = P(x = 73 \text{ or } 74 \text{ or } 75 \text{ or } 76)$$

$$= 0.17 + 0.20 + 0.18 + 0.12 = 0.67$$

H

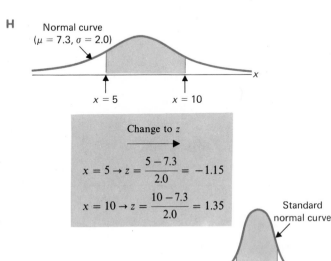

Normal curve
($\mu = 7.3$, $\sigma = 2.0$)

$x = 5$ $x = 10$

Change to z

$x = 5 \to z = \dfrac{5 - 7.3}{2.0} = -1.15$

$x = 10 \to z = \dfrac{10 - 7.3}{2.0} = 1.35$

Standard normal curve

$z = -1.15$ $z = 1.35$

$$0.9115 - 0.1251 = 0.7864$$

I a) $z = \dfrac{110 - 100}{10} = 1$ b) $z = \dfrac{115 - 100}{10} = 1.5$

c) $z = \dfrac{70 - 100}{10} = -3$

J a) $z = \dfrac{x - 100}{10}$

$x = 10z + 100 = 10(3) + 100 = 130$

b) $z = \dfrac{x - 100}{10}$

$x = 10z + 100 = 10(0) + 100 = 100$

c) $z = \dfrac{x - 100}{10}$

$x = 10z + 100 = 10(-1.5) + 100 = 85$

K   a) 65.5, 70.5   b) 60.5, 75.5

L

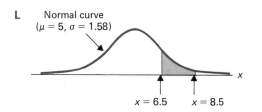

Normal curve
($\mu = 5$, $\sigma = 1.58$)

$x = 6.5$ $x = 8.5$

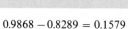

Change to z

$x = 6.5 \rightarrow z = \dfrac{6.5 - 5}{1.58} = 0.95$

$x = 8.5 \rightarrow z = \dfrac{8.5 - 5}{1.58} = 2.22$

$0.9868 - 0.8289 = 0.1579$

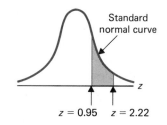

Standard
normal curve

$z = 0.95$ $z = 2.22$

M   a)

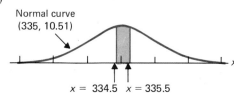

Normal curve
(335, 10.51)

$x = 334.5$ $x = 335.5$

Change to z

$x = 334.5 \rightarrow z = \dfrac{334.5 - 335}{10.51} = -0.05$

$x = 335.5 \rightarrow z = \dfrac{335.5 - 335}{10.51} = 0.05$

$0.5199 - 0.4801 = 0.0398$

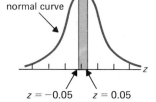

Standard
normal curve

$z = -0.05$ $z = 0.05$

b)

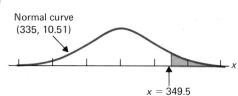

Normal curve
(335, 10.51)

$x = 349.5$

Change to z

$x = 349.5 \rightarrow z = \dfrac{349.5 - 335}{10.51} = 1.38$

$1 - 0.9162 = 0.0838$

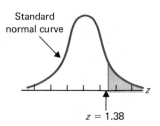

Standard
normal curve

$z = 1.38$

The sampling
distribution of the mean

6.1 Random samples

We are now ready to show how we can calculate probabilities for $\bar{x}$, using areas under normal curves. Mathematically speaking, this requires finding the probability distribution of the random variable $\bar{x}$, which is so important that it is given a special name.

DEFINITION

> The probability distribution of $\bar{x}$ is called the **sampling distribution of the mean.**

First, we need to mention the concept of a *random sample*. This concept can be better understood if we consider that the main reason we take a sample in the first place is *to make inferences* (*educated guesses*) *about a characteristic of a population, based on data obtained from a sample of the population.* For instance, we might be interested in making an inference about the mean tar content, μ, of all cigarettes of a certain brand, based on the (sample) mean tar content, $\bar{x}$, of 30 such cigarettes. Or we might want to gain some insight about the mean height, μ, of all adult U.S. males by looking at the mean height, $\bar{x}$, of a sample of 250 such males.

Clearly, then, we would like our sample to be **representative**. That is, we would like our sample to reflect, as well as possible, the relevant characteristic(s) of the population. For example, it would not make much sense to use the mean height of a sample of 250 basketball players, to make an inference (educated guess) about the mean height of all adult U.S. males. Nor would it be reasonable to try to estimate the mean income of residents of California, by looking at the mean income of a sample of residents of Beverly Hills.

There is a very famous example where a *nonrepresentative* sample led to an incorrect inference. In 1936, before the presidential election, the *Literary Digest* magazine conducted an opinion poll of the voting population. That is, its survey team took a sample of the voting population and asked each individual in the sample whether they would vote for Roosevelt or for Landon. Based on the results of this sample, the magazine predicted (inferred) that Landon would win easily. The election results: Roosevelt won by a landslide. What happened? Some people think the sample was misleading, because the magazine obtained its sample from people who owned a car or had a telephone and, at the time, only the more well-to-do people owned a car or had a telephone. Others believe that the sample was misleading because response was by return mail, and thus self-selected (see Bryson, M. C., "The Literary Digest Poll," *American Statistican*, **30**, 4, 1976).

Whatever the reason, the sample taken by the *Literary Digest* was obviously *not* representative.

Our goal in sampling is to obtain a representative sample. For only then can we expect to make reasonable generalizations (inferences) concerning a population, based on the information obtained from a sample of the population.

One of the most common methods of sampling is called **simple random sampling** or, more briefly, **random sampling.** *In random sampling, each possible sample is equally likely to be the one selected.* Let's look at an example.

◖ In Example 2 of Chapter 5 we considered a *population* of heights of the five starting   Example **1**
players on a basketball team. We mentioned there that this is not a very realistic
example, because for such a small population there would really be no need to sample.
For instance, if we were interested in the (population) mean height μ, we would
undoubtedly just measure all five players, and compute μ exactly. However, this
population of heights will serve as a concrete illustration of what we are talking about.

The population of heights of the five players is given below:

Player	1	2	3	4	5
Height (in.)	67	73	75	76	84

Suppose we decide to take a sample of size $n = 2$. There are ten possible samples of
size 2. They are listed in Table 6.1.

Table **6.1**

Sample	1	2	3	4	5	6	7	8	9	10
Players selected	1 & 2	1 & 3	1 & 4	1 & 5	2 & 3	2 & 4	2 & 5	3 & 4	3 & 5	4 & 5
Heights (in.)	67 & 73	67 & 75	67 & 76	67 & 84	73 & 75	73 & 76	73 & 84	75 & 76	75 & 84	76 & 84

One method we could use to obtain a *random* sample of size 2 is to write each of the
five heights on a separate slip of paper, throw the five slips of paper into a box, shake the
box, and then, while blindfolded, pick two of the slips of paper out. This would ensure
that we are taking a *random* sample of size 2. That is, it would ensure that each of the ten
possible samples of size 2 has the *same probability of being the one selected*—a
probability of $\frac{1}{10}$.

If we consider samples of size $n = 4$, then there are five possible outcomes:

Sample	1	2	3	4	5
Players selected	1, 2, 3, 4	1, 2, 3, 5	1, 2, 4, 5	1, 3, 4, 5	2, 3, 4, 5
Heights (in.)	67, 73, 75, 76	67, 73, 75, 84	67, 73, 76, 84	67, 75, 76, 84	73, 75, 76, 84

In this case, if we take a *random* sample of size 4 (say, by picking four of the slips of
paper out of the box), then each of the five possible samples of size 4 has probability $\frac{1}{5}$ of
being the one selected. ◗

Since the statistical inference techniques you will study in this book require that we
deal with (simple) random samples, from now on we will assume that, when we refer to a
sample of a population, we mean a **random sample**.

EXERCISES Section **6.1**

In Exercises 1 through 3, you are given a population and a sample size n.

a) List all of the possible samples of size n for the given population.
b) Find the probability that any particular sample is randomly selected.

1 Population is the numbers 1, 2, 3, 4, and $n = 2$.

2 Population is 1, 2, 3, 4, 5, 6, and $n = 2$.

3 Population is 1, 2, 3, 4, 5, and $n = 3$.

4 Each of the following samples is *not* representative. Explain why, in your own words.
  a) A sample of incomes from 30 dentists in Seattle is used to represent all incomes in the state of Washington.
  b) The weights of the members of the UCLA football team are used as a sample of the weights of all UCLA students.
  c) A survey of the political opinions of 30 voters in the retirement community of Sun City, Arizona, is used as a sample of the opinions of all Americans.

5 In this section we mentioned that you could select two heights at random from the five heights given in Example 1 by picking numbered slips of paper from a box (blindfolded).
  a) Find another selection procedure that would ensure that two heights are selected at random.
  b) Give an example of a method of selecting two heights that is *not* random.

6 a) List all ten possible samples of size 3 from the five heights given in Example 1.
  b) In a random sampling procedure for obtaining three heights, what is the probability of choosing a particular sample of size 3?

7 Four debaters of equal ability qualify to be taken to a tournament, but there is only enough money to take two of them. The team's coach decides to select the two debaters for the tournament at random.
  a) Describe a selection procedure for the coach to use.
  b) The four debaters are Smith, Lopez, Jones, and Flores. What is the probability that Smith and Lopez are chosen to go?

6.2 The mean and standard deviation of $\bar{x}$

The next preliminary step for describing the sampling distribution of the mean is to learn how to find the mean and standard deviation of the random variable $\bar{x}$. This is necessary in order to use normal-curve methods to find probabilities for $\bar{x}$.

We will look at an example where we calculate the mean and standard deviation of $\bar{x}$. This example is extremely oversimplified, and would not arise in practice. However, it does illustrate, simply and concretely, the points we want to make.

Example **2** ◖ Consider again the illustration of Example 1. The *population* consists of the heights of the starting five players on a basketball team:

Player	1	2	3	4	5
Height (in.)	67	73	75	76	84

Using our usual formulas, we find that $\mu = 75.0$ and $\sigma = 5.48$ for this population.

Now, let's consider *random* samples of size $n = 2$. There are ten possible samples of size 2. These ten samples are listed below, along with their $\bar{x}$-values (see the table on page 199).

Sample	$\bar{x}$	Sample	$\bar{x}$
67, 73	70.0	73, 76	74.5
67, 75	71.0	73, 84	78.5
67, 76	71.5	75, 76	75.5
67, 84	75.5	75, 84	79.5
73, 75	74.0	76, 84	80.0

Table 6.2
$\bar{x}$ values for samples of size $n = 2$

Since each sample has probability $\frac{1}{10}$ of being the one selected, the probability distribution of $\bar{x}$ is:

$\bar{x}$	$P(\bar{x})$
70.0	0.1
71.0	0.1
71.5	0.1
74.0	0.1
74.5	0.1
75.5	0.2	(Two of the ten samples yield $\bar{x} = 75.5$)
78.5	0.1
79.5	0.1
80.0	0.1

Table 6.3
Probability distribution of $\bar{x}$ for samples of size 2 ($n = 2$)

Using the formulas we learned in Chapter 4 (see page 154), we can find the mean and standard deviation of the random variable $\bar{x}$. To keep things straight, *we use the symbol $\mu_{\bar{x}}$ to denote the mean of $\bar{x}$, and $\sigma_{\bar{x}}$ to denote the standard deviation of $\bar{x}$.*

$$\mu_{\bar{x}} = \sum \bar{x} P(\bar{x})$$
$$= (70.0) \cdot (0.1) + (71.0) \cdot (0.1) + \cdots + (80.0) \cdot (0.1)$$
$$= 75.0$$

and

$$\sigma_{\bar{x}} = \sqrt{\sum (\bar{x} - \mu_{\bar{x}})^2 P(\bar{x})}$$
$$= \sqrt{(70.0 - 75.0)^2 \cdot (0.1) + (71.0 - 75.0)^2 \cdot (0.1) + \cdots + (80.0 - 75.0)^2 \cdot (0.1)}$$
$$\doteq 3.35$$

Note that $\mu_{\bar{x}} = 75.0$, which is the same as μ. That is, at least for samples of size $n = 2$, the mean of $\bar{x}$ is the same as the population mean. (We'll consider σ later.) In other words, for samples of size $n = 2$, $\mu_{\bar{x}} = \mu$.

Is this a coincidence? Let's try samples of size $n = 4$. There are five possible samples of size 4. These five samples, along with their $\bar{x}$-values, are given in Table 6.4.

Table 6.4	Sample	$\bar{x}$
	67, 73, 75, 76	72.75
	67, 73, 75, 84	74.75
	67, 73, 76, 84	75.00
	67, 75, 76, 84	75.50
	73, 75, 76, 84	77.00

Since each of the five samples has probability $\frac{1}{5}$ of being the one selected, the probability distribution of $\bar{x}$ is:

Table 6.5	$\bar{x}$	$P(\bar{x})$
	72.75	0.2
	74.75	0.2
	75.00	0.2
	75.50	0.2
	77.00	0.2

We computed the mean and standard deviation of $\bar{x}$ and found that $\mu_{\bar{x}} = 75.0$ and $\sigma_{\bar{x}} \doteq 1.37$. So, again $\mu_{\bar{x}} = \mu$.

The results of the above example are no accident.

> Suppose a random sample of size n is taken from a population with mean μ. Then the *mean of* $\bar{x}$ always equals the mean of the population (regardless of sample size). That is,
>
> $$\mu_{\bar{x}} = \mu$$

Exercise **A**   Using the heights 72, 75, 76, 77 (whose mean is $\mu = 75.0$),

a) List all possible samples of size 2,
b) Find $\bar{x}$ for each sample,
c) Find the probability distribution of $\bar{x}$,
d) Find $\mu_{\bar{x}}$,
e) Find $\sigma_{\bar{x}}$.

We now know that the mean of $\bar{x}$ is just the same as the population mean; $\mu_{\bar{x}} = \mu$. What about the standard deviation of $\bar{x}$? This is a little more complicated. Let's return to Example 2.

◀ In the table below we have included the information obtained in Example 2. In addition we computed $\mu_{\bar{x}}$ and $\sigma_{\bar{x}}$ for random samples of size 3, and have also given these values in the table below.

Example **3**

Sample size (n)	$\mu_{\bar{x}}$	$\sigma_{\bar{x}}$
2	75.0	3.35
3	75.0	2.24
4	75.0	1.37

Recall also, that the population parameters are $\mu = 75.0$ and $\sigma \doteq 5.48$. Now we know that $\mu_{\bar{x}}$ is always equal to μ, regardless of sample size but, as we see from the table above, $\sigma_{\bar{x}}$ is *not* always the same. In fact it appears that $\sigma_{\bar{x}}$ gets smaller as we deal with larger samples. There *is* a formula for computing $\sigma_{\bar{x}}$ exactly, in this special case of a small finite population. The formula is discussed in Exercise 10, but we won't be using it because the populations we consider in this text will be very large or infinite and a simpler formula will be more than adequate. ▶

Suppose a random sample of size n is taken from a population with standard deviation σ. Then the *standard deviation of $\bar{x}$* is (approximately) equal to the standard deviation of the population divided by the square root of the sample size. That is,

$$\sigma_{\bar{x}} \doteq \frac{\sigma}{\sqrt{n}} \, ^*$$

If you try the formula $\sigma_{\bar{x}} \doteq \sigma/\sqrt{n}$ in Example 3, you won't get very good results. This is because the population in Example 3 is unrealistically small. In any realistic problem we consider in this text, and in almost any practical problem you will ever encounter, the formula $\sigma_{\bar{x}} \doteq \sigma/\sqrt{n}$ *will be* extremely accurate.

◀ In a large city, the (population) mean height of adult males is $\mu = 68''$, with a standard deviation of $2.5''$. Suppose a random sample of n adult males is taken (from this city). Let $\bar{x}$ be the (sample) mean height of the adult males selected. Compute $\mu_{\bar{x}}$ and $\sigma_{\bar{x}}$, if

Example **4**

a) $n = 10$,
b) $n = 30$,
c) $n = 100$.

---

* The formula $\sigma_{\bar{x}} \doteq \sigma/\sqrt{n}$ is, in fact, exact if we are sampling from a finite population *with replacement* (i.e., if an item is replaced after it is sampled, so that it can possibly appear several times in the final sample). It is also exact if we are sampling from an infinite population. But if we are sampling from a finite population *without replacement* (as in Example 3), then the exact formula is

$$\sigma_{\bar{x}} = \frac{\sigma}{\sqrt{n}} \cdot \sqrt{\frac{N-n}{N-1}},$$

where N is the population size. This is discussed in Exercise 10.

Solutions   a) $\mu_{\bar{x}} = \mu = 68''$, $\sigma_{\bar{x}} \doteq \dfrac{\sigma}{\sqrt{n}} = \dfrac{2.5}{\sqrt{10}} \doteq 0.79''$

b) $\mu_{\bar{x}} = \mu = 68''$, $\sigma_{\bar{x}} \doteq \dfrac{\sigma}{\sqrt{n}} = \dfrac{2.5}{\sqrt{30}} \doteq 0.46''$

c) $\mu_{\bar{x}} = \mu = 68''$, $\sigma_{\bar{x}} \doteq \dfrac{\sigma}{\sqrt{n}} = \dfrac{2.5}{\sqrt{100}} = 0.25''$ ▶

Exercise **B**   A brand of steel-belted radial tires has a (population) mean tire life of $\mu = 40,000$ miles and a standard deviation of $\sigma = 2,500$ miles. A random sample of n such tires is selected, and each tire in the sample is tested for mileage. If $\bar{x}$ is the (sample) mean tire life of the n tires tested, find the mean and standard deviation of the random variable $\bar{x}$ where:
a) $n = 8$,
b) $n = 16$,
c) $n = 100$.

EXERCISES Section **6.2**

1 The values of four game show prizes are given below (in dollars).

Prize	A	B	C	D
Value (x)	150	50	0	800

a) Find μ and σ for this population of prize values. (Use the formulas

$$\mu = \frac{\Sigma x}{N} \quad \text{and} \quad \sigma = \sqrt{\frac{\Sigma(x-\mu)^2}{N}}$$

from Section 3.7).

A contestant in the show picks *two* letters *at random* to decide which prizes he will win (he does not know their values).
b) List all samples of size 2.
c) Find $\bar{x}$ for each sample.
d) Give the probability distribution of $\bar{x}$.
e) Find $\mu_{\bar{x}}$ and $\sigma_{\bar{x}}$, using part (d).

2 a) Answer parts (b) through (e) of Exercise 1 for samples of size 3 instead of 2.
b) Compare the values $\mu_{\bar{x}}$ and $\sigma_{\bar{x}}$ for samples of size 2 and 3.

3 A man dressed as a rabbit (at a costume party) is allowed to reach into a grab bag and take two prizes. There are five prizes in the bag. Their values (in dollars) are:

Prize	Book	Pen	Empty box	Desk set	Comic glasses with false nose
Value	$2	$1	$0	$5	$3

If the prizes are similarly wrapped and his selections are at random, the probability distribution of values (for selecting *one* prize) is:

Value x	2	1	0	5	3
Probability $P(x)$	0.2	0.2	0.2	0.2	0.2

a) Find μ and σ for this probability distribution.
b) List all samples of size 2.
c) Find $\bar{x}$ for each sample.
d) Give the probability distribution of $\bar{x}$.
e) Find $\mu_{\bar{x}}$ and $\sigma_{\bar{x}}$, using part (d).

4 Answer parts (b) through (e) of Exercise 3 for samples of size 4 instead of 2.

In Exercises 5 through 8, you will be given the values of μ and σ for a *large* population. Find the values of $\mu_{\bar{x}}$ and $\sigma_{\bar{x}}$ for each given value of n.

5 $\mu = 16$, $\sigma = 0.7$
  a) $n = 36$ b) $n = 50$ c) $n = 72$

6 $\mu = 4.61$, $\sigma = 2.4$
  a) $n = 30$ b) $n = 49$ c) $n = 61$

7 $\mu = 138$, $\sigma = 5$
  a) $n = 32$ b) $n = 64$ c) $n = 76$

8 $\mu = 63.5$, $\sigma = 0.27$
  a) $n = 35$ b) $n = 55$ c) $n = 81$

9 For the game-show contestant in Exercise 1, $\mu_{\bar{x}}$ is the same for two prizes or four prizes—the mean of his average prize amount is unchanged. Why is he better off with four prizes?

10 The exact formula for $\sigma_{\bar{x}}$ when $\bar{x}$ is obtained from a sample of size n, without replacement, from a population of size N is

$$\sigma_{\bar{x}} = \frac{\sigma}{\sqrt{n}} \sqrt{\frac{N-n}{N-1}}$$

For the population in Example 2, $N = 5$ and $\sigma = 5.48$. Find $\sigma_{\bar{x}}$ for the given values of n, using the *exact* formula given above.
a) $n = 2$ b) $n = 3$ c) $n = 4$
d) Compare the values of $\sigma_{\bar{x}}$ obtained here with those listed in Example 3.

11 a) Prove that if $n \leqslant 0.05N$,

$$0.97 \leqslant \sqrt{\frac{N-n}{N-1}} \leqslant 1.$$

b) Use the result from part (a) to show that, in the situation of Exercise 10, $\sigma/\sqrt{n}$ is a very good approximation to $\sigma_{\bar{x}}$ if $n \leqslant 0.05N$.

12 In Exercise 14 of Section 4.9, we showed that for two *independent* random variables x and y, the mean and variance of $x + y$ are given by

$\mu_{x+y} = \mu_x + \mu_y$ (Mean of sum = sum of means)
$\sigma_{x+y}^2 = \sigma_x^2 + \sigma_y^2$ (Variance of sum = sum of variances)

We also showed, in Exercise 16 of Section 4.9, that $\mu_{kx} = k\mu_x$ and $\sigma_{kx}^2 = k^2\sigma_x^2$, for any constant k. These formulas can be used to derive some of the results given in this section. If we are taking a random sample of size n from a population, we can think of the n elements we pick as described by n independent random variables $x_1, \ldots, x_n$, where

$x_1 = $ first element sampled,
$\vdots \qquad \vdots$
$x_n = $ last element sampled.

If the population from which we are sampling has mean μ and variance σ^2, each random variable has mean μ and variance σ^2:

$$\mu_{x_i} = \mu, \qquad \sigma_{x_i}^2 = \sigma^2, \qquad i = 1, \ldots, n.$$

The sample mean $\bar{x}$ obtained from $x_1, \ldots, x_n$ is

$$\bar{x} = \frac{\sum x}{n} = \frac{x_1 + \cdots + x_n}{n}$$

Use the formulas and information given above to prove that

a) $\mu_{\bar{x}} = \mu$

b) $\sigma_{\bar{x}}^2 = \dfrac{\sigma^2}{n}$ (which implies $\sigma_{\bar{x}} = \sigma/\sqrt{n}$).

Finally we are ready to discuss how to find probabilities for $\bar{x}$. We said that we would be able to use areas under normal curves to find probabilities for $\bar{x}$. But *there's a catch!* Let's look at an example, returning once more to our *population* of heights of the five starting players on a basketball team.

The central-limit theorem and the sampling distribution of the mean 6.3

Example **5**

Player	1	2	3	4	5
Height (in.)	67	73	75	76	84

In Example 2 (page 199), Table 6.3, we found the probability distribution of $\bar{x}$ for samples of size $n = 2$. We repeat it as Table 6.6.

$\bar{x}$	70.0	71.0	71.5	74.0	74.5	75.5	78.5	79.5	80.0
$P(\bar{x})$	0.1	0.1	0.1	0.1	0.1	0.2	0.1	0.1	0.1

Table **6.6**

The histogram for this probability distribution is given in Fig. 6.1.

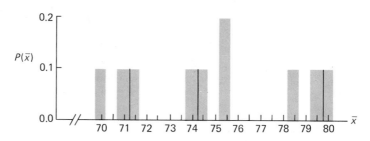

Figure 6.1
Histogram for $\bar{x}$ (sample size $n = 2$).

It is pretty clear from Fig. 6.1 that the histogram for $\bar{x}$ is not at all bell-shaped, and that there is no way we could use areas under a normal curve to calculate (even approximately) probabilities for $\bar{x}$. (Compare with Fig. 5.6 of Section 5.4, page 176.)

Now all along we have promised that we would be able to calculate probabilities for $\bar{x}$ (at least approximately) by using areas under normal curves. But Example 5 makes it clear that we won't *always* be able to do this. What's going on? Well, as we said, there's a catch. The catch is that, generally, we will be able to use areas under normal curves to calculate probabilities for $\bar{x}$ *only if the sample size is relatively large*. The rule of thumb is that we can use areas under normal curves to find probabilities for $\bar{x}$ as long as the sample size is at least 30 (that is, $n \geqslant 30$).

Central-limit theorem

Suppose a random sample of size n (where n *is at least* 30) is taken from a population. Then $\bar{x}$ is (approximately) *normally distributed* with mean $\mu_{\bar{x}} = \mu$ and standard deviation $\sigma_{\bar{x}} \doteq \sigma/\sqrt{n}$. That is, probabilities for $\bar{x}$ can be found approximately by using areas under the normal curve with parameters μ and $\sigma/\sqrt{n}$.

A few comments are in order. First of all, even though there is a restriction on sample size ($n \geqslant 30$), the central limit theorem is truly amazing. The amazing thing is that, as long as $n \geqslant 30$, $\bar{x}$ is approximately normally distributed *no matter what the population looks like*.* This is not only amazing, but it is crucial for using $\bar{x}$ to make inferences about μ because, more often than not, we *won't* know what the population looks like.

Example 6 A brand of thread has a (population) mean breaking strength of $\mu = 25$ lbs, with a standard deviation of $\sigma = 0.5$ lbs. A random sample of 50 such pieces of thread is tested for breaking strength. What is the probability that the (sample) mean breaking strength, $\bar{x}$, of the 50 pieces of thread will be between 24.9 and 25.1 lbs?

The problem is to find $P(24.9 < \bar{x} < 25.1)$. The sample size is $n = 50$ (which is at least 30). So by the central-limit theorem, $\bar{x}$ is approximately normally distributed, and we can find probabilities for $\bar{x}$ using areas under the normal curve with parameters μ

* We emphasize that "$n \geqslant 30$" is a "rule of thumb." In general, the more "nonnormal" the population, the larger n must be in order to use the normal curve to find probabilities for $\bar{x}$.

and $\sigma/\sqrt{n}$. Now, $\mu = 25$ and $\sigma/\sqrt{n} = 0.5/\sqrt{50} \doteq 0.07$. Therefore, to find the approximate value of $P(24.9 < \bar{x} < 25.1)$, all we have to is find the area under the normal curve (with parameters 25 and 0.07) between 24.9 and 25.1.

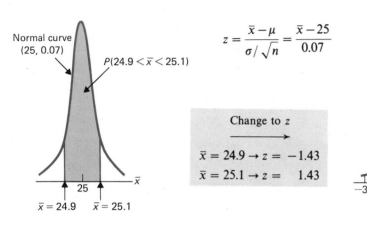

$$z = \frac{\bar{x} - \mu}{\sigma/\sqrt{n}} = \frac{\bar{x} - 25}{0.07}$$

Change to z

$\bar{x} = 24.9 \rightarrow z = -1.43$

$\bar{x} = 25.1 \rightarrow z = 1.43$

Area $= 0.9236 - 0.0764 = 0.8472$

So,

$$P(24.9 < \bar{x} < 25.1) \doteq 0.8472$$

An interesting way to interpret this result is as follows: Suppose we test a random sample of 50 pieces of thread: Then there is about an 85 percent chance that the sample mean breaking strength ($\bar{x}$) will be within 0.1 lb of the true mean breaking strength of $\mu = 25$ lbs. ▶

The weight, at birth, of newborn babies has a (population) mean of $\mu = 7.2$ lbs and a standard deviation of $\sigma = 2.0$ lbs. If a random sample of 80 newborn babies is weighed, what is the probability that their mean weight ($\bar{x}$) will be between 7 and 7.5 lbs?

Exercise **C**

We have seen that, in general, if we want to use normal curves to find probabilities for $\bar{x}$, then we need to have a sample of size at least 30 ($n \geqslant 30$). There is, however, one notable exception to this rule.

Suppose a random sample of size n is taken from a *population* that is itself *normally distributed*. Then $\bar{x}$ is normally distributed (with mean $\mu_{\bar{x}} = \mu$ and standard deviation $\sigma_{\bar{x}} = \sigma/\sqrt{n}$), *regardless of the sample size n*. That is, if the population is normally distributed then probabilities for $\bar{x}$ can be found using areas under the normal curve with parameters μ and $\sigma/\sqrt{n}$, regardless of sample size.

Example **7**   ◖ IQ's are *normally distributed* with mean 100 and standard deviation 10. Suppose we take a random sample of size 16 from the population. What is the probability that $\bar{x}$ will be between 95 and 105?

Solution   In this illustration we have $n = 16$, $\mu = 100$, and $\sigma = 10$. *Since the population* (IQ) *is normally distributed*, we know *that $\bar{x}$ is normally distributed* (*regardless of sample size*), and we can find probabilities for $\bar{x}$ using areas under the normal curve with parameters $\mu = 100$ and $\sigma/\sqrt{n} = 10/\sqrt{16} = 2.5$. Therefore, to determine $P(95 < \bar{x} < 105)$, we need to find the area under the normal curve (with parameters 100 and 2.5) between 95 and 105:

$$z = \frac{\bar{x} - \mu}{\sigma/\sqrt{n}} = \frac{\bar{x} - 100}{2.5}$$

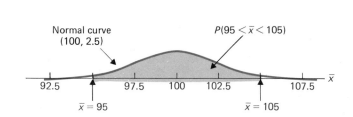

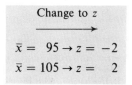

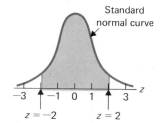

$$\text{Area} = 0.9772 - 0.0228 = 0.9544$$

Consequently,

$$P(95 < \bar{x} < 105) = 0.9544$$

So when we take a random sample of only 16 IQ's, we have better than a 95 percent chance of $\bar{x}$ being within 5 of the true mean of 100. ◗

Exercise **D**   The values of homes in a certain large community are (approximately) *normally distributed* with mean $\mu = \$48,500$ and standard deviation $\sigma = \$5,000$. A real-estate broker, *who doesn't know this*, wants to get some information concerning the mean value of a home in this community. She decides to take a random sample of ten homes and have each of the homes appraised. Then she plans to use the mean value ($\bar{x}$) of these ten homes as her estimate for the actual mean value μ. Find the probability that her estimate ($\bar{x}$) will be within \$2,000 of the true mean.

In Chapters 7 and 8 it will be necessary for us to consider the "standardized" version of $\bar{x}$; that is

$$\frac{\bar{x} - \mu_{\bar{x}}}{\sigma_{\bar{x}}}$$

Since $\mu_{\bar{x}} = \mu$ and $\sigma_{\bar{x}} = \sigma/\sqrt{n}$, this is the same as

$$\frac{\bar{x} - \mu}{\sigma/\sqrt{n}}$$

In Section 5.5 (page 182) we learned that, if we standardize a *normally distributed* random variable, then the resulting random variable has the *standard* normal distribution. Therefore, we can summarize the results of Section 6.2 and the present section as follows:

Suppose we take a random sample of size *n* from a population with mean μ and standard deviation σ. Then the random variable

$$\frac{\bar{x} - \mu}{\sigma/\sqrt{n}}$$

1. has approximately the *standard* normal distribution if $n \geqslant 30$, regardless of the distribution of the population;
2. has exactly the *standard* normal distribution if the population itself is normally distributed, regardless of sample size.

In other words, if *either* $n \geqslant 30$ *or* the population is normally distributed, then probabilities for the random variable

$$z = \frac{\bar{x} - \mu}{\sigma/\sqrt{n}}$$

can be found (at least approximately) using areas under the *standard* normal curve.

We conclude this section by illustrating the *sampling distribution of the mean* (i.e., the probability distribution of $\bar{x}$) by means of some pictures. We do this first for a case where the population itself is normally distributed, and then for a case where the population itself is *not* normally distributed.

As an illustration of the sampling distribution of the mean in the case where the population itself is normally distributed, we consider IQ's, which are normally distributed with mean $\mu = 100$ and standard deviation $\sigma = 10$. The normal curve for the population of IQ's is shown in Fig. 6.2.

Now the population here is itself normally distributed, so we know that, if we take a random sample of size *n* from the population, then $\bar{x}$ is normally distributed, *regardless of sample size*, with $\mu_{\bar{x}} = \mu = 100$ and $\sigma_{\bar{x}} = \sigma/\sqrt{n} = 10/\sqrt{n}$.

In Figs. 6.3 through 6.5 we illustrate the normal curves for $\bar{x}$ for $n = 3, 10,$ and 30.

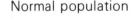

Normal population

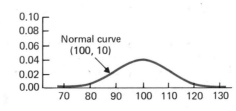

Figure 6.2
Normal curve for population of IQ's.

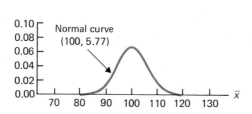

Figure 6.3
Normal curve for $\bar{x}$, when $n = 3$. Here $\mu_{\bar{x}} = \mu = 100$ and $\sigma_{\bar{x}} = \sigma/\sqrt{n} = 10/\sqrt{3} = 5.77$.

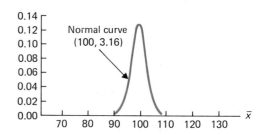

Figure 6.4
Normal curve for $\bar{x}$ when $n = 10$. Here $\mu_{\bar{x}} = \mu = 100$ and $\sigma_{\bar{x}} = \sigma/\sqrt{n} = 10/\sqrt{10} = 3.16$.

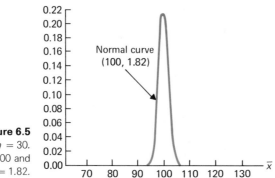

Figure 6.5
Normal curve for $\bar{x}$ when $n = 30$. Here $\mu_{\bar{x}} = \mu = 100$ and $\sigma_{\bar{x}} = \sigma/\sqrt{n} = 10/\sqrt{30} = 1.82$.

Since $\sigma_{\bar{x}} = \sigma/\sqrt{n}$, the larger the sample size, the smaller the standard deviation of $\bar{x}$. This fact is illustrated by Figs. 6.3 through 6.5 and means that *the larger the sample size, the "more likely" $\bar{x}$ will be close to μ.* To illustrate this fact we have computed the probability that $\bar{x}$ is within 5 of the true population mean of $\mu = 100$ for $n = 3$, 10, and 30.

Sample size (n)	$P(95 < \bar{x} < 105)$
3	0.6156
10	0.8860
30	0.9940

Nonnormal population   To illustrate the sampling distribution of the mean in the case where the population itself is *not* normally distributed, we consider the duration of telephone conversations within a given city. The population of telephone-conversation durations has what is called an *exponential distribution*. (Don't worry too much about this. We are using it only for illustration.)

Specifically, suppose in a given city that the mean duration of a telephone call is 5 minutes; that is, $\mu = 5$. Using calculus, we can obtain probabilities for this population by computing areas under the curve shown in Fig. 6.6, called an *exponential curve*. Moreover, it turns out that the standard deviation, σ, of the population is also 5.

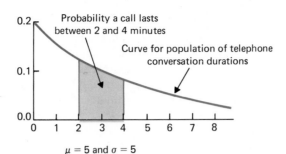

Figure 6.6
Curve for population of telephone-conversation durations, with $\mu = 5$ and $\sigma = 5$

If we take a random sample of n phone calls, then $\bar{x}$, in this case, denotes the average duration of the n calls. Figures 6.7, 6.8, and 6.9 show the curves for $\bar{x}$ for $n = 3$, 10, and 30. (You are not expected to know how we got these curves.)

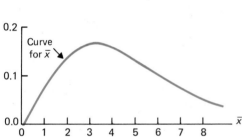

Figure 6.7
Curve for $\bar{x}$ when $n = 3$. Here $\mu_{\bar{x}} = \mu = 5$ and $\sigma_{\bar{x}} = \sigma/\sqrt{n} = 5/\sqrt{3} = 2.89$.

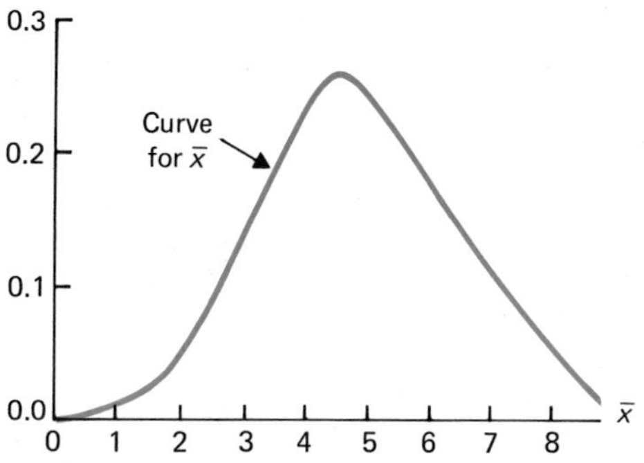

Figure 6.8
Curve for $\bar{x}$ when $n = 10$. Here $\mu_{\bar{x}} = \mu = 5$ and $\sigma_{\bar{x}} = \sigma/\sqrt{n} = 5/\sqrt{10} = 1.58$.

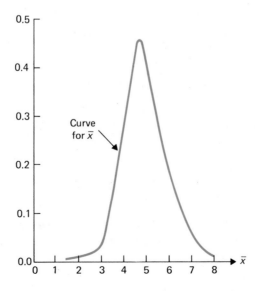

Figure 6.9
Curve for $\bar{x}$ when $n = 30$. Here $\mu_{\bar{x}} = \mu = 5$ and $\sigma_{\bar{x}} = \sigma/\sqrt{n} = 5/\sqrt{30} = 0.91$.

As we see by looking at the curve for the population of telephone-conversation durations (Fig. 6.6), this population is certainly *not* normally distributed. But according to the central-limit theorem, $\bar{x}$ is approximately normally distributed for $n \geqslant 30$. This is borne out pictorially by the three curves for $\bar{x}$. The first curve for $\bar{x}$ ($n = 3$) certainly does not look like a normal curve. The second curve for $\bar{x}$ ($n = 10$) is beginning to look more like a normal curve, and the third curve for $\bar{x}$ ($n = 30$) definitely looks like a normal curve.

EXERCISES Section **6.3**

1 Suppose a population has mean $\mu = 65$ and standard deviation $\sigma = 15$. If we take a random sample of size 36 from this population, what is the probability that the sample mean, $\bar{x}$, is between

a) 62 and 66?   b) 64 and 68?   c) 60 and 70?

2 Suppose a population has a mean $\mu = 12$ and standard deviation $\sigma = 2$. If we take a random sample of size 64 from this population, what is the probability that the sample mean, $\bar{x}$, is between

a) 11.5 and 14?   b) 11.6 and 12.4?   c) 11.9 and 12.2?

3 Suppose a population has a mean of $\mu = 36$ and a standard deviation of $\sigma = 6.2$. If we take a random sample of size 50 from this population, what is the probability that the sample mean, $\bar{x}$, is between

a) 35 and 37?   b) 35.4 and 38?   c) 30 and 40?

4 Suppose the population in Exercise 3 has a standard deviation of $\sigma = 3.1$. Find the new probabilities for parts (a), (b), and (c) of Exercise 3.

5 The length of the Western Rattlesnake is *normally* distributed with a mean of $\mu = 3.5$ feet and a standard deviation of $\sigma = 0.17$ foot. What is the probability that a sample of 15 snakes in a den has a mean length between 3.45 and 3.56 feet?

6 The number of young in a litter of minks is (approximately) normally distributed with a mean of $\mu = 7$ and a standard deviation of $\sigma = 1$. What is the probability that the litters of ten minks on a mink farm have a mean, $\bar{x}$, of between 6 and 8 young?

7 A graph in the August 1977 issue of *Runner's World* shows that the times of the finishers in the New York City 10-km run are normally distributed with a mean of $\mu = 61$ minutes and a standard deviation of $\sigma = 9$ minutes. If a random sample of 16 runners is selected, what is the probability that the mean, $\bar{x}$, of their finishing times is between 60 and 63 minutes?

8 The May 1977 issue of *Scientific American* states that the batting averages of all major league players are approximately normally distributed with a mean of $\mu = .270$ and a standard deviation of $\sigma = .015$. What is the probability that the mean, $\bar{x}$, of the batting averages of a random sample of nine players is between .269 and .271?

9 Why were we permitted to use sample sizes smaller than 30 in Exercises 5 through 8?

CHAPTER REVIEW

Key Terms		
	sampling distribution of the mean	**mean of $\bar{x}$**
	representative sample	**standard deviation of $\bar{x}$**
	simple random sampling	$\mu_{\bar{x}}$
	random sample	$\sigma_{\bar{x}}$
		central-limit theorem

simple random sampling:

In (simple) random sampling, each possible sample (of the given sample size) is equally likely to be the one selected.

mean of $\bar{x}$:

$$\mu_{\bar{x}} = \mu \qquad (\mu = \text{population mean})$$

standard deviation of $\bar{x}$:

$$\sigma_{\bar{x}} \doteq \frac{\sigma}{\sqrt{n}} \qquad (\sigma = \text{population standard deviation, } n = \text{sample size})$$

sampling distribution of the mean:

If a random sample of size n is taken from a population with mean μ and standard deviation σ, then $\bar{x}$ has mean $\mu_{\bar{x}} = \mu$ and standard deviation $\sigma_{\bar{x}} \doteq \sigma/\sqrt{n}$, and

1. if $n \geqslant 30$, $\bar{x}$ is approximately normally distributed, regardless of the distribution of the population;

2. if the population is itself normally distributed, $\bar{x}$ is exactly normally distributed, regardless of sample size.

   In terms of standardized random variables this means that the random variable

$$z = \frac{\bar{x} - \mu}{\sigma/\sqrt{n}}$$

has *approximately* the standard normal distribution if $n \geqslant 30$, and has *exactly* the standard normal distribution if the population itself is normally distributed.

You should be able to

1 Explain what is meant by a (*simple*) *random sample*.

2 Compute the *mean* ($\mu_{\bar{x}}$) and *standard deviation* ($\sigma_{\bar{x}}$) *for the random variable* $\bar{x}$, knowing the sample size (n), the population mean (μ), and the population standard deviation (σ).

3 State and use the *central-limit theorem*.

4 *Find probabilities for $\bar{x}$ (if $n \geqslant 30$), by looking at areas under normal curves.*

5 *Find probabilities for $\bar{x}$ (when sampling from a normal population) by looking at areas under normal curves.*

REVIEW TEST

1 Classify each sampling procedure below as "random" or "not random."
   a) A college student is hired to interview a random sample of all 200,000 voters in her town. She stays on campus and interviews 100 students in the cafeteria.
   b) A pollster wants to interview a random sample of 20 gas-station managers in Denver. He pastes a list of all such managers on his wall, closes his eyes, and tosses a dart at the list 20 times. He interviews the people whose names the dart hits.

2 A population consists of the numbers 1, 2, 3, 4, 5, 6. For this population $\mu = 3.5$ and $\sigma = 1.71$.

   a) What is $\mu_{\bar{x}}$?
   b) Which is larger, σ or $\sigma_{\bar{x}}$?

3 For a very large population, $\mu = 50$ and $\sigma = 100$. Find $\mu_{\bar{x}}$ and $\sigma_{\bar{x}}$ if $\bar{x}$ is obtained from a sample of size
   a) $n = 36$ b) $n = 81$,   c) $n = 100$

4 Suppose a population has mean $\mu = 50$ and standard deviation $\sigma = 18$. If we take a random sample of size $n = 64$,
   a) What is $\mu_{\bar{x}}$?   b) What is $\sigma_{\bar{x}}$?
   c) The distribution of $\bar{x}$ is approximately _____
   d) Find the probability that $46 \leqslant \bar{x} \leqslant 53$.

FURTHER READINGS

KRUSKAL, W., *et al.* (Eds.), *Statistics by Example: Detecting Patterns.* Reading, Massachusetts: Addison-Wesley, 1973. See "Grocery shopping and the Central-Limit Theorem" by Samuel Zahl.

LAPIN, L., *Statistics: Meaning and Method.* New York: Harcourt Brace Jovanovich, Inc., 1975. Section 5.5 and all of Chapter 6 combine to give a nice overview of sampling, the normal distribution, and the central-limit theorem.

ANSWERS TO REINFORCEMENT EXERCISES

A

Heights 72, 75, 76, 77

(a) through (c)

Sample	$\bar{x}$	$P(\bar{x})$
72, 75	73.5	1/6
72, 76	74.0	1/6
72, 77	74.5	1/6
75, 76	75.5	1/6
75, 77	76.0	1/6
76, 77	76.5	1/6

d) $\mu_{\bar{x}} = (73.5)\left(\dfrac{1}{6}\right) + \cdots + (76.5)\left(\dfrac{1}{6}\right) = \dfrac{450}{6} = 75$

e) $\sigma_{\bar{x}}^2 = (73.5 - 75)^2\left(\dfrac{1}{6}\right) + (74 - 75)^2\left(\dfrac{1}{6}\right)$

$\qquad + (74.5 - 75)^2\left(\dfrac{1}{6}\right) + (75.5 - 75)^2\left(\dfrac{1}{6}\right)$

$\qquad + (76 - 75)^2\left(\dfrac{1}{6}\right) + (76.5 - 75)^2\left(\dfrac{1}{6}\right)$

$\qquad = \dfrac{7}{6} \doteq 1.17$

$\qquad \sigma_{\bar{x}} \doteq \sqrt{1.17} \doteq 1.08$

B   a) $\mu_{\bar{x}} = 40{,}000$; $\sigma_{\bar{x}} \doteq \dfrac{2500}{\sqrt{8}} \doteq 883.9$

   b) $\mu_{\bar{x}} = 40{,}000$; $\sigma_{\bar{x}} \doteq \dfrac{2500}{\sqrt{16}} = 625$

   c) $\mu_{\bar{x}} = 40{,}000$; $\sigma_{\bar{x}} \doteq \dfrac{2500}{\sqrt{100}} = 250$

C $\mu = 7.2$, $\sigma = 2.0$, $\sigma_{\bar{x}} = 2/\sqrt{80} \doteq 0.22$.

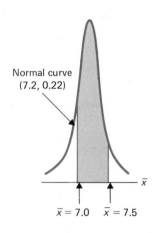

Normal curve
(7.2, 0.22)

$\bar{x} = 7.0$ $\bar{x} = 7.5$

$$z = \frac{\bar{x} - 7.2}{0.22}$$

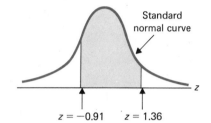

Change to z

$\bar{x} = 7.0 \to z \doteq -.91$

$\bar{x} = 7.5 \to z \doteq 1.36$

Standard
normal curve

$z = -0.91$ $z = 1.36$

$$P(7.0 < \bar{x} < 7.5) = P(-0.91 < z < 1.36)$$
$$= 0.9131 - 0.1814 = 0.7317$$

D $\mu = 48,500$, $\sigma_{\bar{x}} = 5000/\sqrt{10} \doteq 1581.14$. For $\bar{x}$ to be within $2000 of the true mean, we must have $46,500 < \bar{x} < 50,500$. We need to find $P(46500 < \bar{x} < 50500)$:

$$z = \frac{\bar{x} - 48500}{1581.14}$$

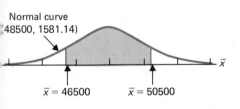

Normal curve
(48500, 1581.14)

$\bar{x} = 46500$ $\bar{x} = 50500$

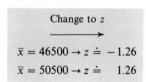

Change to z

$\bar{x} = 46500 \to z \doteq -1.26$

$\bar{x} = 50500 \to z \doteq 1.26$

Standard
normal curve

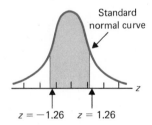

$z = -1.26$ $z = 1.26$

$$P(46500 < \bar{x} < 50500) = P(-1.26 < z < 1.26)$$
$$= 0.8962 - 0.1038 = 0.7924$$

Estimating a population mean

7.1 Confidence intervals

In Chapters 5 and 6 we learned two important properties about the random variable $\bar{x}$:

1. In Section 5.1, we saw that it is reasonable to use a sample mean ($\bar{x}$) to estimate a population mean (μ). Specifically, the *law of large numbers* states that: If a "large" random sample is taken from a population, then it is "likely" that $\bar{x}$ will be "close" to μ.
2. In Section 6.3 we discussed the *central-limit theorem*, which states that, if a random sample of size n ($n \geqslant 30$) is taken from a population with mean μ and standard deviation σ, then $\bar{x}$ is (approximately) normally distributed and we can find probabilities for $\bar{x}$ using areas under the normal curve with parameters μ and $\sigma/\sqrt{n}$.

The reason for studying properties of sample means ($\bar{x}$) is so that we can use $\bar{x}$ to make inferences about μ. The next example shows how we can use the properties of $\bar{x}$ to obtain information about an *unknown* population mean μ.

Example 1 ◖ A statistician is interested in gaining information about the population mean height, μ, of all adult males in a large city.

From past experience, she knows that the population standard deviation is about $\sigma = 2.50$ in.* If she takes a random sample of 30 adult males living in the city, then what is the probability that the sample mean height $\bar{x}$ will be within one inch of the population mean height μ?

Solution We want to find the probability that $\bar{x}$ will be within one inch of μ; that is, $P(\mu - 1 < \bar{x} < \mu + 1)$. Since the sample size is $n = 30$, we know from the central-limit theorem that $\bar{x}$ is approximately normally distributed, and we can find probabilities for $\bar{x}$ using areas under the normal curve with parameters μ (which we don't know) and $\sigma/\sqrt{n} = 2.50/\sqrt{30} = 0.46$.

Thus to find $P(\mu - 1 < \bar{x} < \mu + 1)$, we need only find the area under the normal curve (with parameters μ and 0.46) between $\mu - 1$ and $\mu + 1$. See Fig. 7.1.

The difference between this problem and the previous problems we have solved is that here we don't know the values of $\mu - 1$ and $\mu + 1$. But we can still solve the problem in the usual way—by standardizing. That is, we look at

$$z = \frac{\bar{x} - \mu}{\sigma/\sqrt{n}} = \frac{\bar{x} - \mu}{0.46}$$

The z-score for $\bar{x} = \mu - 1$ is

$$z = \frac{(\mu - 1) - \mu}{0.46} = \frac{-1}{0.46} \doteq -2.17$$

---

* She might know this from her previous work with heights of adult males, or she could have done a preliminary study to estimate σ (you will learn how to estimate σ in Chapter 9). The main reason that we are assuming she knows σ, in this example, is to retain simplicity. In Section 7.3 we shall deal with such problems when σ is *not* known (which is usually the case in practice).

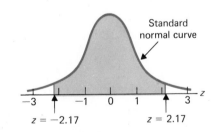

$$P(\mu - 1 < \overline{x} < \mu + 1) \quad \textbf{Figure 7.1}$$

Normal curve
$(\mu, 0.46)$

and the z-score for $\overline{x} = \mu + 1$ is

$$z = \frac{(\mu + 1) - \mu}{0.46} = \frac{1}{0.46} \doteq 2.17.$$

So we get the following:

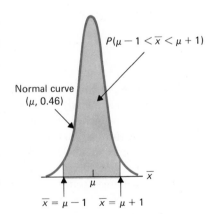

$$z = \frac{\overline{x} - \mu}{0.46}$$

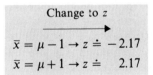

Change to z

$\overline{x} = \mu - 1 \rightarrow z \doteq -2.17$

$\overline{x} = \mu + 1 \rightarrow z \doteq \quad 2.17$

Standard
normal curve

$z = -2.17 \qquad z = 2.17$

$$\text{Area} = 0.9850 - 0.0150 = 0.9700$$

Consequently,

$$P(\mu - 1 < \overline{x} < \mu + 1) \doteq 0.97$$

In words, the probability is about 0.97 that $\overline{x}$ will be within one inch of μ. ▶

Exercise A   In Example 1, suppose that a random sample is taken of $n = 50$ adult males. Find the probability that the sample mean height $\bar{x}$ will be within 0.5 inch of the unknown population mean μ.

We will now continue with Example 1, since we need to make a few additional comments.

Example 1   ◖ In Example 1 we saw that if a random sample of 30 adult males is taken, then the
(continued)   probability is 0.97 that $\bar{x}$ will be within one inch of μ. In symbols,

$$P(\mu - 1 < \bar{x} < \mu + 1) = 0.97$$

Now, to say, "$\bar{x}$ will be within one inch of μ" is the same as saying "μ will be within one inch of $\bar{x}$". So, we can rewrite the above equation as

$$P(\bar{x} - 1 < \mu < \bar{x} + 1) = 0.97*$$

In other words, we know that:

If a random sample of 30 adult males is taken, then the probability that the interval from $\bar{x} - 1$ to $\bar{x} + 1$ will contain μ is 0.97.

It is very important to realize that in this last statement (and similar such statements) the random variable is $\bar{x}$, not μ.

Suppose now, for example, that when the statistician actually does take the random sample and computes $\bar{x}$, she gets $\bar{x} = 67.0$ inches. Then

$$\bar{x} - 1 = 66.0 \quad \text{and} \quad \bar{x} + 1 = 68.0.$$

Since she knows that, in the long run, 97 percent of such intervals will contain μ, she can be 97 percent *confident* that μ lies somewhere between 66.0 and 68.0 inches. Therefore, the interval from 66.0 to 68.0 inches is called a 97-*percent confidence interval* for μ.

If the statistician's sample of 30 adult males had yielded $\bar{x} = 66.4$ inches (instead of 67.0 inches), then she would get a different 97-percent confidence interval. In that case, the interval from 65.4 ($= 66.4 - 1$) to 67.4 ($= 66.4 + 1$) inches would be the 97-percent confidence interval for μ. The statistician would be 97-percent confident that μ is somewhere between 65.4 and 67.4 inches. This shows that things change from sample to sample. The 97-percent confidence interval the statistician actually gets depends on the value of $\bar{x}$ obtained (which in turn depends upon the sample she selects).

It is important to remember that the 97-percent confidence interval the statistician actually gets may or may not contain μ. But she can be 97 percent confident that it does. ◗

---

* We could also provide a direct proof that the algebraic statement "$\mu - 1 < \bar{x} < \mu + 1$" is the same as

$$\text{"}\bar{x} - 1 < \mu < \bar{x} + 1\text{"}.$$

You will be asked to do this in the exercises.

Suppose that the statistician's sample of 30 adult males yielded $\bar{x} = 68.2$ inches. Fill in the blanks: "The interval from ____ to ____ inches is a 97-*percent confidence interval* for μ." Interpret the statement in quotes (with the blanks filled in).

Exercise **B**

◖ In Exercise A you were asked to solve the following problem: If a random sample is taken of 50 adult males, find the probability that $\bar{x}$ will be within 0.5 inch of μ. The answer is about 0.85. That is,

Example **2**

$$P(\mu - 0.5 < \bar{x} < \mu + 0.5) \doteq 0.85$$

or, equivalently,

$$P(\bar{x} - 0.5 < \mu < \bar{x} + 0.5) \doteq 0.85$$

In other words, if a random sample of size 50 is taken, then the probability that the interval from $\bar{x} - 0.5$ to $\bar{x} + 0.5$ will contain μ is (about) 0.85.

Find the 85-percent confidence interval for μ, in case the (sample) mean height of the 50 adult males sampled turns out to be

a) $\bar{x} = 67.5$ inches   b) $\bar{x} = 68.1$ inches

Since

Solutions

$$P(\bar{x} - 0.5 < \mu < \bar{x} + 0.5) = 0.85$$

we get the following:

a) If $\bar{x} = 67.5$ inches, then $\bar{x} - 0.5 = 67.0$ and $\bar{x} + 0.5 = 68.0$, and consequently, the interval from 67.0 to 68.0 inches is the 85-percent confidence interval for μ. We can be 85 percent confident that μ is somewhere between 67.0 and 68.0 inches.

b) If $\bar{x}$ turns out to be 68.1 inches, then the interval from 67.6 to 68.6 inches is the 85 percent confidence interval for μ. That is, we can be 85 percent confident that μ is somewhere between 67.6 and 68.6 inches. ◗

In the final example of this section, we want to emphasize that the confidence interval we get may or may not contain the population mean μ.

◖ The manufacturer of a certain brand of cigarette is interested in obtaining information concerning the (population) mean tar content, μ, of all such cigarettes. He decides to take a random sample of 30 *cigarettes*, and use their sample mean tar content, $\bar{x}$, to get a confidence interval for μ. From a preliminary study, he knows that $\sigma \doteq 0.5$ mg. †

Example **3***

If you do the calculations described in Example 1, you will find that

$$P(\mu - 0.1 < \bar{x} < \mu + 0.1) \doteq 0.73$$

---

* This example is optional.

† As we said before, you will see, in Chapter 9, how he could get this information.

or, equivalently,

$$P(\bar{x} - 0.1 < \mu < \bar{x} + 0.1) \doteq 0.73$$

Consequently, if a random sample of 30 cigarettes is tested, then the probability that the interval from $\bar{x} - 0.1$ to $\bar{x} + 0.1$ will contain μ is about 0.73. So, once the random sample of 30 cigarettes is actually tested and $\bar{x}$ is computed, a 73-percent confidence interval can be constructed; namely $\bar{x} - 0.1$ to $\bar{x} + 0.1$.

In order to illustrate that μ may or may not be contained in the 73-percent confidence interval, we simulated the experiment 20 times using $\mu = 11$ mg. (Of course, in reality we won't know μ, for if we *knew* $\mu = 11$ mg, there would be no need to use $\bar{x}$ to estimate μ.)

Table 7.1

Sample	$\bar{x}$	73 % confidence interval $\bar{x} - 0.1$ to $\bar{x} + 0.1$	Is μ in the confidence interval?
1	10.88	10.78 to 10.98	No
2	10.98	10.88 to 11.08	Yes
3	10.92	10.82 to 11.02	Yes
4	10.91	10.81 to 11.01	Yes
5	10.96	10.86 to 11.06	Yes
6	11.08	10.98 to 11.18	Yes
7	10.98	10.88 to 11.08	Yes
8	10.97	10.87 to 11.07	Yes
9	10.84	10.74 to 10.94	No
10	11.20	11.10 to 11.30	No
11	10.84	10.74 to 10.94	No
12	10.97	10.87 to 11.07	Yes
13	10.89	10.79 to 10.99	No
14	11.03	10.93 to 11.13	Yes
15	11.14	11.04 to 11.24	No
16	11.03	10.93 to 11.13	Yes
17	10.91	10.81 to 11.01	Yes
18	10.99	10.89 to 11.09	Yes
19	11.04	10.94 to 11.14	Yes
20	10.92	10.82 to 11.02	Yes

(Graphical display of the confidence intervals plotted on number lines with scale marks at 10.8, 10.9, 11.0, 11.1, 11.2, with μ indicated at 11.0.)

From Table 7.1, we see that in 14 out of the 20 samples of size 30, μ was in the 73-percent confidence interval. In other words, in 70 % (14/20) of the 20 samples of size 30, μ was in the 73-percent confidence interval. If, instead of 20 samples of size 30, we had looked at, say, 1000 samples of size 30, then we would find that the percentage of these 1000 samples for which μ was in the 73-percent confidence interval would be very close indeed to 73 percent. Thus, we can be 73 percent confident that any computed 73-percent confidence interval will actually contain μ.

1 A psychologist wished to find out the mean IQ, μ, of adults in a midwestern suburb. From past experience she knows that $\sigma \doteq 10$ for these IQ's. She takes a random sample of 50 adult IQ scores in the city.

   a) What is the probability that $\bar{x}$ will be within two points of μ? [*Hint*: refer to the argument used in Example 1.]

   b) Suppose the psychologist finds that $\bar{x} = 110$. Use your answer from (a) to fill in the blanks. The interval from _108_ to _112_ is a ____ percent confidence interval for μ.

   c) Interpret the statement in part (b)

2 Answer Exercise 1 for the new sample size $n = 100$.

3 A quality-control engineer in a bakery-goods plant needs to find the mean weight, μ, of potato-chip bags that are packed by a machine. He knows from experience that $\sigma = 0.1$ (ounce) for this machine. He takes a sample of $n = 36$ bags and finds the sample mean weight $\bar{x}$.

   a) What is the probability that $\bar{x}$ will be within 0.03 ounce of μ?

   b) Suppose the engineer finds $\bar{x} = 9.98$. Use your answer from (a) to fill in the blanks. The interval from _9.95_ to _10.01_ is a _99_ percent confidence interval for μ.

   c) Interpret the statement in (b).

4 Answer Exercise 3 for the new sample size $n = 64$.

5 The personnel director of a supermarket chain has every new employee take an arithmetic test. She would like to know μ, the mean time for completion of this test by new employees. She actually times a random sample of 40 new employees on the test and computes $\bar{x}$. If $\sigma = 2$ minutes for this test:

   a) What is the probability that $\bar{x}$ will be within one-half minute of μ?

   b) Suppose that the director finds $\bar{x} = 10$. Use your answer from (a) to fill in the blanks. The interval from ____ to ____ is a ____ percent confidence interval for μ.

   c) Interpret the statement in (b).

6 (*Point estimators*) A statement such as "we are 85 percent confident that $67.0 < \mu < 68.0$" is called an *interval estimate* of μ. An interval estimate gives an idea of the margin of error involved in estimating a population mean, μ, from a sample

mean $\bar{x}$. The calculated number $\bar{x}$ by itself is called a point estimate of μ. (A *point estimate* of μ is a single number, which we calculate from a sample to estimate μ.) For example, the interval estimate $67.0 < \mu < 68.0$ given above was derived from a sample with a point estimate of $\bar{x} = 67.5$ for μ.

   a) Which estimate gives a more accurate description of our knowledge of μ—the point estimate of 67.5 or the interval estimate $67.0 < \mu < 68.0$?

Interval estimates depend on point estimates. To find a confidence interval for μ in this section, it was first necessary to calculate the point estimate $\bar{x}$, since $\bar{x}$ is at the center of the confidence interval. However, there is more than one kind of point estimate for a population mean. Computing $\bar{x}$ from a sample is not the only way to estimate μ. Below are three different possible point estimators for μ.

   i) $\bar{x}$ = sample mean

   ii) Md = sample median

   iii) x_L = mean of two smallest numbers in sample.

Suppose, for example, that our population consisted of the four numbers 1, 2, 3, 5. The true mean for this population is $\mu = 2.75$. Suppose that you were taking samples of size 3 from this population and obtained the sample $\{1, 2, 3\}$. Then

$$\bar{x} = \frac{1+2+3}{3} = 2$$

$$Md = 2$$

$$x_L = \frac{1+2}{2} = 1.5$$

   b) List all possible samples of size 3 from the population $\{1, 2, 3, 5\}$.

   c) Find $\bar{x}$ for each sample.

   d) Find Md for each sample.

   e) Find x_L for each sample.

In Chapter 6, we stated that $\mu_{\bar{x}}$ (the mean of all possible sample means, $\bar{x}$) was equal to μ: $\mu_{\bar{x}} = \mu$. This states an important property of the sample mean $\bar{x}$. Most individual sample means $\bar{x}$ are not equal to μ but, *on the average*, $\bar{x}$ estimates μ accurately. Since $\mu_{\bar{x}} = \mu$, the point estimator $\bar{x}$ is called *unbiased*. An *unbiased estimator* is one whose sample values average out to equal μ.

f) Find $\mu_{\bar{x}}$, the mean of all sample means in part (c), and verify that $\mu_{\bar{x}} = \mu$.

g) Find the mean of all sample medians, Md, in part (d). Is the sample median an unbiased estimator of μ for this particular population?

h) Find the mean of all x_L values from part (e). Is x_L an unbiased estimator of μ for this population?

i) If you had to choose between $\bar{x}$ and Md to estimate μ, which would you choose? (Look at all the possible values of these estimators from (c) and (d) and decide which one seems to give you a better chance of obtaining a value close to μ.)

7 Prove that the inequality

$$\mu - a < \bar{x} < \mu + a$$

is equivalent to

$$\bar{x} - a < \mu < \bar{x} + a$$

7.2 Specifying the confidence level

In the previous section we learned how to find confidence intervals for a population mean μ, based on the information obtained from a sample mean $\bar{x}$. In the examples there, we specified the sample size and the form of the confidence interval and then, from these specifications, computed the confidence. However, it is often desirable to specify the confidence beforehand. An example will make this point clear.

Example 4 ◖ A telephone company in a large city is interested in obtaining information concerning the mean length, μ, of telephone conversations. A preliminary study indicated that the standard deviation of such phone conversations is about $\sigma = 4.4$ minutes.*

The company monitored the lengths of a random sample of $n = 100$ phone conversations. The (sample) mean length of these 100 phone conversations turned out to be $\bar{x} = 5.8$ minutes (that is, 5 minutes and 48 seconds). Knowing that $\bar{x} = 5.8$ minutes, find a 95-percent confidence interval for μ.

In this question (unlike all questions considered previously) *the confidence is specified beforehand*: we want a 95-percent confidence interval. The solution to this problem requires, in a sense, the "reverse" of the procedure used in solving Example 1. Before we can solve this problem we need to learn how to use Table I "in reverse." We will now see how to do this, and then we will return to solve Example 4. ◗

Example 5 ◖ Find the z-value for which the area under the standard normal curve to the *right* of this value is 0.0250.

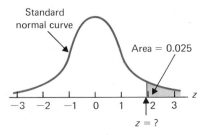

Solution To find the z-value in question, first notice that since the total area under the curve is 1,

---

* We shall deal with problems where σ is unknown in the next section.

the *unshaded* portion of the figure has area $1 - 0.0250 = 0.9750$. So, to solve the problem all we need to do is find the z-value that has an area of 0.9750 to its *left*.

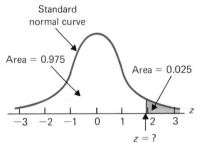

Standard normal curve

Area = 0.975

Area = 0.025

If we use Table 7.2 or Table I in the appendix, we find that the area under the standard normal curve to the *left* of 1.96 is 0.9750 (search the *body* of the table for 0.9750, and then note that the corresponding z-value is 1.96). Thus, the required z-value is $z = 1.96$.

$z = ?$

Table 7.2

z	0	1	2	3	4	5	6	7	8	9
⋮	·	·	·	·	·	·	·	·	·	·
1.5	.9332	.9345	.9357	.9370	.9382	.9394	.9406	.9418	.9429	.9441
1.6	.9452	.9463	.9474	.9484	.9495	.9505	.9515	.9525	.9535	.9545
1.7	.9554	.9564	.9573	.9582	.9591	.9599	.9608	.9616	.9625	.9633
1.8	.9641	.9649	.9656	.9664	.9671	.9678	.9686	.9693	.9699	.9706
1.9	.9713	.9719	.9726	.9732	.9738	.9744	.9750	.9756	.9761	.9767
2.0	.9772	.9778	.9783	.9788	.9793	.9798	.9803	.9808	.9812	.9817
2.1	.9821	.9826	.9830	.9834	.9838	.9842	.9846	.9850	.9854	.9857
2.2	.9861	.9864	.9868	.9871	.9875	.9878	.9881	.9884	.9887	.9890
2.3	.9893	.9896	.9898	.9901	.9904	.9906	.9909	.9911	.9913	.9916
2.4	.9918	.9920	.9922	.9925	.9927	.9929	.9931	.9932	.9934	.9936
⋮	·	·	·	·	·	·	·	·	·	·

We will use the notation

$$z_{0.025}$$

to denote the z-value for which the area under the standard normal curve to its right is 0.025. So, as we have just seen, $z_{0.025} = 1.96$. See Fig. 7.2 ▶

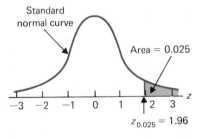

Standard normal curve

Area = 0.025

$z_{0.025} = 1.96$

Figure 7.2

◀ Find $z_{0.33}$. That is, find the z-value for which the area under the standard normal curve to its *right* is 0.33.

Example **6**

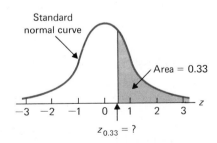

Standard normal curve

Area = 0.33

$z_{0.33} = ?$

The unshaded portion of the figure has area $1 - 0.33 = 0.67$. If we use Table I, we find that the area under the standard normal curve to the *left* of $z = 0.44$ is 0.67.

Solution

Consequently the area under the standard normal curve to the *right* of $z = 0.44$ is 0.33. That is, $z_{0.33} \doteq 0.44$. See Fig. 7.3 ▶

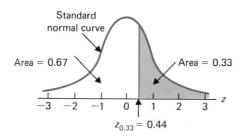

Figure 7.3

Exercise **C**   Find $z_{0.015}$. That is, find the z-value for which the area under the standard normal curve to its *right* is 0.015.

Example **7**   ◀ Suppose z has the standard normal distribution. Find the z-value, c, so that $P(-c < z < c) = 0.95$.

Solution   Since probabilities for z can be found by using areas under the standard normal curve, we can illustrate our problem pictorially as follows:

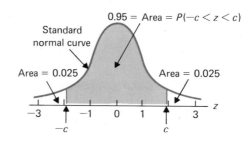

Each of the unshaded portions in the figure must have area 0.025 [since $(1 - 0.95)/2 = 0.025$]. So $c = z_{0.025}$. Using Table I, we find that $z_{0.025} = 1.96$. Consequently, $c = 1.96$. In other words,

$$P(-1.96 < z < 1.96) = 0.95 \text{ ▶}$$

Example **8**   ◀ Suppose z has the standard normal distribution. Find the z-value, c, so that $P(-c < z < c) = 0.90$.

Solution   The procedure follows that of Example 7. See Fig. 7.4.

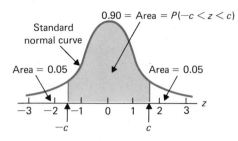

Figure 7.4

Each of the unshaded portions of the figure must have area 0.05 [since $(1 - 0.90)/2 = 0.05$]. So $c = z_{0.05}$. From Table I we find that the area under the standard normal curve to the *left* of 1.64 is 0.9495, and so the area to the *right* of 1.64 is 0.0505 (this is as close to 0.05 as we can get by using Table I). So $z_{0.05} \doteq 1.64$. Consequently, $c = 1.64$. That is,

$$P(-1.64 < z < 1.64) = 0.90 \quad \blacktriangleright$$

Suppose z has the standard normal distribution. Find c so that $P(-c < z < c) = 0.99$.   Exercise **D**

The results of Examples 7 and 8 and Exercise D will be particularly important in our work with confidence intervals. We summarize these results below.

Suppose z has the standard normal distribution. Then

1. $P(-1.64 < z < 1.64) = 0.90$ (that is, $z_{0.05} = 1.64$)
2. $P(-1.96 < z < 1.96) = 0.95$ (that is, $z_{0.025} = 1.96$)
3. $P(-2.58 < z < 2.58) = 0.99$ (that is, $z_{0.005} = 2.58$)

We now return to solving Example 4.

◀ A telephone company in a large city is interested in obtaining information   Example **9** concerning the mean duration, μ, of telephone conversations. A preliminary study indicated that the standard deviation of such phone conversations is about $\sigma = 4.4$ minutes.

The company monitored the durations of a random sample of $n = 100$ phone conversations. The (sample) mean duration of these 100 phone conversations turned out to be $\bar{x} = 5.8$ minutes.

Based on this information, find a 95-percent confidence interval for the mean duration, μ, of all such phone conversations.

We know $n = 100$. By the central-limit theorem, $\bar{x}$ is (approximately) normally   Solution distributed with mean $\mu_{\bar{x}} = \mu$ (which we don't know) and standard deviation

$$\sigma_{\bar{x}} \doteq \frac{\sigma}{\sqrt{n}} = \frac{4.4}{\sqrt{100}} = 0.44$$

Consequently (see page 207), the random variable

$$z = \frac{\bar{x} - \mu}{\sigma / \sqrt{n}} = \frac{\bar{x} - \mu}{0.44}$$

has approximately the *standard normal distribution*. We want a 95-*percent confidence interval* for μ. In Example 7 we showed that

$$P(-1.96 < z < 1.96) = 0.95 \quad (95\%)$$

(that is, $z_{0.025} = 1.96$). Since, in this case,

$$z = \frac{\bar{x} - \mu}{0.44},$$

we conclude that

$$P\left(-1.96 < \frac{\bar{x} - \mu}{0.44} < 1.96\right) = 0.95$$

We want to rewrite the expression on the left of this last equation so that "μ is in the middle." This requires some algebra. The steps for doing this are as follows:

$$-1.96 < \frac{\bar{x} - \mu}{0.44} < 1.96$$

$$-1.96 < \frac{\mu - \bar{x}}{0.44} < 1.96*$$

$$-1.96(0.44) < \mu - \bar{x} < 1.96(0.44)$$

$$\bar{x} - 1.96(0.44) < \mu < \bar{x} + 1.96(0.44)$$

Consequently, we can rewrite

$$P\left(-1.96 < \frac{\bar{x} - \mu}{0.44} < 1.96\right) = 0.95$$

as

$$P(\bar{x} - 1.96(0.44) < \mu < \bar{x} + 1.96(0.44)) = 0.95$$

In other words (since $1.96(0.44) \doteq 0.86$), we have

$$P(\bar{x} - 0.86 < \mu < \bar{x} + 0.86) = 0.95.$$

That is, if we take a random sample of size $n = 100$, then the *probability that the interval from $\bar{x} - 0.86$ to $\bar{x} + 0.86$ will contain μ is 0.95.*

---

* This first step uses the fact that if a number, b, is between -1.96 and 1.96, then so is $-b$.

Since the results of the monitoring of the 100 phone conversations gave an $\bar{x} = 5.8$ minutes,

$$\bar{x} - 0.86 = 5.8 - 0.86 = 4.94$$
$$\bar{x} + 0.86 = 5.8 + 0.86 = 6.66$$

Consequently, *the interval from 4.94 to 6.66 minutes is a 95-percent confidence interval for μ.* The phone company can be 95-percent confident that the mean duration, μ, of (all) phone conversations is somewhere between 4.94 and 6.66 minutes. ▶

Following the same procedure (step by step) used in the previous example, solve the following problem: Suppose the phone company monitors the duration of a random sample of 50 phone conversations, and finds $\bar{x} = 5.12$ minutes. Find a 90-percent confidence interval for μ (again, assume $\sigma = 4.4$ minutes).

Exercise **E**

EXERCISES Section **7.2**

1 Find $z_{0.305}$.

2 Find $z_{0.695}$.

3 Find $z_{0.01}$ (approximately).

4 Find $z_{0.1}$ (approximately).

5 Find $z_{0.02}$ (approximately).

In Exercises 6 through 9, z has the standard normal distribution.

6 Find the z-value, c, so that $P(-c < z < c) = 0.98$.

7 Find the z-value, c, such that $P(-c < z < c) = 0.85$.

8 Find the z-value, c, such that $P(-c < z < c) = 0.995$.

9 Find the z-value, c, such that $P(-c < z < c) = 0.5$.

10 The psychologist in Exercise 1 of Section 7.1 wishes to find a 95-percent confidence interval for the mean IQ of the population she is studying. She knew that $\sigma = 10$, took a sample of size $n = 50$, and found $\bar{x} = 110$. Follow the procedure of Example 9 and find the 95-percent confidence interval for this data.

11 The quality-control engineer in Exercise 3 of Section 7.1 wishes to find a 90-percent confidence interval for the mean weight μ of his potato-chip bags. He took a sample of size $n = 36$ and found $\bar{x} = 9.98$. Use $\sigma = 0.10$ and find a 90-percent confidence interval for him. Follow the procedure of Example 9.

In Example 9, we were asked to find a 95-percent confidence interval. The number 0.95 is called the **confidence level** (or **confidence coefficient**). In statistics, it is customary to write the number 0.95 as $1 - 0.05$. The number that is subtracted from 1 (0.05 in this case) to get the confidence level is denoted by the Greek letter, α. For a 90-percent confidence interval, the confidence level is 0.90 and $\alpha = 0.10$.

7.3

Confidence intervals for means—large samples

Find the confidence level and α in case we are dealing with
a) 99-percent confidence intervals,
b) 98-percent confidence intervals.

Exercise **F**

We now present a quick and simple procedure for finding confidence intervals for μ.

Procedure

Procedure for finding confidence intervals for μ based on $\bar{x}$:
Assumption: (1) Sample size is at least 30 ($n \geqslant 30$), and (2) σ is known.

Step 1. If the desired confidence level is $1 - \alpha$, use Table I to

$$\text{Find } z_{\alpha/2}$$

Step 2. The desired confidence interval for μ is

$$\bar{x} - z_{\alpha/2} \cdot \frac{\sigma}{\sqrt{n}} \quad \text{to} \quad \bar{x} + z_{\alpha/2} \cdot \frac{\sigma}{\sqrt{n}}$$

where $z_{\alpha/2}$ is found in Step 1, n is the sample size, and $\bar{x}$ is computed from the actual sample data obtained.

Let's apply this procedure to a couple of examples.

Example **10**   A manufacturer of a certain brand of cigarette was interested in obtaining information concerning the mean tar content, μ, of all such cigarettes. A preliminary study indicated that $\sigma = 0.5$ mg.* The manufacturer decided to test a random sample of 30 cigarettes. He found that the (sample) mean tar content of these 30 cigarettes was 10.88 mg. Based on this data, find a 90-percent confidence interval for μ.

Solution

To use the procedure given above, we first need to check the assumptions:
1.   Sample size is at least 30? Yes, $n = 30$, here.
2. σ is known? Yes, $\sigma = 0.5$ mg.

We can now apply the procedure set forth above.

1.   The desired confidence level is $0.90 = 1 - 0.10$. So here $\alpha = 0.10$. Using Table I, we find that

$$z_{\alpha/2} = z_{0.05} = 1.64$$

2.   Since $z_{\alpha/2} = 1.64$ (from Step 1), $n = 30$, $\sigma = 0.5$, and $\bar{x} = 10.88$, the 90-percent confidence interval for μ is

$$\bar{x} - z_{\alpha/2} \cdot \frac{\sigma}{\sqrt{n}} \quad \text{to} \quad \bar{x} + z_{\alpha/2} \cdot \frac{\sigma}{\sqrt{n}}$$

or

$$10.88 - 1.64 \left(\frac{0.5}{\sqrt{30}} \right) \quad \text{to} \quad 10.88 + 1.64 \left(\frac{0.5}{\sqrt{30}} \right)$$

or

$$10.73 \text{ to } 11.03$$

---

* We will show how to deal with problems where σ is unknown after the next example.

The manufacturer can be 90 percent confident that the mean tar content, μ, of the brand of cigarettes he sells is somewhere between 10.73 and 11.03 mg. ◗

◖ A statistician at a large university is interested in estimating the mean IQ, μ, of the students attending. His previous experience indicates that σ is about 10. He decides to take a random sample of 90 students, and finds that their (sample) mean IQ is 112.75. Based on this data, find a 95-percent confidence interval for the mean IQ, μ, of all students attending the university.

Example **11**

To use the procedure on page 228, we first need to check the assumptions:
1. Sample size is at least 30? Yes, here $n = 90$.
2. σ is known? Yes, $\sigma = 10$.

Solution

We can now apply the procedure to get the confidence interval.

1. The desired confidence level is $0.95 = 1 - 0.05$. So here, $\alpha = 0.05$. By Table I,

$$z_{\alpha/2} = z_{0.025} = 1.96$$

2. Since $z_{\alpha/2} = 1.96$, $n = 90$, $\sigma = 10$, and $\bar{x} = 112.75$, the 95-percent confidence interval for μ is

$$\bar{x} - z_{\alpha/2} \cdot \frac{\sigma}{\sqrt{n}} \quad \text{to} \quad \bar{x} + z_{\alpha/2} \cdot \frac{\sigma}{\sqrt{n}},$$

or

$$112.75 - 1.96 \left(\frac{10}{\sqrt{90}} \right) \quad \text{to} \quad 112.75 - 1.96 \left(\frac{10}{\sqrt{90}} \right)$$

or

$$110.68 \text{ to } 114.82$$

Thus the statistician can be 95 % confident that the mean IQ of all students attending the university is somewhere between 110.68 and 114.82. ◗

Use the data in Example 11 to find a 90-percent confidence interval for μ.

Exercise **G**

The procedure given on page 228 for finding confidence intervals for μ assumes that the population standard deviation σ is known. In most applications, however, σ will *not* be known. One way to deal with this problem is to do a "pilot" (preliminary) study to estimate σ.* Fortunately, *there is* a way to find confidence intervals when σ is unknown, *without* doing a preliminary study.

---

* We will see how to estimate σ in Chapter 9.

If we don't know the *population* standard deviation σ (we usually won't), then *we use the sample standard deviation s in place of σ.* This is acceptable because, for large samples ($n \geqslant 30$), *the value of s is extremely likely to be a good approximation to σ.* The mathematical consequence of this is the following. By the central-limit theorem, if $n \geqslant 30$, then the random variable

$$\frac{\bar{x} - \mu}{\sigma / \sqrt{n}}$$

has approximately the standard normal distribution. Since the sample standard deviation s is extremely likely to be close to σ, it turns out that the random variable

$$\frac{\bar{x} - \mu}{s / \sqrt{n}}$$

(we replaced σ by s) also has approximately the standard normal distribution. We repeat this fact below for emphasis.

> If we take a random sample of size n ($n \geqslant 30$) from a population with mean μ, then the random variable
>
> $$z = \frac{\bar{x} - \mu}{s / \sqrt{n}}$$
>
> has (approximately) the *standard normal distribution.*

What all this boils down to is that if we want to find a confidence interval for μ but we don't know σ, then (if $n \geqslant 30$), we can use the same procedure as given on page 228, *but we use s instead of σ.*

Procedure

Procedure for finding confidence intervals for μ based on $\bar{x}$ when σ is unknown:
Assumption: Sample size is at least $30 (n \geqslant 30)$.

Step 1. If the desired confidence level is $1 - \alpha$, use Table I to

$$\text{Find } z_{\alpha/2}$$

Step 2. The desired confidence interval for μ is

$$\bar{x} - z_{\alpha/2} \cdot \frac{s}{\sqrt{n}} \qquad \text{to} \qquad \bar{x} + z_{\alpha/2} \cdot \frac{s}{\sqrt{n}}$$

where $z_{\alpha/2}$ is found in Step 1, n is the sample size, and $\bar{x}$ and s are computed from the actual sample data obtained.

Example **12**

❡ An economist is interested in obtaining information concerning the mean income (μ) of the residents of a certain community. She decides to take a random sample of 250 residents to estimate μ. After taking a random sample of 250, she computes $\bar{x}$ and s. Her results are that

$$\bar{x} = \$18,216.66 \qquad \text{and} \qquad s = \$4,048.32.$$

Based on this information, find a 90-percent confidence interval for μ.

The sample size here is $n = 250$ ($\geqslant 30$). Since we don't know σ, we use the procedure just given.　　Solution

1. The desired confidence level is $0.90 = 1 - 0.10$. So here $\alpha = 0.10$. Using Table I we find that

$$z_{\alpha/2} = z_{0.05} \doteq 1.64$$

2. Now, $n = 250$, and we are given that $\bar{x} = 18{,}216.66$ and $s = 4{,}048.32$. Also, from Step 1, $z_{\alpha/2} = 1.64$. Thus, the 90-percent confidence interval for μ is

$$\bar{x} - z_{\alpha/2} \cdot \frac{s}{\sqrt{n}} \quad \text{to} \quad \bar{x} + z_{\alpha/2} \cdot \frac{s}{\sqrt{n}}$$

or

$$18{,}216.66 - 1.64 \left(\frac{4{,}048.32}{\sqrt{250}} \right) \quad \text{to} \quad 18{,}216.66 + 1.64 \left(\frac{4{,}048.32}{\sqrt{250}} \right)$$

or

$$17{,}796.76 \quad \text{to} \quad 18{,}636.56.$$

In other words, we can be 90-percent confident that the mean income (μ) of the residents of the community is somewhere between \$17,796.76 and \$18,636.56. ▶

Suppose a statistician took a random sample of 200 newborn babies, and found　Exercise **H** $\bar{x} = 7.52$ lbs., $s = 1.85$ lbs. Based on this data, find a 90-percent confidence interval for μ, the mean weight of newborn babies.

　　In many cases, the first problem of a statistician is to determine how large a sample needs to be taken in order to guarantee a specific amount of accuracy in estimating μ by $\bar{x}$. This problem will be taken up in the advanced exercises at the end of this section.

EXERCISES Section **7.3**

In Exercises 1 through 8, find the desired confidence interval for μ based on the given $\bar{x}$ and the given value of σ or s, for the given sample size n.

1　Find a 95-percent confidence interval for the mean length, μ, of western rattlesnakes, if a random sample of 75 such snakes yielded a sample mean length of $\bar{x} = 3.5$ feet. Assume that the population standard deviation of such lengths is $\sigma = 0.17$ foot.

2　Suppose the sample size in Exercise 1 is doubled, so that $n = 150$, but that $\bar{x}$ remains the same.

3　Suppose the sample size in Exercise 1 is quadrupled, so that $n = 300$, but that $\bar{x}$ remains the same.

4　A city planner is interested in the mean living area, μ, of one-family houses in her city. A preliminary study indicates that $\sigma = 25$ square feet. A random sample of the sizes of 30 houses resulted in $\bar{x} = 2250$ square feet. Based on this information find a 95-percent confidence interval for μ.

5　Ninety-two boys in the eighth grade of a *large* junior high school have a mean weight of 112.3 lbs, and a standard deviation of 11.6 lbs. Use this data to find a 95-percent confidence interval for μ, the mean weight of all eighth-grade boys in the school.

6 A manufacturer of nails takes a sample of the lengths of 50 nails. Find the 99-percent confidence interval for μ, the mean length of all such nails, if $\bar{x} = 1.75$ in., and $s = 0.1$ in.

7 The mean length, $\bar{x}$, of 40 giant squids is 47.5 feet, and the sample standard deviation is 2.5 feet. Find a 98-percent confidence interval for μ, the mean length of all giant squids.

8 A random sample of 35 half-gallons of a certain brand of nonfat milk yielded a mean net volume of 64.00 fluid ounces, with a sample standard deviation of 0.05 ounce. Based on this data, find a 98-percent confidence interval for the mean net volume, μ, of all half-gallons of nonfat milk of this brand.

In Exercises 9 through 13, find the confidence interval for mean IQ (μ) that would be obtained from the given values of $\bar{x}$ and s, based on a random sample of size 90.

9 Find the 98-percent confidence interval for μ, if $\bar{x} = 100.58$ and $s = 8.72$.

10 Find the 90-percent confidence interval for μ, if $\bar{x} = 100.58$ and $s = 8.72$.

11 Find the 90-percent confidence interval for μ, if $\bar{x} = 110.09$ and $s = 8.16$.

12 Find the 75-percent confidence interval for μ, if $\bar{x} = 105.61$ and $s = 7.65$.

13 Find the 97-percent confidence interval for μ, if $\bar{x} = 98.48$ and $s = 8.39$.

14 A population has $\sigma = 10$. Find a 95-percent confidence interval for μ based on:

a) A sample mean of $\bar{x} = 20$ obtained from a sample of size $n = 100$.

b) A sample mean of $\bar{x} = 20$ obtained from a sample of size $n = 400$.

c) A sample mean of $\bar{x} = 20$ obtained from a sample of size $n = 900$.

d) What is the relationship between the length of a confidence interval and the size of the sample from which it was obtained? Comment on the practical implications of this relationship.

The following exercises explain how sample size is related to the length of the confidence interval.

15 The first situation considered in this chapter was one in which large samples were taken from a population with known standard deviation σ. For such a population,

$$z = \frac{\bar{x} - \mu}{\sigma / \sqrt{n}}$$

is approximately a standard normal random variable, and $P(-z_{\alpha/2} < z < z_{\alpha/2}) = 1 - \alpha$. Use these facts to derive the confidence-interval formula:

$$P\left(\bar{x} - z_{\alpha/2} \cdot \frac{\sigma}{\sqrt{n}} < \mu < \bar{x} + z_{\alpha/2} \cdot \frac{\sigma}{\sqrt{n}}\right) = 1 - \alpha.$$

16 a) What is the length of the confidence interval given in Exercise 15?

b) How does this length change as sample size increases?

c) What must you do to sample size if you wish to cut the length of that confidence interval in half?

17 The quantity that determines the accuracy in estimating μ by $\bar{x}$ is

$$E = z_{\alpha/2} \cdot \frac{\sigma}{\sqrt{n}}$$

and is called the *maximum error of the estimate* (E is just half the length of the confidence interval). If we specify the confidence level ($1 - \alpha$) and the maximum error of the estimate (E), then we can determine the sample size required by solving for n in the above equation. Do the necessary algebra to show that

$$n = \frac{z_{\alpha/2}^2 \cdot \sigma^2}{E^2}$$

[*Note*: The formula for n usually doesn't yield a whole number. To get the required sample size, *round up* to the next whole number.]

18 A statistician is interested in estimating the (population) mean weight, μ, of newborn babies. From previous experience, she knows that $\sigma \doteq 2$ lbs. How large a sample must she take in order to be 95 percent confident that μ is within 0.5 lb of her calculated value of $\bar{x}$? (Use the results of Exercise 17.)

19 A biologist wishes to estimate the mean length of western rattlesnakes. From previous experience, she estimates that $\sigma \doteq 0.17$ foot. How large a sample must she take if she wishes to be 95 percent confident that μ is within 0.05 foot of her sample mean $\bar{x}$? (Use the results of Exercise 17.)

20 Use the result of Exercise 17 to show the following: For a fixed confidence level it is necessary to (approximately) *quadruple* the sample size, in order to *double* the accuracy of the estimate.

In this section we will learn how to find confidence intervals for proportions. Essentially the procedure is just a special case of what we did in Section 7.3.

Before the 1976 Presidential election, an independent researcher wished to estimate the proportion (percentage) of Arizona voters who would vote for Jimmy Carter. He took a random sample of 500 voters. He then used the proportion of these 500 voters who said they would vote for Carter to construct a 95-percent confidence interval for the proportion of all Arizona voters who would vote for Carter. We will now explain how to obtain such confidence intervals. Let p be the actual (population) proportion of voters who would vote for Jimmy Carter. Then the problem is to find a 95-percent confidence interval for p.

Let $\hat{p}$ be the proportion of the *sampled* voters who said they would vote for Carter. Note that $\hat{p}$ is a random variable. Its value depends on chance—namely, on the voters selected in the sample. For example, if 200 of the 500 sampled voters said they would vote for Carter, then $\hat{p} = 200/500 = 0.40$. On the other hand, if 250 of the 500 sampled voters said they would vote for Carter, then $\hat{p} = 250/500 = 0.50$.

The procedure for finding confidence intervals for p, based on $\hat{p}$, relies on the following fact, which is a consequence of the central-limit theorem.

> If p is the population proportion, and $\hat{p}$ is the sample proportion, based on a random sample of size n, then the random variable
>
> $$z = \frac{\hat{p} - p}{\sqrt{\hat{p}(1 - \hat{p})/n}} \quad *$$
>
> has (approximately) the standard normal distribution. The approximation is "good" if both $n\hat{p}$ and $n(1 - \hat{p})$, the observed number of "success" and "failures," respectively, are at least 5.

Using the above statement and reasoning similar to that in Section 7.2, we can get the following procedure for finding confidence intervals for proportions.

Procedure for finding confidence intervals for p (the population proportion) based on $\hat{p}$ (the sample proportion)

1. Check that the observed number of successes and failures are both at least 5 (if they aren't, you should not use this procedure).
2. If the desired confidence level is $1 - \alpha$, use Table I to

   Find $z_{\alpha/2}$

3. The desired confidence interval for p is

$$\hat{p} - z_{\alpha/2} \sqrt{\frac{\hat{p}(1 - \hat{p})}{n}} \quad \text{to} \quad \hat{p} + z_{\alpha/2} \sqrt{\frac{\hat{p}(1 - \hat{p})}{n}}$$

   where $z_{\alpha/2}$ is found in Step 2, n is the sample size, and $\hat{p}$ is computed from the actual sample data obtained.

---

* The random variable $(\hat{p} - p)/\sqrt{\hat{p}(1 - \hat{p})/n}$ is essentially just the random variable $(\bar{x} - \mu)/(s/\sqrt{n})$ when specialized to the situation of proportions.

Example 13 ◀ We now apply this procedure to find a 95-percent confidence interval for the (population) proportion of Arizona voters who would actually vote for Jimmy Carter.

The researcher took a random sample of 500 Arizona voters ($n = 500$). Of these 500 voters, 202 said they would vote for Carter.

1. Number of successes = 202; number of failures = $500 - 202 = 298$. Both are at least 5.
2. Since the desired confidence level is 0.95, $\alpha = 0.05$. Using Table I, we find that

$$z_{\alpha/2} = z_{0.025} = 1.96$$

3. Now, $n = 500$ and $\hat{p} = 202/500 = 0.404$. Thus, the 95-percent confidence interval for p is

$$\hat{p} - z_{\alpha/2}\sqrt{\frac{\hat{p}(1-\hat{p})}{n}} \quad \text{to} \quad \hat{p} + z_{\alpha/2}\sqrt{\frac{\hat{p}(1-\hat{p})}{n}}$$

or

$$0.404 - 1.96\sqrt{\frac{0.404(1-0.404)}{500}} \quad \text{to} \quad 0.404 + 1.96\sqrt{\frac{0.404(1-0.404)}{500}}$$

or

$$0.361 \quad \text{to} \quad 0.447$$

In other words, the researcher was 95 percent confident that the proportion, p, of Arizona voters who would vote for Jimmy Carter was somewhere between 0.361 (36.1%) and 0.447 (44.7%).* ▶

Example 14 ◀ A statistician working for a television company is interested in obtaining information concerning the percentage (proportion) of U.S. families who own a color T.V. He takes a random sample of 250 families, and finds that 165 of these 250 families own a color T.V. Based on this data, find a 90-percent confidence interval for the proportion, p, of U.S. families owning a color T.V.

Solution
1. Number of successes = 165. Number of failures = $250 - 165 = 85$. Both are at least 5.
2. Since the desired confidence level is 0.90, $\alpha = 0.10$. Using Table I, we find that

$$z_{\alpha/2} = z_{0.05} = 1.64$$

3. Now $n = 250$ and $\hat{p} = 165/250 = 0.66$. Thus, the 90-percent confidence interval for p is

$$\hat{p} - z_{\alpha/2}\sqrt{\frac{\hat{p}(1-\hat{p})}{n}} \quad \text{to} \quad \hat{p} + z_{\alpha/2}\sqrt{\frac{\hat{p}(1-\hat{p})}{n}}$$

---

* Actually, about 41.4% of the Arizona voters voted for Jimmy Carter.

or

$$0.66 - 1.64 \sqrt{\frac{0.66(1-0.66)}{250}} \quad \text{to} \quad 0.66 + 1.64 \sqrt{\frac{0.66(1-0.66)}{250}}$$

or

$$0.61 \quad \text{to} \quad 0.71$$

That is, the statistician is 90 percent confident that the proportion of U.S. families owning a color T.V. is somewhere between 0.61 (61%) and 0.71 (71%). ◗

Suppose that, in Example 14, the statistician takes a random sample of 500 families, and finds that 315 of these families own a color T.V. Based on this data, find a 95-percent confidence interval for the proportion, p, of U.S. families owning a color T.V.   Exercise ❙

EXERCISES Section **7.4**

1 A statistician working for an automobile company is interested in obtaining information concerning the percentage of U.S. families who own only one automobile. She takes a random sample of 300 families and finds that 147 of these 300 families own only one car. Based on this data, find a 95-percent confidence interval for the proportion, p, of U.S. families owning only one car.

2 The statistician in Exercise 1 also found that 99 out of another sample of 300 families owned two or more cars. Based on this data, find a 95-percent confidence interval for the proportion, p, of U.S. families owning two or more automobiles.

3 A distributor of T.V. sets in the North Central states is interested in obtaining information concerning the percentage of North Central families who own T.V. sets. He takes a random sample of 350 families and finds that 336 own T.V. sets. Based on this data, find a 90-percent confidence interval for the proportion, p, of North Central families who own T.V.'s.

4 A manufacturer of air-conditioning units is interested in obtaining information concerning the percentage of families in the South who own central air-conditioning units. He takes a random sample of 300 families, and finds that 90 of these 300 families own central air-conditioning units. Based on this data, find a 98-percent confidence interval for the proportion, p, of Southern families owning central air-conditioning units.

5 The manufacturer in Exercise 4 also found that 114 out of another sample of 300 families own room air-conditioners. Based on the data, find a 98-percent confidence interval for the proportion, p, of Southern families owning room air-conditioners.

6 A transportation planner in the Northeast is interested in obtaining information concerning the percentage of workers in the Northeast who drive to work. She takes a random sample of 250 workers, and finds that 60 of these 250 workers drive to their jobs. Based on this data, find a 90-percent confidence interval for the proportion, p, of northeastern workers who drive to work.

7 A statistician working for the government is interested in obtaining information concerning the percentage of U.S. families that took at least one overnight trip in 1978. She takes a random sample of 350 families, and finds that 252 of these 350 families made at least one overnight trip. Based on this data, find a 95-percent confidence interval for the proportion, p, of families who made at least one overnight trip.

8 A statistician working for a company that manufactures dishwashers is interested in obtaining information concerning the percentage of U.S. households with dishwashers. She takes a random sample of 400 households and finds that 152 of the 400 households have dishwashers. Based on this data, find a 95-percent confidence interval for the proportion, p, of U.S. households with dishwashers.

9 A pharmaceutical company is interested in testing the effectiveness of a new drug developed to cure a disease. The company has the drug administered to a random sample of 900 people with the disease. The conclusion is that the drug is effective on 747 of the 900 people sampled. Based on this data, find a 98-percent confidence interval for p, the actual proportion of people with the disease who would be cured by the drug.

10 A frozen-food manufacturer is interested in obtaining information concerning whether or not people will like the company's new frozen yogurt dessert. Their statistician gives the dessert to a random sample of 250 people, and finds that 195 would buy the yogurt dessert. Based on this data, find a 90-percent confidence interval for p, the actual proportion of all people familiar with the dessert who would buy it.

11 A 95-percent confidence interval for p always has the form

$$\hat{p} - 1.96 \sqrt{\frac{\hat{p}(1-\hat{p})}{n}} \quad \text{to} \quad \hat{p} + 1.96 \sqrt{\frac{\hat{p}(1-\hat{p})}{n}}$$

The maximum error of the estimate for such a confidence interval is

$$E = 1.96 \sqrt{\frac{\hat{p}(1-\hat{p})}{n}}$$

A politician running for mayor in a city with 50,000 voters knows these formulas and wishes to determine the sample size to use in a political poll to estimate p, the proportion of voters who will vote for her. She is in a two-candidate race, and is sure that $\hat{p}$ will not turn out to be less than 0.4 or more than 0.6.

a) Calculate $\hat{p}(1-\hat{p})$ for the extreme cases where $\hat{p} = 0.4$ or 0.6, and for the more realistic case where $\hat{p}$ is 0.5. Is there much difference in these values?

b) Decide on a numerical value of $\hat{p}(1-\hat{p})$ that can be substituted into the formula above for planning studies of the relation between n and E.

c) Make the substitution described in (b), and compute (approximate) values of n necessary to have a maximum error E less than or equal to

  i) 0.03   ii) 0.02   iii) 0.01

d) A newspaper poll based on a random sample of 400 voters has announced that the candidate is ahead 56 % to 44 %. Use the newspaper's data to calculate a 95-percent confidence interval for the candidate, and use this interval to decide whether she can be fairly sure that she is going to win.

e) It will cost the candidate $1.00 per voter surveyed for her own poll. Her funds are running low, and she wants to spend more than $5,000 only if absolutely necessary. What sample size from (c) would you advise her to use in her poll? (Assume that she still wants to have 95 percent confidence.)

Note. This problem is simplified, and does not correspond exactly to the analyses done by pollsters. But it does give you some idea of the problems involved in voter surveys—especially the problem of newspaper headlines that announce "Smith leading 51 % to 49 %."

7.5 The Student's *t*-distribution

In the previous sections of this chapter, we learned how to find confidence intervals for μ, when dealing with *large* random samples ($n \geq 30$). However, there are many cases where large samples are either unavailable, extremely expensive, or for some reason, simply undesirable.

For instance, suppose we wanted to estimate (find a confidence interval for) the mean lifetime of a certain type of lightbulb. The lightbulbs selected in the sampling are used until they are burned out. So the process of sampling, in this case, actually involves destroying the items in the sample, since the lightbulbs are no good after they are tested. In this particular situation, the fact that "sampling" involves actually destroying the items sampled is not of great concern. Not only is it easy to obtain *large* samples, but such samples are relatively inexpensive.

On the other hand, suppose we wanted to estimate the mean life of a certain brand of tire. Again, sampling actually involves destroying the items tested—and tires are more expensive than lightbulbs. In this case, it is desirable to use small random samples, if possible.

We will have more to say about *large vs. small* random samples a little later. But right now, let's look at an example.

◀ The owner of a retail tire shop is interested in a new line of tires. The manufacturer of this new line of tires has made the claim that the mean life, μ, of these tires is 40,000 miles. However, the owner of the shop is somewhat skeptical, and would like to obtain some firsthand information concerning the mean life of these tires.   Example **15**

To get this information he decides to test, for himself, the life of 16 such tires. He then plans to use the mean life, $\bar{x}$, of these 16 tires to get a 95-percent confidence interval for the true mean life, μ. Based on the results of this test he will then decide whether or not to purchase this new line of tires. ▶

Note that, in Example 15, σ is not given and is therefore *unknown*. Up to this point, the only procedure we have for finding confidence intervals for μ, when σ is *unknown*, is the one on page 230 of Section 7.3. However, the method there required that the sample size be at least 30 (that is, $n \geqslant 30$). Since, in the above example, the sample size is only 16 ($n = 16$), we can't use the procedure of Section 7.3. Consequently, it appears that, in order to solve problems such as Example 15 (where σ is unknown and $n < 30$), we need to develop a new procedure.*

The procedure of Section 7.3 is based on the fact that, *if the sample size is at least* 30 ($n \geqslant 30$), then the random variable

$$z = \frac{\bar{x} - \mu}{s/\sqrt{n}}$$

has (approximately) *the standard normal distribution* (see page 230) and so we can find probabilities for $(\bar{x} - \mu)/(s/\sqrt{n})$ using areas under the standard normal curve. However, if the sample size is *not* at least 30 (that is, if $n < 30$), then we *cannot* use the standard normal curve to find probabilities for $(\bar{x} - \mu)/(s/\sqrt{n})$. Consequently, in order to develop a procedure for finding confidence intervals for μ when σ is unknown and $n < 30$, we first need to know *how we can find probabilities for the random variable* $(\bar{x} - \mu)/(s/\sqrt{n})$ *when* $n < 30$.

In general, the answer is not at all simple because there is not just one answer. However, we *can* give an answer, *if the population itself is approximately normally distributed*. So we make the following requirement for use in this text.

> When obtaining confidence intervals for μ, based on small random samples ($n < 30$), we require that the *population itself be approximately normally distributed*.

Now, if the population itself is approximately normally distributed, how can we find probabilities for the random variable

$$\frac{\bar{x} - \mu}{s/\sqrt{n}}?$$

---

* We shall concentrate on the case where σ is not known since, in practice, it usually isn't. At the end of this chapter we will make a few remarks about the case where σ is known.

If $n < 30$, we *can't* use the *standard normal curve* to find probabilities for $(\bar{x} - \mu)/(s/\sqrt{n})$. However, a man named W. S. Gosset found probability curves that can be used. These curves are called Student's *t*-curves, or simply **t-curves**.* The shape of a *t*-curve depends on the sample size n. If the sample size is n, then we identify the *t*-curve in question by saying that it is the *t*-curve with $n - 1$ *degrees of freedom*.†

> Suppose we take a random sample of size n from a *population that is approximately normally distributed* with mean μ. Then the random variable
>
> $$t = \frac{\bar{x} - \mu}{s/\sqrt{n}}$$
>
> has the (*Student's*) *t-distribution with $n - 1$ degrees of freedom*. That is, probabilities for this random variable can be found using areas under the *t*-curve with $n - 1$ degrees of freedom.

For convenience we will often write d.f. $= n - 1$ to indicate that we are dealing with $n - 1$ degrees of freedom.

Example **16**   ◖ From past experience it is known that tire life (for a particular line of tire) is *normally distributed*. Suppose the tire-shop owner in Example 15 decides to take a *random sample of* 16 of the new line of tires, and determine the lives of these 16 tires. Let $\bar{x}$ be the (sample) mean life of the 16 tires, and s the (sample) standard deviation of their lives. Then the random variable

$$t = \frac{\bar{x} - \mu}{s/\sqrt{n}} = \frac{\bar{x} - \mu}{s/\sqrt{16}}$$

has the *t*-distribution with $16 - 1 = 15$ degrees of freedom. Probabilities for $t = (\bar{x} - \mu)/(s/\sqrt{16})$ can be found using areas under the *t*-curve with d.f. $= 15$. ◗

Exercise **J**   Suppose that, in Example 16, the tire dealer takes a random sample of 20 tires. What curve would you use, in order to find probabilities for the random variable

$$t = \frac{\bar{x} - \mu}{s/\sqrt{20}}?$$

We will now look at some specific examples of *t*-curves, and then examine some general properties of such curves. As we said, there is a different *t*-curve for each number of degrees of freedom. But all the *t*-curves look quite alike, and are very similar to the

---

* W. S. Gosset (1876–1936) did much of his research while working for the Guiness Brewery of Ireland. The company prohibited its employees from publishing research. Consequently, Gosset presented his findings using the pen name "Student."

† We will not go into the mathematical concepts involved in defining degrees of freedom, but simply use degrees of freedom as a method for identifying the appropriate *t*-curve.

standard normal curve. For illustration, we have drawn in Fig. 7.5 the standard normal curve along with two *t*-curves. One *t*-curve has 3 degrees of freedom, and the other has 12 degrees of freedom.

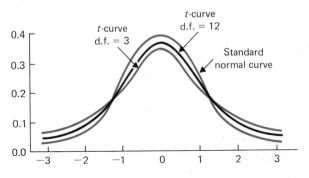

Figure 7.5

As illustrated by Fig. 7.5, *t*-curves have the following properties:

1. The total area under any *t*-curve is 1.
2. *t*-curves are *symmetric* about the vertical line passing through 0.
3. *t*-curves extend indefinitely in both directions.
4. As the number of degrees of freedom (or sample size) gets larger, *t*-curves begin to look more and more like the standard normal curve.

In Section 5.2, we learned how to find areas under the standard normal curve by using Table I. We will now see how to find areas under *t*-curves by using Table II.

For our purposes, among which is determining confidence intervals for means, we do not need a complete *t*-table (like the standard normal table) for each *t*-curve. There are only certain areas under *t*-curves that will be important for us to know. In Table 7.3 we have reproduced part of the Student's *t*-distribution table which can be found in Table II in the appendix. We will illustrate the use of this table.

Table 7.3

d.f.	$t_{0.10}$	$t_{0.05}$	$t_{0.025}$	$t_{0.01}$	$t_{0.005}$
$\vdots$	$\vdots$	$\vdots$	$\vdots$	$\vdots$	$\vdots$
12	1.36	1.78	2.18	2.68	3.05
→ 13	1.35	(1.77)	2.16	2.65	3.01
14	1.35	1.76	2.14	2.62	2.98
15	1.34	1.75	2.13	2.60	2.95
16	1.34	1.75	2.12	2.58	2.92
$\vdots$	$\vdots$	$\vdots$	$\vdots$	$\vdots$	$\vdots$

Example **17**   ◖ Suppose that we are considering a *t*-curve with 13 degrees of freedom (d.f. = 13). A typical problem will be to find the *t*-value for which the area under the *t*-curve to the *right* of this value is 0.05. See Fig. 7.6.

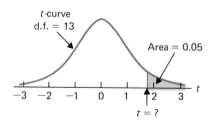

Figure 7.6

To find the *t*-value in question we use Table 7.3 (or Table II in the appendix). The lefthand column of the table gives us the number of degrees of freedom. In this case, d.f. = 13, so we concentrate on the row of the table with d.f. = 13. If you go across this row until you are under the column headed $t_{0.05}$, then you will see the number 1.77 (see Table 7.3). This is the desired *t*-value. That is, the area under the *t*-curve (with d.f. = 13) to the *right* of 1.77 is 0.05. See Fig. 7.7.

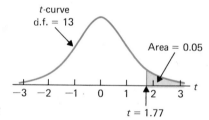

Figure 7.7

We will use the notation $t_{0.05}$ to denote the *t*-value for which the area under the *t*-curve to its *right* is 0.05. So, as we have just seen, for a *t*-curve with d.f. = 13, $t_{0.05} = 1.77$. ◗

Exercise **K**   For a *t*-curve with d.f. = 14, find $t_{0.025}$. That is, find the *t*-value for which the area under the *t*-curve to the *right* of this value is 0.025. Sketch a picture illustrating your work.

Now that we have learned to read the Student's *t*-distribution table, we can consider further uses.

Example **18**   ◖ For a *t*-curve with d.f. = 16, find the *t*-value for which the area under the curve to its *left* is 0.025. See Fig. 7.8.

Solution   Since *t*-curves are symmetric about the vertical line passing through 0, it follows that the desired *t*-value is $t = -t_{0.025}$. See Fig. 7.9.

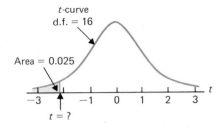

Figure 7.8

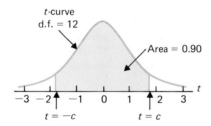

Figure 7.9

From Table 7.3, or Table II in the appendix, we find that, for a *t*-curve with d.f. = 16, $t_{0.025} = 2.12$. Thus, $t = -t_{0.025} = -2.12$. That is, the area under the *t*-curve to the left of -2.12 is 0.025. By the way, since the total area under a *t*-curve is 1, the unshaded portion of Fig. 7.9 has area 0.95 (since $1 - 0.025 - 0.025 = 0.95$). ▶

For a *t*-curve with d.f. = 12, find the *t*-value for which the area under the *t*-curve to the   Exercise **L**
left of this value is 0.05. Draw a picture similar to Fig. 7.8.

◀ For a *t*-curve with d.f. = 12, find the *t*-value, *c*, for which the area under the curve   Example **19**
between $t = -c$ and $t = c$ is 0.90.

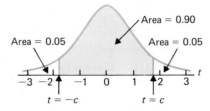

Since the shaded area in the above picture is to be 0.90, this means that each of the two   Solution
unshaded areas must be 0.05 (since $(1 - 0.90)/2 = 0.05$).

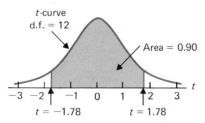

t-curve
d.f. = 12

Area = 0.90

$t = -1.78$ $t = 1.78$

So we must have $c = t_{0.05}$. From Table 7.3, we see that for a *t*-curve with d.f. = 12, $t_{0.05} = 1.78$. Consequently, $c = 1.78$. That is, the area under the *t*-curve between -1.78 and 1.78 is 0.90.

Just as for normal curves, areas under *t*-curves represent probabilities. So, for instance, we see from this example that if a random variable *t* has the *t*-distribution with d.f. = 12, then $P(-1.78 < t < 1.78) = 0.90$. ▶

Exercise M For a *t*-curve with d.f. = 16, find the *t*-value, *c*, for which the area under the curve between $t = -c$ and $t = c$ is 0.95. Illustrate your work with a picture.

Exercise N Suppose *t* has the Student's *t*-distribution with d.f. = 22. Find *c* so that
$$P(-c < t < c) = 0.99.$$
Illustrate your work with a picture.

In most of our work with confidence intervals, we will stick to three basic levels of confidence; namely, 90 %, 95 %, and 99 %. The corresponding formulas for these are listed below:

1. **90 % confidence:** $P(-t_{0.05} < t < t_{0.05}) = 0.90$

2. **95 % confidence:** $P(-t_{0.025} < t < t_{0.025}) = 0.95$

3. **99 % confidence:** $P(-t_{0.005} < t < t_{0.005}) = 0.99$

EXERCISES Section **7.5**

In each of Exercises 1 through 5, a study is described in which a researcher wants to find a confidence interval for the population mean, μ. In order to accomplish this, the researcher must be able to compute probabilities for the random variable

$$\frac{\bar{x} - \mu}{s/\sqrt{n}}.$$

In each case, identify the curve the researcher should use to calculate probabilities for that random variable. (Assume the population in each question is approximately normally distributed.)

1 A psychologist wants to find a confidence interval for the mean IQ, μ, of airline pilots by using a random sample of $n = 14$ IQ's.

2 A doctor is interested in the mean weight, μ, of all fourth-grade boys, and wants to find a confidence interval for μ based on a random sample of $n = 12$ weights.

3 An exercise physiologist wants to use a random sample of size $n = 20$ to find a confidence interval for the mean, μ, of reaction times for fast-pitch softball players.

4 A reading specialist wants to find a confidence interval for the mean number of words read per minute by fourth-grade girls, based on a random sample of 18 such reading speeds.

5 A quality-control engineer, working for a potato-chip company, needs to find a confidence interval for the mean weight, μ, of all packages of potato chips produced by the company. He decides to take a random sample of the weights of 16 such packages.

6 Which of the following looks most like the standard normal curve?

a) a *t*-curve with d.f. = 14   b) a *t*-curve with d.f. = 23
c) a *t*-curve with d.f. = 1

In Exercises 7 through 11, the value of d.f. is given.
a) Find the *t*-value.
b) Draw a picture showing the area under the *t*-curve to the *right* of that *t*-value.

7 d.f. = 19. Find $t_{0.10}$.

8 d.f. $= 16$. Find $t_{0.05}$.

9 d.f. $= 22$. Find $t_{0.01}$.

10 d.f. $= 27$. Find $t_{0.025}$.

11 d.f. $= 3$. Find $t_{0.005}$.

In Exercises 12 through 16, the value of d.f. is given.

a) Find the indicated t-value.

b) Draw a picture showing the area under the t-curve to the *left* of that t-value.

12 d.f. $= 14$. Find $t = -t_{0.01}$.

13 d.f. $= 20$. Find $t = -t_{0.025}$.

14 d.f. $= 4$. Find $t = -t_{0.05}$.

15 d.f. $= 17$. Find $t = -t_{0.10}$.

16 d.f. $= 28$. Find $t = -t_{0.005}$.

17 What is the area to the right of the t-value in Exercise 15?

In Exercises 18 through 22, the value of d.f. is given.

a) Find the indicated t-value.

b) Draw a picture of the indicated area.

18 d.f. $= 21$, area between $-t$ and t is 0.90.

19 d.f. $= 6$, area between $-t$ and t is 0.98.

20 d.f. $= 18$, area between $-t$ and t is 0.80.

21 d.f. $= 25$, area between $-t$ and t is 0.95.

22 d.f. $= 20$, area between $-t$ and t is 0.99.

In Exercises 23 through 27, t has the Student's t-distribution and the value for d.f. is given. Find c so that $P(-c < t < c)$ equals the given value.

23 d.f. $= 11$, $P(-c < t < c) = 0.90$

24 d.f. $= 27$, $P(-c < t < c) = 0.95$

25 d.f. $= 13$, $P(-c < t < c) = 0.99$

26 d.f. $= 18$, $P(-c < t < c) = 0.80$

27 d.f. $= 8$, $P(-c < t < c) = 0.98$

Confidence intervals for means—small samples 7.6

We are now ready to discuss the procedure for finding confidence intervals for μ based on $\bar{x}$, when the sample size is less than 30 ($n < 30$). We will illustrate the procedure with an example. But first let's review the important fact that we learned in Section 7.5.

> If we take a random sample of size n from a *population* that is *approximately normally distributed* with mean μ, then the random variable
>
> $$t = \frac{\bar{x} - \mu}{s/\sqrt{n}}$$
>
> has the *Student's t-distribution with d.f.* $= n - 1$. That is, probabilities for t can be found using areas under the t-curve with d.f. $= n - 1$.*

Using this fact, we can find confidence intervals for μ based on $\bar{x}$ for samples of size less than 30 ($n < 30$).

◀ A real-estate broker is interested in obtaining information concerning the mean value (μ) of the homes in a certain community. Her past experience tells her that the home values are (approximately) normally distributed.

Example **20**

---

* We should emphasize that, if the population is normally distributed, then $(\bar{x} - \mu)/(s/\sqrt{n})$ has the Student's t-distribution with d.f. $= n - 1$, regardless of the value of n. However, if $n \geqslant 30$, $(\bar{x} - \mu)/(s/\sqrt{n})$ is also (approximately) normally distributed (see page 230). So, in this latter case, we can use the procedure on page 230 for finding confidence intervals. It is only when $n < 30$ that we *need* to assume that the population is normally distributed, so that we can use t-curves to find probabilities for $(\bar{x} - \mu)/(s/\sqrt{n})$. If $n \geqslant 30$, we can use the standard normal curve to find probabilities for $(\bar{x} - \mu)/(s/\sqrt{n})$, regardless of whether or not the population is normally distributed.

To estimate μ (the mean value), she takes a *random sample of* 10 *home values*. She then computes $\bar{x}$ and s. Her computations show that

$$\bar{x} = \$50,473.00$$
$$s = \$ \; 3,623.97.$$

Based on this data, find a 90-percent confidence interval for μ.

Solution   The population (of home values) is normally distributed. The sample size is $n = 10$. Therefore, we know that the random variable

$$t = \frac{\bar{x} - \mu}{s/\sqrt{n}} = \frac{\bar{x} - \mu}{s/\sqrt{10}}$$

has the Student's t-distribution with $10 - 1 = 9$ degrees of freedom.

Now, we want a 90-percent confidence interval. By (1) on page 242,

$$P(-t_{0.05} < t < t_{0.05}) = 0.90 \quad (90\%)$$

Using Table II in the appendix, we find that

$$t_{0.05} = 1.83 \quad \text{for d.f.} = 9$$

So,

$$P(-1.83 < t < 1.83) = 0.90$$

In our case,

$$t = \frac{\bar{x} - \mu}{s/\sqrt{10}}$$

Consequently,

$$P\left(-1.83 < \frac{\bar{x} - \mu}{s/\sqrt{10}} < 1.83\right) = 0.90$$

Doing some algebra, as on page 226, we can rewrite this last equation as

$$P\left(\bar{x} - 1.83\frac{s}{\sqrt{10}} < \mu < \bar{x} + 1.83\frac{s}{\sqrt{10}}\right) = 0.90$$

So the probability is 0.90 that the interval from $\bar{x} - 1.83(s/\sqrt{10})$ to $\bar{x} + 1.83(s/\sqrt{10})$ will contain μ. That is, once we compute the values $\bar{x}$ and s for the random sample actually obtained, then the range,

$$\bar{x} - 1.83\frac{s}{\sqrt{10}} \quad \text{to} \quad \bar{x} + 1.83\frac{s}{\sqrt{10}}$$

will be the 90-percent confidence interval for μ.

The broker's sample yielded the values $\bar{x} = 50{,}473$ and $s = 3{,}623.97$. Using these values we get

$$\bar{x} - 1.83 \frac{s}{\sqrt{10}} = 50{,}473 - 1.83 \left(\frac{3{,}623.97}{\sqrt{10}} \right) \doteq 48{,}375.82$$

$$\bar{x} + 1.83 \frac{s}{\sqrt{10}} = 50{,}473 + 1.83 \left(\frac{3{,}623.97}{\sqrt{10}} \right) \doteq 52{,}570.18$$

So the broker can be 90 percent confident that the mean value (μ) of the homes in the community is somewhere between

$$\$48{,}375.82 \quad \text{and} \quad \$52{,}570.18 \quad \blacktriangleright$$

If we study Example 20 carefully, we can write down the procedure for finding confidence intervals for μ based on $\bar{x}$, when the sample size is less than 30.

Assumption: **Normal population***   Procedure for finding confidence intervals for

Step 1. If the desired confidence level is $1 - \alpha$, use Table II in the appendix, to   confidence intervals for μ based on $\bar{x}$ for samples of size n less than 30

$$\text{Find } t_{\alpha/2} \quad \text{for d.f.} = n - 1$$

Step 2. The desired confidence interval is

$$\bar{x} - t_{\alpha/2} \cdot \frac{s}{\sqrt{n}} \quad \text{to} \quad \bar{x} + t_{\alpha/2} \cdot \frac{s}{\sqrt{n}}$$

where $t_{\alpha/2}$ is found in Step 1, n is the sample size, and $\bar{x}$ and s are computed from the actual sample data obtained.

◀ An administrator at a large university is interested in obtaining information   Example **21** concerning the one-way driving distance of commuter students. He takes a random sample of 25 commuter students, and finds that their average one-way driving distance is $\bar{x} = 10$ km, with a (sample) standard deviation of $s = 3$ km. Using this data, find a 99-percent confidence interval for μ (the mean one-way driving distance of all commuter students at the university). Assume the one-way driving distances are normally distributed.

We use the procedure given above. Here $n = 25$.   Solution

1. Since the desired confidence level is 0.99, $\alpha = 0.01$. Using Table II, we find that for d.f. $= 25 - 1 = 24$,

$$t_{\alpha/2} = t_{0.005} = 2.80.$$

2. Since $t_{\alpha/2} = 2.80$, $\bar{x} = 10$, $s = 3$, and $n = 25$, the 99-percent confidence interval is

$$\bar{x} - t_{\alpha/2} \cdot \frac{s}{\sqrt{n}} \quad \text{to} \quad \bar{x} + t_{\alpha/2} \cdot \frac{s}{\sqrt{n}}$$

---

* We will use the phrase "normal population" as an abbreviated form for "the population is normally distributed."

or

$$10 - 2.80 \frac{3}{\sqrt{25}} \quad \text{to} \quad 10 + 2.80 \frac{3}{\sqrt{25}}$$

or

$$8.32 \quad \text{to} \quad 11.68$$

So, the administrator can be 99 percent confident that μ (the mean one-way driving distance) is somewhere between 8.32 km and 11.68 km. ▶

Exercise **O**   Suppose that the administrator in Example 21 took a random sample of 29 commuter students, and found $\bar{x} = 11.1$ and $s = 2.8$. Based on this data, find a 95-percent confidence interval for μ.

Example **22**   ◀ Let us now return to the tire-life illustration of Example 15. Recall that a tire-shop owner is interested in obtaining information regarding the mean life (μ) of a new line of tire. The manufacturer of the tire claims that, on the average, his new line of tire will get 40,000 miles (that is, he claims μ is 40,000 miles). The tire-shop owner wants to test, for himself, the life of this new line of tire.

So he decides to take a random sample of 16 such tires, and performs a mileage test on them. The results of his testing are as follows:

Table 7.4	Tire	Life (miles)	Tire	Life (miles)
	1	43,725	9	39,783
	2	40,652	10	44,652
	3	37,732	11	38,740
	4	41,868	12	39,385
	5	44,473	13	39,686
	6	43,097	14	44,019
	7	37,396	15	40,220
	8	42,200	16	40,742

Using this data, find a 95-percent confidence interval for μ (the mean life of the new line of tire). Assume that tire life is normally distributed.

Solution   We use the procedure given above. Here $n = 16$.

1. Since the desired confidence level is 0.95, $\alpha = 0.05$. Using Table II, we find that for d.f. $= 16 - 1 = 15$,

$$t_{\alpha/2} = t_{0.025} = 2.13$$

2. The 95-percent confidence interval is

$$\bar{x} - 2.13 \frac{s}{\sqrt{16}} \quad \text{to} \quad \bar{x} + 2.13 \frac{s}{\sqrt{16}}$$

Using the data in Table 7.4 along with our formulas for computing $\bar{x}$ and s, we find that

$$\bar{x} = \frac{1}{n}\Sigma x \doteq 41{,}148.13$$

and

$$s = \sqrt{\frac{n\Sigma x^2 - (\Sigma x)^2}{n(n-1)}} \doteq 2{,}360.32$$

Consequently,

$$\bar{x} - 2.13\,\frac{s}{\sqrt{16}} = 41{,}148.13 - 2.13\left(\frac{2{,}360.32}{\sqrt{16}}\right) \doteq 39{,}891.26$$

$$\bar{x} + 2.13\,\frac{s}{\sqrt{16}} = 41{,}148.13 + 2.13\left(\frac{2{,}360.32}{\sqrt{16}}\right) \doteq 42{,}405.00$$

In other words, the tire-shop owner can be 95 percent confident that μ (the mean life of the new line of tire) is somewhere between 39,891.26 and 42,405.00 miles. So it appears that the tire manufacturer's claim that $\mu = 40{,}000$ miles may very well be correct. ▶

We should make one final comment. In this section, we wrote down the procedure for finding confidence intervals for μ when $n < 30$ and we are sampling from a population that is normally distributed.

The procedure we gave assumes that the population standard deviation (σ) is *unknown*. However, if, for some reason, σ *is* known, then we can just use the procedure given on page 228 of Section 7.3 since we learned in Section 6.3 that if a sample of size n is taken from a *normal* population with mean μ and standard deviation σ, then the random variable

$$\frac{\bar{x} - \mu}{\sigma/\sqrt{n}}$$

has the *standard normal distribution, regardless of the size of the sample.*

In other words *if* σ is known, use the procedure given on page 228 of Section 7.3. If σ is unknown, use the procedure given on page 245 of this section.

EXERCISES Section **7.6**

1 A city planner working on bikeways is interested in obtaining information on local bicycle commuters. She designs a questionnaire, and one of the questions asks how many minutes it takes the rider to pedal from home to his or her destination each day. A random sample of 25 bicycle riders reveals that the sample mean time $\bar{x} = 18$ minutes and $s = 4$ minutes. Find a 90-percent confidence interval for μ, the mean one-way bicycling time for commuters in the city. (Assume the one-way bicycling time is normally distributed.)

2 A preschool center is considering the establishment of a nursery school in the southeast section of a city. Among other things, information concerning the mean income μ of the families in the area will be helpful in the decision-making process. A random sample of 20 family incomes gives $\bar{x} = \$20{,}358$ and $s = \$2{,}954.62$. Using the assumption that incomes are approximately normally distributed, find a 95-percent confidence interval for μ.

3 An agricultural agent in Yuma, Arizona, is interested in obtaining information concerning the price that farmers in the area received for their lettuce crops over a period of time. He took a random sample of prices paid to 15 farmers and found $\bar{x} = \$138.36$ per ton and $s = \$3.08$. Find a 98-percent confidence interval for the mean price, μ, per ton. (Assume that the price per ton is normally distributed.)

4 A public transportation official in a metropolitan area is interested in obtaining the mean μ of monthly transportation costs (both public and private) for four-person households in the area. From a random sample of 18 four-person households, the sample mean cost, $\bar{x}$, is calculated to be $\bar{x} = \$91.83$ per month and $s = \$8.95$. Find a 99-percent confidence interval for μ. (Assume that the monthly transportation costs for four-person households is normally distributed.)

5 A researcher in public school administration is interested in obtaining information on the mean, μ, of administrative costs for school districts in the U.S. She takes a random sample of costs in 22 school districts in the country and calculates that $\bar{x} = \$2,276$ (in millions of dollars), and $s = \$85$ (in millions of dollars). Find a 95-percent confidence interval for μ. (Assume that administration costs of school districts in the U.S. are normally distributed.)

6 A high-school newspaper reporter is interested in obtaining information on the mean cost, μ, of a Saturday night date for male seniors at his school. He takes a random sample of 17 such costs from male seniors, and finds that $\bar{x} = \$14.21$, and $s = \$2.86$. Find a 99-percent confidence interval for μ. (Assume that such costs are normally distributed.)

7 A social worker is interested in determining the mean weekly cost, μ, of babysitting for working mothers in her city. She interviews a random sample of 14 working mothers, and calculates that $\bar{x} = \$36.73$, and $s = \$3.91$. Find a 95-percent confidence interval for μ. (Assume that such costs are normally distributed.)

8 A random sample of 20 aluminum alloy bolts from a manufacturer can withstand a mean of $\bar{x} = 1854$ lbs of stress with a standard deviation of $s = 30.9$. Find a 98-percent confidence interval for μ, the true mean stress the bolts can withstand. (Assume that the stresses such bolts can take are normally distributed.)

9 A manufacturer of rubber bands claims that his rubber bands stretch to 18 inches before breaking. A random sample of 15 rubber bands yields the following data set. Using this data, find a 95-percent confidence interval for μ, the mean length to which

such rubber bands will stretch before breaking. (Assume that such lengths are normally distributed.)

18.6	17.5	17.4	18.0	17.3
18.0	17.6	17.9	18.2	18.5
18.7	17.5	17.0	18.4	17.7

10 A wildlife biologist measures 17 red foxes in her area. Their lengths, in inches, are given below. Using this data, find a 95-percent confidence interval for μ, the mean length of all red foxes in her area. (Assume that these lengths are normally distributed.)

37.4	36.9	37.8	37.6	37.9	36.6
37.4	37.9	36.5	37.7	39.1	38.0
37.8	37.4	38.4	37.7	36.9	

11 The random variable

$$t = \frac{\bar{x} - \mu}{s/\sqrt{n}}$$

is used for analysis of small samples from normal populations when σ is unknown. If the population standard deviation σ is known, then we can use the random variable

$$z = \frac{\bar{x} - \mu}{\sigma/\sqrt{n}}$$

which has the standard normal distribution—even if the *sample size is small.*

a) Rework Exercise 7 under the assumption that σ is known, and $\sigma = 3.91$.

b) Compare the confidence interval from (a) with the one found in Exercise 7.

c) What is the general effect on a $(1 - \alpha)$-level confidence interval if σ is known and a z-statistic can be used instead of a t-statistic?

12 Use the fact that

$$P\left[-t_{\alpha/2} < \frac{\bar{x} - \mu}{s/\sqrt{n}} < t_{\alpha/2} \right] = 1 - \alpha$$

to derive the $(1 - \alpha)$-level confidence interval; that is,

$$P\left(\bar{x} - t_{\alpha/2} \cdot \frac{s}{\sqrt{n}} < \mu < \bar{x} + t_{\alpha/2} \cdot \frac{s}{\sqrt{n}} \right) = 1 - \alpha.$$

sample mean
law of large numbers
central-limit theorem
confidence interval for μ
$z_{\alpha/2}$
standard normal distribution
confidence level $(1 - \alpha)$
sample standard deviation

sample proportion
$\hat{p}$
normally distributed population
t-curves
degrees of freedom (d.f.)
Student's t-distribution
$t_{\alpha/2}$

Key Terms

central-limit theorem:

Formulas and Key Facts

If a random sample of size $n \geqslant 30$ is taken from a population with mean μ and standard deviation σ, then the random variable $\bar{x}$ *is approximately normally distributed with* $\mu_{\bar{x}}$ $= \mu$ *and* $\sigma_{\bar{x}} \doteq \sigma/\sqrt{n}$.

sample mean (ungrouped data)

$$\bar{x} = \frac{\sum x}{n} \qquad (n = \text{sample size})$$

sample standard deviation (ungrouped data):

$$s = \sqrt{\frac{n\sum x^2 - (\sum x)^2}{n(n-1)}}$$

formula for finding confidence intervals for a population proportion, p:

$$\hat{p} - z_{\alpha/2} \cdot \sqrt{\frac{\hat{p}(1-\hat{p})}{n}} \qquad \text{to} \qquad \hat{p} + z_{\alpha/2} \cdot \sqrt{\frac{\hat{p}(1-\hat{p})}{n}}$$

($\hat{p}$ = sample proportion, $(1 - \alpha)$ = confidence level, n = sample size.)
Assumption: The observed number of successes and failures are both at least 5.

Student's t-distribution:

If a random sample of size n is taken from a *normally distributed population* with mean μ, then the random variable

$$t = \frac{\bar{x} - \mu}{s/\sqrt{n}}$$

has the *Student's t-distribution with d.f.* $= n - 1$. That is, probabilities for t can be found using areas under the t-curve with d.f. $= n - 1$.

Table 7.5 summarizes the procedures for finding confidence intervals for μ. $[(1 - \alpha) = \text{confidence level.}]$

Table 7.5 Confidence intervals for μ	Sample size assumption	Population assumption	σ known?	Table	Confidence interval		
	$n \geqslant 30$	None	Yes	Normal	$\bar{x} - z_{\alpha/2} \cdot \dfrac{\sigma}{\sqrt{n}}$	to	$\bar{x} + z_{\alpha/2} \cdot \dfrac{\sigma}{\sqrt{n}}$
	$n \geqslant 30$	None	No	Normal	$\bar{x} - z_{\alpha/2} \cdot \dfrac{s}{\sqrt{n}}$	to	$\bar{x} + z_{\alpha/2} \cdot \dfrac{s}{\sqrt{n}}$
	None	Normal	Yes	Normal	$\bar{x} - z_{\alpha/2} \cdot \dfrac{\sigma}{\sqrt{n}}$	to	$\bar{x} + z_{\alpha/2} \cdot \dfrac{\sigma}{\sqrt{n}}$
	None	Normal	No	t, d.f. $= n - 1$	$\bar{x} - t_{\alpha/2} \cdot \dfrac{s}{\sqrt{n}}$	to	$\bar{x} + t_{\alpha/2} \cdot \dfrac{s}{\sqrt{n}}$

You should be able to

1 Use all of the formulas and key facts.

2 Find the probability that $\bar{x}$ will be within a specified distance, d, from μ if you know n and σ. That is, given n, d, and σ you should be able to compute

$$P(\mu - d < \bar{x} < \mu + d)$$

[see, e.g., Example 1 where $n = 30$, $d = 1$, and $\sigma = 2.50$].

3 Use the standard normal table (Table I) to find $z_{\alpha/2}$ for any given value of α.

4 Use the Student's t-distribution table (Table II) to find $t_{\alpha/2}$, for d.f. $= n - 1$.

REVIEW TEST

1 a) Find $z_{0.02}$
   b) Find $z_{0.05}$
   c) Find $z_{0.5}$
   d) Find $z_{0.9}$
   e) If z has the standard normal distribution, find the value, c, such that $P(-c < z < c) = 0.9$.

2 A doctor is collecting data on the heights of male college students involved in an intramural sport. Past experience leads him to estimate that $\sigma = 2.4$ inches for this population. He collects a random sample of $n = 64$ heights and finds that $\bar{x} = 70$ inches. Find a 95-percent confidence interval for μ, the population mean.

3 A planner who is studying a large suburban area wishes to estimate the mean income of the residents. She obtains family income data from a random sample of 100 households. The mean of these 100 family incomes is $\bar{x} = \$30,000$ and the standard deviation is $s = \$4000$. Find a 98-percent confidence interval for the true mean income, μ, of all homeowners in this tract.

4 A city councilman wishes to know whether residents of his city favor use of city funds to build bike lanes. He sponsors a poll in which 400 voters are asked their opinion, and 280 of those voters say that they favor the building of bike lanes with city funds. Find a 95-percent confidence interval for p, the actual proportion of all voters favoring the building of bike lanes.

5 a) Find $t_{0.05}$ for d.f. = 12
   b) Find $t_{0.01}$ for d.f. = 25

c) For d.f. = 18, find c such that $P(-c < t < c) = 0.90$
d) For d.f. = 18, find c such that $P(-c < t < c) = 0.95$

6 Sixteen calculator battery packs (all the same brand and model) were tested to see how long they would work without recharging. They lasted a mean time of $\bar{x} = 178$ minutes with a standard deviation of $s = 8$ minutes. Find a 95-percent confidence interval for the true mean life of battery packs of this brand and model. (Assume that such lifetimes are normally distributed.)

FURTHER READINGS

NETER, J., W. WASSERMAN, and G. A. WHITMORE, *Applied Statistics*. Boston: Allyn and Bacon, 1978. "Estimation of Population Means" covers some more advanced topics in estimation at a level only slightly higher than the level of this book.

SNEDECOR, G., and W. COCHRAN, *Statistical Methods*. Ames, Iowa: Iowa State University Press, 1967 (Sixth Ed.). The first chapter of this book introduces the reader to statistics by describing the construction of a confidence interval for the percentage of farmers in Boone County, Iowa, who spray for corn borers. This presentation unifies the work of the statistician, covering the process from sampling to estimation. There is further material on confidence intervals for means in Chapters 2 and 3.

WALPOLE, R., and R. MYERS, *Probability and Statistics for Engineers and Scientists*. New York: Macmillan Publishing Company, 1978 (Second Ed.). Pages 212–218 cover Bayesian methods of estimation. The mathematical level is above that of this book, but the material should interest a student with a calculus background.

ANSWERS TO REINFORCEMENT EXERCISES

A To find: $P(\mu - 0.5 < \bar{x} < \mu + 0.5)$

First calculate: $\dfrac{\sigma}{\sqrt{n}} = \dfrac{2.50}{\sqrt{50}} = 0.35$

Then look at: $z = \dfrac{\bar{x} - \mu}{\sigma/\sqrt{n}} = \dfrac{\bar{x} - \mu}{0.35}$

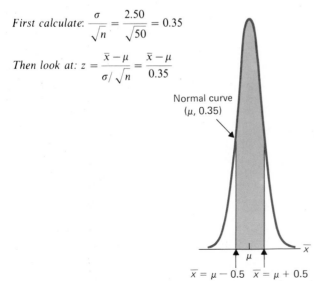

$\bar{x} = \mu - 0.5 \rightarrow z = \dfrac{(\mu - 0.5) - \mu}{0.35} = -1.43$

$\bar{x} = \mu + 0.5 \rightarrow z = \dfrac{(\mu + 0.5) - \mu}{0.35} = 1.43$

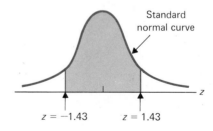

Area $= 0.9236 - 0.0764 = 0.8472$

$P(\mu - 0.5 < \bar{x} < \mu + 0.5) = 0.8472$

B   67.2 to 69.2

C

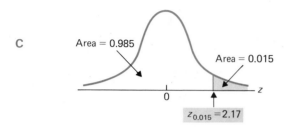

D

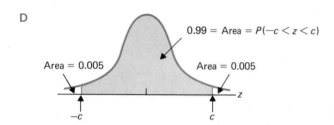

From the picture, $c = z_{0.005}$. From Table I the area to the left of 2.58 is very close to 0.995, so $z_{0.005} \doteq 2.58$. Thus, $c \doteq 2.58$ and $P(-2.58 < z < 2.58) \doteq 0.99$.

E $n = 50$, $\mu_{\bar{x}} = \mu$

$$\sigma_{\bar{x}} \doteq \frac{\sigma}{\sqrt{n}} = \frac{4.4}{\sqrt{50}} \doteq 0.62 \qquad z = \frac{\bar{x} - \mu}{\sigma/\sqrt{n}} = \frac{\bar{x} - \mu}{0.62}$$

$$P(-1.64 < z < 1.64) = 0.90$$

or

$$P\left(-1.64 < \frac{\bar{x} - \mu}{0.62} < 1.64\right) = 0.90$$

Algebra: $-1.64 < \dfrac{\bar{x} - \mu}{0.62} < 1.64$

$$-1.64 < \frac{\mu - \bar{x}}{0.62} < 1.64$$

$$-1.64(0.62) < \mu - \bar{x} < 1.64(0.62)$$

$$\bar{x} - 1.64(0.62) < \mu < \bar{x} + 1.64(0.62)$$

Consequently,

$$P\left(-1.64 < \frac{\bar{x} - \mu}{0.62} < 1.64\right) = 0.90$$

can be rewritten as

$$P(\bar{x} - 1.64(0.62) < \mu < \bar{x} + 1.64(0.62)) = 0.90.$$

Since $1.64(0.62) \doteq 1.02$, we then have

$$P(\bar{x} - 1.02 < \mu < \bar{x} + 1.02) = 0.90.$$

If $\bar{x} = 5.12$, then $\bar{x} - 1.02 = 4.10$ and $\bar{x} + 1.02 = 6.14$. So the interval from 4.10 to 6.14 minutes is a 90-percent confidence interval for μ.

F   a) The confidence level is 0.99 and $\alpha = 0.01$.
     b) The confidence level is 0.98 and $\alpha = 0.02$.

G   1. $\alpha = 0.10$, $z_{\alpha/2} = 1.64$

     2. $n = 90$, $\sigma = 10$, $\bar{x} = 112.75$

$$\bar{x} - z_{\alpha/2} \cdot \frac{\sigma}{\sqrt{n}} = 112.75 - 1.64 \, \frac{10}{\sqrt{90}} = 111.02$$

$$\bar{x} + z_{\alpha/2} \cdot \frac{\sigma}{\sqrt{n}} = 112.75 + 1.64 \, \frac{10}{\sqrt{90}} = 114.48$$

The 90-percent confidence interval for μ is 111.02 to 114.48.

H   1. $\alpha = 0.10$, $z_{\alpha/2} = 1.64$
     2. $n = 200$, $\bar{x} = 7.52$, $s = 1.85$

$$\bar{x} - z_{\alpha/2} \cdot \frac{s}{\sqrt{n}} = 7.52 - 1.64 \, \frac{1.85}{\sqrt{200}} = 7.31$$

$$\bar{x} + z_{\alpha/2} \cdot \frac{s}{\sqrt{n}} = 7.52 + 1.64 \, \frac{1.85}{\sqrt{200}} = 7.73$$

The 90-percent confidence interval for μ is 7.31 to 7.73.

I $n = 500$, $\hat{p} = \dfrac{315}{500} = 0.63$

     1. Number of successes $= 315$, number of failures $= 185$.
     2. $\alpha = 0.05$, $z_{\alpha/2} = z_{0.025} = 1.96$

3. $\hat{p} - z_{\alpha/2} \sqrt{\dfrac{\hat{p}(1 - \hat{p})}{n}} = 0.63 - 1.96 \sqrt{\dfrac{(0.63)(0.37)}{500}} \doteq 0.59$

$$\hat{p} + z_{\alpha/2} \sqrt{\frac{\hat{p}(1 - \hat{p})}{n}} = 0.63 + 1.96 \sqrt{\frac{(0.63)(0.37)}{500}} \doteq 0.67$$

The 95-percent confidence interval for p is 0.59 to 0.67.

J   The t-curve with d.f. $= 19$.

K $\quad t_{0.025} = 2.14$

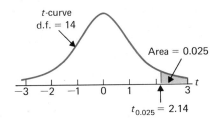

L $\quad -t_{0.05} = -1.78$

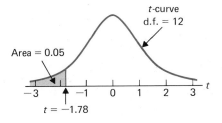

M $\quad c = 2.12$

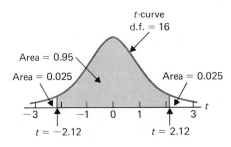

N $\quad c = 2.82$

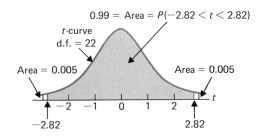

O $\quad \alpha = 0.05 \qquad\qquad\qquad \bar{x} = 11.1$

$\qquad$ d.f. $= 29 - 1 = 28 \qquad\quad s = 2.8$

1. $t_{\alpha/2} = t_{0.025} = 2.05$

2. $\bar{x} - 2.05 \dfrac{s}{\sqrt{29}} = 11.1 - 2.05 \dfrac{2.8}{\sqrt{29}} = 10.03$

$\qquad \bar{x} + 2.05 \dfrac{s}{\sqrt{29}} = 11.1 + 2.05 \dfrac{2.8}{\sqrt{29}} = 12.17$

95-percent confidence interval: 10.03 to 12.17

Hypothesis tests for means

8.1 Introduction

In Chapter 7 we learned how to make an "educated guess" (inference) about a population mean, μ, by looking at the sample mean, $\bar{x}$, of a random sample of the population. In this chapter we will basically be looking at the reverse procedure. Namely, *we* (or someone) *will make an "educated guess" or* **hypothesis** *about μ, and then use $\bar{x}$ to make inferences concerning our hypothesis.* That is, we will use $\bar{x}$ to decide whether or not we believe that the hypothesis concerning μ is correct.

For example, the mayor of a city might claim that the mean income of all residents of the city is $\mu = \$18,000$. We will learn how to use the mean income, $\bar{x}$, of a random sample of residents of the city to decide whether or not we believe the claim made by the mayor that $\mu = \$18,000$.

Example 1 ◖ A wheat farmer has used a certain brand of fertilizer for the past twenty years. His experience with this fertilizer indicates that the mean yield using it was 40 bushels per acre.

Just recently the company that manufactured this brand of fertilizer went out of business and the farmer was forced to use a new fertilizer this season. He is, of course, interested in knowing *whether or not* the use of this new fertilizer will result in the *same* mean yield of wheat. Let μ be the actual mean yield of wheat, in bushels per acre, resulting from using the new fertilizer. In statistical terminology the farmer wants to *test the hypothesis that $\mu = 40$ bushels per acre against the hypothesis that $\mu \neq 40$ bushels per acre.* The hypothesis that $\mu = 40$ bushels per acre is called the **null hypothesis**. We write this compactly as:

$$\text{Null hypothesis:} \quad \mu = 40$$

The alternative that $\mu \neq 40$ bushels per acre is called the **alternative hypothesis**. We write this as:

$$\text{Alternative hypothesis:} \quad \mu \neq 40^* \;◗$$

Example 2 ◖ The manufacturer of a new model car claims that a "typical" car gets 35 mpg (miles per gallon). An independent consumer agency is somewhat skeptical of this claim, and thinks the gas mileage may very well be *less than* 35 mpg. In other words, the consumer agency wants to *test the hypothesis that the mean gas mileage, μ, equals 35 mpg against the hypothesis that $\mu < 35$ mpg.* So, in this case the null hypothesis is

$$\text{Null hypothesis:} \quad \mu = 35 \text{ mpg}$$

and the alternative hypothesis is

$$\text{Alternative hypothesis:} \quad \mu < 35 \text{ mpg}$$

The consumer agency plans to justify its opinion by doing mileage tests on a random sample of 30 ($n = 30$) of the manufacturer's new model cars. Following this, its

---

* We will discuss methods of choosing the null and alternative hypotheses at the end of Section 8.2.

statisticians will calculate the sample mean gas mileage, $\bar{x}$, of the 30 cars sampled. Using $\bar{x}$ they will then make their decision concerning the hypothesis test.

The question now is, just how will the statisticians use $\bar{x}$ to make their decision? The idea is simply this: We know that $\bar{x}$ should be approximately equal to μ.* So if, in fact, $\mu = 35$ (that is, the null hypothesis is true), then we expect $\bar{x}$ to be "about" 35 mpg. To put it another way, *if $\bar{x}$ is "too much smaller" than 35 mpg, then we would tend to doubt that $\mu = 35$ mpg, and be more willing to believe that $\mu < 35$ mpg.* The question now is "*How much smaller is 'too much smaller'?*"

If for instance, the mean gas mileage of the sample of 30 cars tested were 15 mpg (that is, $\bar{x} = 15$ mpg), then we would probably doubt that the null hypothesis is true. But, for example, if the mean gas mileage of the 30 cars tested were 34.8 mpg (that is, $\bar{x} = 34.8$ mpg), then we would probably be willing to believe that the discrepancy between $\bar{x} = 34.8$ and the null hypothesis value of $\mu = 35$ is due to the fact that only a *sample* of all the cars is being taken. (Remember that, although generally $\bar{x}$ is "close" to μ, it will rarely *equal* μ.)

The point of the preceding paragraph is that if the sample mean differs by a lot from the hypothesized value of the population mean (in the null hypothesis), then we would probably reject the null hypothesis. On the other hand, if the sample mean differs by only a little from the hypothesized value of the population mean, then we would probably be willing to accept the null hypothesis as being reasonable. The problem in making a decision arises when the difference between the sample mean and the hypothesized value of the population mean (in the null hypothesis) is neither very small nor very large.

So our question remains: "*How much smaller is 'too much smaller'?*" We need to establish a "cutoff point." That is, how small will $\bar{x}$ have to be before we are *unwilling to believe* that the null hypothesis is true (that is, $\mu = 35$ mpg), and decide in favor of the alternative hypothesis (that is, $\mu < 35$ mpg)? Mathematically speaking we want to find a number d, so that if $\bar{x} < d$, then we will reject the null hypothesis that $\mu = 35$ mpg. (See Fig. 8.1.)

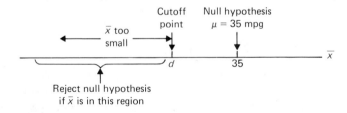

Figure 8.1

Before we discuss how to find the cutoff point d, it will be worthwhile to consider some possible outcomes of the consumer agency's sampling.

Suppose, for example, that the consumer agency's sampling of 30 cars yields an $\bar{x} = 34.8$ mpg with a sample standard deviation of $s = 1.45$ mpg. *If the null hypothesis*

---

* According to the law of large numbers (see page 161), if the sample size is "large" it is likely that $\bar{x}$ will be "close" to μ.

that $\mu = 35$ *mpg is true*, then we can compute the probability of getting an $\bar{x}$ of 34.8 mpg or less. We can do this computation because we know (see page 230) that

$$z = \frac{\bar{x} - \mu}{s/\sqrt{n}} = \frac{\bar{x} - 35}{1.45/\sqrt{30}} = \frac{\bar{x} - 35}{0.265}$$

has approximately the standard normal distribution. Now $\bar{x} = 34.8$ corresponds to

$$z = \frac{34.8 - 35}{0.265} \doteq -0.75,$$

so

$$P(\bar{x} \leqslant 34.8) \doteq P(z \leqslant -0.75) = 0.2266.$$

Thus, if the true mean *is* $\mu = 35$, then we can expect a value of $\bar{x}$ of 34.8 or less over 22 percent of the time. This is not strong evidence against the null hypothesis that $\mu = 35$ mpg.

On the other hand, suppose that the consumer agency's sampling of 30 cars yields an $\bar{x} = 34$ mpg with a sample standard deviation of $s = 1.45$ mpg. As above, if the null hypothesis that $\mu = 35$ mpg is true, then

$$z = \frac{\bar{x} - \mu}{s/\sqrt{n}} = \frac{\bar{x} - 35}{1.45/\sqrt{30}} = \frac{\bar{x} - 35}{0.265}$$

has approximately the standard normal distribution. Now, $\bar{x} = 34$ corresponds to

$$z = \frac{34 - 35}{0.265} \doteq -3.77$$

and therefore

$$P(\bar{x} \leqslant 34) \doteq P(z \leqslant -3.77) \doteq 0.0001.*$$

Thus, if the true mean *is* $\mu = 35$, then we would expect a value of $\bar{x}$ of 34 or less only about 0.01 percent of the time. So, either

1. The null hypothesis is true, and an *extremely unlikely* event occurred, or
2. The null hypothesis is false.

In such a case we would surely select (2) as the more reasonable choice. Consequently, the fact that $\bar{x}$ is 34 does constitute pretty strong evidence that the null hypothesis $\mu = 35$ mpg is false.

The above two computations indicate how we will decide on the "cutoff point," d. Basically, we must decide on how unlikely a value of $\bar{x}$ we will tolerate before we feel compelled to reject the null hypothesis. Generally statisticians use 1, 5, or 10 percent as limits for how unlikely a value of $\bar{x}$ they will tolerate, before rejecting the null hypothesis.

---

* The value $z = -3.77$ is not included in Table I. We used a different table to get $P(z \leqslant -3.77)$.

Suppose, for example, we decide on 5 percent (0.05). Then we choose the cutoff point d so that, *if the null hypothesis were true*, values of $\bar{x}$ less than d would occur only 5 percent of the time. Figure 8.2 illustrates the situation.

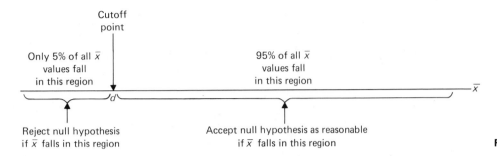

Cutoff
point

Only 5% of all $\bar{x}$
values fall
in this region

95% of all $\bar{x}$
values fall
in this region

$\bar{x}$

d

Reject null hypothesis
if $\bar{x}$ falls in this region

Accept null hypothesis as reasonable
if $\bar{x}$ falls in this region

Figure 8.2

In the language of probability this means that, *if the null hypothesis is true*,

$$P \text{ (rejecting null hypothesis)} = P(\bar{x} < d) = 0.05 \quad (5\%)$$

The number 0.05 (5%) is called the **significance level** of the test and is generally denoted by the Greek letter α. So in the above example, $\alpha = 0.05$ (5%).

If we decide on a 10 percent significance level ($\alpha = 0.10$), then we choose the cutoff point d so that, if the null hypothesis were true, values of $\bar{x}$ less than d would occur only 10 percent of the time. That is, we choose d so that, *if the null hypothesis is true*,

$$P \text{ (rejecting null hypothesis)} = P(\bar{x} < d) = 0.10$$

Figure 8.3 illustrates this case.

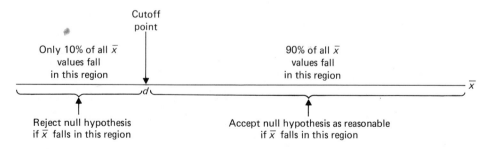

Cutoff
point

Only 10% of all $\bar{x}$
values fall
in this region

90% of all $\bar{x}$
values fall
in this region

$\bar{x}$

d

Reject null hypothesis
if $\bar{x}$ falls in this region

Accept null hypothesis as reasonable
if $\bar{x}$ falls in this region

Figure 8.3

As we shall see in the next section, it will not actually be necessary to find the cutoff point d explicitly, because we will carry out our hypothesis test by changing to z, the standard normal random variable. ◗

EXERCISES Section **8.1**

1 A manufacturer claims that the long-life battery the company produces lasts a mean of $\mu = 36$ months. A consumer agency is skeptical and thinks that $\mu < 36$ months. The agency wants to perform the hypothesis test:

Null hypothesis: $\mu = 36$
Alternative hypothesis: $\mu < 36$

The agency tests a random sample of 50 batteries and determines that the sample standard deviation of the lifetimes of these 50 batteries is $s = 3.1$ months. Would you be inclined to reject the null hypothesis if the sample mean life of these 50 batteries turned out to be

a) $\bar{x} = 35.67$? b) $\bar{x} = 34.90$?

[Use the techniques of Example 2 to answer these questions. That is, for (a) calculate $P(\bar{x} \leqslant 35.67)$ under the assumption that, in fact, $\mu = 36$. Similarly for (b).]

2 A teacher at a large state university thinks that the students in the College Algebra course study, on the average, more than six hours per week. She wants to test the hypotheses:

Null hypothesis: $\mu = 6$
Alternative hypothesis: $\mu > 6$,

where μ is the actual mean of the number of hours studied per week by College Algebra students. A random sample of 35 students yielded an $\bar{x} = 6.61$ hours with $s = 1.5$ hours.

a) Use the technique of Example 2 to calculate $P(\bar{x} \geqslant 6.61)$ under the null hypothesis $\mu = 6$.

b) Would you be inclined to reject the null hypothesis?

3 A rancher feels that his cattle have a mean weight of $\mu = 2000$ pounds. He wants to test the hypotheses:

Null hypothesis: $\mu = 2000$
Alternative hypothesis: $\mu \neq 2000$

He selects 40 cattle at random and finds that they have a mean weight of $\bar{x} = 1957.4$ pounds with $s = 90.6$ pounds.

a) Use the technique of Example 2 to find $P(\bar{x} \leqslant 1957.4)$ under the null hypothesis $\mu = 2000$.

b) Would you be inclined to reject the null hypothesis?

4 The methods we used in Section 8.1 are based on the assumption that $\bar{x}$ is (approximately) normally distributed. How large must the sample size be, if we are to be reasonably sure that this assumption is satisfied?

5 When performing a hypothesis test, it is possible to choose samples that lead us to erroneous conclusions. The errors that can be made are of two kinds.

Type I: The null hypothesis is true, but we accept the alternative hypothesis.

Type II: The alternative hypothesis is true, but we accept the null hypothesis (as being reasonable).*

The situation can be summarized in the chart below:

Hypothesis that is accepted	Hypothesis that is true	
	Null	Alternative
Null	Correct conclusion	Type II error
Alternative	Type I error	Correct conclusion

Describe in your own words what each of the following would mean for the teacher in Exercise 2:

a) Type I error b) Type II error

c) Correct conclusion (this can happen in two ways).

6 The rancher in Exercise 3 wishes to test the null hypothesis "$\mu = 2000$" against the alternative "$\mu \neq 2000$". Describe his conclusion as a *Type I error*, *Type II error*, or *correct* if:

a) He accepts the null hypothesis when, in fact, the mean is not 2000.

b) He concludes that the mean weight is not 2000, and, in fact, it isn't.

c) He concludes that $\mu \neq 2000$, but the mean weight actually is 2000.

---

* The material on Type I and Type II errors in this and later exercises will be covered in detail in Chapter 14.

In this section we will discuss, in detail, the procedure for setting up a hypothesis test concerning the mean, μ, of a population when the sample size is large. Before we begin, recall that the **significance level, α,** of a hypothesis test is the probability of rejecting the null hypothesis when it is actually true. We usually take $\alpha = 0.01$, 0.05, or 0.10.

In designing our hypothesis-testing procedure we can use the *standard normal distribution* because we know from Section 7.3 that if we take a random sample of size n ($n \geqslant 30$) from a population with mean μ, then the random variable

$$z = \frac{\bar{x} - \mu}{s/\sqrt{n}}$$

has approximately the *standard normal distribution*.

◖ Let's return, once again, to the gas mileage illustration of Example 2. The manufacturer of a new model car claims that the mean gas mileage, μ, of all his cars is 35 mpg. The consumer agency wants to test the hypotheses

Null hypothesis: $\mu = 35$ mpg
Alternative hypothesis: $\mu < 35$ mpg

The agency will base its decision on taking a random sample of 30 cars ($n = 30$), and finding the (sample) mean gas mileage, $\bar{x}$, of the 30 cars selected at random.

Moreover, the agency has decided to use a *significance level* of $\alpha = 0.05$. In other words, the agency wants the test designed so that, *if the null hypothesis is true* (that is, if $\mu = 35$ mpg), *then*

$$P(\text{rejecting null hypothesis}) = 0.05.$$

Now remember that the agency will reject the null hypothesis if $\bar{x}$ is "too small." However, *if the null hypothesis is true*, then the random variable

$$z = \frac{\bar{x} - \mu}{s/\sqrt{n}} = \frac{\bar{x} - 35}{s/\sqrt{30}}$$

has approximately the standard normal distribution. Since small values of $\bar{x}$ lead to small values of z and vice versa, to say that $\bar{x}$ is "too small" is the same as saying that z is "too small." Consequently, we need to find the cutoff point c, so that we will reject the null hypothesis if $z < c$. Since the significance level here is 0.05, this means that we want to find c so that $P(z < c) = 0.05$. Here c is called the **critical value** of the test, and the set of values for z that will cause us to reject the null hypothesis (in this case, $z < c$) is called the **rejection region** or **critical region**.

But by Table I, $P(z < -1.64) = 0.0505$. (This is as close to 0.05 as we can get by using Table I.) So the *critical value c* in this case is

$$c \doteq -1.64.$$

(The *rejection region* is that part of the horizontal axis under the shaded area in Fig. 8.4.)

Setting up a   **8.2**
hypothesis test:
large samples

Example **3**

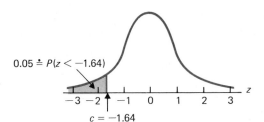

$0.05 \doteq P(z < -1.64)$

Figure 8.4

$c = -1.64$

Putting together all that we have done above, we can now state precisely how to carry out the hypothesis test: *From the sample mean $\bar{x}$ and sample standard deviation s of the gas mileage of the 30 cars selected at random, compute*

$$z = \frac{\bar{x} - 35}{s/\sqrt{30}}$$

If $z < -1.64$, then reject the null hypothesis. If $z \geq -1.64$ then accept the null hypothesis as being reasonable (see Fig. 8.5).

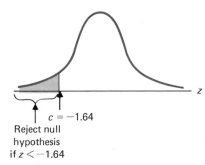

$c = -1.64$

Reject null
hypothesis
if $z < -1.64$

Figure 8.5

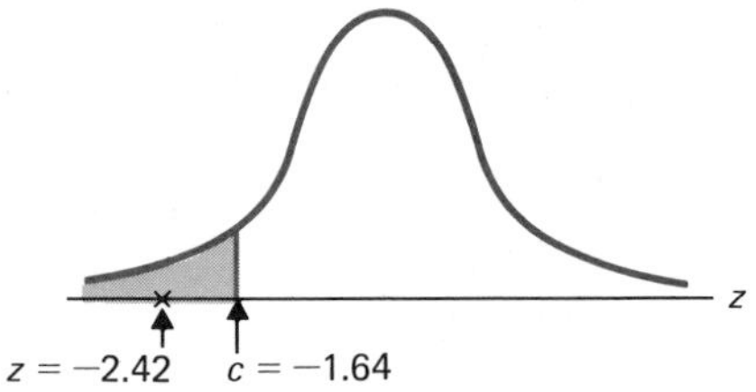

$z = -2.42$ $c = -1.64$

Figure 8.6

The results of the consumer agency's random sampling of 30 cars was

$$\bar{x} = 34.36 \text{ mpg}$$
$$s = 1.45 \text{ mpg}$$

So

$$z = \frac{\bar{x} - 35}{s/\sqrt{30}} = \frac{34.36 - 35}{1.45/\sqrt{30}} = -2.42$$

Since $z < -1.64$, we *reject* the null hypothesis and conclude that $\mu < 35$ mpg. (See Fig. 8.6) ▸

The hypothesis test in Example 3 is an example of what is called a **one-tailed** test. The reason for this terminology is that the null hypothesis is rejected only when $\bar{x}$ is "too far away" in *one direction* (in Example 3, "too much smaller") from the value of μ

given in the null hypothesis. Hypothesis tests in which the null hypothesis is rejected when $\bar{x}$ is "too far away" in *either direction* from the value of μ given in the null hypothesis are called **two-tailed** tests.

In this section and the next we will concentrate on *one-tailed* tests. In Section 8.4 we will discuss *two-tailed* tests. One-tailed tests are of two possible types. The first type is illustrated in Example 3. It takes the form:

Null hypothesis: $\mu = \mu_0$
Alternative hypothesis: $\mu < \mu_0$

In Example 3, $\mu_0 = 35$ mpg. Generally, μ_0 represents the value of μ given in the null hypothesis.

The second type of one-tailed test takes the form:

Null hypothesis: $\mu = \mu_0$
Alternative hypothesis: $\mu > \mu_0$

An illustration of this type of one-tailed test is given in Example 4.

◀ A tire dealer has been purchasing a certain brand of tire that he knows from experience will give good service for 35,000 miles, on the average. A different tire manufacturer has sent a representative to talk to the tire dealer. That representative tells the dealer that his company's brand of tire costs only slightly more than the brand the dealer is presently purchasing, and will *outperform* it. The dealer decides that it will be worth it for him to purchase the new brand of tire if the representative's claim is correct. He wants to do the following hypothesis test:

Null hypothesis: $\mu = 35,000$ mi,
Alternative hypothesis: $\mu > 35,000$ mi,

where μ is the mean tire life of the new brand of tire (here $\mu_0 = 35,000$). If the dealer can *reject* the null hypothesis, he will purchase the new brand of tire. ▶

Example **4**

An important question in setting up a hypothesis test is how to choose the hypotheses. That is, how does one decide what the null hypothesis should be and what the alternative hypothesis should be? Unfortunately there is no simple answer to this question. The choices of null and alternative hypotheses are somewhat subjective. These choices depend on such things as who is the tester, and what the tester wants to "prove."

In most examples and problems you consider in this text it should be fairly clear how to make such choices—but there is often room for disagreement. As you proceed through this chapter you should gain a better understanding of how to choose the null and alternative hypotheses. However, in order to help you on your way, we offer the following *guidelines*.

Choosing the hypotheses in a hypothesis test

1. *Null hypothesis:* In this text the null hypothesis (for tests concerning a population mean) will *always* be of the form

$$\mu = \mu_0,$$

(that is, μ *equals* a certain value) and never of the form $\mu < \mu_0$, $\mu > \mu_0$, or $\mu \neq \mu_0$. For instance, in Example 2 the null hypothesis is $\mu = 35$ mpg, and in Example 1 it is $\mu = 40$ bushels per acre. Consequently, the choice of the null hypothesis is almost always clear.

2. *Alternative hypothesis*: The choice for the alternative hypothesis can be broken down into two cases.

a) *One-tailed test*: If the tester believes, suspects, or wants to "prove" that the true value of μ is *smaller* than a value μ_0, then the *alternative hypothesis* will be $\mu < \mu_0$. For instance, in Example 2 a manufacturer claims that the mean gas mileage, μ, of all his new-model cars is 35 mpg. The consumer agency suspects that the mean gas mileage, μ, may very well be less (*smaller*) than 35 mpg. Thus, the *alternative hypothesis* is $\mu < 35$ mpg (that is, $\mu < \mu_0$).

On the other hand, if the tester believes, suspects, or wants to "prove" that the true value of μ is *larger* than a value μ_0, then the *alternative hypothesis* will be $\mu > \mu_0$. For instance, in Example 4, a tire dealer wants to know whether a new brand of tire has a mean tire life, μ, that is in excess of (*larger* than) 35,000 miles. If he can "prove" this, he will purchase the new brand of tire. Therefore, the *alternative hypothesis* is $\mu > 35,000$ miles (that is, $\mu > \mu_0$).

b) *Two-tailed test*: If the tester has no suspicion or indication whether the true value of μ is smaller or larger than a value μ_0, then the *alternative hypothesis* will be $\mu \neq \mu_0$. For instance, in Example 1 a wheat farmer is using a new fertilizer, since the brand he previously used is no longer available. From his past experience, he knows that the fertilizer he previously used resulted in a mean yield of 40 bushels per acre. The farmer does not know what mean wheat yield, μ, will result from using the new fertilizer, but he would certainly like to have some idea of whether or not it will give the same mean wheat yield of 40 bushels per acre as the fertilizer he used previously. Since the farmer has no suspicion or indication of whether the true mean wheat yield, μ, is smaller or larger than 40 bushels per acre, he might choose to do a hypothesis test with the *alternative hypothesis* $\mu \neq 40$ bushels per acre.*

Remember, the above paragraphs are just *guidelines* for choosing the null and alternative hypotheses in a hypothesis test.

Finally we should like to make the following comments. We said before that we usually use significance levels of $\alpha = 0.01, 0.05$, or 0.10. Since the significance level is the probability of rejecting a true null hypothesis, it is "unlikely" that the null hypothesis will be rejected when it is actually true. Consequently, *if we can reject the null hypothesis in a hypothesis test, then we can be reasonably confident that the alternative hypothesis is true.* On the other hand, if we *cannot* reject the null hypothesis, this does not prove that the null hypothesis is true, but simply means that it is reasonable. Therefore, when we say "accept the null hypothesis," we really mean "accept the null hypothesis as being a reasonable hypothesis" or "the null hypothesis cannot be rejected."

---

* This example is a good illustration of the point that the choice of hypotheses in a hypothesis test is subjective. The farmer might also wish to do a one-tailed test here.

EXERCISES Section **8.2**

1   A manufacturer orders thread with a breaking strength of 25 pounds. After the shipment arrives the manufacturer wishes to test the breaking strength of the thread in the shipment using a null hypothesis of $\mu = 25$ and an alternative hypothesis of $\mu < 25$.

  a) Write the formula for z that would be used in testing this hypothesis with a sample mean $\bar{x}$ from a sample of size 50.

  b) Find c so that $P(z < c) = 0.01$.

  c) Draw a figure similar to Fig. 8.5 to show which z-values would lead the manufacturer to reject the null hypothesis, at the 0.01 significance level.

  d) Suppose a sample of size 50 gives $\bar{x} = 24.9$ and $s = 0.4$. Would this lead you to reject the null hypothesis at the 0.01 significance level?

2   The national mean score on a reading test administered to third-graders is 3.0. A group of parents suspects that scores in their district are two low and wishes to compare the mean score, μ, achieved in the district against the national mean score. Their null hypothesis is $\mu = 3.0$, and their alternative hypothesis is $\mu < 3.0$.

  a) Write the formula for z that would be used to test this hypothesis from a sample mean $\bar{x}$ for a sample size of 66.

  b) Find c such that $P(z < c) = 0.05$.

  c) Draw a figure similar to Fig. 8.5 to show which z-values would lead the parents' group to reject the null hypothesis at the 0.05 significance level.

  d) Suppose the sample of 66 of their third-graders gives $\bar{x} = 2.9$ and $s = 0.6$. Would this lead you to reject the null hypothesis at the 0.05 significance level?

3   The mean hourly wage of a building trades journeyman in the U.S. is \$9.50. Union leaders in a southwest city wish to compare their mean hourly wage, μ, against the national average. Their null hypothesis is $\mu = 9.5$, and the alternative hypothesis is $\mu < 9.5$.

  a) Write the formula for z that would be used to test this hypothesis from a sample mean $\bar{x}$ for a sample size of 75.

  b) Find c such that $P(z < c) = 0.10$.

  c) Draw a figure similar to Fig. 8.5 to show which values would lead the union leaders to reject the null hypothesis at the 0.10 significance level.

  d) Suppose a sample size of 75 gives $\bar{x} = 9.38$ and $s = 0.40$. Would this lead you to reject the null hypothesis at the 0.10 significance level?

4   A type of air-conditioning unit consumes (on the average) 7.2 kilowatts of electricity per day. A consumer group interested in lowering energy costs wishes to test a new type of air conditioner against the other model. Their null hypothesis for the new type is $\mu = 7.2$, and their alternative hypothesis is $\mu > 7.2$ because they suspect the new units are *less energy-efficient*.

  a) Write the formula for z that would be used to test this hypothesis from a sample mean $\bar{x}$ for a sample size of 30.

  b) Find c such that $P(z > c) = 0.025$.

  c) Draw a figure similar to Fig. 8.5 to show which values would lead the consumer group to reject the null hypothesis at the 0.025 significance level.

  d) Suppose a sample size of 30 air conditioners gives $\bar{x} = 7.3$ and $s = 0.2$. Would this lead you to reject the null hypothesis at the 0.025 significance level?

5   The national mean salary for high-school teachers in the U.S. in 1976 was \$12,500. The teachers in a district suspected that they were *below* this mean. What are the null and alternative hypotheses they would have chosen to test in order to try to prove that their mean salary μ was lower?

6   The intermediate budget for a four-person family in the U.S. was \$15,318 in 1975. Economists think that budgets are higher now. What are the null and alternative hypotheses they would choose to test in order to try to prove that present budgets are *higher*?

7   The mean public-school expenditure per pupil in the U.S. in 1976 was \$1,388. A group of administrators in California think that California may be below that national average. What are the null and alternative hypotheses they would test in order to decide whether California is *below* that national average?

8   Consumer credit outstanding in October 1978 was estimated to be \$2,144 per capita. Economists believe the amount is still rising. What are the null and alternative hypotheses they would test to decide whether current consumer credit is *higher*?

9   In Exercise 5 of Section 8.1, we defined Type I error to be the error of rejecting the null hypothesis when it is actually true.

  a) If a hypothesis test has significance level α, what is the probability of a Type I error in that test?

  b) What is the probability of a Type I error in the hypothesis test in Exercise 1?

  c) What is the probability of a Type I error in the hypothesis test in Exercise 3?

10 *Probability of Type II error.* In Exercise 5 of Section 8.1 we
defined Type II error to be the error of accepting the null
hypothesis when, in fact, it is actually false. If the null
hypothesis is false, we can find the probability of a Type II error
only if we know the actual (true) value of μ. We will illustrate
below the computations involved in finding the probability of a
Type II error.

The mean cost of a hospital stay nationally in 1974 was $995. A
hospital administrator in Ohio decided to take a random
sample of $n = 50$ hospital stays in that state to test whether the
mean cost, μ, in Ohio was below the national mean. Specifically,
he wanted to test the hypotheses:

Null hypothesis: $\mu = \$995$
Alternative hypothesis: $\mu < \$995$

We will look at the case where "$\mu = \$995$" (the null hypothesis)
is false, because the true value of μ is $980.

a) Find the critical value , c, for the above hypothesis test, if the
significance level is chosen to be $\alpha = 0.05$.
b) From previous experience, the administrator knows that
$\sigma = \$30$. Using this fact and the fact that the sample size is
$n = 50$, find the cutoff value d for this test. That is, find the
value d such that the null hypothesis will be rejected if $\bar{x} < d$,
and it will be accepted if $\bar{x} \geq d$. (See Fig. 8.7.)

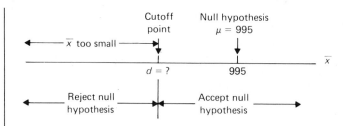

Figure 8.7

[*Hint:* $d = 995 + c(30/\sqrt{50})$. Why?]

c) If μ is actually $980, is the null hypothesis true or false?
d) If μ is actually $980 and you perform the hypothesis test,
what is the probability that $\bar{x}$ is at least d? That is, what is
the probability of accepting the null hypothesis, if in
fact $\mu = \$980$. [*Recall:* $\sigma = 30$ and $n = 50$.]
e) What is the probability of a Type II error for this test, if μ is
actually $980?
f) What is the probability of a Type I error for this test, if μ is
actually $980?

8.3 Hypothesis tests for a single mean— one-tailed tests for large samples

In this section we will summarize the procedure for doing a one-tailed hypothesis test
for μ. If we study Example 3 carefully, then we can write down a procedure for doing
one-tailed hypothesis tests. Remember that *we are assuming that the sample size is
at least 30 ($n \geq 30$).*

Procedure

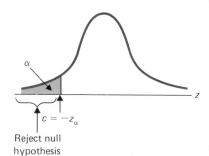

α

$c = -z_\alpha$

Reject null
hypothesis

Procedure for performing a hypothesis test of the form:
Null hypothesis: $\mu = \mu_0$
Alternative hypothesis: $\mu < \mu_0$

Assumption: Sample size is at least 30 ($n \geq 30$).

1. Decide on the *significance level*, α. Usually $\alpha = 0.01$, 0.05, or 0.10.
2. The *critical value* is
$$c = -z_\alpha$$
(See the figure at the left.) Use Table I to find z_α.
3. Compute the value of
$$z = \frac{\bar{x} - \mu_0}{s/\sqrt{n}}$$
where μ_0 is the value for μ given in the null hypothesis, n is the sample size, and $\bar{x}$
and s are the sample mean and sample standard deviation.
4. If $z < c$, *reject* the null hypothesis.

For the other type of one-tailed test the procedure is very similar.

Procedure

Procedure for performing a hypothesis test of the form:
Null hypothesis: $\mu = \mu_0$
Alternative hypothesis: $\mu > \mu_0$

Assumption: Sample size is at least 30 ($n \geqslant 30$).

1. Decide on the *significance level*, α.
2. The *critical value* is

$$c = z_\alpha$$

   (See the figure at the right.) Use Table I to find z_α.
3. Compute the value of

$$z = \frac{\bar{x} - \mu_0}{s/\sqrt{n}}$$

4. If $z > c$, *reject* the null hypothesis.

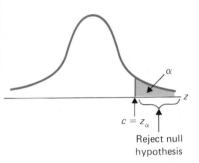

Remark: In some cases the population standard deviation, σ, will be known. In such cases *always* use σ instead of s in Step 3.

◀ The tire dealer of Example 4 decides that he will test 40 (10 sets) of the new brand of tire. He will then base his decision (whether or not to purchase the new brand of tire) on the outcome of his results with these 40 tires. Statistically speaking, he wants to perform the following hypothesis test:

Example **5**

Null hypothesis: $\mu = 35{,}000$ mi,
Alternative hypothesis: $\mu > 35{,}000$ mi,

where μ is the mean tire life of the new brand of tire. If the dealer can *reject* the null hypothesis, then he will purchase the new brand of tire. This, of course, is because rejection of the null hypothesis indicates that the new brand of tire will outperform the brand he presently sells.

   The dealer's test on the random sample of 40 tires of the new brand results in a sample mean life of $\bar{x} = 36{,}720.36$ miles, with a sample standard deviation of $s = 2{,}390$ miles. Since $n = 40$ ($\geqslant 30$), we can use the procedure given above to do the hypothesis test.

1. Decide on the *significance level*, α. The dealer wants to be very sure that he does not reject the null hypothesis if it is actually true, since if he rejects the null hypothesis, then he will purchase the new brand of tire, which costs him more than the old brand. And if the null hypothesis is actually true, the new brand does *not* outperform the old, less expensive brand.

What all this means is that he wants to make the significance level, α, very small. For remember, α *is the probability of rejecting the null hypothesis when it is actually true*. So the dealer decides on a significance level of 0.01. That is,

$$\alpha = 0.01$$

This means that the probability is only 0.01 that he will decide to purchase the new brand of tire when in fact the new brand does *not* outperform the old brand.

2. The *critical value* is $c = z_\alpha = z_{0.01}$. From Table I we find that $z_{0.01} \doteq 2.33$. Therefore,

$$c = 2.33$$

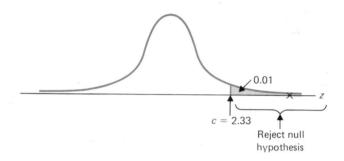

Figure 8.8

3. Compute the value of z:

$$z = \frac{\bar{x} - \mu_0}{s/\sqrt{n}}$$

We have $\mu_0 = 35{,}000$, $n = 40$, $\bar{x} = 36{,}720.36$, and $s = 2{,}390$. So

$$z = \frac{36{,}720.36 - 35{,}000}{2{,}390/\sqrt{40}} \doteq 4.55$$

We have marked this value of z with a cross in the above figure.

4. Since $z > c$ the dealer *rejects* the null hypothesis and will purchase the new brand. ◗

Exercise A   Perform the hypothesis test in Example 5 using the significance level $\alpha = 0.05$, if the results of the dealer's test on the 40 tires turned out to be $\bar{x} = 35{,}840.65$ miles and $s = 2420.38$ miles.

Example 6   ◖ A soft-drink manufacturer sells a "one-liter" bottle of soda. The FDA is concerned that the manufacturer may be "short-changing" the customer. Specifically, the manufacturer claims that the mean content, μ, of the bottles of soda he produces is one liter. The FDA is concerned that μ is less than one liter, and therefore decides to perform the hypothesis test:

Null hypothesis: $\mu = 1$ liter
Alternative hypothesis: $\mu < 1$ liter

at the 5 percent significance level.

A random sample of 100 bottles of soda is selected by the FDA, which then measures precisely the contents of each bottle. The results are

$$\bar{x} = 0.997 \text{ liters}$$

$$s = 0.021 \text{ liters}$$

Since $n = 100$ (≥ 30) the first procedure given above on page 266 can be used.

1. The *significance level* is $\alpha = 0.05$.
2. The critical value is $c = -z_\alpha = -z_{0.05}$. From Table I we find that $z_{0.05} \doteq 1.64$, so $c = -1.64$. See Fig. 8.9.
3. Compute the value of z:

$$z = \frac{\bar{x} - \mu_0}{s/\sqrt{n}} = \frac{0.997 - 1}{0.021/\sqrt{100}} \doteq -1.43$$

This value of z is marked with a cross on the figure at the right.
4. Since $z \geq c$, the FDA *cannot* reject the null hypothesis.

In other words, on the basis of the data, the FDA *cannot* reject, at the 0.05 significance level, the hypothesis that μ is one liter. ▶

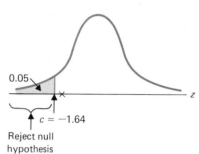

0.05

$c = -1.64$

Reject null
hypothesis

Figure 8.9

Perform the hypothesis test in Example 6 using the significance level $\alpha = 0.01$, if the data are $\bar{x} = 0.992$ and $s = 0.023$, for a sample of size $n = 90$.

Exercise **B**

EXERCISES Section **8.3**

For Exercises 1 through 7,

a) state the null hypothesis,
b) state the alternative hypothesis,
c) find the critical value,
d) draw a picture,
e) calculate the value of z,
f) state whether or not the null hypothesis should be rejected.

1 Consumer credit outstanding in October 1978 was estimated to be \$2,144 per person nationally. A local banker thought that the mean, μ, for the people in his city was *below* the national mean. He randomly sampled 50 people and found $\bar{x} = \$1,989$ and $s = \$278$. Perform an appropriate hypothesis test using the significance level $\alpha = 0.05$.

2 The best-selling glue can withstand a mean of 2,500 pounds of stress before its bonds break. A competitor claims, in a T.V. commercial, that its glue outperforms the best-seller. A consumer agency suspects that the new glue may be *weaker* than the best-selling glue, and tests 70 samples of the glue advertised in the commercial. The agency finds that $\bar{x} = 2481$ pounds and $s = 151.14$ pounds. Perform an appropriate hypothesis test using the significance level $\alpha = 0.025$.

3 A consumer protection group is interested in testing a brand of cereal because of the suspicion that the cereal contains *less* than the advertised (mean of) 18 ounces per box. They took a random sample of 40 boxes of the cereal. They found $\bar{x} = 17.98$ and $s = 0.11$. Perform an appropriate hypothesis test at the significance level $\alpha = 0.1$.

4 A politician running for office in California in 1976 thought that there might be an issue in per capita school expenditures. She thought that California's per capita expense μ might be *lower* than the national average. In 1976 the national mean was $314. A random sample of 40 districts in California gave $\bar{x} = \$311$ and $s = \$13$. Perform an appropriate hypothesis test at the significance level $\alpha = 0.1$.

5 In 1977, car dealers of a particular type of car sold a mean of 14 cars per dealership per month. The manufacturers of the car are interested in testing the results of their current advertising campaign to sell *more* cars. They took a sample of 90 dealerships. The information yielded $\bar{x} = 16$ cars per dealership and $s = 1$. Perform an appropriate hypothesis test at the significance level $\alpha = 0.05$.

6 In Arizona in 1974, the mean yield of cotton grown by a particular farmer from Gila Bend was 1,178 pounds per acre. The farmer was interested in finding out whether the new type of insecticide he used the next year *increased* his mean yield over the previous year. He sampled 30 acres, and found $\bar{x} = 1180$, and $s = 20.6$. Perform an appropriate hypothesis test at the significance level $\alpha = 0.025$.

7 The mean cost of a hospital stay nationally in 1974 was $996. A hospital administrator in Ohio sampled 75 hospital stays in that state in order to determine whether the mean cost, μ, in her state was *below* the national mean. She found $\bar{x} = \$967$ and $s = \$29.28$. Perform an appropriate hypothesis test at the significance level $\alpha = 0.05$.

8 A curriculum coordinator was told to decide whether or not to implement a new educational method in algebra classes at his district schools. The cost of the new method was more than the traditional one (currently in use), but was projected to improve performance. The coordinator knew that classes using the traditional method had a mean score of 80 on a standardized final exam. He decided to try the new method and use exam results from it to perform the test:

Null hypothesis: $\mu = 80$ (The new method is the same as the old.)

Alternative hypothesis: $\mu > 80$ (The new method is better.)

$\alpha = 0.05$

Suppose that he tried the new method in his district, gave the standardized final to the students taught by it, and found that the exam results were $\bar{x} = 80.5$, $s = 6$.

a) What conclusion would he reach if his sample size was
   i) $n = 36$? ii) $n = 64$? iii) $n = 900$?

b) Find a 95-percent confidence interval for μ if $n = 900$, $\bar{x} = 80.5$, and $s = 6$.

c) If $n = 900$, which gives the coordinator more useful information, a 95-percent confidence interval or the result of a hypothesis test with $\alpha = 0.05$?

9 In Exercise 10 of Section 8.2 we showed how to find the probability of a Type II error in a hypothesis test if we are given the actual value of μ. Suppose that we plan to use a sample of size $n = 100$ to perform the educational test in Exercise 8, and that we are sure that $\sigma = 6$ even before we collect our data.

a) Find the critical value, c, that would be used in this test if $\alpha = 0.05$.

b) Find the probability of a Type II error if the actual value of μ is 81. (It might be helpful for you to review Exercise 10 of Section 8.2.)

c) Find the probability of a Type II error if the actual value of μ is
   i) 82.0 ii) 82.5 iii) 82.6 iv) 82.7

d) What happens to the probability of a Type II error as the actual value of μ becomes larger? What is the practical importance of this to the coordinator?

8.4 Hypothesis tests for a single mean— two-tailed tests for large samples

In Sections 8.2 and 8.3 we discussed hypothesis tests that are one-tailed; that is, the alternative hypothesis is of the form $\mu < \mu_0$ or $\mu > \mu_0$. These are called *one-tailed* tests because the null hypothesis is rejected only when $\bar{x}$ is "too far away" in *one direction* from the value of μ given in the null hypothesis.

In this section we will discuss *two-tailed* tests, in which the null hypothesis is rejected when $\bar{x}$ is "too far away" in *either direction* from the value of μ given in the null hypothesis. They take the form:

Null hypothesis: $\mu = \mu_0$

Alternative hypothesis: $\mu \neq \mu_0$

The procedure for doing two-tailed tests is basically the same as that for one-tailed tests, except that in two-tailed tests there are *two critical values*, instead of one. As an illustration of a two-tailed test, we return to the wheat farmer in Example 1.

◀ Let μ be the actual mean yield of wheat, in bushels per acre, resulting from using the new fertilizer. The farmer wants to test the hypotheses: Example **7**

Null hypothesis: $\mu = 40$ bushels/acre
Alternative hypothesis: $\mu \neq 40$ bushels/acre.

To actually carry out this test the farmer decides to take a random sample of 100 yields of one-acre plots of wheat on which the new fertilizer was applied. If the sample mean yield, $\bar{x}$, of the 100 acres selected is "too far away" *in either direction* from 40 bushels per acre, then he will reject the null hypothesis (see Fig. 8.10).

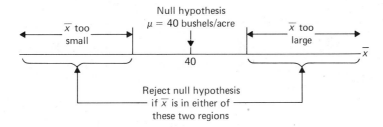

Figure **8.10**

As in one-tailed tests, how far away is "too far away" will depend on the significance level, α. The farmer decides to use a significance level of $\alpha = 0.05$. That is, if *the null hypothesis is true*, then he wants P(rejecting null hypothesis) = 0.05. But if the null hypothesis is true, then

$$z = \frac{\bar{x} - \mu}{s/\sqrt{n}} = \frac{\bar{x} - 40}{s/\sqrt{100}}$$

has approximately the standard normal distribution. So, to say that $\bar{x}$ is either too small or too large is the same as saying that z is either too small or too large. In other words, we need to find a number c so that if either $z < -c$ or $z > c$, then we will reject the null hypothesis. Since the significance level is 0.05, we want

$$P(z < -c \quad \text{or} \quad z > c) = 0.05$$

(or, equivalently, $P(z > c) = 0.025$). See Fig. 8.11.

From Fig. 8.11 we see that, in this case ($\alpha = 0.05$),

$$c = z_{0.025} = 1.96$$

In summary then, the procedure for doing this hypothesis test at the 0.05 significance level is: *From the sample mean, $\bar{x}$, and the sample standard deviation, s,*

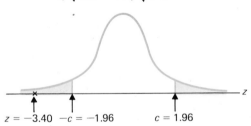

Figure 8.11

of the wheat yield of the 100 *one-acre plots selected at random, compute*

$$z = \frac{\bar{x} - 40}{s/\sqrt{100}}$$

If either $z < -1.96$ or $z > 1.96$, then reject the null hypothesis.

Now, it turned out that the sample mean yield of the 100 one-acre plots selected was $\bar{x} = 38.2$ bushels per acre, with a sample standard deviation of $s = 5.3$ bushels per acre. Thus,

$$z = \frac{\bar{x} - 40}{s/\sqrt{100}} = \frac{38.2 - 40}{5.3/\sqrt{100}} \doteq -3.40$$

Since $z = -3.40$, we see that $z < -1.96$, and consequently we *reject* the null hypothesis. The farmer concludes that the new fertilizer *does not* give the same mean yield of 40 bushels per acre as given by the old fertilizer.

The following procedure for performing two-tailed hypothesis tests can be obtained by studying Example 7. Remember that we are assuming that the sample size is at least 30.

Procedure

Procedure for performing a hypothesis test of the form:
Null hypothesis: $\mu = \mu_0$
Alternative hypothesis: $\mu \neq \mu_0$

Assumption: Sample size is at least 30 ($n \geqslant 30$).

1. Decide on the *significance level*, α. Usually $\alpha = 0.01, 0.05$, or 0.10.
2. The *critical values* are

$$-c = -z_{\alpha/2} \qquad \text{and} \qquad c = z_{\alpha/2}$$

Use Table I to find $z_{\alpha/2}$.

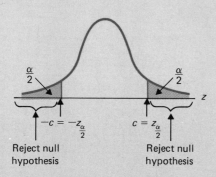

Reject null hypothesis   Reject null hypothesis

3. Compute the value of

$$z = \frac{\bar{x} - \mu_0}{s/\sqrt{n}}$$

4. If either $z < -c$ *or* $z > c$, *reject* the null hypothesis.

◀ The mayor of a city claims that the mean annual income, μ, of all the residents of his   Example **8**
city is \$18,000. A retail chain is thinking about locating several of its stores in the city.
Before it does, however, it wants to see whether the claim made by the mayor is
accurate. They decide to take a random sample of 250 incomes of the residents of the
city in order to test the hypotheses,

Null hypothesis: $\mu = \$18,000$
Alternative hypothesis: $\mu \neq \$18,000$

at the 0.05 significance level.

The results of the random sample of 250 incomes are

$$\bar{x} = \$18,225.15 \qquad \text{and} \qquad s = \$4,125.17$$

Use the procedure above, along with this data, to perform the hypothesis test. (*Note.*
Here $n = 250$ ($\geqslant 30$) so the procedure can be used.)

1. $\alpha = 0.05$.
2. The *critical values* here are $-c = -z_{\alpha/2} = -z_{0.025} = -1.96$ and $c = z_{\alpha/2} = z_{0.025} = 1.96$.

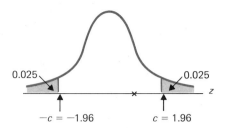

3. Compute the value of z:

$$z = \frac{\bar{x} - \mu_0}{s/\sqrt{n}} = \frac{18,225.15 - 18,000}{4,125.17/\sqrt{250}} \doteq 0.86$$

We have marked this value of z with a cross on the above figure.

4. Since $z = 0.86$, the retail chain *cannot* reject the null hypothesis that the mean income, μ, of all residents of the city is $18,000.

Exercise **C** Do the hypothesis test in Example 8, if a random sample of size 400 yielded an $\bar{x} = \$17,331.05$ and an $s = \$3920.62$. Use $\alpha = 0.05$.

EXERCISES Section **8.4**

For Exercises 1 through 8,
a) state the null hypothesis,
b) state the alternative hypothesis,
c) find the critical values,
d) draw a picture,
e) calculate the value of z, and
f) state whether or not the null hypothesis should be rejected.

1 A dairy cooperative has a machine for filling half-gallon cartons (64 fluid ounces) with milk. To ensure that, on the average, not too much or not too little milk is dispensed into each carton, the quality-control personnel randomly sampled 75 cartons filled by the machine. These 75 cartons were found to contain $\bar{x} = 63.97$ fluid ounces of milk and $s = 0.24$ fluid ounce. Perform an appropriate hypotheses test at the significance level $\alpha = 0.05$.

2 According to the *Statistical Abstract of the United States*, 1976, the average sentence in months imposed on criminals convicted in 1970 was 38.6 months. A law enforcement official was interested in knowing whether this figure had changed in 1975. A random sample of 100 prisoners sentenced in 1975 revealed $\bar{x} = 39.8$ and $s = 54$ months. Perform an appropriate hypothesis test at the significance level $\alpha = 0.1$.

3 The mean income of American families in 1977 was $18,264. An economist in Virginia took a random sample of 85 wage earners and calculated their mean income at $\bar{x} = \$18,671$ and $s = \$1986$. Perform an appropriate hypothesis test, at the significance level $\alpha = 0.05$, to determine whether the families in Virginia have incomes with $\mu = \$18,264$.

4 The economist in Exercise 3 also took a random sample of incomes of 85 residents of Washington, D.C. The results were $\bar{x} = \$17,751$ and $s = \$1758$. Perform an appropriate hypothesis test at the significance level $\alpha = 0.05$ to determine whether the residents of Washington, D.C., have incomes with $\mu = \$18,264$.

5 The estimated mean for personal savings accounts in the U.S. in 1976 was $454. A banker samples 120 residents of his area to see if the area's mean μ equals the national mean. The results yielded $\bar{x} = \$438$ and $s = \$101$. Perform an appropriate hypothesis test at the significance level $\alpha = 0.1$.

6 Do the hypothesis test in Exercise 5, if a sample size of 60 yielded $\bar{x} = \$442$ and $s = \$101$. (Use $\alpha = 0.1$.)

7 In 1974, an average of 1082 immigrants per day were admitted to the U.S. An immigration official was interested in comparing the mean influx μ in 1975 to that of 1974. He randomly selected 45 days and found $\bar{x} = 1059$ and $s = 81$. Perform an appropriate hypothesis test at the significance level $\alpha = 0.01$.

8 Do the hypothesis test in Exercise 7, at the significance level $\alpha = 0.01$, if a random sample of 45 days yielded $\bar{x} = 1049$ and $s = 63$.

9 a) Compute a 95-percent confidence interval for μ, the true mean volume of the contents of the milk cartons studied in Exercise 1.

b) The mean used in the null hypothesis of Exercise 1 was $\mu = 64$. Does 64 lie within the confidence interval found in (a)?

c) Compute a 99-percent confidence interval for μ, the true mean number of immigrants per day in Exercise 8.

d) The mean used in the null hypothesis of Exercise 8 was $\mu = 1082$. Does 1082 lie within the confidence interval found in (c)?

e) (*Hypothesis testing based on confidence intervals*) Based on your observations in (a) through (d) above, complete the following statements concerning a hypothesis test of "$\mu = \mu_0$" against "$\mu \neq \mu_0$" at the level α and based on sample data $\bar{x}$ and s:

i) If μ_0 lies in the $(1 - \alpha)$-level confidence interval for μ computed from $\bar{x}$ and s, (*accept*, *reject*) the null hypothesis.

ii) If μ_0 lies outside the $(1 - \alpha)$-level confidence interval for μ computed from $\bar{x}$ and s, (*accept*, *reject*) the null hypothesis.

Hypothesis tests concerning proportions follow basically the same pattern as those for means. The essential fact used in constructing hypothesis tests for proportions follows from the *central-limit theorem* (see Section 6.3, page 204). We state this fact below.

Hypothesis tests **8.5** concerning proportions for large samples

> If p is the population proportion, and $\hat{p}$ is the sample proportion, based on a sample of size n, then the random variable
>
> $$z = \frac{\hat{p} - p}{\sqrt{p(1 - p)/n}}$$
>
> has approximately the *standard normal distribution*. The approximation is good if both np and $n(1 - p)$ are at least 5.*

Using the fact stated above, we can perform hypothesis tests for proportions in the same manner as we did for means. The procedure is illustrated in the following example.

◀ A political incumbent received 58 percent of the vote during the last election. He feels that his four years in office have been good ones, and believes that his popularity has increased. To obtain relevant information concerning his belief, he decides to take a random sample of 300 voters to test the hypotheses:

Example **9**

---

* The random variable

$$\frac{\hat{p} - p}{\sqrt{p(1 - p)/n}}$$

is just the familiar $(\bar{x} - \mu)/(\sigma/\sqrt{n})$, when specialized to the situation of proportions.

Null hypothesis: $p = 0.58$
Alternative hypothesis: $p > 0.58$

where p is the actual (population) proportion of the vote he will receive in the upcoming election.

If $\hat{p}$ is the sample proportion of the 300 voters sampled who plan to vote for him, then $\hat{p}$ should be close to p.* So, if the null hypothesis ($p = 0.58$) is true, then $\hat{p}$ should be about 0.58. If $\hat{p}$ is "too much larger" than 0.58, then he can reject the null hypothesis, and conclude that $p > 0.58$ (i.e., his popularity has increased).

Now, if the *null hypothesis* is true, then

$$z = \frac{\hat{p} - p}{\sqrt{p(1-p)/n}} = \frac{\hat{p} - 0.58}{\sqrt{0.58(1-0.58)/300}}$$

has (approximately) the standard normal distribution.† So to say $\hat{p}$ is "too large" is the same as saying z is "too large." Consequently, we need to find the critical value, c, so that if $z > c$, then we will reject the null hypothesis. As usual we can determine c once we decide on the significance level, α.

The politician decides to use a significance level of $\alpha = 0.10$. So $c = z_{0.10} \doteq 1.28$. Therefore, if $z > 1.28$, *then he will reject the null hypothesis.*

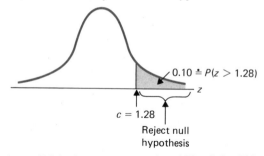

The result of the politician's survey was that 179 of the 300 voters sampled said they intended to vote for him. So the sample proportion $\hat{p}$ is equal to

$$\hat{p} = \frac{179}{300} \doteq 0.597$$

Consequently,

$$z = \frac{\hat{p} - 0.58}{\sqrt{0.58(1-0.58)/300}} = \frac{0.597 - 0.58}{\sqrt{0.58(1-0.58)/300}} \doteq 0.60$$

---

* This follows from the law of large numbers (see page 161). Also refer to the relative-frequency interpretation of probability (page 149).
† Note that both np and $n(1 - p)$ are at least 5, so that the normal approximation applies. In fact, $np = 300(0.58) = 174$ and $n(1 - p) = 300(1 - 0.58) = 126$.

Since $z \leqslant 1.28$, we cannot reject the null hypothesis. In other words, based on the survey of 300 voters, the politician is unable to conclude, at the 0.10 significance level, that his popularity has increased. ◗

Below we summarize the procedure illustrated in Example 9 for performing a hypothesis test concerning a proportion.

Procedure for performing a hypothesis test of the form: Procedure
Null hypothesis: $p = p_0$
Alternative hypothesis: $p > p_0$

1. Check that both np_0 and $n(1 - p_0)$ are at least 5.
2. Decide on the *significance level, α.*
3. The *critical value* is

$$c = z_\alpha$$

Use Table I to find z_α.

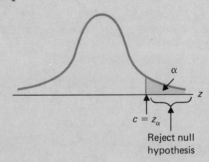

4. Compute the value of

$$z = \frac{\hat{p} - p_0}{\sqrt{p_0(1 - p_0)/n}}$$

where p_0 is the value given for p in the null hypothesis, n is the sample size, and $\hat{p}$ is the sample proportion.
5. If $z > c$, *reject* the null hypothesis.

Just as for hypothesis tests concerning means, the procedure for performing a hypothesis test concerning a proportion when the alternative hypothesis is of the form $p < p_0$ or $p \neq p_0$ is similar to that given for the alternative $p > p_0$.

The next example is an illustration of a *two-tailed* test concerning a proportion.

◗ On a roulette wheel there are 38 numbers, of which 18 are red, 18 are black, and 2   Example **10**
are green. If the wheel is "true," the probability of the ball landing on red is

$$p = \frac{18}{38} \doteq 0.474$$

A gambler has been observing a certain roulette wheel, and wonders if he can find an imperfection in the wheel so that he can improve his odds of winning. Specifically, he wants to test whether the probability of the ball landing on red is 0.474:

Null hypothesis: $p = 0.474$
Alternative hypothesis: $p \neq 0.474$

The gambler observes 200 spins of the wheel and finds that, out of these 200 trials, the ball lands on red 97 times. Using this data, perform the hypothesis test at the 0.05 significance level.

1. Check that both np_0 and $n(1 - p_0)$ are at least 5:

$$np_0 = 200(0.474) = 94.8,$$

and

$$n(1 - p_0) = 200(1 - 0.474) = 105.2$$

2. Decide on the *significance level*, α. We will use $\alpha = 0.05$.
3. For a two-tailed test there are *two critical values*. Namely, $-c = -z_{\alpha/2}$ and $c = z_{\alpha/2}$. In this case $\alpha = 0.05$. So, we need to find $z_{\alpha/2} = z_{0.025}$. From Table I, we find that $z_{0.025} = 1.96$. The critical values are

$$-c = -1.96 \qquad \text{and} \qquad c = 1.96$$

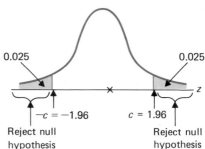

4. Compute the value of z:

$$z = \frac{\hat{p} - p_0}{\sqrt{p_0(1 - p_0)/n}}$$

We have $p_0 = 0.474$, $n = 200$, and

$$\hat{p} = \frac{97}{200} = 0.485$$

Consequently,

$$z = \frac{0.485 - 0.474}{\sqrt{0.474(1 - 0.474)/200}} \doteq 0.31$$

We have marked this value of z with a cross on the figure on page 278.

5.   So $z = 0.31$, and we *cannot* reject the null hypothesis.

In other words, on the basis of this information, the gambler must assume that the ball is landing on red about as often as it should with a "true" wheel. ◗

Perform the hypothesis test in Example 10 using the data $n = 300$ and $\hat{p} = 0.411$.   Exercise **D**

EXERCISES Section **8.5**

In Exercises 1 through 9,

a) calculate np_0,
b) calculate $n(1 - p_0)$,
c) state the null hypothesis,
d) state the alternative hypothesis,
e) find the critical value(s),
f) draw a picture,
g) calculate the z-value,
h) state whether or not the null hypothesis should be rejected.

1   In 1974, when a political incumbent ran for city council, 63.8 percent of the voters in the city thought that repair of city streets was an important issue. Since then, repairs have been effected in some areas of the city. The councilman wondered if attitudes on this issue had *changed*. In preparing for his 1978 campaign, the councilman polled 500 randomly selected voters. He found that 291 of these voters still thought repair of city streets was an important issue in the upcoming election. Perform an appropriate hypothesis test at the significance level $\alpha = 0.05$.

2   An incumbent U.S. Senator was running for another term. She was interested in knowing whether the percentage of registered Democrats in her state had *stayed the same*. In the previous election in which she had been involved, 52.8 percent of the registered voters in her state were Democrats. A recent poll indicated that, out of a random sample of 700 registered voters, 380 were registered Democrats. Perform the appropriate hypothesis test at the 0.05 significance level.

3   In 1974, the delinquent debt rate was 3.13 percent of the installment debt rate. A banker was interested in whether the new year had brought a *decrease* in this rate. He randomly sampled 500 debtors and found that 14 were delinquent. Perform an appropriate hypothesis test at the significance level $\alpha = 0.01$.

4   Should you perform the above hypothesis test in Exercise 3, if the sample size is 140?

5   At the beginning of the fall semester, a poll showed that 44 percent of the students in dormitories at a large university wanted loud music turned off at 10 o'clock on nights before school days. Proponents of the music curfew conducted a second poll in November hoping to show that the percentage had *increased*. They randomly sampled 120 people and found that 63 people favored the music curfew. Perform an appropriate hypothesis test at the significance level $\alpha = 0.05$.

6   Perform the above hypothesis test in Exercise 5, if the significance level is 0.01.

7   A drug was once shown to be 98 percent effective in curing a disease. After some years, however, medical researchers began to suspect that the drug had become *less effective* due to a mutation in the strain causing the disease. (The new strain appeared to be more resistant to the drug.) Medical researchers tested 1000 people with the disease and 930 were cured. Perform an appropriate hypothesis test at the significance level $\alpha = 0.05$.

8   In 1973, 45.7 percent of college faculty members in the U.S. earned over \$15,600. A female professor interested in determining whether women were participating equally in the better-paying jobs randomly sampled 300 female faculty members and found that 75 earned over \$15,600. Perform an appropriate hypothesis test at the significance level $\alpha = 0.05$.

9   In 1974, there were 1438 burglaries in a city with a population of 100,000. Before the end of the following year a law enforcement officer, interested in determining if the burglary rate was *increasing*, randomly sampled 400 people. The results yielded six people who had been burglarized. Perform an appropriate hypothesis test at the significance level $\alpha = 0.05$.

(*Using the exact distribution of $\hat{p}$.*) The theory given in this section applies only to large samples. In Exercises 10 and 11 we will look at a small-sample situation.

<u>10</u> A student is interested in the outcome of an election. The papers claim that 60 percent of all voters favor the incumbent and 40 percent favor her opponent. The student plans to take a random sample of only $n = 8$ voters.

a) Can we use the normal approximation? (Compute np and $n(1 - p)$, using $p = 0.6$.)

b) Complete the table below, using $p = 0.6$.

Number of voters for incumbent x	Probability $P(x)$	Proportion of sample for incumbent $\hat{p} = x/8$
0		
1		
2		
3		
4		
5		
6		
7		
8		

c) The probabilities of the x-values follow the binomial distribution with $n = 8$ and $p = 0.6$. What is the probability distribution of $\hat{p}$?

d) Find $P(\hat{p} \leqslant 0.25)$, the probability that at most 25 percent of the student's small sample favor the incumbent, if the true value of p is 0.6.

<u>11</u> Suppose that the student in Exercise 10 felt that the estimate of 60 percent for the incumbent was too high and wished to perform a test of the form:

Null hypothesis: $p = 0.6$
Alternative hypothesis: $p < 0.6$

where p is the true proportion of voters who will vote for the incumbent.

a) If his sample contained only two voters for the incumbent, then $\hat{p} = 0.25$. Using the result from part (d) of Exercise 10 about $P(\hat{p} \leqslant 0.25)$, what conclusion would you make for this hypothesis test at the level
   i) $\alpha = 0.05$?   ii) $\alpha = 0.025$?   iii) $\alpha = 0.01$?

b) If you disregarded the fact that np_0 and $n(1 - p_0)$ were less than 5, and used the normal-approximation methods of this section, what conclusion would you make for this hypothesis test at the level
   i) $\alpha = 0.05$?   ii) $\alpha = 0.025$?   iii) $\alpha = 0.01$?

c) Compare the conclusions for (a) and (b). Discuss the results.

8.6 Hypothesis tests for differences between means for independent samples —large samples

Up to this point we have considered hypothesis tests for a single population mean. In this section we will discuss hypothesis tests that allow us to compare the values of *two* population means. Our test will be based on the information we obtain from taking two *independent* random samples from the two populations under consideration. By *independent* samples we mean that the selection of one sample has no effect on the selection of the other. For example, in comparing two toothpastes, we could select two groups of 100 people each, and let one group use the first toothpaste, and the other group use the second. We could then compare the results of using each of the two toothpastes by comparing each of the two groups as a whole.

As opposed to independent samples, we may have *dependent* or *paired* samples. This is the type of sampling that occurs in "before and after" experiments. For example, in comparing the two toothpastes, we could select a single group of 100 people. The group could then use the first toothpaste for a while, and then switch to the second. We could then measure the difference in results, person by person, and use this data as a basis for comparing the two toothpastes.

In this section we will consider hypothesis tests for differences between means for *independent samples*. In the next section, hypothesis tests for differences between means for paired (dependent) samples will be studied.

Example 11 ◖ A manufacturer of a new toothpaste claims that his toothpaste is more effective than the leading brand. That is, the manufacturer claims that the mean number of cavities using his brand will be fewer than the mean number of cavities using the leading brand.

For ease in reference, let us call the new brand, Brand I, and the leading brand, Brand II. Let μ_1 be the mean number of cavities using Brand I, and μ_2 the mean number using Brand II. Then we want to perform the hypothesis test:

Null hypothesis: $\mu_1 = \mu_2$
Alternative hypothesis: $\mu_1 < \mu_2$

As usual, we will base our decision on the results of sampling. In this case we take two (independent) random samples of individuals. The first group uses Brand I, and the second, Brand II. Following the use of the two toothpastes, the (sample) mean number of cavities will be computed for each group. Here $\bar{x}_1$ denotes the sample mean for Brand I, and $\bar{x}_2$ the sample mean for Brand II. If $\bar{x}_1$ is "too much" smaller than $\bar{x}_2$, then we will reject the null hypothesis in favor of the alternative hypothesis. Just as in our previous hypothesis tests, we need to decide how much smaller is "too much" smaller. ▶

We shall return to this example in a moment. But first we present the essential fact needed in performing such a hypothesis test. This fact is a consequence of the central-limit theorem.

Let $\bar{x}_1$ be the sample mean of a random sample of size n_1 from a population with mean μ_1. Let $\bar{x}_2$ be the sample mean of a random sample of size n_2 from a population with mean μ_2. Also assume that the two samples are independent. If n_1 and n_2 are both at least 30, then the random variable

$$z = \frac{(\bar{x}_1 - \bar{x}_2) - (\mu_1 - \mu_2)}{\sqrt{(s_1^2/n_1) + (s_2^2/n_2)}}$$

has approximately the *standard normal distribution*. Here s_1 and s_2 are the sample standard deviations for the respective samples.*

◀ Now, *if the null hypothesis is true* (that is, $\mu_1 = \mu_2$), then by the above,

$$z = \frac{\bar{x}_1 - \bar{x}_2}{\sqrt{(s_1^2/n_1) + (s_2^2/n_2)}}$$

Example **11**
(continued)

has approximately the *standard normal distribution* (provided n_1 and n_2 are both at least 30). So to say that $\bar{x}_1$ is "too much" smaller than $\bar{x}_2$ (that is, $\bar{x}_1 - \bar{x}_2$ is "too small"), is the same as saying that z is "too small."

---

* The quantity $\sqrt{(s_1^2/n_1) + (s_2^2/n_2)}$ is used as an *estimate* for the standard deviation of the random variable $\bar{x}_1 - \bar{x}_2$. The *actual* standard deviation of $\bar{x}_1 - \bar{x}_2$ is $\sqrt{(\sigma_1^2/n_1) + (\sigma_2^2/n_2)}$. In Exercise 10 you will be asked to verify that the random variable

$$\frac{(\bar{x}_1 - \bar{x}_2) - (\mu_1 - \mu_2)}{\sqrt{(s_1^2/n_1) + (s_2^2/n_2)}}$$

has approximately the standard normal distribution.

Consequently, we need to find the critical value c, so that if $z < c$, then we will reject the null hypothesis. As usual we can determine c once we decide on the significance level, α. For this problem we shall use $\alpha = 0.05$. Then we need to find c so that $P(z < c) = 0.05$. Using Table I, $c \doteq -1.64$. See Fig. 8.12.

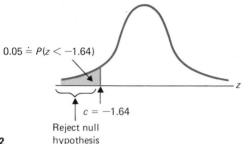

$0.05 \doteq P(z < -1.64)$

$c = -1.64$

Reject null
hypothesis

Figure 8.12

The hypothesis test can now be carried out as follows: *Using the values $\bar{x}_1, \bar{x}_2, s_1$, and s_2 obtained from the sampling, compute*

$$z = \frac{\bar{x}_1 - \bar{x}_2}{\sqrt{(s_1^2/n_1) + (s_2^2/n_2)}}$$

If $z < -1.64$, then reject the null hypothesis.

For this problem the following data were obtained:

Brand I: $n_1 = 100$, $\bar{x}_1 = 0.88$, $s_1 = 0.96$
Brand II: $n_2 = 150$, $\bar{x}_2 = 1.12$, $s_2 = 0.94$

Thus,

$$z = \frac{\bar{x}_1 - \bar{x}_2}{\sqrt{(s_1^2/n_1) + (s_2^2/n_2)}} = \frac{0.88 - 1.12}{\sqrt{[(0.96)^2/100] + [(0.94)^2/150]}} \doteq -1.95$$

Since $z < -1.64$, we *reject* the null hypothesis and conclude that $\mu_1 < \mu_2$. That is, based on the testing of the toothpastes, we can conclude that the new toothpaste (Brand I) is more effective in fighting cavities than the leading brand (Brand II). ▶

Below we will summarize the procedure used in Example 11.

Procedure

Procedure for performing a hypothesis test of the form:
Null hypothesis: $\mu_1 = \mu_2$
Alternative hypothesis: $\mu_1 < \mu_2$

Assumptions: (1) Both samples are of size at least 30, and (2) the samples are independent.

1. Decide on the *significance level, α.*
2. The *critical value* is

$$c = -z_\alpha$$

Use Table I to find z_α. (See the figure at the top of page 283.)

3. Compute the value of

$$z = \frac{\bar{x}_1 - \bar{x}_2}{\sqrt{(s_1^2/n_1) + (s_2^2/n_2)}}$$

where

n_1 = sample size from the first population,
$\bar{x}_1$ = sample mean from the first population,
s_1 = sample standard deviation from the first population,

and n_2, $\bar{x}_2$, s_2 are the corresponding values from the second population.
4. If $z < c$, reject the null hypothesis.

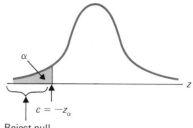

Reject null
hypothesis

A scientist working for a company that produces lightbulbs believes she has developed a new type of filament that will prolong the life of the lightbulbs. The new filament costs slightly more to make than the filament presently in use. So the president of the company does not want to use the new filament unless there is sufficient evidence to indicate that the new filament definitely increases the life of the lightbulbs.

Example **12**

The president decides to consult with a statistician in order to determine whether to use the new type of filament. The statistician lets μ_1 denote the (actual) mean life of the bulbs using the present filament, and μ_2 the mean life of the bulbs using the newly developed filament. He wants to perform the test:

Null hypothesis: $\mu_1 = \mu_2$
Alternative hypothesis: $\mu_1 < \mu_2$

If the statistician can reject the null hypothesis, then he will tell the president that the new filament is superior.

The statistician decides to take random samples of 100 lightbulbs using the presently used filament, and 100 lightbulbs equipped with the new type of filament. The results of his testing are as follows:

$$\bar{x}_1 = 1214.52 \qquad \bar{x}_2 = 1260.09$$
$$s_1 = 119.86 \qquad s_2 = 123.57$$
$$n_1 = 100 \qquad n_2 = 100$$

Since both samples are of size at least 30 ($n_1 = 100$, $n_2 = 100$), and the samples are obviously independent, we can perform the hypothesis test using the procedure described above.

1. Decide on the *significance level*, α. Since the president does not want to use the new filament unless there is strong evidence that it is superior, the statistician chooses $\alpha = 0.01$.
2. The *critical value* is $c = -z_\alpha = -z_{0.01} = -2.33$.
3. $z = \dfrac{\bar{x}_1 - \bar{x}_2}{\sqrt{(s_1^2/n_1) + (s_2^2/n_2)}} = \dfrac{1214.52 - 1260.09}{\sqrt{[(119.86)^2/100] + [(123.57)^2/100]}} = -2.65$
4. So $z = -2.65$ and $c = -2.33$. Consequently, $z < c$, and we *reject* the null hypothesis.

In other words, the statistician will report to the president that there is strong evidence that the new filament improves the life of the lightbulbs. ▶

Exercise E   Perform the hypothesis test in Example 12 using the following data: $\bar{x}_1 = 1217.64$, $s_1 = 119.86$, $n_1 = 100$, $\bar{x}_2 = 1258.85$, $s_2 = 123.57$, and $n_2 = 150$. Use $\alpha = 0.01$.

For a one-tailed test with alternative hypothesis $\mu_1 > \mu_2$, the procedure is the same as that described on page 282 except that the critical value is $c = z_\alpha$ and the rejection region is $z > c$.

We can also perform *two-tailed* tests for the difference of two means. The procedure is the same as that for one-tailed tests except that we use *two* critical values, instead of one.

Example 13   ◀ An education professor at a large university was interested in trying a new method of instruction that involves more student participation than the classical lecture method.

To compare the two methods of instruction, she taught two sections of the same course. The mean final grades in each of the two sections were as follows:

Section 1 (lecture method)	*Section* 2 (new method)
$\bar{x}_1 = 72.3$	$\bar{x}_2 = 74.6$
$s_1 = 9.1$	$s_2 = 8.8$
$n_1 = 35$	$n_2 = 44$

Test, at the 0.05 significance level, whether or not the methods are equally effective.

Solution   Let μ_1 be the mean grade for all students who have taken or will take the course by the lecture method. Let μ_2 be the corresponding mean for the new method. We want to test:

Null hypothesis: $\mu_1 = \mu_2$
Alternative hypothesis: $\mu_1 \neq \mu_2$

$\alpha = 0.05$

We can use the two-tailed analogue of the procedure described on page 282, since both samples are of size at least 30 ($n_1 = 35$, $n_2 = 44$), and the samples are obviously independent.

1. We will use $\alpha = 0.05$.
2. For a two-tailed test there are *two critical values.* Namely, $-c = -z_{\alpha/2}$ and $c = z_{\alpha/2}$. In this case $\alpha = 0.05$. By Table I, $z_{\alpha/2} = z_{0.025} = 1.96$. So the critical values are $-c = -1.96$ and $c = 1.96$. See Fig. 8.13.
3. $z = \dfrac{\bar{x}_1 - \bar{x}_2}{\sqrt{(s_1^2/n_1) + (s_2^2/n_2)}} = \dfrac{72.3 - 74.6}{\sqrt{[(9.1)^2/35] + [(8.8)^2/44]}} \doteq -1.13$

We have marked this value of z with a cross on Fig. 8.13.

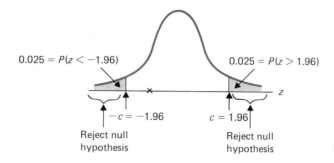

$0.025 = P(z < -1.96)$ $0.025 = P(z > 1.96)$

z

$-c = -1.96$ $c = 1.96$

Reject null   Reject null
hypothesis   hypothesis   **Figure 8.13**

4. So $z = -1.13$ and we *cannot* reject the null hypothesis.

In other words, on the basis of the information obtained from the two sections, there is insufficient evidence to conclude that there is a difference in effectiveness in the two methods of teaching. ▶

Perform the hypothesis test in Example 13 using the following data: Exercise **F**

Lecture method	New method
$\bar{x}_1 = 72.3$	$\bar{x}_2 = 78.9$
$s_1 = 9.1$	$s_2 = 8.8$
$n_1 = 35$	$n_2 = 32$

Hypothesis tests for differences between two proportions are similar to the tests described above for two means. The details for tests concerning two proportions will be covered in the exercises.

EXERCISES Section **8.6**

In Exercises 1 through 8,
a) state the null hypothesis,
b) state the alternative hypothesis,
c) draw a picture,
d) compute z,
e) state whether or not the null hypothesis should be rejected.

1 A mathematics professor at a large university read some research literature on teaching techniques. He was led to think that if he used mastery testing in College Algebra he could increase learning by the students. He taught two sections of the same course. He set up an experiment where mastery testing was done in one class and not in the other. The mean final exam grade in the mastery testing section (Section 2) with 50 students was $\bar{x}_2 = 82.1$, with $s_2 = 8.2$. The mean final exam grade in the traditional section (Section 1) with 56 students was $\bar{x}_1 = 73.4$, with $s_1 = 10.8$. Test, at the 0.05 significance level, whether the mastery testing method is *more effective* than the professor's traditional method.

2 A researcher at a college of agriculture was interested in determining whether a larger crop could be harvested by introducing sterilized males of an insect pest to control the pest population instead of using insecticides. Eighty different one-acre plots in the same county were used for the experiment. The $n_2 = 40$ plots where sterilized males were used had a mean yield of $\bar{x}_2 = 446$ bushels per acre with $s_2 = 17.8$ bushels per acre. The yield on the $n_1 = 40$ plots that were sprayed was $\bar{x}_1 = 438$ bushels per acre with $s_1 = 16.4$ bushels per acre. Test, at the 0.05 significance level, whether the method of using sterilized males is *more effective* than spraying with insecticide.

3 A regional sales manager is interested in studying the effectiveness of a new training program aimed at *increasing* sales. The manager chose two sales offices for a test. One office received the training; the other office did not. Office 2, which received the training program, had 51 salespeople. The mean of the total amount of sales per person for the month at this office

was $\bar{x}_2 = \$3,229$, with $s_2 = \$106.18$. The 47 people in Office 1 received no training. The mean of the total amount of sales per person for the month in Office 1 was $\bar{x}_1 = \$3,197$ with $s_1 = \$101.93$. Determine, at the 0.10 significance level, whether the training program *increases sales effectiveness*.

4 Once a week for five weeks, a counselor lectured a group of 31 juniors in a high-school social studies class on how to study American history. Another class of 30 students did not receive the lectures. The head of the social studies department was interested in determining whether the mean final exam grade would be *higher* for students who received the lectures. The group that received the lectures, Group 2, had a mean grade $\bar{x}_2 = 79.7$, with $s_2 = 7.2$. The other group, Group 1, had a mean grade $\bar{x}_1 = 76.5$, with $s_1 = 6.8$. Determine at the 0.05 significance level, whether the lectures are likely to produce *better* final grade results for similar classes.

5 The head of the department in Exercise 4 was interested in determining whether students who receive the lectures study a *longer* amount of time than students who don't. A survey indicated that, for the 31 students in Group 2, $\bar{x}_2 = 1.3$ hours per week, with $s_2 = 0.15$. For the 30 students in Group 1, $\bar{x}_1 = 1.55$, with $s_1 = 0.5$. Determine, at the 0.05 significance level, whether students who receive the lectures are likely to study *longer*.

6 A supplier of roofing materials wanted to test whether there was a *difference* in roof life between homes in the snow belt and homes in the sun belt. A group of 30 homes with shake roofs in Aspen, Colorado, were compared with a group of 30 similarly roofed homes in Scottsdale, Arizona. The results were:

Aspen	Scottsdale
$\bar{x}_1 = 15.9$ years	$\bar{x}_2 = 18.2$ years
$s_1 = 3.3$ years	$s_2 = 0.8$ year
$n_1 = 30$	$n_2 = 30$

Determine, at the 0.10 significance level, whether there is a difference in roof life between the two climates.

7 Researchers in obesity wanted to test the effectiveness of dieting and exercise against dieting without exercise. A group of 73 patients were divided into two groups. Group 1, numbering 37 patients, was put on a program of dieting and exercise. Group 2, numbering 36 patients, dieted only. The results in weight loss per person after two months are summarized below.

Diet and exercise group	Diet only group
$\bar{x}_1 = 16.8$ pounds	$\bar{x}_2 = 17.1$ pounds
$s_1 = 0.9$ pound	$s_2 = 1.0$ pound
$n_1 = 37$	$n_2 = 36$

Determine, at the 0.05 significance level, whether there is a *difference* between the two treatments.

8 The researchers in Exercise 7 were also interested in the change in the waistlines of the patients in the two groups. They took the waist measurements of each group before the experiment. After two months they measured the groups again. The results of the mean loss per person per group is summarized below:

Diet and exercise	Diet only
$\bar{x}_1 = 4.0$ inches	$\bar{x}_2 = 3.5$ inches
$s_1 = 1.0$ inch	$s_2 = 0.5$ inch
$n_1 = 37$	$n_2 = 36$

Determine, at the 0.05 significance level, whether the diet-and-exercise method is *superior* to the diet-only method in reducing waistlines.

9 (*Confidence intervals for difference between two means.*) A $(1 - \alpha)$-level confidence interval for the difference of two population means can be obtained by taking independent random samples of sizes n_1 and n_2 from the two populations and writing the inequality

$$(\bar{x}_1 - \bar{x}_2) - z_{\alpha/2}\sqrt{(s_1^2/n_1) + (s_2^2/n_2)} \leqslant \mu_1 - \mu_2$$
$$\leqslant (\bar{x}_1 - \bar{x}_2) + z_{\alpha/2}\sqrt{(s_1^2/n_1) + (s_2^2/n_2)}$$

a) Find a 95-percent confidence interval for the difference between mean weight loss due to dieting and exercise (μ_1) and mean weight loss due to diet alone (μ_2) in Exercise 7.

b) Use your results from (a) and Exercise 7 to complete the following statement:

A hypothesis test of $\mu_1 = \mu_2$ against $\mu_1 \neq \mu_2$, based on data from two independent random samples, will lead to acceptance of the null hypothesis $\mu_1 = \mu_2$ at the level α, if and only if the number ____ lies in the $(1 - \alpha)$-level confidence interval for $\mu_1 - \mu_2$.

10 In this exercise you will be asked to verify that the random variable

$$\frac{(\bar{x}_1 - \bar{x}_2) - (\mu_1 - \mu_2)}{\sqrt{(s_1^2/n_1) + (s_2^2/n_2)}}$$

has (approximately) the standard normal distribution, if two independent random samples of sizes n_1 and n_2 (both at least 30) are taken from populations with means μ_1 and μ_2, respectively.

a) From Section 6.2 we know that $\mu_{\bar{x}} = \mu$. Use this fact, along with $\mu_{x+y} = \mu_x + \mu_y$ (Exercise 14, Section 4.9) and $\mu_{kx} = k\mu_x$ (Exercise 16, Section 4.9), to show that $\mu_{\bar{x}_1 - \bar{x}_2} = \mu_1 - \mu_2$.

b) From Section 6.2 we know that $\sigma_{\bar{x}} \doteq \sigma/\sqrt{n}$. Use this fact, along with $\sigma_{x+y}^2 = \sigma_x^2 + \sigma_y^2$ for independent random variables (Exercise 14, Section 4.9) and $\sigma_{kx}^2 = k^2 \sigma_x^2$ (Exercise 16, Section 4.9), to show that

$$\sigma_{\bar{x}_1 - \bar{x}_2} \doteq \sqrt{(\sigma_1^2/n_1) + (\sigma_2^2/n_2)}.$$

c) Assuming that the difference between two independent normally distributed random variables is normally distributed, use the results from (a) and (b), and the central-limit theorem, to show that

$$\frac{(\bar{x}_1 - \bar{x}_2) - (\mu_1 - \mu_2)}{\sqrt{(s_1^2/n_1) + (s_2^2/n_2)}}$$

has (approximately) the standard normal distribution if both n_1 and n_2 are at least 30.

11 (*Hypothesis tests for the difference between two proportions.*) Hypothesis tests for the difference between two proportions are very similar to those for the difference between two means. Such tests are based on the following fact, which follows from the central-limit theorem. Suppose two independent random samples of sizes n_1 and n_2 are taken from two populations with population proportions p_1 and p_2. Let $\hat{p}_1$ and $\hat{p}_2$ be the sample proportions. Then for relatively large sample sizes n_1 and n_2, the random variable

$$z = \frac{(\hat{p}_1 - \hat{p}_2) - (p_1 - p_2)}{\sqrt{[p_1(1 - p_1)/n_1] + [p_2(1 - p_2)/n_2]}}$$

has (approximately) the standard normal distribution.

a) If the null hypothesis is $p_1 = p_2$ (that is, the population proportions are equal), then show that the above random variable becomes

$$z = \frac{\hat{p}_1 - \hat{p}_2}{\sqrt{p(1 - p)[(1/n_1) + (1/n_2)]}}$$

where p is the common population proportion.

We cannot use the random variable in (a) as a test statistic since p is *unknown*. Since the null hypothesis is $p_1 = p_2 = p$, the best estimate of p is not obtained by using $\hat{p}_1$ or $\hat{p}_2$ separately, but by "pooling" these values by taking their "weighted average," the weights being in proportion to the sample sizes. This weighted average of $\hat{p}_1$ and $\hat{p}_2$ is

$$p^* = \frac{n_1 \hat{p}_1 + n_2 \hat{p}_2}{n_1 + n_2}$$

Thus the test statistic that is used with a null hypothesis of $p_1 = p_2$ is

$$z = \frac{\hat{p}_1 - \hat{p}_2}{\sqrt{p^*(1 - p^*)[(1/n_1) + (1/n_2)]}}$$

b) A political pollster is collecting data to measure voter opinion on an upcoming referendum. Specifically he is interested in whether the voter opinion is the *same* in two key districts. He obtains the following data:

Will Vote:

District number	Yes	No	Total
11	20	30	50
15	43	57	100

Let p_1 be the actual proportion of District 11 that will vote *yes*, and p_2 be the actual proportion of District 15 that will vote *yes*. Using the test statistic given above, perform the hypothesis test:

Null hypothesis: $p_1 = p_2$
Alternative hypothesis: $p_1 \neq p_2$

Use $\alpha = 0.10$.

In the previous section we discussed hypothesis tests for differences between means for *independent* samples. The method presented there does *not* apply to situations such as "before and after" experiments. For, in such situations, the samples are *dependent*. In fact, in such cases, instead of having two random samples, there is *one* random sample of pairs.

A team of medical researchers has developed a new exercise program that they feel will be helpful in reducing hypertension (high blood pressure). In order to test this, the

Hypothesis tests 8.7 for differences between means for paired samples —large samples

Example **14**

team selected 35 hypertensive individuals and subjected them to the exercise program. The results of the tests are given in Table 8.1.

Person	Before	After	Difference, (x)	Person	Before	After	Difference, (x)
1	97	86	−11	19	98	96	−2
2	95	97	2	20	95	98	3
3	93	93	0	21	97	93	−4
4	96	89	−7	22	94	96	2
5	98	94	−4	23	94	90	−4
6	97	98	1	24	91	91	0
7	93	96	3	25	93	85	−8
8	96	88	−8	26	97	89	−8
9	95	95	0	27	93	95	2
10	98	92	−6	28	94	89	−5
11	94	98	4	29	97	89	−8
12	95	90	−5	30	93	96	3
13	95	91	−4	31	93	87	−6
14	98	92	−6	32	94	88	−6
15	95	97	2	33	96	89	−7
16	95	92	−3	34	95	94	−1
17	95	93	−2	35	92	92	0
18	97	92	−5				

Table 8.1
Diastolic blood pressure

Now, the idea of the test is this. We can think of the 35 values of x (differences) given in Table 8.1 as a random sample of all such possible differences. If μ is the (*population*) *mean of all such possible differences*, then, if the exercise program has no effect on lowering diastolic blood pressure, we should have $\mu = 0$. On the other hand, if the exercise program, in fact, has the effect of lowering diastolic blood pressure, then we should have $\mu < 0$. Consequently, we want to perform the hypothesis test:

Null hypothesis: $\mu = 0$
Alternative hypothesis: $\mu < 0$.

So we are right back to the same type of hypothesis test as described on page 266 of Section 8.3. We follow the procedure given there. (*Note*: Since the sample size is at least 30 ($n = 35$, here), the procedure applies.)

1. We will use $\alpha = 0.05$.
2. The *critical value* is $c = -z_\alpha = -z_{0.05} = -1.64$.
3. Compute the value of z:

$$z = \frac{\bar{x} - \mu_0}{s/\sqrt{n}}$$

We have $\mu_0 = 0$ and $n = 35$, and from the "Difference" columns of Table 8.1, we find that $\bar{x} = -2.80$ and $s = 4.06$. Thus,

$$z = \frac{-2.80 - 0}{4.06/\sqrt{35}} \doteq -4.08$$

4. Since $z < c$, we *reject* the null hypothesis.

Consequently, the research team can safely recommend that the exercise program has the effect of lowering diastolic blood pressure. ▶

In Example 14, test the hypothesis that the exercise program lowers the diastolic blood pressure by more than 2. That is, perform the hypothesis test: Exercise **G**

Null hypothesis: $\mu = -2$
Alternative hypothesis: $\mu < -2$

Use $\alpha = 0.05$.

Before we present another example, we want to emphasize that *hypothesis tests for differences between means for paired samples are done in the same way as hypothesis tests for a single mean.* The only additional work required is to convert each piece of paired data into a single piece of data by taking the difference. [E.g., the first piece of paired data in Table 8.1 is 97 and 86. This is converted into the single value of -11 ($= 86 - 97$).]

◀ A manufacturer of an oil additive claims that, on the average, the use of his product   Example **15**
increases the gas mileage of a car by 1.5 mpg. A consumer agency is skeptical about this claim, and decides to perform a test to check it out.

The agency first found the gas mileage of 40 cars without using the oil additive. Following this they determined the gas mileage of the same 40 cars using the oil additive. The results are given in Table 8.2. (See the top of page 290.)

The consumer agency wants to see whether the (mean) *difference* μ, in gas mileage, after adding the oil additive, is less than 1.5 mpg. So they want to test:

Null hypothesis: $\mu = 1.5$
Alternative hypothesis: $\mu < 1.5$

Since $n = 40$ ($\geqslant 30$) we can use the usual procedure.

1. The consumer agency wants to be careful not to publicly dispute the manufacturer's claim unless there is very strong evidence against it. Therefore, the agency chooses a small value of α:

$$\alpha = 0.01$$

This means that the probability that the agency will dispute the manufacturer's claim, when his claim is actually true, is only 0.01.

Table 8.2

Car	Without additive (mpg)	With additive (mpg)	Difference, (x)	Car	Without additive (mpg)	With additive (mpg)	Difference, (x)
1	15	17	2	21	31	33	2
2	25	26	1	22	25	26	1
3	23	23	0	23	13	14	1
4	12	14	2	24	11	11	0
5	17	19	2	25	17	19	2
6	21	23	2	26	19	19	0
7	24	24	0	27	21	21	0
8	10	12	2	28	26	28	2
9	12	13	1	29	12	12	0
10	28	31	3	30	11	11	0
11	13	14	1	31	18	18	0
12	11	12	1	32	19	21	2
13	15	16	1	33	16	17	1
14	27	29	2	34	24	24	0
15	31	32	1	35	34	34	0
16	21	22	1	36	18	18	0
17	17	18	1	37	20	20	0
18	19	21	2	38	10	11	1
19	12	15	3	39	17	17	0
20	24	25	1	40	22	22	0

2. The *critical value* is $c = -z_\alpha = -z_{0.01} = -2.33$.
3. Compute the value of z:

$$z = \frac{\bar{x} - \mu_0}{s/\sqrt{n}}$$

We have $\mu_0 = 1.5$ and $n = 40$, and from the "Difference" columns of Table 8.2 we find that $\bar{x} = 1.03$ and $s = 0.92$. Thus

$$z = \frac{1.03 - 1.5}{0.92/\sqrt{40}} = -3.23$$

4. Since $z < c$, we *reject* the null hypothesis.

In other words, based on the data, the consumer agency can dispute the manufacturer's claim that the oil additive has the effect of raising gas mileage, on the average, by 1.5 mpg. ▶

Exercise H   In Example 15, perform the hypothesis test if the oil additive manufacturer claims only that the additive will, on the average, raise gas mileage by 1.25 mpg. Use $\alpha = 0.05$, and the sample data in Table 8.2.

In Exercises 1 through 7,

a) state the null hypothesis,
b) state the alternative hypothesis,
c) draw a picture,
d) calculate z,
e) state whether or not the null hypothesis should be rejected.

1 A typing teacher wished to test whether or not a new method of teaching typing would *increase* typing speed at the rate of more than one word per minute per week. He recorded speed in words per minute of his 31 students. After one week of instruction, based on the new method, he recorded the students' typing speed again. He calculated the difference (x) between the final speed and the beginning speed for each student. The mean of these differences was $\bar{x} = 1.1$, with $s = 3.1$. At the 0.05 significance level, determine whether the new method increases typing speed by more than one word per minute.

2 The developers of the new instructional method for teaching typing in Exercise 1 also claim that errors would be reduced during the first weeks of training. The teacher of the 31 students recorded the number of errors per page per student on a pretest. At the end of one week of the new instructional method, he recorded the number of errors per page per student on a post-test. For each student, he calculated x, the difference between the number of errors after instruction and the number of errors before instruction. He found $\bar{x} = -1.1$, with $s = 2.3$. At the 0.05 significance level, determine whether the new method of instruction decreases errors.

3 A social worker recorded the weekly salaries of 50 unskilled workers who were about to begin a training program. Two months after they finished their training he recorded their salaries again. After finding x, the change (after minus before) in salary for each individual, he calculated $\bar{x} = 50.13$ and $s = 8.41$. At the 0.05 significance level, does it appear that the training program will improve salaries of unskilled workers?

4 A chemistry professor was interested in whether reviewing a test was valuable to the learning process. To test this she gave a class of 34 students an organic chemistry test. The scores are listed below in the columns headed "1". After a review of the test, a second test was given a few days later. The scores from the second test are listed in Columns 2. At the significance level $\alpha = 0.05$, does the review procedure appear to be valuable to the learning process?

1	2	1	2
46	51	51	59
64	70	64	65
89	92	55	50
52	61	51	53
50	58	43	46
60	61	65	79
55	50	52	60
58	68	41	38
61	62	37	43
63	70	58	60
53	54	58	63
77	86	27	22
70	66	69	76
55	56	68	74
54	61	36	38
42	45	85	77
59	65	37	41

5 An exercise physiologist measured the heart rate of 30 people. The rates are listed below in Columns 1. The people were then placed on a long, slow distance (LSD) running program. One year later their heart rates were measured again. These rates are listed below in Columns 2. Based on this sample, does the running program appear to reduce heart rates? Use $\alpha = 0.05$.

1	2	1	2
68	67	75	75
76	75	76	73
74	74	71	69
71	74	72	71
71	69	75	71
72	70	72	70
75	71	68	68
83	77	74	72
75	71	74	75
74	74	71	70
76	73	77	73
77	68	76	73
78	71	75	71
75	72	68	67
75	74	73	72

6 A pediatrician began testing the cholesterol levels of her young patients. She was alarmed to find that a large number had levels over 200. A list of the readings of 40 high-level patients is given below in Columns 1. She developed an education program on lowering cholesterol levels and began instructing the parents of her patients. After two months she tested the children's cholesterol levels again. The results are given in Columns 2. At the 0.05 significance level, is the education program effective in lowering cholesterol levels?

1	2	1	2
210	212	224	224
217	210	201	203
208	210	214	214
215	213	217	218
202	200	205	204
209	208	208	208
207	203	221	219
210	199	216	209
221	218	212	214
218	214	209	204
205	206	204	207
208	208	203	204
213	212	216	218
217	221	213	213
206	208	219	218
219	215	217	216
211	213	208	208
219	219	206	205
221	221	214	219
204	198	211	214

2	3	2	3
212	206	224	216
210	208	203	193
210	193	214	209
213	205	218	215
200	196	204	205
208	194	208	197
203	194	219	209
199	191	209	191
218	209	214	199
214	205	204	196
206	191	207	190
208	198	204	195
212	194	218	218
221	204	213	200
208	195	218	210
215	203	216	207
213	203	208	202
219	208	205	195
221	211	219	216
198	195	214	210

7 After reviewing the results of her efforts, the pediatrician in Exercise 6 decided she needed to get the parents of her patients more involved. She tested the cholesterol level of the parents of the 40 children. Most had readings well above 200. She told them that everyone in the family needed to have a reading of below 200. The whole family was placed on a low-cholesterol diet; the parents were again given instruction on diet, and shown films. After two months she took readings on the whole family. She compared the children's current readings with the readings from Columns 2 in Exercise 6. The newest readings are listed below in Columns 3. At the 0.05 significance level, does this new program appear to be effective in reducing the cholesterol levels of children?

<u>8</u> It is possible to confuse the tests for dependent samples (this section) and independent samples (Section 8.6). This can indeed cause serious problems in results. Suppose, for example, that you collected the data of Exercise 4 (organic chemistry test scores) and mistakenly decided to use the methods of Section 8.6. The means and variances for the two columns of data are:

Column 1	Column 2
$\bar{x}_1 = 56.03$	$\bar{x}_2 = 59.41$
$s_1^2 = 181.42$	$s_2^2 = 212.92$

a) Perform a test of the null hypothesis $\mu_1 = \mu_2$ against the alternative $\mu_1 < \mu_2$ using the methods of Section 8.6. (Use $\alpha = 0.05$.)

b) Compare the results of this (incorrect) test with the actual results of Exercise 4.

9 (*Confidence intervals for differences between means in two dependent samples.*) A $(1-\alpha)$-level confidence interval for the difference between means in paired dependent samples can be obtained using the sample mean difference, $\bar{x}$, and the sample standard deviation of all differences, s. The inequality that gives the confidence interval is

$$\bar{x} - z_{\alpha/2}\frac{s}{\sqrt{n}} < \mu < \bar{x} + z_{\alpha/2}\frac{s}{\sqrt{n}}$$

Here we assume that $n \geqslant 30$, and μ represents the true mean difference between pairs.

a) Find a 95-percent confidence interval for μ, the mean difference between test scores, with and without review, in Exercise 4.

b) Discuss the relation between these confidence intervals and hypothesis tests using the same data.

Setting up a hypothesis test: small samples   **8.8**

Up to this point we have discussed how to perform hypothesis tests for means when dealing with *large random samples* ($n \geqslant 30$). However, as we mentioned in Section 7.5, there are many instances in which large samples are either unavailable, extremely expensive, or for some other reason, simply undesirable.

For instance, suppose we are testing a type of car bumper to determine its effectiveness in protecting a car in a front-end collision. Testing the bumper might actually involve driving cars into a brick wall. Obviously, we would like to use a small sample here, because of the costs involved in such testing.

We would also use small samples if we were testing a new drug that may possibly have harmful side effects, or if we were testing a new teaching procedure in which the number of students in the class is small. (In the latter instance, we have no control over sample size, even if we were willing to "pay for it.")

In Section 8.2 we learned that, when dealing with *large samples*, we can base our hypothesis tests on the standard normal random variable z. However, in Chapter 7 we learned that for small samples ($n < 30$) the random variable

$$\frac{\bar{x} - \mu}{s/\sqrt{n}}$$

does *not* have the standard normal distribution. But *if we assume that the population from which we are sampling is approximately normally distributed, then the random variable*

$$t = \frac{\bar{x} - \mu}{s/\sqrt{n}}$$

has the Student's t-distribution with $n-1$ degrees of freedom. Consequently, when considering *normally distributed populations*, we can perform hypothesis tests for means using small random samples in the same way as was done for large random samples. The only difference is that the test will be designed using the random variable t instead of the random variable z.

We can look at the *large sample* versus *small sample and normal population* situation in two different ways.

Firstly, if we know that the population is approximately normally distributed, then we can take advantage of this knowledge by performing hypothesis tests using small samples. But if we do not know whether or not the population is approximately normally distributed, or if we know in fact that it isn't, then we must resort to large samples.*

On the other hand, if large samples are readily available and, say, inexpensive to obtain, we do not have to worry about whether or not the underlying population is normally distributed. We can simply use the large-sample tests described earlier in this chapter.

For emphasis we repeat, once more, the fundamental fact used in performing *small sample* hypothesis tests for means.

> Suppose we take a random sample of size n from a population that is (approximately) *normally distributed* with mean μ. Let $\bar{x}$ and s be the sample mean and sample standard deviation, respectively. Then the random variable
>
> $$ t = \frac{\bar{x} - \mu}{s/\sqrt{n}} $$
>
> has the *Student's t-distribution with $n-1$ degrees of freedom*. That is, probabilities for t can be found by looking at areas under the t-curve with d.f. $= n - 1$.

In the above paragraph note that we have required that the population be *approximately* normally distributed. That is, we require that the distribution of the population be "close" to a normal distribution. A natural question is "How close is close enough?" We will discuss such questions in detail in Chapter 14. For now we merely mention that, when dealing with tests concerning *means*, the normality assumption is not extremely critical as long as the population does not deviate *too much* from normality.

It might be useful for you to go back and briefly review Section 7.5. The most important thing for you to review will be how to use the Student's t-distribution table (Table II).

EXERCISES Section **8.8**

1 Find the t-value for which the area under the t-curve to the *right* of this value is 0.01, if d.f. $= 16$ (that is, find $t_{0.01}$ for d.f. $= 16$).

2 Find the t-value for which the area under the t-curve to the *right* of this value is 0.10, if d.f. $= 21$ (that is, find $t_{0.10}$ for d.f. $= 21$).

3 Find the t-value for which the area under the t-curve to the *left* of this value is 0.05, if d.f. $= 19$.

4 Find the t-value for which the area under the t-curve to the *left* of this value is 0.01, if d.f. $= 26$.

5 Find the t-value for which the area under the curve between $-t$ and t is 0.95, if d.f. $= 8$.

6 What is the distribution of $(\bar{x} - \mu)/(\sigma/\sqrt{n})$, if $\bar{x}$ is based upon a random sample of size n from a normal population?

---

* It is also possible that a nonparametric test could be used in order to retain a small sample size. Such tests will be discussed in Chapter 13.

In Sections 8.3 and 8.4 we gave the procedure for performing a hypothesis test for a single mean for large random samples. The procedure for *small* random samples is the same except for two things:

Hypothesis tests **8.9**
for a single mean
for small samples

1. The *population* from which we are sampling *must be* (approximately) *normally distributed*, and
2. We use the *random variable t* instead of the random variable *z*.

Procedure for performing a hypothesis test of the form:
Null hypothesis: $\mu = \mu_0$
Alternative hypothesis: $\mu > \mu_0$

Procedure

Assumption: **Normal population**

1. Decide on the *significance level, α.*
2. The *critical value* is

$$c = t_\alpha, \qquad \text{d.f.} = n - 1$$

(*n* is the sample size). Use Table II to find t_α.

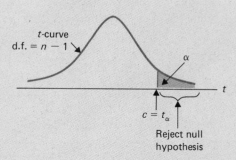

3. Compute the value of

$$t = \frac{\bar{x} - \mu_0}{s / \sqrt{n}}$$

4. If $t > c$, *reject* the null hypothesis.

The procedure for a one-tailed test of the form

Null hypothesis: $\mu = \mu_0$
Alternative hypothesis: $\mu < \mu_0$

is similar to the above, except that the critical value is $c = -t_\alpha$, and we reject the null hypothesis if $t < c$.

For a two-tailed test the procedure is as usual.

Procedure

Procedure for performing a hypothesis test of the form:
Null hypothesis: $\mu = \mu_0$
Alternative hypothesis: $\mu \neq \mu_0$

Assumption: Normal population

1. Decide on the *significance level*, α.
2. The *critical values* are

$$-c = -t_{\alpha/2} \quad \text{and} \quad c = t_{\alpha/2}, \quad \text{d.f.} = n-1$$

Use Table II, to find $t_{\alpha/2}$.

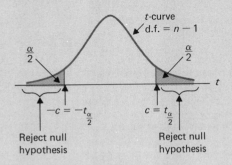

3. Compute the value of

$$t = \frac{\bar{x} - \mu_0}{s/\sqrt{n}}$$

4. *Reject* the null hypothesis if either $t < -c$ or $t > c$.

We illustrate these procedures in the following examples.

Example **16** ◖ An automobile manufacturer is testing a new bumper that he feels will reduce the cost of repairs in front-end collisions at low speeds. Experience with the presently used bumper indicates that at 10 mph the cost of repair resulting from a front-end collision has a mean of $325.

Let μ be the mean cost of repair resulting from a front-end collision at 10 mph with the *new* bumper and assume that such costs are normally distributed. Then the manufacturer wants to test the hypotheses:

Null hypothesis: $\mu = \$325$
Alternative hypothesis: $\mu < \$325$.

To perform the test, the manufacturer equips 15 of his cars with the new bumper, and has these 15 cars undergo front-end collisions at 10 mph. Following this, the cars are repaired, and the cost of repair to each car is recorded. The results are as follows.

The sample mean cost of repair to the 15 cars is $\bar{x} = \$291.70$ with a sample standard deviation of $s = \$19.82$. Perform the hypothesis test at the 0.05 significance level.

1. $\alpha = 0.05$

2. The *critical value* is $c = -t_\alpha = -t_{0.05}$, for d.f. $= 15 - 1 = 14$. Using Table II we find that $t_{0.05} = 1.76$ for d.f. $= 14$. So

$$c = -1.76$$

See Fig. 8.14.

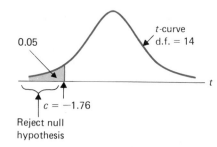

0.05

t-curve
d.f. = 14

$c = -1.76$

Reject null
hypothesis

Figure 8.14

3. $t = \dfrac{\bar{x} - \mu_0}{s/\sqrt{n}} = \dfrac{291.70 - 325}{19.82/\sqrt{15}} \doteq -6.51$

4. Since $t < c$, we *reject* the null hypothesis.

In other words, on the basis of the testing with the new bumper, the manufacturer can conclude that the new bumper will lower the cost of repairs for front-end collisions at 10 mph. ▶

Do the hypothesis test in **Example 16** with the same data, except take $\bar{x} = \$315.86$.　Exercise **I**

◀ A beer manufacturer puts out a 16-oz can of beer. He wants to make sure that the machine being used to fill the cans is working properly. That is, he wants to see whether the mean volume, μ, of beer put into the can *is* 16 fluid oz.

To check this he takes a random sample of 20 cans of beer and determines the (net) volume of each of the 20 cans of beer. His findings are $\bar{x} = 16.02$ fluid oz and $s = 0.18$ fluid oz. Assuming that the amount of beer put in a can is *normally distributed*, test the hypotheses:

Example **17**

Null hypothesis:　$\mu = 16$
Alternative hypothesis:　$\mu \neq 16$.

1. Decide on the *significance level*, α. We will use $\alpha = 0.10$.
2. The *critical values* are $-c = -t_{\alpha/2}$ and $c = t_{\alpha/2}$ for d.f. $= 20 - 1 = 19$. Since $\alpha = 0.10$, $t_{\alpha/2} = t_{0.05}$, and by Table II, $t_{0.05} = 1.73$ for d.f. $= 19$. So $-c = -1.73$ and $c = 1.73$. See Fig. 8.15.

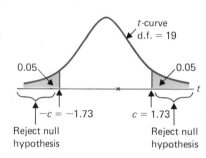

t-curve
d.f. = 19

0.05　　　0.05

$-c = -1.73$　　$c = 1.73$

Reject null
hypothesis

Reject null
hypothesis

Figure 8.15

3. $t = \dfrac{\bar{x} - \mu_0}{s/\sqrt{n}} = \dfrac{16.02 - 16}{0.18/\sqrt{20}} \doteq 0.50$

    This value of t is marked with a cross in Fig. 8.15
4. Since $t = 0.50$, we cannot reject the null hypothesis.

Therefore, on the basis of this sampling, the manufacturer will assume that his can-filling machine is working properly. ◗

Exercise J   Suppose that, in Example 17, $\bar{x} = 16.08$.
a) What conclusions can you draw from the hypothesis test, using $\alpha = 0.10$?
b) What if the significance level is $\alpha = 0.05$?

EXERCISES Section **8.9**

In Exercises 1 through 7 assume that the population is normally distributed.

a) State the null hypothesis.
b) State the alternative hypothesis.
c) Find the critical value(s).
d) Draw a picture.
e) Calculate t.
f) State whether or not the null hypothesis should be rejected.

1   A retailer wishes to test a large shipment of batteries from a new supplier. The supplier claims that the batteries have a mean life of 36 months. The retailer wants to determine whether the mean life, μ, of the batteries she received is different. A test of 20 batteries randomly sampled from the shipment shows that $\bar{x} = 33$, with $s = 9$. At the 0.05 significance level, determine whether the mean life of the shipment is *different* from what the supplier claims.

2   Do the hypothesis test in Exercise 1 with $\bar{x} = 31$ and all other data unchanged.

3   A manufacturer of lightbulbs has a bulb which, on the average, burns for 2000 hours. The research department developed a new bulb which it claims outlasts the older type. A test of 25 new bulbs yields $\bar{x} = 2011$, with $s = 35.6$. At the 0.05 significance level, test whether the new bulb is *longer-burning*, on the average.

4   Do the hypothesis test in Exercise 3 with the same data, but take $\alpha = 0.1$.

5   From past experience a sugar cane farmer knew he was getting an average of 33.8 tons of sugar cane per acre on his farm. He decided to test a new fertilizer on $n = 20$ similar one-acre plots. His yield was $\bar{x} = 35.8$ tons per acre, with $s = 4.2$. At the 0.05 significance level, does the new fertilizer appear to *increase* the mean yield?

6   Do the hypothesis test in Exercise 5 with the same data, but take $\bar{x} = 35.2$.

7   A dog food manufacturer packs 50-pound bags of dog food. He wanted to be sure that on the average the bags do contain 50 pounds. He weighed 25 randomly selected bags, and found that $\bar{x} = 49.8$ pounds with $s = 0.5$. At the 0.05 significance level, test the hypothesis that $\mu = 50$ pounds.

8   Do the hypothesis test in Exercise 7 with the same data, except take $\bar{x} = 49.9$ pounds.

9   What is the population of interest in
a) Exercise 5?   b) Exercise 7?

10 a) Suppose that you are performing the lightbulb test of Exercise 3 and *mistakenly* conclude that the variable

$$t = \frac{\bar{x} - \mu}{s/\sqrt{n}}$$

has the standard normal distribution. What critical value (z-value) would you use for the hypothesis test of Exercise 3, in which $\alpha = 0.05$?
b) What critical value (t-value) was actually used in that hypothesis test?
c) Does the mistaken use of z instead of t make you more or less likely to reject the null hypothesis?

11 If the sampled population is normally distributed, then the random variable

$$z = \frac{\bar{x} - \mu}{\sigma/\sqrt{n}}$$

has the standard normal distribution. Suppose that you wish to perform the hypothesis test of Exercise 1 and you know that the battery life is normally distributed with $\sigma = 9$.

a) What critical values would be used for z if you wish to do your test at the level $\alpha = 0.05$?

b) What critical values were actually used for t in the situation of Exercise 1 where σ was not assumed to be known and $\alpha = 0.05$?

c) If σ is known and you mistakenly use a t-value, will you be more or less likely to reject the null hypothesis?

12 Suppose that you are performing the hypothesis test of Exercise 3, and mistakenly use a z-value instead of a t-value, as in Exercise 10. Show that the actual significance level of the resulting test will be greater than 0.05.

13 Suppose that you are performing the hypothesis test of Exercise 11 with known $\sigma = 9$, and use t-values instead of z-values as critical values. What will the significance level of the resulting test be?

In Section 8.6 we learned how to perform hypothesis tests for differences between means for independent samples in the case of *large random samples* (i.e., both samples are of size at least 30). The fundamental fact used for such tests was given on page 281. Namely, suppose $\bar{x}_1$ and $\bar{x}_2$ are sample means based on samples of sizes n_1 and n_2, respectively, from populations with means μ_1 and μ_2, respectively. If the samples are independent, and *if n_1 and n_2 are both at least* 30, then the random variable

$$\frac{(\bar{x}_1 - \bar{x}_2) - (\mu_1 - \mu_2)}{\sqrt{(s_1^2/n_1) + (s_2^2/n_2)}}$$

Small-sample 8.10 hypothesis tests for differences between means for independent samples— equal variances

has approximately the standard normal distribution.

Now, if we are *not* dealing with large random samples, then the situation changes somewhat. Just as in the case of a single mean, we can use the Student's t-distribution, but we need to consider two cases separately. Assuming that both populations are approximately normally distributed, the first case is when the *population* variances (σ_1^2 and σ_2^2) are equal, and the second case is when they are not equal. Since usually we won't know the population variances, you might wonder how we can decide whether or not the population variances are equal. Sometimes we can decide by performing a hypothesis test:

Null hypothesis: $\sigma_1^2 = \sigma_2^2$
Alternative hypothesis: $\sigma_1^2 \neq \sigma_2^2$

You will learn how to do such a hypothesis test in Chapter 9, but for now, we will see what we do once we have decided whether or not to believe that the population variances are equal.

In this section we assume that we are dealing with the case where the population variances are equal. Specifically, the assumptions we make are the following:

1. *Independent random samples* are taken from two populations.
2. Both populations are (approximately) *normally distributed.*
3. The two *population variances are equal.**

---

* We are assuming that the population variances are unknown, since they usually are. If it happened that they were known, then we could base our test of the difference of the means on the fact that the random variable

$$\frac{(\bar{x}_1 - \bar{x}_2) - (\mu_1 - \mu_2)}{\sqrt{(\sigma_1^2/n_1) + (\sigma_2^2/n_2)}}$$

has the standard normal distribution. We will explore this in the exercises.

Now, recall that when we are sampling from a single population, we use the sample standard deviation, s, as our estimate of the population standard deviation, σ. When taking two independent random samples from two populations with the same population standard deviation (i.e., same variance), it turns out that the best estimate for the common standard deviation is

$$s_p = \sqrt{\frac{(n_1 - 1)s_1^2 + (n_2 - 1)s_2^2}{n_1 + n_2 - 2}}$$

Here s_1 and s_2 are the sample standard deviations obtained by sampling from the first and second populations, respectively. Also, of course, n_1 and n_2 are the sizes of the samples taken from the first and second populations, respectively. The reason for the subscript p in s_p is that the estimate above given for the standard deviation is obtained by "pooling" the two sample standard deviations s_1 and s_2. The derivation of the formula for the "pooled" sample standard deviation, s_p, will be considered in the exercises.

In order to carry out the hypothesis tests of this section we will need the following fact:

> If both populations are normally distributed and $\sigma_1^2 = \sigma_2^2$, then the random variable
>
> $$t = \frac{(\bar{x}_1 - \bar{x}_2) - (\mu_1 - \mu_2)}{s_p \sqrt{(1/n_1) + (1/n_2)}}$$
>
> has the *Student's t-distribution with $n_1 + n_2 - 2$ degrees of freedom.*

With this knowledge we can now proceed in the usual manner with our hypothesis tests. [*Note:* In most of our work the null hypothesis will be $\mu_1 = \mu_2$ ($\mu_1 - \mu_2 = 0$). So *if the null hypothesis is true*, then, by what we said above,

$$t = \frac{\bar{x}_1 - \bar{x}_2}{s_p \sqrt{(1/n_1) + (1/n_2)}}$$

has the *Student's t-distribution with d.f. $= n_1 + n_2 - 2$.*]

Procedure

> *Procedure for performing a hypothesis test of the form:*
> *Null hypothesis:* $\mu_1 = \mu_2$
> *Alternative hypothesis:* $\mu_1 < \mu_2$
>
> *Assumptions:* (1) Independent samples, (2) normal populations, (3) equal population variances
>
> 1. Decide on the *significance level, α.*
> 2. The *critical value* is
>
> $$c = -t_\alpha, \qquad \text{d.f.} = n_1 + n_2 - 2$$
>
> Use Table II to find t_α.

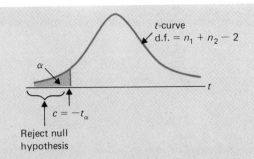

3. Compute the value of

$$t = \frac{\bar{x}_1 - \bar{x}_2}{s_p \sqrt{(1/n_1) + (1/n_2)}} \quad \text{with} \quad s_p = \sqrt{\frac{(n_1 - 1)s_1^2 + (n_2 - 1)s_2^2}{n_1 + n_2 - 2}}$$

where n_1, $\bar{x}_1$, s_1 are, respectively, the sample size, sample mean, and sample standard deviation from the first population, and n_2, $\bar{x}_2$, s_2 are those for the second population.

4. If $t < c$, *reject the null hypothesis.*

◀ A manufacturer of automobile products is interested in comparing a newly   Example **18**
developed wax with the wax he is presently producing. Specifically, he wishes to
perform the hypothesis test:

Null hypothesis: $\mu_1 = \mu_2$
Alternative hypothesis: $\mu_1 < \mu_2$

$\alpha = 0.05$

Here μ_1 is the mean effectiveness time for the wax presently being produced and μ_2
is the mean effectiveness time for the newly developed wax.

In order to make the comparison, five new cars are given applications of the
present wax, and another five new cars are given applications of the newly developed
wax. The cars are then exposed to the same environmental conditions, and the length
of time of effectiveness of the wax is measured for each car. The results are given in
Table 8.3 (time measured in days).

Time of effectiveness for presently used wax	Time of effectiveness for newly developed wax	Table **8.3**
89	94	
86	91	
83	88	
87	92	
89	87	

Using this data we can compute the sample means and sample standard deviations. They are

Presently used wax	Newly developed wax
$n_1 = 5$	$n_2 = 5$
$\bar{x}_1 = 86.80$	$\bar{x}_2 = 90.40$
$s_1 = 2.49$	$s_2 = 2.88$

The manufacturer's experience with waxes indicates that the length of time of effectiveness for waxes is *normally distributed*, and that *variances* for different waxes are about *the same*. Since the two samples here are obviously *independent*, we can use the procedure described on page 300.

1. $\alpha = 0.05$.
2. The critical value is $c = -t_\alpha = -t_{0.05}$ for d.f. $= n_1 + n_2 - 2 = 5 + 5 - 2 = 8$. Using Table II, we see that for d.f. $= 8$, $t_{0.05} = 1.86$. Thus, $c = -1.86$. See Fig. 8.16.

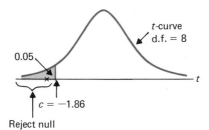

0.05

t-curve
d.f. = 8

t

$c = -1.86$

Reject null
hypothesis

Figure 8.16

3. Compute the value of

$$t = \frac{\bar{x}_1 - \bar{x}_2}{s_p\sqrt{(1/n_1) + (1/n_2)}}$$

with

$$s_p = \sqrt{\frac{(n_1 - 1)s_1^2 + (n_2 - 1)s_2^2}{n_1 + n_2 - 2}}$$

We have $n_1 = 5$, $\bar{x}_1 = 86.80$, $s_1 = 2.49$, $n_2 = 5$, $\bar{x}_2 = 90.40$, and $s_2 = 2.88$. So,

$$s_p = \sqrt{\frac{(5 - 1)(2.49)^2 + (5 - 1)(2.88)^2}{5 + 5 - 2}} \doteq 2.69$$

Consequently,

$$t = \frac{86.80 - 90.40}{2.69\sqrt{\frac{1}{5} + \frac{1}{5}}} \doteq -2.12$$

We have marked this value of t with a cross in Fig. 8.16.

4. Since $t < c$, we *reject* the null hypothesis.

In other words, based on the data, we conclude that the newly developed wax has a longer mean effectiveness time than the presently used wax. ◗

Perform the hypothesis test in Example 18 with the same $\bar{x}$'s and s's, but assume that $n_1 = 4$ and $n_2 = 3$ and use $\alpha = 0.10$.   Exercise **K**

Hypothesis tests in which the alternative hypothesis is of the form $\mu_1 > \mu_2$ are handled in the same way, except the critical value is $c = t_\alpha$, and we reject the null hypothesis if $t > c$. For *two-tailed* tests the procedure is basically the same except that there are *two critical values*; namely, $-c = -t_{\alpha/2}$ and $c = t_{\alpha/2}$.

In this section we learned how to perform small-sample hypothesis tests for the difference between two means for normal populations when the population variances are *equal* and the samples are *independent*. In cases where the population variances are *unequal*, a different test is required. We shall discuss this test in the advanced exercises. Also discussed there are small-sample hypothesis tests for the difference between means for *paired* samples. This is the small-sample version of the test discussed in Section 8.7.

EXERCISES Section **8.10**

In Exercises 1 through 6, assume that the samples are independent, the populations are normally distributed, and the population variances are equal.

a) State the null hypothesis.
b) State the alternative hypothesis.
c) Find the critical value(s).
d) Draw a picture.
e) Calculate t.
f) State whether or not the null hypothesis should be rejected.

1 A highway official wished to test whether a new paint for striping roads is longer-lasting than the paint in current use. Fifteen stripes of each paint were run across a highway. The number of months for the life of each paint are given below

Current paint	New paint
$\bar{x}_1 = 12$	$\bar{x}_2 = 14$
$s_1 = 1$	$s_2 = 1$
$n_1 = 15$	$n_2 = 15$

At the 0.05 significance level, do a hypothesis test to determine whether the new paint appears to *last longer* on the average.

2 In a packing plant, a machine packs cartons with jars. A sales person claims a new machine will pack the jars faster. The packing results (in seconds) of the two machines for 10 cartons are given in the next table.

New machine	Old machine
$\bar{x}_1 = 43$	$\bar{x}_2 = 44$
$s_1 = 1.5$	$s_2 = 1.6$
$n_1 = 10$	$n_2 = 10$

At the 0.05 significance level, do a hypothesis test to determine whether the new machine *packs faster* on the average.

3 The owner of two boutiques was interested in comparing the business done by the two stores. The sales for 12 randomly selected weeks yielded the following results.

Store 1	Store 2
$\bar{x}_1 = \$4327.64$	$\bar{x}_2 = \$4138.67$
$s_1 = \$ 392.43$	$s_2 = \$ 425.82$
$n_1 = 12$	$n_2 = 12$

At the 0.05 significance level, determine whether the mean sales are different at the two stores.

4 A research botanist is interested in whether or not there is a difference in the growth of a plant if a particular amount of a chemical is added to the soil. Two trays of 14 plants each are prepared. Tray 1 has the chemical mixed in with its soil, but Tray 2 does not. After the plants have been grown in a controlled environment, measurements are made of their heights. The results (in centimeters) are given in Table 8.4.

Table 8.4

Tray 1 (with additive)	Tray 2 (without additive)
15.9	16.2
16.9	14.8
14.6	14.5
18.1	15.8
18.2	15.4
18.6	14.8
14.5	16.3
16.8	16.3
14.6	17.1
16.7	18.5
14.1	16.5
17.5	18.6
16.1	18.3
15.6	18.2

At the 0.05 significance level, determine whether there is a difference in the growth of plants in the two different soils.

5 A professor obtained the grade point averages of the graduate students in her class. She divided the class into two groups. Group 1, married students, and Group 2, unmarried students. The grade point averages are listed below.

Group 1	Group 2
3.4	3.2
2.9	3.6
3.8	3.1
3.2	3.6
3.1	3.2
3.9	3.4
	2.6
	3.5

On the basis of this sample, is there reason to believe that there is a difference in the grade point averages of married and unmarried students? (Use $\alpha = 0.05$)

6 In 1973, a sample of ten farms in Iowa yielded $\bar{x}_1 = 107.0$ bushels per acre of corn, with $s_1 = 2.2$ bushels. In California, nine randomly selected farms yielded $\bar{x}_2 = 105.0$ bushels per acre, with $s_2 = 2.1$ bushels. At the 0.05 significance level, did Iowa farms yield more bushels per acre than California farms?

7 The methods of this section are designed for situations in which (1) the normal population variances σ_1^2 and σ_2^2 are unknown, and (2) we are willing to assume that $\sigma_1^2 = \sigma_2^2$. If the population variances are *known* and equal to a single common variance σ^2, we may use as our statistic for a test of $\mu_1 = \mu_2$:

$$z = \frac{\bar{x}_1 - \bar{x}_2}{\sigma\sqrt{(1/n_1) + (1/n_2)}},$$

which has the standard normal distribution. Suppose that the engineers at the packing plant in Exercise 2 are willing to assume (from prior studies) that $\sigma_1 = \sigma_2 = 1.4$ for both machines being studied. Use the z-statistic to perform the required hypothesis test if $n_1 = n_2 = 10$, $\bar{x}_1 = 43$ and $\bar{x}_2 = 44$. (Use $\alpha = 0.05$)

8 If you wish to test the null hypothesis $\mu_1 = \mu_2$ for two normal populations with *known* variances σ_1^2 and σ_2^2, the statistic to use is:

$$z = \frac{\bar{x}_1 - \bar{x}_2}{\sqrt{(\sigma_1^2/n_1) + (\sigma_2^2/n_2)}}$$

Show that if $\sigma_1^2 = \sigma_2^2 = \sigma^2$, this z-statistic is identical with the one given in Exercise 7.

9 The formula for the "pooled" sample standard deviation is

$$s_p = \sqrt{\frac{(n_1 - 1)s_1^2 + (n_2 - 1)s_2^2}{n_1 + n_2 - 2}}$$

The idea behind this formula is that if the two sampled populations have the same variance (that is, $\sigma_1^2 = \sigma_2^2$), then we should be able to get a better estimate of this common population variance by "pooling" the sample data obtained from both populations. Thus, the common variance, σ^2, is estimated not by s_1^2 or s_2^2 separately, but by a "weighted" average of s_1^2 and s_2^2. It is reasonable to "weight" according to sample size, since large sample size should lead, on the average, to a more accurate estimate of σ^2. It turns out that the proper way to "weight" is according to *degrees of freedom*. Show that the "weighted" average of s_1^2 and s_2^2, weighted proportionally to $n_1 - 1$ and $n_2 - 1$, respectively, is s_p^2.

In this section we learned how to perform small-sample hypothesis tests for the difference between two means when (1) the samples are *independent*, (2) both populations are *normally distributed*, and (3) the population variances are *equal*. If the population variances are *unequal* (and unknown), the following test statistic is *often* used to test the null hypothesis $\mu_1 = \mu_2$:

$$t = \frac{\bar{x}_1 - \bar{x}_2}{\sqrt{(s_1^2/n_1) + (s_2^2/n_2)}}$$

with degrees of freedom given by

$$\text{d.f.} = \frac{[(s_1^2/n_1) + (s_2^2/n_2)]^2}{(s_1^2/n_1)^2[1/(n_1 - 1)] + (s_2^2/n_2)^2[1/(n_2 - 1)]}$$

rounded to the nearest integer.

Using the test statistic described above, perform an appropriate hypothesis test for the problem below. (Assume independent samples, normal populations, and *unequal* population variances.)

10 A sociologist, interested in comparing life styles, randomly sampled a group of commuters on the East Coast and a group on the West Coast. He then determined the number of miles that they traveled one way to work. The results are given below.

East Coast	West Coast
$\bar{x}_1 = 15.6$	$\bar{x}_2 = 25.3$
$s_1 = 5.1$	$s_2 = 15.8$
$n_1 = 37$	$n_2 = 25$

At the 0.05 significance level, is there any difference in the (mean) commuting distance of East and West Coast commuters?

In Exercise 11 we will apply the small-sample analogue of the test discussed in Section 8.7. That is, we will perform a small-sample hypothesis test for the difference between means for *paired* samples.

When performing a hypothesis test for differences between means for paired data using a small ($n < 30$) random sample, we simply apply the techniques of Section 8.9 to the *differences* of the pairs sampled. However, if the test is to apply, the *population of all such differences must be (approximately) normally distributed.* In other words, to test the null hypothesis that the mean difference, μ, is $\mu = \mu_0$ we use the statistic

$$t = \frac{\bar{x} - \mu_0}{s/\sqrt{n}} \quad (\text{d.f.} = n - 1)$$

where $\bar{x}$ is the sample mean *difference* of the paired data and s is the sample standard deviation of the *differences.* Using the test statistic described above, perform an appropriate hypothesis test for the problem below.

11 The owner of a health spa advertises that after one month of his fitness program a person will lose (on the average) more than 10 pounds.

A group of three couples decides to enroll in the program. Moreover, they plan to stay in the program if it appears that the owner's claim is true. They record their weights just before beginning the program, and then again after one month. The data are given below:

Person	Before	After	Weight loss (x)
1	110	102	8
2	163	151	12
3	124	106	18
4	185	170	15
5	137	126	11
6	208	196	12

To test the owner's claim, let μ be the (population) mean weight loss resulting from one month of the fitness program and perform the hypothesis test:

Null hypothesis: $\mu = 10$
Alternative hypothesis: $\mu > 10$

(Assume that the population of such weight differences is normally distributed.) Use $\alpha = 0.05$.

CHAPTER REVIEW

Key Terms

population mean (μ)
sample mean ($\bar{x}$)
population standard deviation (σ)
sample standard deviation (s)
inference
hypothesis
null hypothesis
alternative hypothesis
significance level (α)
critical values
rejection region

one-tailed test
two-tailed test
population proportion (p)
sample proportion ($\hat{p}$)
independent samples
paired samples
normally distributed populations
t-distribution
degrees of freedom (d.f.)
pooled sample standard
   deviation (s_p)

Formulas and Key Facts

sampling distribution of the mean for large samples:

If a random sample of size $n \geqslant 30$ is taken from a population with mean μ, then the random variable

$$z = \frac{\bar{x} - \mu}{s/\sqrt{n}}$$

has approximately the *standard normal distribution*.

sampling distribution of a proportion:

If p is the population proportion and $\hat{p}$ is the sample proportion based on a random sample of size n, and if both np and $n(1-p)$ are at least 5, then the random variable

$$z = \frac{\hat{p} - p}{\sqrt{p(1-p)/n}}$$

has approximately the *standard normal distribution*.

sampling distribution for the difference between two means for large samples:

If two independent random samples of sizes $n_1 \geqslant 30$ and $n_2 \geqslant 30$ are taken from two populations with means μ_1 and μ_2, then the random variable

$$z = \frac{(\bar{x}_1 - \bar{x}_2) - (\mu_1 - \mu_2)}{\sqrt{(s_1^2/n_1) + (s_2^2/n_2)}}$$

has approximately the *standard normal distribution*.

sampling distribution of the mean for random samples from a normally distributed population:

If a random sample of size n is taken from a normally distributed population with mean μ, then the random variable

$$t = \frac{\bar{x} - \mu}{s/\sqrt{n}}$$

has the *Student's t-distribution with* d.f. $= n - 1$.

sampling distribution for the difference between the means of two normally distributed populations with equal variances:

If two independent random samples of sizes n_1 and n_2 are taken from two normally distributed populations with means μ_1 and μ_2 and equal variances, then the random variable

$$t = \frac{(\bar{x}_1 - \bar{x}_2) - (\mu_1 - \mu_2)}{s_p\sqrt{(1/n_1) + (1/n_2)}}$$

has the *Student's t-distribution* with d.f. $= n_1 + n_2 - 2$. Here

$$s_p = \sqrt{\frac{(n_1 - 1)s_1^2 + (n_2 - 1)s_2^2}{n_1 + n_2 - 2}}$$

In Table 8.5 we give a brief summary of the hypothesis-testing procedures covered in this chapter. For detailed procedures, see the text.

Table 8.5
Hypothesis tests for means and proportions

Type	Assumptions	Null hypothesis	Alternative hypothesis	Test statistic	Rejection region
Single mean (large sample)	$n \geqslant 30$	$\mu = \mu_0$	$\mu > \mu_0$ $\mu < \mu_0$ $\mu \neq \mu_0$	$z = \dfrac{\bar{x} - \mu_0}{s/\sqrt{n}}$	$z > z_\alpha$ $z < -z_\alpha$ $z < -z_{\alpha/2}$ or $z > z_{\alpha/2}$
Two means (large samples)	(1) $n_1 \geqslant 30$, $n_2 \geqslant 30$ (2) Independent samples	$\mu_1 = \mu_2$	$\mu_1 > \mu_2$ $\mu_1 < \mu_2$ $\mu_1 \neq \mu_2$	$z = \dfrac{\bar{x}_1 - \bar{x}_2}{\sqrt{(s_1^2/n_1) + (s_2^2/n_2)}}$	$z > z_\alpha$ $z < -z_\alpha$ $z < -z_{\alpha/2}$ or $z > z_{\alpha/2}$
Single proportion	$np_0 \geqslant 5$ $n(1 - p_0) \geqslant 5$	$p = p_0$	$p > p_0$ $p < p_0$ $p \neq p_0$	$z = \dfrac{\hat{p} - p_0}{\sqrt{p_0(1 - p_0)/n}}$	$z > z_\alpha$ $z < -z_\alpha$ $z < -z_{\alpha/2}$ or $z > z_{\alpha/2}$
Single mean (small sample)	Normal population	$\mu = \mu_0$	$\mu > \mu_0$ $\mu < \mu_0$ $\mu \neq \mu_0$	$t = \dfrac{\bar{x} - \mu_0}{s/\sqrt{n}}$ (d.f. $= n - 1$)	$t > t_\alpha$ $t < -t_\alpha$ $t < -t_{\alpha/2}$ or $t > t_{\alpha/2}$
Two means (small samples)	(1) Independent samples (2) Normal populations (3) Equal variances	$\mu_1 = \mu_2$	$\mu_1 > \mu_2$ $\mu_1 < \mu_2$ $\mu_1 \neq \mu_2$	$t = \dfrac{\bar{x}_1 - \bar{x}_2}{s_p\sqrt{(1/n_1) + (1/n_2)}}$ (d.f. $= n_1 + n_2 - 2$)	$t > t_\alpha$ $t < -t_\alpha$ $t < -t_{\alpha/2}$ or $t > t_{\alpha/2}$

You should be able to

1 State the *null and alternative hypotheses* in a hypothesis test.

2 Perform *large sample* hypothesis tests of the form:
   a) *Single mean* †
      Null hypothesis: $\mu = \mu_0$
      Alternative hypothesis: $\mu > \mu_0$ or $\mu < \mu_0$ or $\mu \neq \mu_0$
   b) *Two means*
      Null hypothesis: $\mu_1 = \mu_2$
      Alternative hypothesis: $\mu_1 > \mu_2$ or $\mu_1 < \mu_2$ or $\mu_1 \neq \mu_2$
   c) *Single proportion*
      Null hypothesis: $p = p_0$
      Alternative hypothesis: $p > p_0$ or $p < p_0$ or $p \neq p_0$

3 Perform small-sample hypothesis tests of the form:
   Null hypothesis: $\mu = \mu_0$
   Alternative hypothesis: $\mu > \mu_0$ or $\mu < \mu_0$ or $\mu \neq \mu_0$
   using data obtained from a random sample from a *normal population*.

4 Perform small-sample hypothesis tests of the form:
   Null hypothesis: $\mu_1 = \mu_2$
   Alternative hypothesis: $\mu_1 > \mu_2$ or $\mu_1 < \mu_2$ or $\mu_1 \neq \mu_2$

   using data obtained from two *independent* random samples from two *normal populations* with *equal variances*.

REVIEW TEST

1 In 1976, the mean salary for American wives who worked full time was $7321. An inner-city political leader in a large Western city suspected that the working wives in his community had a *lower* mean income μ than that for all American wives. He took a random sample of $n = 100$ working wives in his city and found $\bar{x} = \$6625$ with $s = \$2000$. Use this data to perform the hypothesis test

   Null hypothesis: $\mu = 7321$
   Alternative hypothesis: $\mu < 7321$

   (Use $\alpha = 0.01$.)

2 In 1978, the mean selling price of homes in a Southwestern suburb was $59,000. A realtor wished to show that the mean selling price μ had *risen* in 1979. She took a random sample of $n = 70$ transactions in 1979 and found $\bar{x} = 65000$ with $s = 5000$. Use this data to perform an appropriate hypothesis test for the realtor with $\alpha = 0.05$.

3 A high-school district claimed (on the basis of data collected in the last five years) that 80 percent of all parents of students in the district were satisfied with the schools there. A teacher felt that this figure should be *lower* and decided to do a statistical test. His null hypothesis was $p = 0.8$.

   a) What should his alternative hypothesis be?
   b) The teacher took a random sample of $n = 40$ parents and found that 30 of them were satisfied with the district schools. Use this data to test the null and alternative hypotheses above with $\alpha = 0.01$.

4 The owner of two travel agencies wished to do a quick check on whether the mean prices of airline tickets sold at the two agencies last month were the same. She took random samples from the sales records of each agency. The figures were:

Adventure Travel	Escapetours, Inc.
$\bar{x}_1 = 178.30$	$\bar{x}_2 = 181.20$
$s_1 = 40.10$	$s_2 = 39.90$
$n_1 = 36$	$n_2 = 36$

   Perform a two-tailed hypothesis test for the owner at the level $\alpha = 0.10$.

5 A nutrition specialist was interested in the effects of a new diet. The diet's inventor claimed that his diet has reduced the weights of thousands of American females by a mean of $\mu = 20$ pounds. The nutritionist recruited a sample of $n = 50$ women and had them try the diet. He then computed the differences, x, between the before and after weights of the women, and found $\bar{x} = 18$, $s = 6$.

---

† This includes tests from paired samples.

Use this data to test the hypotheses:

Null hypothesis: $\mu = 20$
Alternative hypothesis: $\mu < 20$

(Use $\alpha = 0.05$.)

6   A real-estate dealer claims (in a brochure) that the mean cost of air conditioning in July for a three-bedroom house in a Southwestern city is $\mu = \$65$. A skeptical homeowner takes a random sample of bills for $n = 15$ such homes for the month of July. Her results are $\bar{x} = \$78$ and $s = \$6$. Use this data to test the null hypothesis "$\mu = 65$" against the alternative "$\mu > 65$" with $\alpha = 0.01$. (You may assume that the costs are approximately normally distributed.)

7   The homeowner in Problem 6 knew that most homes in her city were built by the two major builders in her city—Quality Homes and Smith Brothers. She decided to test whether the mean air-conditioning costs for homes from the two builders are the same. In August, she took independent random samples of eight homes from each of the two builders. Her results were:

Quality Homes	Smith Brothers
$n_1 = 8$	$n_2 = 8$
$\bar{x}_1 = \$81$	$\bar{x}_2 = \$74$
$s_1^2 = \$32$	$s_2^2 = \$37$

Assuming that costs are normally distributed and the population variances are equal, perform an appropriate hypothesis test with $\alpha = 0.05$.

FURTHER READINGS

DIXON, W., and F. MASSEY, *Introduction to Statistical Analysis.* New York: McGraw-Hill, 1969. Chapters 7 and 8 cover hypothesis testing. Particularly helpful are pages 114–123, which give a concise survey of the many different statistics that can be used to test hypotheses concerning the means of two populations.

GILBERT, N., *Statistics.* Philadelphia: W. B. Saunders, 1976. Chapter 11 of this text covers hypothesis testing on an elementary level, but includes material on Type I and Type II errors, which in this chapter we covered only in the exercises. [We *will* cover this material in detail in Chapter 14.]

MENDENHALL, W., and L. OTT, *Understanding Statistics. North Scituate, Massachusetts*:Duxbury Press, 1976. Chapter 7 of this text is similar in topic coverage to Chapter 11 of the Gilbert text.

Readings from Contemporary Problems in Statistics (Bernhardt Lieberman, ed.). London: Oxford University Press, 1971. Section 2 of this book contains a number of papers attacking the general principle of hypothesis testing. It is not easy reading, but should be of interest to those who want to see what some researchers believe to be wrong with the principle of hypothesis testing.

ROMANO, A., *Applied Statistics for Science and Industry.* Boston: Allyn and Bacon, 1977. Chapter 3 surveys hypothesis testing for large and small samples, with special attention to Type I and Type II error. (The level is higher than in this text.) Of special interest is Section 3.12, which gives a brief introduction to Bayesian decision analysis—an alternative to the classical hypothesis testing covered here.

ANSWERS TO REINFORCEMENT EXERCISES

A   *Null hypothesis:* $\mu = 35{,}000$ miles
Alternative hypothesis: $\mu > 35{,}000$ miles

1. $\alpha = 0.05$

2. $c = z_{0.05} = 1.64$

3. $z = \dfrac{35{,}840.65 - 35{,}000}{2420.38 / \sqrt{40}} \doteq 2.20$

4. Reject the null hypothesis.

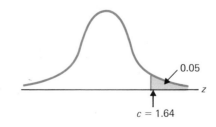

B   *Null hypothesis:* $\mu = 1$ liter
      Alternative hypothesis: $\mu < 1$ liter

1. $\alpha = 0.01$
2. $c = -z_{0.01} = -2.33$

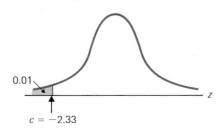

0.01

$c = -2.33$

3. $z = \dfrac{0.992 - 1}{0.023/\sqrt{90}} = -3.30$
4. Reject the null hypothesis.

C   *Null hypothesis:* $\mu = \$18{,}000$
      Alternative hypothesis: $\mu \neq \$18{,}000$

1. $\alpha = 0.05$
2. $-c = -z_{0.025} = -1.96;\ c = z_{0.025} = 1.96$

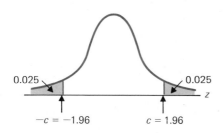

0.025 0.025

$-c = -1.96$ $c = 1.96$

3. $z = \dfrac{17{,}331.05 - 18{,}000}{3{,}920.62/\sqrt{400}} \doteq -3.41$
4. Reject the null hypothesis.

D   *Null hypothesis:* $p = 0.474$
      Alternative hypothesis: $p \neq 0.474$
1. $np_0 = 300(0.474) = 142.2$,
   $n(1 - p_0) = 300(1 - 0.474) = 157.8$
2. $\alpha = 0.05$

3. $-c = -z_{0.025} = -1.96;\ c = z_{0.025} = 1.96$
4. $\hat{p} = 0.411$

$$z = \frac{0.411 - 0.474}{\sqrt{0.474(1 - .474)/300}} \doteq -2.19$$

5. Reject the null hypothesis.

E   *Null hypothesis:* $\mu_1 = \mu_2$
      Alternative hypothesis: $\mu_1 < \mu_2$

1. $\alpha = 0.01$
2. $c = -z_{0.01} = -2.33$

3. $z = \dfrac{1217.64 - 1258.85}{\sqrt{[(119.86)^2/100] + [(123.57)^2/150]}} \doteq -2.63$

4. Reject the null hypothesis.

F   *Null hypothesis:* $\mu_1 = \mu_2$
      Alternative hypothesis: $\mu_1 \neq \mu_2$

1. $\alpha = 0.05$
2. $-c = -z_{0.025} = -1.96;\ c = z_{0.025} = 1.96$

3. $z = \dfrac{72.3 - 78.9}{\sqrt{[(9.1)^2/35] + [(8.8)^2/32]}} \doteq -3.02$

4. Reject the null hypothesis.

G   *Null hypothesis:* $\mu = -2$
      Alternative hypothesis: $\mu < -2$

1. $\alpha = 0.05$
2. $c = -z_{0.05} = -1.64$

3. $z = \dfrac{-2.80 - (-2)}{4.06/\sqrt{35}} \doteq -1.17$

4. The null hypothesis cannot be rejected.

H   *Null hypothesis:* $\mu = 1.25$
      Alternative hypothesis: $\mu < 1.25$

1. $\alpha = 0.05$
2. $c = -z_{0.05} = -1.64$

3. $z = \dfrac{1.03 - 1.25}{0.92/\sqrt{40}} \doteq -1.51$

4. The null hypothesis cannot be rejected.

I *Null hypothesis:* $\mu = 325$
   Alternative hypothesis: $\mu < 325$

1. $\alpha = 0.05$
2. $c = -t_{0.05}$, d.f. = 4; $c = -1.76$

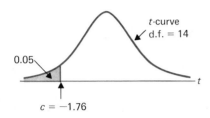

3. $t = \dfrac{315.86 - 325}{19.82/\sqrt{15}} \doteq -1.79$
4. Reject the null hypothesis.

J a) *Null hypothesis:* $\mu = 16$
   Alternative hypothesis: $\mu \neq 16$

1. $\alpha = 0.10$
2. $t_{\alpha/2} = t_{0.05} = 1.73$ for d.f. = 19; $-c = -1.73$ and $c = 1.73$

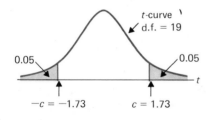

3. $t = \dfrac{16.08 - 16}{0.18/\sqrt{20}} \doteq 1.99$
4. Reject the null hypothesis.

b) *Null hypothesis:* $\mu = 16$
   Alternative hypothesis: $\mu \neq 16$

1. $\alpha = 0.05$
2. $t_{\alpha/2} = t_{0.025} = 2.09$ for d.f. = 19; $-c = -2.09$ and $c = 2.09$

3. $t = \dfrac{16.08 - 16}{0.18/\sqrt{20}} \doteq 1.99$

4. The null hypothesis cannot be rejected.

K *Null hypothesis:* $\mu_1 = \mu_2$
   Alternative hypothesis: $\mu_1 < \mu_2$

1. $\alpha = 0.10$
2. $t_{0.10} = 1.48$ for d.f. = $4 + 3 - 2 = 5$; $c = -t_{0.10} = -1.48$

3. $s_p = \sqrt{\dfrac{(4-1)(2.49)^2 + (3-1)(2.88)^2}{4+3-2}} \doteq 2.65$

$t = \dfrac{86.80 - 90.40}{2.65\sqrt{\frac{1}{4} + \frac{1}{3}}} \doteq -1.78$

4. Reject the null hypothesis.

Inferences concerning variances

9.1 Introduction—why we study variances

In this chapter we will look at methods for testing hypotheses and finding confidence intervals for variances. Researchers study variances in situations where variation must be controlled.

For example, a manufacturer who is setting up production lines to manufacture bolts knows that each bolt manufactured may be *slightly* different from all others. But even if he is willing to accept some variation in bolt size, he cannot tolerate *too much* variation. He must look at variance, and be sure that it is kept from getting too large. Educators who design tests also attempt to control variation in scores. On some tests, such as college entrance examinations, test designers will try to keep the variance from getting too small or too large. On other tests, such as minimum competency tests, testers will attempt to make the variance as small as possible.

In this section, we will give examples of manufacturing and test-design situations that illustrate the necessity for studying variances. In the following sections, we will show how statistical inferences are made in such situations.

Example 1 ◀ A manufacturer needs to produce bolts approximately 10 millimeters in diameter to fit into a circular hole 10.4 millimeters in diameter. Anyone who has ever put together a child's bicycle, a barbecue grill, or a metal bookcase would hope that the manufacturer gets things right: Bolts that are more than 10.4 mm in diameter won't fit into the holes;

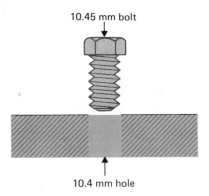

10.45 mm bolt

10.4 mm hole

those that are much less than 10.0 mm won't stay in place to keep the components together!

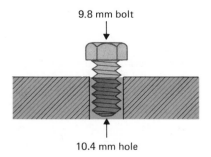

9.8 mm bolt

10.4 mm hole

The manufacturer sets up two trial production lines. He is aiming for a 10-mm bolt, and each production line *does* produce bolts whose diameters are *normally distributed* with mean $\mu = 10$ mm. He is also interested in saving money. The second production line is less costly to run, but has a larger variation in output. The normal distribution for the first line has $\sigma = 0.1$ mm, while the normal distribution for the second line has $\sigma = 0.3$ mm.*

The manufacturer stages a trial run for each line, and has a quality-control engineer take a random sample of 20 bolts from each line. These bolts are then very carefully measured. The diameters of the bolts from the two lines are given in Table 9.1 (to the nearest hundredth of a millimeter).

It seems pretty clear that the manufacturer is due for trouble with Production line 2. Three of the bolts won't fit at all, and one came very close to being too large. Production line 1, on the other hand, seems to present no problems.

Production line 1 $\mu = 10, \sigma = 0.1$	Production line 2 $\mu = 10, \sigma = 0.3$	**Table 9.1** Diameter of bolts (in mm)
10.12	10.38 ⟵ Too close for comfort	
10.02	10.07	
9.92	9.77	
10.06	10.10	
10.15	10.45	
10.10	10.31	
9.91	9.74	
10.07	10.22	
9.99	9.98	
10.16	10.47	
9.96	9.87	
9.98	9.94	Won't fit
9.99	9.97	
10.13	10.04	
10.01	10.02	
10.02	10.07	
10.00	10.01	
10.10	10.31	
10.16	10.48	
10.03	10.08	

Since the bolt diameters from the two production lines are normally distributed, it is not hard to see why this happened. We will compute the probability that a bolt is too large, $P(x > 10.4)$, for each production line.

* We are assuming the manufacturer knows the information given in this paragraph for the sake of illustrating a point. In practice this information must be obtained by using statistical inference methods. Such methods will be discussed in Sections 9.2 and 9.3.

Production line 1: $\mu = 10.0, \sigma = 0.1$.
$x = 10.4$ corresponds to $z = (10.4 - 10)/0.1 = 4$, and thus

$$P(x > 10.4) = P(z > 4) = 1 - 0.99997 = 0.00003 = \frac{3}{100,000} \;^*$$

Production line 2: $\mu = 10.0, \sigma = 0.3$.
$x = 10.4$ corresponds to $z = (10.4 - 10)/0.3 \doteq 1.33$. Therefore,

$$P(x > 10.4) = P(z > 1.33) = 1 - 0.9082 = 0.0918 \doteq \frac{9}{100}$$

In the long run, this means that for Production line 1 we would expect only three out of every 100,000 bolts to be too large, but that for Production line 2 we would expect an average of nine out of every 100 bolts to be too large. Production line 2 may be less costly, but its variation is too large. This illustrates an important point. *The manufacturer wishes to keep production-line variance small.* ▶

In actual practice a manufacturer will not be able to set up production lines with *known* variances as we just did for the sake of an example. Instead, a manufacturer will start out with some idea of the variance that is needed, set up production lines, sample their output, and attempt to determine whether one of the lines produces bolts whose diameters have a reasonably small variance. The following example illustrates this situation.

Example **2** ◀ A manufacturer needs to produce bolts approximately 10 mm in diameter to fit in a hole of diameter 10.4 mm. The manufacturer has two trial production lines. From past experience with such production lines, the manufacturer knows that each line will produce bolts whose diameters are *normally distributed.*

He calls in a statistician. She first takes random samples of bolts from both production lines. Then using tests on *means* (see Chapter 8), she finds that she is able to assure the manufacturer that each line does, in fact, produce bolts whose mean diameter is about 10 mm.

However, as we have seen, the manufacturer also needs to be concerned about the *variance* of bolt diameters. First he needs to know what an appropriate variance would be; and secondly, he must know whether either (or both) of his production lines have this appropriate variance.

To answer the first question, the statistician does some normal-curve calculations (as in Example 1) for a number of different possible values of σ. She then tells the manufacturer, "If you can keep $\sigma \doteq 0.1$ for the production line you use, then, on the average, only three bolts out of every 100,000 will be too large. Aim for $\sigma \doteq 0.1$." In other words, $\sigma^2 \doteq 0.01$ is an appropriate variance of bolt diameters for the manufacturer's production line.

Now to the manufacturer's second question. Do either (or both) of his production lines have an appropriate bolt-diameter variance of $\sigma^2 \doteq 0.01$? In order to answer this

* We used a table giving more accuracy than Table I to get this value.

question the statistician decides to first take a random sample of 20 bolt diameters from each production line. She obtains the data shown in Table 9.2.

Production line 1		Production line 2	
9.91	10.05	10.48	10.26
9.97	10.18	10.07	9.73
9.84	10.06	9.89	10.29
9.97	9.98	10.38	9.97
10.18	9.91	9.50	10.38
10.08	10.07	9.95	9.94
10.03	9.98	9.81	10.14
10.02	10.10	9.87	10.17
9.88	9.99	10.13	10.17
10.03	9.97	10.03	10.09

Table 9.2
Sample diameters (in mm)

Based on this data, the statistician can get an *estimate* of each production line's variance by computing the *sample* variance. Using the usual formula (see page 58) she obtains

$$\text{Production line 1} \qquad \text{Production line 2}$$
$$s^2 = 0.008 \qquad\qquad s^2 = 0.058$$

Since the manufacturer is aiming for a variance of $\sigma^2 \doteq 0.01$, it *appears* from this data that Production line 1 *is* suitable, but that Production line 2 isn't.* However, the statistician would not simply *guess* on the basis of this data; she would perform a hypothesis test, or find a confidence interval. In this case, for example, the statistician might wish to do a test of the following kind for each of the production lines:

Null hypothesis: $\sigma^2 = 0.01$
Alternative hypothesis: $\sigma^2 > 0.01$

If the null hypothesis can be rejected, this would indicate that the production line output is too varied. If the null hypothesis cannot be rejected, this would indicate that the production line might be acceptable. ▶

Suppose you sample from a production line, and obtain the following bolt diameters.   Exercise **A**

9.81	9.72	9.98	10.17	9.79
9.93	10.25	10.09	10.26	9.96
10.04	10.21	9.87	10.10	10.23
10.10	10.15	9.73	10.14	9.46

a) Calculate $\bar{x}$, s^2, and s.

---

* You could probably come to the same conclusion by looking at the table of sample diameters, and recalling that the bolts must fit into a hole of diameter 10.4 mm.

b) What guess might you make about using this production line, if you needed bolts with a mean diameter of 10 mm and a variance of about 0.01 mm^2?

c) What hypothesis test would you wish to perform here regarding μ? regarding σ^2?

The statistician in Example 2 might also wish to calculate a 95-percent confidence interval for σ^2 for each of the production lines, and use this information to evaluate the production lines. At present, we can neither perform a hypothesis test nor find a confidence interval for σ^2, since we do not know what the probability distribution of s^2 looks like. In the next few sections, we will introduce the probability distribution used for s^2, and show how to set up hypothesis tests and confidence intervals for σ^2.

Before moving on to the next section, we will give two more examples of situations in which attempts are made to control variation.

Example 3 ◀ In the year 1976, more than one million candidates for college admission took the Scholastic Aptitude Test (SAT). This test, provided by the Educational Testing Service of Princeton, New Jersey, is used by many colleges to measure the verbal and mathematical abilities of prospective students. Student scores are reported on a scale that ranges from a low of 200 to a high of 800. This scale was introduced in 1941. At that time, the mean of 1941 students was arbitrarily set at 500 and the standard deviation at 100. Since 1941 a process called *equating* has been used to ensure that scores on new editions of the test maintain their old meaning—a 600 in 1960 should indicate nearly the same level of ability as a 600 today. In recent years there has been a great deal of discussion of downward trends in mean scores. No single reason for mean-score drops has been identified, but it is clear that scores have dropped. Below we list means and standard deviations for all test takers in five test years between 1956 and 1976.

	SAT Verbal		SAT Math	
Test year	Mean	SD	Mean	SD
1956–57	473	105	496	111
1960–61	474	108	495	109
1966–67	467	109	495	110
1970–71	454	109	487	113
1976–77	429	108	471	117

As you can plainly see, the means have dropped—average performance is certainly down. However, standard deviations have never moved very far from the 1941 value of 100. This is important. If the standard deviation of all SAT verbal scores became very small, there would be very little variation in test scores. All students would have nearly identical scores, and the tests would be useless to colleges that intend to use them for selection purposes. With a standard deviation that is close to 100 every year, users of the SAT can be more confident that scores are spread apart nearly as they have been in the past. Test designers really do construct aptitude tests in such a way that standard deviations do not become too large or too small.* ▶

◀ Another kind of test has received a good deal of attention in the last few years—the *minimum competency test*. Many states have passed laws requiring that students pass some kind of basic-skills test covering arithmetic, reading, and writing. The goal of such basic-skill tests is to ensure that every school graduate can write a letter, read a contract, and figure prices. The tests cover only the most basic skills, and the test designers would like students to be able to do almost everything on them—i.e., they would like all school graduates to have enough basic skill to score nearly 100 percent. When this happens, data variation will be very small. For an educator running a basic-skills program, the main objective is to make test-score variance small. This is quite different from the objective of the SAT test designers. The SAT designers design aptitude tests that spread students' scores out year after year with almost the same variance. The basic-skills program manager takes a group of students with a large variance in basic-skills scores and trains them so that their score variance will be made much smaller when they take a similar test a few months later. ▶

Example **4**

EXERCISES Section **9.1**

Note: The exercises in this section are intended only to test your *basic understanding* of how tests concerning variances are set up. You will be asked only to make guesses here. You will see how these hypothesis tests are actually done in Section 9.3.

In Exercises 1 through 5:

a) State the appropriate null hypothesis,

b) State an alternative hypothesis,

c) Calculate s^2,

d) Make a *guess* as to whether or not the null hypothesis should be rejected.

1 A manufacturer is making tubes to be fitted into a receptacle 1.1 cm in diameter. The variance should not exceed 0.01. Below is a sample of size $n = 25$ from the production line.

1.04	1.00	0.94	1.00	1.09
1.02	1.10	1.00	0.93	0.94
1.17	1.00	0.86	0.96	1.14
0.84	1.11	1.19	0.97	1.04
1.00	0.97	0.99	1.09	0.92

2 A shop class attempts to make one-foot-long metal rods. They want to keep the variance at about 0.02. Below is a sample of size $n = 20$ from the class.

0.86	1.14	1.11	0.97	0.82
0.89	0.86	1.14	1.21	0.80
1.05	1.14	1.06	0.98	0.84
1.14	1.24	0.97	1.11	0.90

3 A bottling plant distributes four-liter bottles of bleach. The variance should not exceed 0.02. Below is a sample of 16 bottles from the plant.

3.96	3.97	4.12	3.87
4.14	3.79	4.13	3.72
3.77	4.25	4.09	3.91
4.22	3.84	3.86	4.15

4 A manufacturer needs to produce 10-mm bolts to fit into a hole of diameter 10.4 mm. The machine that is presently used produces bolts with a variance of 0.02. A new machine has come out, and a salesman claims it produces bolts with a variance smaller than 0.02. The manufacturer decides that he will buy the new machine if he can be convinced that it actually does produce bolts with a variance *smaller* than 0.02.

---

* The reader who wishes further information on these tests might read *The College Board Scholastic Aptitude Test*, by James Braswell, in the March 1978 issue of the journal *The Mathematics Teacher*.

A random sample of 15 bolts from the new machine yields the following data on bolt diameters.

9.91	10.03	10.06	9.99	10.08
9.84	9.88	9.91	9.97	10.02
10.18	10.05	9.98	9.97	10.03

5 A company produces a 14-ounce can of stewed tomatoes. Recently the company hired a quality-control engineer. He is interested in the variance in the weight of the cans of stewed tomatoes. The production manager claims that the variance is 0.5, but the quality-control engineer decides he wants to check this claim. Below are the data obtained from a random sample of 10 cans.

12.76	13.34	14.76	13.77	14.09
13.79	13.68	14.36	14.06	13.86

9.2 The χ^2-statistic and the chi-square curve

Let us return to Example 2 where the bolt manufacturer needs to prevent excessive variation of bolt diameters on his production line. As we pointed out in Section 9.1, this manufacturer would like to perform a hypothesis test with:

Null hypothesis: $\sigma^2 = 0.01$
Alternative hypothesis: $\sigma^2 > 0.01$

The rough idea behind this hypothesis test is to estimate σ^2 by calculating s^2, and to reject the null hypothesis if the calculated value of s^2 is "too large." The quantity that a statistician will actually look at here is not simply s^2. A statistician will calculate the quantity χ^2 (pronounced "ky-square," and written "chi-square") given below.

The χ^2-statistic for variance tests

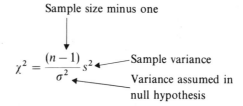

This doesn't change things very much; χ^2 is simply s^2 multiplied by a number. For example, in the case of the production lines, where $n = 20$ and σ^2 is assumed (in the null hypothesis) to be 0.01, we have

$$\chi^2 = \frac{(20-1)}{0.01}s^2 = \frac{19}{0.01}s^2 = 1900s^2$$

Larger values of s^2 result in larger values of χ^2, and smaller values of s^2 result in smaller values of χ^2.

The advantage of considering χ^2 instead of s^2 is quite similar to the advantage of standardizing a normal random variable x. Recall that if x is a normal random variable with mean μ and standard deviation σ, then we can use the standard normal table to find probabilities for x by considering the standardized random variable $z = (x - \mu)/\sigma$. Subtracting μ from x and then dividing by σ allows us to use the *same* probability table to find probabilities for normal random variables x regardless of the values of μ and σ. Similarly, multiplying s^2 by $n - 1$ and then dividing by σ^2 enables us to use the *same* probability table to find probabilities for s^2 regardless of the value of σ^2.

◀ Compute the χ^2-statistic for the sample-bolt data from Production line 1 in Example 2.   Example **5**

On Production line 1, $n = 20$, $\sigma^2 = 0.01$ (value assumed in null hypothesis), and   Solution
$s^2 = 0.008$ (from page 317). So

$$\chi^2 = \frac{(20-1)}{0.01}(0.008) = 1900(0.008) = 15.2 \quad ▶$$

Compute the χ^2-statistic for the sample-bolt data from Production line 2 in Example 2.   Exercise **B**
(Take $\sigma^2 = 0.01$ (the value assumed in the null hypothesis), and use the value of
$s^2 = 0.058$ computed for Production line 2 on page 317.)

The χ^2-statistic is a random variable—its value depends on chance. When you
take different samples, you will obtain different values of χ^2. The way to find probabilities for χ^2 is quite similar to the method for finding probabilities for the random
variable z.

> If a random sample of size n is taken from a population that is *normally distributed*
> with variance σ^2, then probabilities for the random variable
>
> $$\chi^2 = \frac{(n-1)}{\sigma^2}s^2$$
>
> can be found using areas under special curves called *chi-square curves* or
> χ^2-*curves*.

The shape of a chi-square curve depends on the sample size n. Three chi-square
curves are pictured in Fig. 9.1. As with the Student's t-distribution curves (t-curves), the
number $n-1$ is called the degrees of freedom, or d.f. We will use this terminology in
Fig. 9.1.

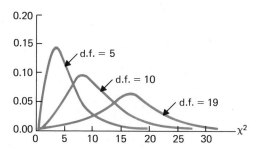

Figure 9.1
Chi-square curves for
d.f. = 5, 10, and 19.

In this figure, we can observe some basic properties of the chi-square curves.

1. Chi-square curves are different for different degrees of freedom.
2. A chi-square curve starts at 0 on the horizontal axis and extends to the right. The
   χ^2-values of interest to us are marked on the horizontal axis.

3. A chi-square curve is not symmetrical. It climbs to its high point rapidly, and comes back to the axis more slowly. It is skewed to the right.

4. The total area under a chi-square curve is 1.

Areas Since probabilities for χ^2 are equal to areas under chi-square curves, we will now show how to find areas under such curves. A part of the chi-square table in this book, Table III, is reproduced below in Table 9.3.

Table 9.3

					Area to the right of χ^2-value					
α d.f.	.995	.99	.975	.95	.90	.10	.05	.025	.01	.005
1	.00	.00	.00	.00	.02	2.71	3.84	5.02	6.63	7.88
2	.01	.02	.05	.10	.21	4.61	5.99	7.38	9.21	10.60
.	.	.	.	.	.	.	.	.	.	.
6	.68	.87	(1.24)	1.64	2.20	10.64	12.59	(14.45)	16.81	18.55
7	.99	1.24	1.69	2.17	2.83	12.02	14.07	16.01	18.48	20.28
.	.	.	.	.	.	.	.	.	.	.
19	6.84	7.63	8.91	10.12	11.65	27.20	30.14	32.85	36.19	38.58
.	.	.	.	.	.	.	.	.	.	.

The values inside the table are χ^2-values, i.e., the values marked off on the horizontal axis for chi-square curves. On top of the table, *the area to the right* of each χ^2-value is given. The column on the left gives the degrees of freedom for the chi-square curve.

For example, we have circled the value 1.24 inside the table. The 6 on the left under d.f. indicates that this χ^2-value has been calculated for a chi-square curve with six degrees of freedom. The 0.975 above 1.24 in the table shows that *with six degrees of freedom, the area under the chi-square curve to the right of 1.24 is 0.975*. This is pictured in Fig. 9.2. Note that the unshaded area to the left of 1.24 is 0.025, since the total area under the curve is 1 and $1 - 0.975 = 0.025$.

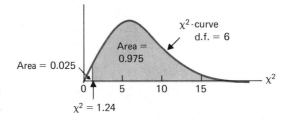

Figure 9.2
Chi-square curve with d.f. = 6.

To find the χ^2-value that has an area of 0.025 to its right for the same curve, we would read down from the value of 0.025 above the table until we reached row 6. For an area of 0.025, $\chi^2 = 14.45$. This is pictured in Fig. 9.3.

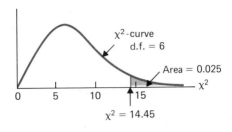

Figure 9.3
Chi-square curve, d.f. $= 6$

Once we know the areas of 0.025 on either end of the curve, we can find the two χ^2-values that divide the area under the curve into a middle 0.95 and an outside 0.05. This is shown in Fig. 9.4.

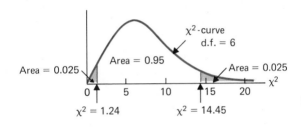

Figure 9.4

For 7 degrees of freedom: Exercise **C**

1. Find the χ^2-value with area 0.05 to its right.
2. Find the χ^2-value with area 0.95 to its right.
3. Using the χ^2-graph below, shade the two outer areas of 0.05, and label the middle segment of area 0.90.

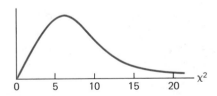

Find the χ^2-values that separate the middle area of 0.90 from the two outer areas of   Exercise **D**
0.05, for a chi-square curve with nine degrees of freedom. Illustrate your work with a picture.

It is useful to use the symbol $\chi^2_{0.05}$ for the χ^2-value with area 0.05 to its right, $\chi^2_{0.025}$ Notation
for the χ^2-value with area 0.025 to its right, and so forth.

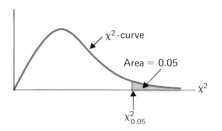

Example **6** ◖ For a χ^2-curve with six degrees of freedom (see Table 9.3),

$$\chi^2_{0.05} = 12.59$$

$$\chi^2_{0.025} = 14.45$$

$$\chi^2_{0.95} = 1.64$$ ◗

Exercise **E**   For a χ^2-curve with seven degrees of freedom, find
a) $\chi^2_{0.10}$ b) $\chi^2_{0.05}$ c) $\chi^2_{0.01}$ d) $\chi^2_{0.995}$ e) $\chi^2_{0.95}$

EXERCISES Section **9.2**

1 For 19 degrees of freedom
   a) Find the χ^2-value with area 0.05 to the right.
   b) Find the χ^2-value with area 0.95 to the right.
   c) Draw a chi-square curve and locate the values from (a) and (b) on it.
   d) Shade the two outer areas of 0.05 and label the middle segment of area 0.90.

2 Find the χ^2-values that separate the middle 0.95 area from the outer 0.025 areas for a chi-square curve with seven degrees of freedom. Illustrate your work with a picture.

3 Find the χ^2-values that separate the middle 0.98 area from the outer 0.01 areas for a chi-square curve with two degrees of freedom. Illustrate your work with a picture.

4 For a χ^2-curve with nineteen degrees of freedom find
   a) $\chi^2_{0.05}$ b) $\chi^2_{0.025}$ c) $\chi^2_{0.005}$ d) $\chi^2_{0.975}$

5 For a χ^2-curve with ten degrees of freedom find
   a) $\chi^2_{0.01}$ b) $\chi^2_{0.95}$ c) $\chi^2_{0.975}$ d) $\chi^2_{0.99}$

6 For a χ^2-curve with two degrees of freedom find
   a) $\chi^2_{0.10}$ b) $\chi^2_{0.025}$ c) $\chi^2_{0.005}$

9.3 Hypothesis tests for a single variance

It is time to return to the bolt-manufacturing problem once more. In Section 9.2, we were forced to leave the problem to do some necessary mathematics. Let us review the important fact we learned there.

Suppose a random sample of size n is taken from a population that is *normally distributed* with variance σ^2. If s^2 is the *sample* variance, then the random variable

$$\chi^2 = \frac{(n-1)}{\sigma^2} s^2$$

has the *chi-square distribution* with d.f. $= n - 1$. That is, probabilities for the random variable χ^2 can be determined using areas under the chi-square curve with d.f. $= n - 1$.

◀ As you may remember, for Production line 2 of the bolt-manufacturing problem, we   Example **7**
want to perform the hypothesis test:

Null hypothesis: $\sigma^2 = 0.01$
Alternative hypothesis: $\sigma^2 > 0.01$

Our sample consists of 20 bolts randomly selected from Production line 2. The
diameters of these bolts are given on page 317; their sample variance is $s^2 = 0.058$. We
have already pointed out that we will compute the χ^2-statistic for this sample:

$$\chi^2 = \frac{(n-1)}{\sigma_0^2}s^2$$

where σ_0^2 is the value for σ^2 given in the null hypothesis (in this case $\sigma_0^2 = 0.01$). Then we
will try to determine whether this χ^2-statistic is "too large." And what, exactly, does "too
large" mean? Just as in our previous hypothesis tests, this depends on our significance
level α. For example, if $\alpha = 0.05$, we reason as follows:

For 19 $(= 20 - 1)$ degrees of freedom, $\chi^2_{0.05} = 30.14$

So if the null hypothesis is true, then the probability that our calculated χ^2-value is
larger than 30.14 is 0.05. In mathematical symbols, $P(\chi^2 > 30.14) = 0.05$. Thus, if the
null hypothesis is true, χ^2-values greater than 30.14 should occur only 5 percent of the
time. We will classify χ^2-values greater than 30.14 as "too large" (see Fig. 9.5). As in
Chapter 8, we will call 30.14 the *critical value* for the test.

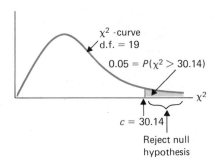

χ^2 -curve
d.f. = 19

$0.05 = P(\chi^2 > 30.14)$

χ^2

$c = 30.14$

Reject null
hypothesis   **Figure 9.5**

We can now perform the hypothesis test:

Null hypothesis: $\sigma^2 = 0.01$
Alternative hypothesis: $\sigma^2 > 0.01$

$\alpha = 0.05.$

Since the sample of 20 bolts from Production line 2 yielded $s^2 = 0.058$, we have

$$\chi^2 = \frac{(n-1)}{\sigma_0^2} s^2 = \frac{(20-1)}{0.01}(0.058) = 110.20$$

Since $\chi^2 > 30.14$ we reject the null hypothesis, and conclude that $\sigma^2 > 0.01$ for Production line 2. ▶

The general procedure for doing a hypothesis test as in Example 7 is given below.

Procedure

Procedure for performing a hypothesis test of the form:
Null hypothesis: $\sigma^2 = \sigma_0^2$
Alternative hypothesis: $\sigma^2 > \sigma_0^2$

Assumption: Normal population

1. Decide on the *significance level*, α.
2. The *critical value* is ·

$$c = \chi_\alpha^2, \qquad \text{d.f.} = n-1$$

Use Table III to find χ_α^2 for d.f. $= n - 1$.

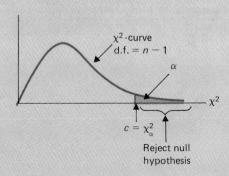

3. Compute the value of

$$\chi^2 = \frac{n-1}{\sigma_0^2} s^2$$

where σ_0^2 is the value given for σ^2 in the null hypothesis, n is the sample size and s^2 is the sample variance.
4. If $\chi^2 > c$, *reject* the null hypothesis.

We illustrate this procedure in the following example.

◀ A teacher in a *large* high school believes that the variance in College Board verbal   Example **8**
scores in his school in 1976–77 is significantly greater than the national figure of 11,664.
He takes a random sample of the scores of 30 students, and finds $s^2 = 19,044$ for this
sample. Use this data to test his hypothesis. Use $\alpha = 0.05$.

Since experience indicates that College Board verbal scores are normally
distributed, we can use the procedure described above to perform the test:

Null hypothesis: $\sigma^2 = 11,664$
Alternative hypothesis: $\sigma^2 > 11,664$

1. $\alpha = 0.05$.
2. The *critical value* is $c = \chi_\alpha^2 = \chi_{0.05}^2$ which equals 42.56 for d.f. $= 30 - 1 = 29$. See
    Fig. 9.6.

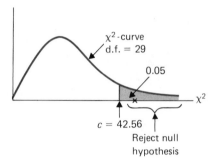

χ^2-curve
d.f. = 29

0.05

χ^2

$c = 42.56$

Reject null
hypothesis

Figure 9.6

3. Compute the value of χ^2:

$$\chi^2 = \frac{(n-1)}{\sigma_0^2} s^2$$

We have $\sigma_0^2 = 11,664$, $n = 30$, and $s^2 = 19,044$. Therefore

$$\chi^2 = \frac{30-1}{11664} \cdot 19044 = 47.35$$

We have marked this value of χ^2 with a cross on Fig. 9.6.
4. Since $\chi^2 > c$, we reject the null hypothesis.

In other words, the teacher can conclude that the variance in College Board verbal
scores in his school is greater than the national average. ▶

Use the bolt-diameter data (page 317) from Production line 1 to test the null hypothe-   Exercise **F**
sis $\sigma^2 = 0.01$ against the alternative hypothesis $\sigma^2 > 0.01$. Use $\alpha = 0.05$.

For a one-tailed test of the form

Null hypothesis: $\sigma^2 = \sigma_0^2$
Alternative hypothesis: $\sigma^2 < \sigma_0^2$

we use the same basic procedure as the one given above on page 326. The only differences are that the critical value is $c = \chi^2_{1-\alpha}$, and we reject the null hypothesis if $\chi^2 < c$.

Similarly, for a two-tailed test of the form

Null hypothesis: $\quad \sigma^2 = \sigma^2_0$
Alternative hypothesis: $\quad \sigma^2 \neq \sigma^2_0$

the procedure is also essentially the same as the one on page 326. However, in this case there are *two* critical values,

$$\chi^2_{1-\alpha/2} \quad \text{and} \quad \chi^2_{\alpha/2}$$

and the null hypothesis is rejected if either $\chi^2 < \chi^2_{1-\alpha/2}$ or $\chi^2 > \chi^2_{\alpha/2}$.

These last two hypothesis tests will be explored in the exercises.

P-values*   We have done our hypothesis tests using tables of critical values. This method was standard before computers and calculators became widely available. However, modern electronic computation will enable you to find the exact area to the right of the value of a χ^2-statistic. For example, our own statistical calculator gave the results shown in Fig. 9.7 for the areas to the right of the χ^2-values for the two production lines.

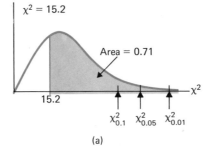
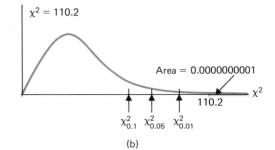

Figure 9.7
(a) Production line 1
(b) Production line 2

These areas are called the **P-values** for the values of our χ^2-statistic. The *P*-value for Production line 1 shows clearly that we could not reject the null hypothesis $\sigma^2 = 0.01$ for Production line 1 at any of the standard levels, 0.10, 0.05, or 0.01. The critical values of χ^2 for tests at these levels have areas far less than 0.71 to their right. But the *P*-value for Production line 2 shows that we would reject the null hypothesis for Production line 2 no matter which of these standard levels we chose.

---

* The remainder of this section is optional

A computer print-out for the same data would not mention the area to the right. It would have statements like these:

Production line 1:

```
SIGNIFICANCE LESS THAN .71
```

Production line 2:

```
SIGNIFICANCE LESS THAN .0000000001.
```

EXERCISES Section **9.3**

In Exercises 1 through 7, assume that the populations are normally distributed, and that the required test is one-tailed with alternative hypothesis of the form $\sigma^2 > \sigma_0^2$.

a) State the null hypothesis,

b) State the alternative hypothesis,

c) Find the critical value,

d) Draw a picture,

e) Compute χ^2,

f) State whether or not the null hypothesis should be rejected.

1 The manufacturer in Exercise 1, Section 9.1, makes tubes to be fitted into a receptor 1.1 cm in diameter. The variance should not exceed 0.01. A random sample of size $n = 25$ yielded $s^2 = 0.008$. Use this information to test the appropriate hypotheses. Use $\alpha = 0.05$.

2. In Exercise 2, Section 9.1, a shop class attempts to make one-foot metal rods. They want to keep the variance of the length at about 0.02. A sample size of $n = 20$ yielded $s^2 = 0.019$. Use this information to test the appropriate hypotheses. Use $\alpha = 0.10$.

3 Suppose in Exercise 2, they want to keep the variance at 0.01. Using the data of Exercise 2, test the appropriate hypotheses (use $\alpha = 0.10$).

4 The bottling plant in Exercise 3, Section 9.1, distributes four-liter bottles of bleach. The variance should not exceed 0.02. A random sample of 16 bottles yielded $s^2 = 0.029$. Use this information to test the appropriate hypotheses. Use $\alpha = 0.05$.

5 Suppose the variance in Exercise 4 was calculated to be $s^2 = 0.04$. Perform the hypothesis test of Exercise 4 using $\alpha = 0.05$.

6 A production line has workers who perform the same task. The job takes each worker an average of $\mu = 15$ minutes. The variance should not exceed $\sigma^2 = 1$. Below is a random sample of times, in minutes, for nine workers.

15.4	14.7	16.7
14.3	15.2	13.5
17.6	14.8	16.3

Use this data to test the appropriate hypotheses. Use $\alpha = 0.05$.

7 A manufacturer of electronic parts makes a component (used in the assemblage of a product) that lasts a mean of $\mu = 4000$ hours of use. The variance of the lifetimes of the component should not exceed 50. Below is a random sample of the lifetimes for eight of these very expensive components.

3993	4002	4007	4007
3994	4007	3993	4012

Use this data to test the appropriate hypotheses. Use $\alpha = 0.05$.

8 (*Tests with alternative* $\sigma^2 < \sigma_0^2$.) A basketball referee claims that he can consistently estimate three seconds without using a watch. To test his consistency, he is asked to do 21 separate estimations while a person with a watch records the actual times. The data is to be used to perform the hypothesis test.

Null hypothesis: $\sigma^2 = 0.1$

Alternative hypothesis: $\sigma^2 < 0.1$

at the significance level $\alpha = 0.1$.

a) If the test leads to rejection of the null hypothesis, this will be taken as a strong indication that the referee is fairly consistent. Explain in your own words why this is so.

b) Perform this hypothesis test using the data $n = 21$ and $s^2 = 0.06$ for the trials. (Remember that your critical value will be $c = \chi_{0.90}^2$, and your rejection region will consist of all χ^2-values *less than c*.)

<u>9</u> In Exercise 4 of Section 9.1 a manufacturer of bolts decided he will buy a new machine if the bolts produced by this new machine have a variance less than 0.02. A random sample of 15 bolts from this machine yielded the following data on bolt diameters.

9.91	10.03	10.06	9.99	10.08
9.84	9.88	9.91	9.97	10.02
10.18	10.05	9.98	9.97	10.03

Perform the appropriate hypothesis test at the 0.05 significance level.

<u>10</u> The manufacturer in Examples 2 and 7 of this chapter is interested in avoiding situations where variance is too large. His test for bolt diameters there had as alternative hypothesis $\sigma^2 > 0.01$—the situation he wished to avoid. This same manufacturer could also have done a test with alternative hypothesis $\sigma^2 < 0.01$.

a) Use the data for Production line 1 given in Example 2 to test the null hypothesis $\sigma^2 = 0.01$ against the alternative hypothesis $\sigma^2 < 0.01$ at the level $\alpha = 0.05$.

b) Compare your results with those obtained using the same data in Exercise F to test the null hypothesis $\sigma^2 = 0.01$ against the alternative hypothesis $\sigma^2 > 0.01$. What is the difference between the two tests?

c) If you were manufacturing a special bolt for a space vehicle in which precision was extremly important, which alternative hypothesis would you use? Why?

d) If you were manufacturing a cheap bolt for a lawn chair, which alternative hypothesis would you use? Why?

<u>11</u> (*Tests with alternative* $\sigma^2 \neq \sigma_0^2$.) The designer of a 100-point aptitude test wishes test scores to have a standard deviation of 10. He gives the test to 25 randomly selected college students in order to test the null hypothesis $\sigma^2 = 100$ against the alternative hypothesis $\sigma^2 \neq 100$ at the level $\alpha = 0.02$.

a) Find the critical values $\chi_{0.01}^2$ and $\chi_{0.99}^2$.

b) What are the acceptance and rejection regions for this test?

c) If the sample data gives $s^2 = 70$, what conclusion will be made?

<u>12</u> In Exercise 5 of Section 9.1, a quality-control engineer took a random sample of 10 cans of stewed tomatoes to check the claim that the variance of all such cans is 0.5. The results of his sampling were the following:

12.76	13.34	14.76	13.77	14.09
13.79	13.68	14.36	14.06	13.86

Perform the appropriate hypothesis test at the 0.05 significance level.

9.4 Confidence intervals for variances

In Chapter 7 we learned how to find confidence intervals for a population mean, μ, based on a sample mean, $\bar{x}$. In this section we will learn how to find confidence intervals for a population variance, σ^2, based on a sample variance, s^2. To construct such confidence intervals we will use the fact (see page 324) that, if a random sample of size n is taken from a population that is normally distributed with variance σ^2, then the random variable

$$\chi^2 = \frac{(n-1)}{\sigma^2} s^2$$

has the chi-square distribution with $n-1$ degrees of freedom.

Example **9** ◖ In Example 8, a teacher took a random sample of the verbal scores of 30 students in his school. The sample variance, s^2, of these scores was 19,044. Use this information to find a 95-percent confidence interval for the (population) variance, σ^2, of College Board verbal scores for all students in the high school. (Assume that these scores are normally distributed.)

Solution In this case $n = 30$, and so

$$\chi^2 = \frac{29}{\sigma^2} s^2$$

has the chi-square distribution with d.f. = 29.

Now, we want a 95-percent confidence interval, so $\alpha = 0.05$. In other words we wish to cut off areas of

$$\frac{\alpha}{2} = 0.025$$

from each end of the χ^2-curve with d.f. = 29. (See Fig. 9.8).

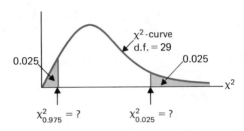

Figure 9.8

From Table III we find that for d.f. = 29,

$$\chi^2_{0.975} = 16.05, \qquad \chi^2_{0.025} = 45.72$$

Consequently,

$$P(16.05 < \chi^2 < 45.72) = 0.95$$

Since, in this case,

$$\chi^2 = \frac{29}{\sigma^2} s^2,$$

we have

$$P\left(16.05 < \frac{29}{\sigma^2} s^2 < 45.72\right) = 0.95$$

Doing some algebra (you will be asked to do this in the exercises), this last expression can be rewritten as

$$P\left(\frac{29s^2}{45.72} < \sigma^2 < \frac{29s^2}{16.05}\right) = 0.95$$

In other words, the probability is 0.95 that the interval from

$$\frac{29s^2}{45.72} \quad \text{to} \quad \frac{29s^2}{16.05}$$

will contain σ^2. Since, in this case, $s^2 = 19,044$, we have

$$\frac{29s^2}{45.72} = \frac{29(19044)}{45.72} = 12079.53$$

and

$$\frac{29s^2}{16.05} = \frac{29(19044)}{16.05} = 34409.72$$

We can be 95-percent confident that σ^2 is somewhere between

$$12{,}079.53 \quad \text{and} \quad 34{,}409.72 \blacktriangleright$$

By studying carefully what was done in Example 9, we obtain the following procedure for finding confidence intervals for σ^2 based on s^2.

Procedure

Procedure for finding confidence intervals for σ^2 based on s^2.

Assumption: Normal population.

1. If the desired confidence level is $1 - \alpha$, use Table III to find

$$\chi^2_{1-\alpha/2} \quad \text{and} \quad \chi^2_{\alpha/2}, \quad \text{for d.f.} = n - 1$$

2. The desired confidence interval for σ^2 is

$$\frac{(n-1)s^2}{\chi^2_{\alpha/2}} \quad \text{to} \quad \frac{(n-1)s^2}{\chi^2_{1-\alpha/2}}$$

We will apply this procedure in the following example.

Example 10   ◖ Find a 90-percent confidence interval for the bolt-diameter variance for bolts produced by Production line 2 in Example 2. The sample variance obtained on page 317 was $s^2 = 0.058$.

Solution   1. The desired confidence level is $0.90 = 1 - 0.10$. So here $\alpha = 0.10$. Since $n = 20$, d.f. $= n - 1 = 19$. Using Table III, we find that, for d.f. $= 19$.

$$\chi^2_{1-\alpha/2} = \chi^2_{0.95} = 10.12 \quad \text{and} \quad \chi^2_{\alpha/2} = \chi^2_{0.05} = 30.14$$

2. The 90-percent confidence interval for σ^2 is

$$\frac{(n-1)s^2}{\chi^2_{\alpha/2}} \quad \text{to} \quad \frac{(n-1)s^2}{\chi^2_{1-\alpha/2}}$$

or

$$\frac{19(0.058)}{30.14} \quad \text{to} \quad \frac{19(0.058)}{10.12}$$

or

$$0.037 \quad \text{to} \quad 0.109$$

In other words, we can be 90 percent confident that the variance in bolt diameters of bolts from Production line 2 is somewhere between 0.037 mm^2 and 0.109 mm^2. ▶

Find a 99-percent confidence interval for the variance in bolt diameters of bolts from Production line 1. Use the sample data on page 317 for Production line 1 ($s^2 = 0.008$ and $n = 20$).   Exercise **G**

EXERCISES Section **9.4**

In Exercises 1 through 9, assume each population is normally distributed.

1 A manufacturer makes nuts to be fitted onto bolts 1.5 cm in diameter. A sample of 30 nuts is measured and found to have a sample variance of $s^2 = 0.009$. Find a 99-percent confidence interval for σ^2.

2 A sunflower-seed packer packages 3.5-ounce packages of shelled sunflower seeds. A sample of 25 package weights yielded $s^2 = 0.05$. Find a 99-percent confidence interval for σ^2.

3 Find a 95-percent confidence interval for σ^2, if $s^2 = 0.009$ and $n = 30$ for the nuts in Exercise 1.

4 Find a 95-percent confidence interval for σ^2, if $s^2 = 0.05$ and $n = 25$ for the packages of sunflower seeds in Exercise 2.

5 A manufacturer of fine watches claims that the company's top-line watch keeps perfect time to within one second every month. A statistician finds the monthly error in seconds for 20 of those watches, and calculates $s^2 = 0.13$ for those errors. Find a 95-percent confidence interval for σ^2.

6 Find a 99-percent confidence interval for σ^2 in Exercise 5.

7 An automobile manufacturer claims that the company's economy car gets 35 miles per gallon in the city. Below is a random sample of the miles per gallon of 10 cars. Find a 95-percent confidence interval for σ^2.

35.3	34.3	35.2	33.6	34.3
33.2	35.3	32.5	34.1	35.2

8 In Exercise 4 of Section 9.1, a manufacturer of bolts was interested in the variance of diameters of bolts produced by a new machine. A random sample of 15 such bolts had the following bolt diameters:

9.91	10.03	10.06	9.99	10.08
9.84	9.88	9.91	9.97	10.02
10.18	10.05	9.98	9.97	10.03

Find a 95-percent confidence interval for the variance, σ^2, of bolt diameters produced by this machine.

9 In Exercise 5 of Section 9.1 a quality-control engineer was interested in the variance, σ^2, of 14-oz cans of stewed tomatoes sold by his company. A random sample of 10 such cans yielded the following data:

12.76	13.34	14.76	13.77	14.09
13.79	13.68	14.36	14.06	13.86

Find a 95-percent confidence interval for σ^2.

10 Use the confidence interval from Exercise 4 to decide what conclusion would be made if a test of the null hypothesis $\sigma^2 = 0.04$ against the alternative hypothesis $\sigma^2 \neq 0.04$ were done with the same data, and $\alpha = 0.05$.

11 Show that the inequality

$$16.05 < \frac{29}{\sigma^2} s^2 < 45.72$$

is equivalent to

$$\frac{29 s^2}{45.72} < \sigma^2 < \frac{29 s^2}{16.05}$$

a) Find

$$P\left(\chi^2_{1-\alpha/2} < \frac{(n-1)s^2}{\sigma^2} < \chi^2_{\alpha/2} \right)$$

b) Use the result from (a) to derive the general formula for a $(1 - \alpha)$-level confidence interval for σ^2.

12 Confidence intervals for μ are symmetric around $\bar{x}$—that is, the two endpoints of the confidence interval are equal distances from $\bar{x}$. The confidence intervals for σ^2 obtained here are not symmetric around s^2. What is the reason for this difference?

9.5 The *F*-statistic and the *F*-curve

In the previous four sections, we have looked at situations involving only one unknown variance. There are many situations in which we might wish to compare *two* unknown variances. One such situation is described in the following example, which deals with minimum-competency testing. In the remainder of this section, we will describe the statistic and the distribution curves used for such variance tests, and then show how the hypothesis test is actually done.

Example 11

◀ Each year at Arizona State University, more than 2000 students register for Intermediate Algebra, and another 2000 register for College Algebra.* These 4000 students are taking algebra to prepare for more advanced courses in which algebra is used: calculus, statistics, economics, and many others. The algebra course director must make sure that students who pass intermediate algebra really have learned to graph a straight line, to factor, and to solve equations. If a student receives a passing grade in algebra and has not learned such basic things, he may fail his next course—and the algebra course director is sure to hear about it.

Since 1974, students in this algebra program have been required to take a quiz on basic skills for each chapter in the textbook. The passing grade on each chapter quiz is 70%; a student who does not pass must restudy and take makeup quizzes until 70% is attained. This kind of testing, known as *mastery testing*, was designed to ensure that all students acquire most of the basic skills. Studies have shown that such mastery testing works fairly well, and would work even better if the eventual passing grade were raised to 100%. However, most algebra teachers are convinced that very few students would register for algebra if a passing grade of 100% were announced.

In the spring of 1976, an education professor who was familiar with these algebra courses made a suggestion for reaching 100% perfection without scaring away all the students. Her idea was to take each student who had passed a quiz with a grade less than 100%, and to require that student to work a number of problems just like the ones he had done incorrectly. The student would work to the 100% level, but only *after* passing the quiz.

The algebra course director liked this idea, and decided to test it. In the fall of 1977, he and the education professor set up an experimental modification of the Intermediate Algebra class. For this modified class, there were drill sheets covering every possible skill on every quiz given. Whenever a student finished a quiz, she was required to work through the drill sheets for every quiz question that had not been answered correctly. The algebra course director randomly split up all the algebra students registered for class at 9:40 into two groups. One of the groups, called the *control group*, was taught the usual algebra class; the other group, called the *experimental group*, was given the new experimental procedure with drill sheets. Both classes covered the same material, and took the same unit quizzes and the same final exam at the same time. The professors agreed to look at final exam scores for the two groups to see if the experiment had improved learning. The professors hoped that the experimental approach would make

---

* At Arizona State University, Intermediate Algebra is a basic algebra course, while College Algebra is a precalculus algebra course.

the average student learn enough algebra to prevent extremely low scores on the final. We could picture what they hoped to find as follows:

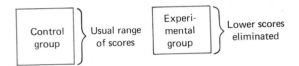

If the experimental group really did not produce low scores, the final exam data from it would be less varied. Using the language of variances:

If the experimental group performed as expected, its final exam scores should have a smaller variance than the final exam scores from the control group. We picture this below.

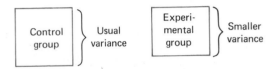

In the third week of December, 1977, the 20 students in the experimental group and the 41 students in the control group took the final exam, which consisted of 40 questions. It was not multiple-choice, but was graded right–wrong with no partial credit. A student's score was just the total number of questions that were answered correctly. The scores were as shown in Table 9.4.

Control (1)			Experimental (2)		**Table 9.4**
36	27	18	36	27	
35	26	17	35	27	
35	26	17	35	26	
33	25	16	35	25	
32	24	15	32	23	
32	24	15	31	23	
31	24	15	30	21	
29	23	15	29	21	
29	20	14	28	21	
28	20	11	28	19	
28	19	10			
28	19	9			
27	18	4			
27	18				
$\bar{x}_1 = 22.4$			$\bar{x}_2 = 27.6$		
$s_1^2 = 61.05$			$s_2^2 = 27.94.$		

The things we had hoped for seem to have happened. The experimental group had fewer low scores, less variance, and a higher mean score. Looking at box-and-whisker diagrams for the data would lead to a similar conclusion (see Fig. 9.9).

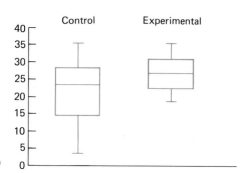

Figure 9.9

There is still a hypothesis test to do here. In fact, this hypothesis test was set up before the data were collected. It is clear that the experimental group in this particular experiment outperformed the control group. However, educators are interested in more than this semester's results. They would like to use this semester's results as support for using the experimental method in the future. They are interested in the *populations* of all students who will be taught by the two methods, and not just the *samples* obtained from this semester's classes. In other words, they are interested in the following hypothesis test for variances.

Null hypothesis: $\sigma_1^2 = \sigma_2^2$
Alternative hypothesis: $\sigma_1^2 > \sigma_2^2$

Here σ_1^2 denotes the (population) variance of *all* final exam scores for all students who have been or will be taught by the control method, and σ_2^2 denotes that for the experimental method. Since s_1^2 estimates σ_1^2, and s_2^2 estimates σ_2^2, it would make sense to test this hypothesis by comparing s_1^2 to s_2^2. In actual practice, statisticians perform this comparison by calculating the quantity $F = s_1^2/s_2^2$. If F is large, s_1^2 is much bigger than s_2^2. If F is small, s_1^2 is much smaller than s_2^2. ▸

F-statistic for comparison of variances

$$F = \frac{s_1^2}{s_2^2}$$

Example **12** ◂ Calculate the F-statistic for the final exam data on page 335.

Solution We have $s_1^2 = 61.05$ and $s_2^2 = 27.94$. Thus,

$$F = \frac{s_1^2}{s_2^2} = \frac{61.05}{27.94} = 2.19 ▸$$

Exercise **H**

At the beginning of the semester, the experimental and control groups in the educational experiment took a 32-question diagnostic test that was much easier than the final. For this test, the following variances were obtained.

Control (1)	Experimental (2)
$s_1^2 = 34.89$	$s_2^2 = 23.10$

Calculate the *F*-statistic for this diagnostic test data.

Just as with the χ^2-statistic, the *F*-statistic depends on sample size. In the case of the *F*-statistic, there are two sample sizes to look at: the sample size n_1 from Group 1, and the sample size n_2 from Group 2. As with the χ^2-statistic, we will subtract one from each of these numbers to obtain *degrees of freedom* for the *F*-statistic. The degrees of freedom for Group 1 is $n_1 - 1$; this number is called the **degrees of freedom for the numerator,** since the variance from Group 1 is the numerator of s_1^2/s_2^2. Group 2 has $n_2 - 1$ degrees of freedom; this number is called the **degrees of freedom for the denominator.**

$$\text{d.f.} = (n_1 - 1, \, n_2 - 1)$$

Degrees of freedom
for numerator

Degrees of freedom
for denominator

Degrees of freedom
for *F*-statistic

◀ Find the degrees of freedom for the *F*-statistic for the final exam data from Example 11 on page 335.

Example **13**

Solution

Since $n_1 = 41$ and $n_2 = 20$, we have

$$\text{d.f.} = (41 - 1, \, 20 - 1) = (40, 19) \; ▶$$

To perform the hypothesis test for the final exam data, we need to be able to determine when $F = s_1^2/s_2^2$ is "too large." As usual, this means that we must find the curves that give probabilities for the *F*-statistic. These curves are called ***F*-curves**. They depend on the degrees of freedom; two of them are pictured in Fig. 9.10.

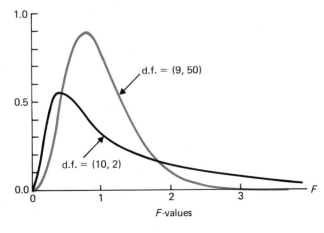

d.f. = (9, 50)

d.f. = (10, 2)

F-values

Figure 9.10
F-curves for d.f. = (10, 2)
and d.f. = (9, 50).

As with chi-square curves, we can detect some basic properties of F-curves from Fig. 9.10.

1. F-curves are different for different degrees of freedom.
2. Each F-curve starts at 0 on the horizontal axis and extends to the right. The F-values of interest are marked on the horizontal axis.
3. An F-curve is *not* symmetrical but is skewed to the right. It climbs to its high point rapidly, and comes back to the axis more slowly.
4. The total area under an F-curve is 1.

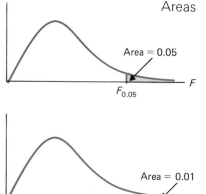

Areas

Areas under F-curves (probabilities for the F-statistic) have also been put in tables. As usual, these tables give areas likely to be used as significance levels: 0.05, 0.025, and 0.01. For F-curves there are entirely separate tables for each area value. This is because we need critical values for each different combination of degrees of freedom for the numerator and degrees of freedom for the denominator. As with the χ^2-statistic, we will use the symbols $F_{0.05}$ for the F-value with area 0.05 to its right, $F_{0.01}$ for the F-value with area 0.01 to its right, and so on. (See Fig. 9.11.)

In this book, we give tables of $F_{0.05}$, $F_{0.025}$, and $F_{0.01}$ for various degrees of freedom. In Table 9.5 we reproduce part of Table VI for $F_{0.05}$-values.

The values of $F_{0.05}$ are displayed inside this table, and the degrees of freedom for numerator and denominator on the top and side. To find $F_{0.05}$ for d.f. = (40, 19), read down from the 40 on top of the table until you reach the row for 19 degrees of freedom. Here, $F_{0.05} = 2.03$; that is, the area under the F-curve to the right of 2.03 is 0.05 if d.f. = (40, 19).

Figure 9.11

Table 9.5

Values of $F_{0.05}$

Degrees of freedom for the numerator

		1	2	3	4	5	6	7	. . .	40
	1	161.4	199.5	215.7	224.6	230.2	234.0	236.8	. . .	251.1
	2	18.51	19.00	19.16	19.25	19.30	19.33	19.35	. . .	19.47
	3	10.13	9.55	9.28	9.12	9.01	8.94	8.89	. . .	8.59
	4	7.71	6.94	6.59	6.39	6.26	6.16	6.09	. . .	5.72
	5	6.61	5.79	5.41	5.19	5.05	4.95	4.88	. . .	4.46
	6	5.99	5.14	4.76	4.53	4.39	4.28	4.21	. . .	3.77
Degrees of freedom for the denominator	7	5.59	4.74	4.35	4.12	3.97	3.87	3.79	. . .	3.34
	8	5.32	4.46	4.07	3.84	3.69	3.58	3.50	. . .	3.04
	9	5.12	4.26	3.86	3.63	3.48	3.37	3.29	. . .	2.83
	⋮	⋮	⋮	⋮	⋮	⋮	⋮	⋮	⋮	⋮
	19	4.38	3.52	3.13	2.90	2.74	2.63	2.54	. . .	(2.03)
	⋮	⋮	⋮	⋮	⋮	⋮	⋮	⋮	⋮	⋮

◀ Using Table IV in the appendix, find   Example **14**
a) $F_{0.01}$ for d.f. = (6, 9)   b) $F_{0.01}$ for d.f. = (10, 8).

a) $F_{0.01} = 5.80$ for d.f. = (6, 9). (Read down the sixth column until you reach the ninth   Solutions
row.)

b) $F_{0.01} = 5.81$ for d.f. = (10, 8). ▶

Use the tables in the appendix to find   Exercise **I**
a) $F_{0.05}$ for d.f. = (7, 7)   b) $F_{0.05}$ for d.f. = (7, 2)
c) $F_{0.025}$ for d.f. = (24, 21)   d) $F_{0.025}$ for d.f. = (15, 18)
e) $F_{0.01}$ for d.f. = (9, 6)   f) $F_{0.01}$ for d.f. = (8, 10)

Now that we have learned how to find areas under *F*-curves, we can perform hypothesis tests for the equality of two variances. The fundamental fact needed to do this is given below.

Suppose two *independent* random samples are taken from two populations that are *normally distributed*. Let

$$n_1 = \text{sample size for the first population,}$$
$$s_1^2 = \text{sample variance for the first population,}$$
$$\sigma_1^2 = \text{population variance for the first population,}$$

and let n_2, s_2^2, σ_2^2 be the corresponding values for the second population. If $\sigma_1^2 = \sigma_2^2$, then the random variable

$$F = \frac{s_1^2}{s_2^2}$$

has the *F*-distribution with d.f. = $(n_1 - 1, n_2 - 1)$. That is, probabilities for the random variable *F* can be determined using areas under the *F*-curve with d.f. = $(n_1 - 1, n_2 - 1)$.

We can now explain how to do a hypothesis test of the form

Null hypothesis: $\sigma_1^2 = \sigma_2^2$
Alternative hypothesis: $\sigma_1^2 > \sigma_2^2$

The idea is this: If the null hypothesis is true, then $\sigma_1^2 = \sigma_2^2$. But s_1^2 is an estimate of σ_1^2, and s_2^2 is an estimate of σ_2^2. So if the null hypothesis is true, then s_1^2 should be about the same as s_2^2 and

$$F = \frac{s_1^2}{s_2^2}$$

should be approximately equal to 1. If *F* is "too much bigger" than 1, then we would reject the null hypothesis in favor of the alternative hypothesis that $\sigma_1^2 > \sigma_2^2$.

As usual "how much bigger is too much bigger" is determined by the significance level, α. For example, if $\alpha = 0.05$, we want to choose the *critical value c* so that $P(F > c) = 0.05$. That is, $c = F_{0.05}$.

We will give the general procedure below, and then illustrate it with an example.

Procedure

Procedure for performing a hypothesis test of the form:

Null hypothesis: $\quad \sigma_1^2 = \sigma_2^2$
Alternative hypothesis: $\quad \sigma_1^2 > \sigma_2^2$

Assumptions: (1) Independent samples, (2) normal populations.

1. Decide on the *significance level*, α.
2. The *critical value* is

$$c = F_\alpha, \qquad \text{d.f.} = (n_1 - 1, n_2 - 1)$$

where n_1 and n_2 are the sample sizes.

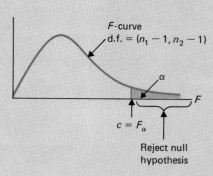

3. Calculate the value of

$$F = \frac{s_1^2}{s_2^2}$$

where s_1^2 and s_2^2 are the sample variances from the first and second populations, respectively.
4. If $F > c$, *reject* the null hypothesis.

We can now perform the hypothesis test concerning two variances for the educational experiment data. We will use a significance level of $\alpha = 0.05$. This means that an F-statistic will be considered "too large" if it is greater than $F_{0.05}$.

Example **15**   ◀ Perform a hypothesis test on the final exam data from Example 11 on page 334 to determine whether the control method produces greater variance in scores than the experimental method. Use $\alpha = 0.05$.

We want to perform the hypothesis test   Solution

Null hypothesis: $\sigma_1^2 = \sigma_2^2$
Alternative hypothesis: $\sigma_1^2 > \sigma_2^2$

The sampling was done in such a way that the two samples are independent. Moreover, educators' experience indicates that test scores are normally distributed. Thus, we can use the procedure given above.

1. We will use $\alpha = 0.05$.
2. The critical value is $c = F_\alpha = F_{0.05}$, which equals 2.03 for d.f. $= (n_1 - 1, n_2 - 1) = (40, 19)$. See Fig. 9.12.
3. From page 335, we know $s_1^2 = 61.05$ and $s_2^2 = 27.94$, so

$$F = \frac{61.05}{27.94} \doteq 2.19$$

We have marked the value of F with a cross on Fig. 9.12

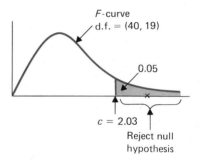

F-curve
d.f. = (40, 19)

0.05

$c = 2.03$

Reject null
hypothesis

Figure 9.12

4. Since $F > c$, we *reject* the null hypothesis and conclude that $\sigma_1^2 > \sigma_2^2$. ◗

If you use a computer, you may be given a *P*-value for your *F*-statistic. The *P*-value   *P*-values
for the *F*-statistic in the last example is slightly larger than 0.03. Thus the area to the right of our computed *F*-statistic is about 0.03 (See Fig. 9.13.) We were able to reject the null hypothesis with $\alpha = 0.05$, but wouldn't have been able to reject it if we had used $\alpha = 0.01$.

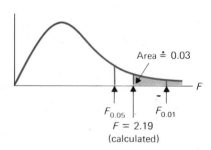

Area $\doteq$ 0.03

F

$F_{0.05}$ $F_{0.01}$

$F = 2.19$
(calculated)

Figure 9.13

Exercise **J**   Use the diagnostic test data given below to test the null hypothesis $\sigma_1^2 = \sigma_2^2$ against the alternative hypothesis $\sigma_1^2 > \sigma_2^2$, for the beginning groups.

Control	Experimental
$s_1^2 = 34.89$	$s_2^2 = 23.10$
$n_1 = 41$	$n_2 = 20$

Use $\alpha = 0.05$. (This data was also discussed in Exercise H.)

On page 340 we gave the procedure for performing a hypothesis test of the form:

Null hypothesis: $\sigma_1^2 = \sigma_2^2$
Alternative hypothesis: $\sigma_1^2 > \sigma_2^2$

Hypothesis tests where the alternative hypothesis is $\sigma_1^2 < \sigma_2^2$ can be done using exactly the same procedure by interchanging the roles of the populations. This will be discussed in the exercises.

We conclude this section by explaining the **procedure** for performing a *two-tailed* hypothesis test for the equality of two variances.* That is, a test of the form

Null hypothesis: $\sigma_1^2 = \sigma_2^2$
Alternative hypothesis: $\sigma_1^2 \neq \sigma_2^2$

One (but not the only) use for such a test is the following. You might recall from Chapter 8 that when we want to perform a *small* sample hypothesis test for the difference between two population means, the *t*-test we use depends on whether or not the population variances are equal. We can sometimes "decide" whether or not the two population variances are equal by performing a two-tailed hypothesis test for the equality of two variances.†

As usual, the procedure for a two-tailed test is basically the same as that for a one-tailed test, except that we need two critical values instead of one. For example, for a two-tailed test for the equality of two variances with $\alpha = 0.05$, we need to cut off areas of 0.025 on each end of the appropriate *F*-curve (see Fig. 9.14). The righthand critical value, $F_{0.025}$, can be found by looking in Table V. The lefthand critical value is $F_{0.975}$. This value cannot be found *directly* in the appendix tables since we don't have a table for values of $F_{0.975}$. However, there is a way to find $F_{0.975}$ by using the tables we have. The way to do this is illustrated in the following example.

---

* The remainder of this section is optional.

† We should emphasize that the *F*-test for equality of variances is extremely sensitive to nonnormality, especially when the sample sizes are unequal. Thus normality is crucial for an *F*-test. On the other hand, the *t*-test is less sensitive to deviations from normality. Consequently, unless one is sure that the populations are normally distributed, it is not recommended that an *F*-test be used to decide whether or not $\sigma_1^2 = \sigma_2^2$ as preliminary to a *t*-test for the equality of two means.

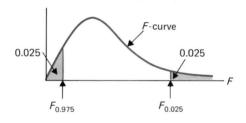

Figure 9.14

◀ Find $F_{0.975}$ for d.f. = (9, 8).   Example **16**

To find $F_{0.975}$ for d.f. = (9, 8), we use the following fact. The value $F_{0.975}$ Solution
for d.f. = (9, 8) is the *reciprocal* of the value $F_{1-0.975} = F_{0.025}$ for d.f. = (8, 9).
(*Note*: we switched the degrees of freedom.) Now from Table V we see that for
d.f. = (8, 9), $F_{0.025} = 4.10$. Therefore, for d.f. = (9, 8),

$$F_{0.975} = \frac{1}{4.10} \doteq 0.24$$

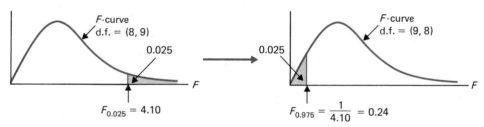

Figure 9.15

We summarize the above work in Fig. 9.15. ▶

Here is another example that illustrates how to find lefthand critical values for an
F-test.

◀ Find $F_{0.95}$ for d.f. = (15, 20).   Example **17**

$F_{0.95}$ for d.f. = (15, 20) is the reciprocal of $F_{1-0.95} = F_{0.05}$ for d.f. = (20, 15). From   Solution
Table VI, for d.f. = (20, 15), $F_{0.05} = 2.33$. Therefore, for d.f. = (15, 20),

$$F_{0.95} = \frac{1}{2.33} = 0.43 \text{ ▶}$$

Find   Exercise **K**
a) $F_{0.975}$ for d.f. = (30, 9),
b) $F_{0.95}$ for d.f. = (12, 40).

Now that we see how to find lefthand critical values, we can give the procedure for
doing a two-tailed hypothesis test for the equality of two variances. Following this we
will illustrate the procedure with an example.

Procedure

Procedure for performing a hypothesis test of the form:

Null hypothesis: $\sigma_1^2 = \sigma_2^2$
Alternative hypothesis: $\sigma_1^2 \neq \sigma_2^2$

Assumptions: (1) Independent samples, (2) normal populations.

1. Decide on the significance level, α.
2. The critical values are

$$F_{1-\alpha/2} \quad \text{and} \quad F_{\alpha/2}, \quad \text{for d.f.} = (n_1 - 1, n_2 - 1)$$

where n_1 and n_2 are the sample sizes.

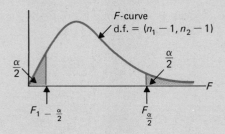

3. Calculate the value of

$$F = \frac{s_1^2}{s_2^2}$$

where s_1^2 and s_2^2 are the sample variances from the two populations.
4. If $F < F_{1-\alpha/2}$ or $F > F_{\alpha/2}$, *reject* the null hypothesis.

Example 18 ◀ An agronomist is interested in comparing the mean corn yields on farms in Iowa and California. A sample of ten farms in Iowa yielded

$$\bar{x}_1 = 107.0 \text{ bushels/acre}$$
$$s_1^2 = 4.84$$

and nine farms in California gave

$$\bar{x}_2 = 105.0 \text{ bushels/acre}$$
$$s_2^2 = 4.41$$

Before she can use a *t*-test to compare the (population) *mean* corn yields on farms in Iowa and California, she needs to decide whether or not she believes that the population *variances* of corn yields are equal. In other words, if σ_1^2 is the variance of corn yield in Iowa, and σ_2^2 is the variance of corn yield in California, then the agronomist wants to test:

Null hypothesis: $\sigma_1^2 = \sigma_2^2$
Alternative hypothesis: $\sigma_1^2 \neq \sigma_2^2$

She decides to perform the test at the 0.10 significance level. Assuming that corn yields on farms in Iowa and California are normally distributed, perform the desired hypothesis test, using the procedure given above.

1. We will use $\alpha = 0.10$.
2. We have $n_1 = 10$ and $n_2 = 9$. Therefore, the critical values are

$$F_{1-\alpha/2} = F_{0.95} \quad \text{and} \quad F_{\alpha/2} = F_{0.05} \quad \text{for d.f.} = (10-1, 9-1) = (9, 8).$$

From Table VI we see that for d.f. = (9, 8),

$$F_{0.05} = 3.39$$

For d.f. = (9, 8), $F_{0.95}$ is the reciprocal of $F_{1-0.95} = F_{0.05}$ for d.f. = (8, 9). From Table VI we see that for d.f. = (8, 9), $F_{0.05} = 3.23$. Therefore, for d.f. = (9, 8),

$$F_{0.95} = \frac{1}{3.23} = 0.31$$

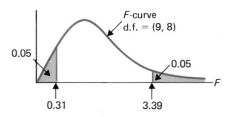

3. From the data, $s_1^2 = 4.84$ and $s_2^2 = 4.41$. Consequently,

$$F = \frac{s_1^2}{s_2^2} = \frac{4.84}{4.41} = 1.10$$

4. Thus we cannot reject the null hypothesis. In other words, the agronomist should assume that the (population) corn-yield variances are equal for farms in Iowa and California. ▶

A banker is interested in discovering whether married males keep their cars (on the average) longer than unmarried males. Random samples of 16 married males and 10 unmarried males yielded the following data on the number of years each kept his last car.   Exercise **L**

Married		Unmarried	
6.3	8.6	1.5	3.7
7.3	6.2	6.2	2.1
7.8	7.2	5.6	5.0
6.5	7.2	6.8	4.7
7.1	5.2	5.1	7.2
6.2	9.3		
7.4	6.3		
7.0	7.5		

Before the banker can use a *t*-test to compare the (population) means for the two groups, he first needs to check whether or not the (population) variances are equal. Assuming that the length of time both married and unmarried males keep their cars is normally distributed, use the data above to test:

Null hypothesis: $\sigma_1^2 = \sigma_2^2$
Alternative hypothesis: $\sigma_1^2 \neq \sigma_2^2$

Use $\alpha = 0.10$

EXERCISES Section **9.5**

In Exercises 1 through 7, use the appropriate appendix tables to find the specified *F*-value.

1 $F_{0.01}$ for d.f. = (20, 25)

2 $F_{0.05}$ for d.f. = (20, 15)

3 $F_{0.01}$ for d.f. = (7, 12)

4 $F_{0.025}$ for d.f. = (9, 4)

5 $F_{0.05}$ for d.f. = (24, 30)

6 $F_{0.05}$ for d.f. = (12, 14)

7 $F_{0.01}$ for d.f. = (10, 11)

In Exercises 8 through 11, assume that the populations are normally distributed and the samples are independent.

a) State the appropriate null hypothesis,
b) State the alternative hypothesis,
c) Find the critical value,
d) Draw a picture,
e) Calculate *F*,
f) State whether or not the null hypothesis should be rejected.

8 An educator wished to test whether stress in the classroom *increases* the variance in test scores. She had two classes in which test-score variance had been the same in the past. In Class 1 (size $n_1 = 21$) she introduced stress by announcing that the test would be long, and by continually calling attention to the limited amount of time left during the test. For Class 2 (size $n_2 = 26$) she gave the test without comment. The results are given below:

Class 1	Class 2
$s_1^2 = 112.36$	$s_2^2 = 54.76$

Perform the appropriate *F*-test using $\alpha = 0.05$.

9 A psychologist was interested in discovering whether workers owning company stock have *less* variance in production output than workers without company stock. She sampled the two groups with the following results.

Workers without stock	Workers with stock
$s_1^2 = 8.0$	$s_2^2 = 6.8$
$n_1 = 21$	$n_2 = 16$

Perform the appropriate *F*-test using $\alpha = 0.05$.

10 A sales manager was interested in determining whether his inexperienced sales people had more variance in total sales receipts than his more experienced sales people. The random-sampling data is given at the top of page 347.

Inexperienced	Experienced
$s_1^2 = 203.49$	$s_2^2 = 163.82$
$n_1 = 8$	$n_2 = 13$

Perform the appropriate F-test using $\alpha = 0.05$.

11 An agronomist planted two varieties of commercially prepared seeds to test whether there was *more* variance in the time to maturity of the older variety of plant than in the newly developed variety. The data from the two groups (in days) is given below:

Old variety			New variety		
39	37	41	37	39	38
42	38	40	39	38	39
40	41	42	39	39	40
37	46	43	40	39	
37					

Perform the appropriate F-test using $\alpha = 0.05$.

Exercises 12 through 18 refer to the *optional* material on a two-tailed F-test.

12 Find $F_{0.975}$ for d.f. = (9, 8).

13 Find $F_{0.975}$ for d.f. = (20, 24).

14 Find $F_{0.95}$ for d.f. = (17, 10).

15 Find $F_{0.99}$ for d.f. = (14, 20).

16 Find $F_{0.99}$ for d.f. = (10, 12).

17 A teacher is going to perform an educational experiment involving two methods of instruction. He plans to use a t-test to decide whether one method of instruction is superior to the other. Before he can do this, he needs to do a preliminary study to check whether or not the population variances are equal. The preliminary study yields the following results:

Method 1	Method 2
$s_1^2 = 63.8$	$s_2^2 = 59.4$
$n_1 = 10$	$n_2 = 9$

Using this data, perform the hypothesis test:

Null hypothesis: $\sigma_1^2 = \sigma_2^2$
Alternative hypothesis: $\sigma_1^2 \neq \sigma_2^2$

at the 5 % significance level (that is, $\alpha = 0.05$).

18 A production manager needs to have the springs in some machines replaced. The spring factory that supplied the company previously is out of business. The manager wants to be certain that the new supply of springs has the same variance in elasticity as the original springs. He has data from a random sample of the original springs when they first arrived. He takes a random sample of 25 of the new springs. The results for the variances are given below.

Original springs	New springs
$s_1^2 = 0.018$	$s_2^2 = 0.020$
$n_1 = 21$	$n_2 = 25$

Perform the appropriate hypothesis test, using $\alpha = 0.05$.

19 Four roommates decided to go on a 1400-calorie-a-day diet, but they were deadlocked over which of two diets to choose. One of them, a statistics student, said she would perform a test to determine whether diet plan I had less variance in total daily calories. The friends would all agree to use diet plan I if it proved to have smaller variance. Below is a sample of the total calories from seven daily menus of both diet plans.

Plan I		Plan II	
1386	1408	1355	1376
1345	1432	1440	1363
1427	1427	1267	1518
1333		1494	

Find s_1^2 *and* s_2^2, and make the appropriate hypothesis test, using $\alpha = 0.05$.

20 A manufacturer of precision tools is interested in determining whether Supplier 1's shipment of ball bearings has less variance in diameter than Supplier 2's. A random sample is taken from each shipment. The diameters are listed below.

Supplier 1		Supplier 2	
0.999	0.999	0.997	0.998
1.000	0.998	0.997	1.001
1.000	0.998	1.000	1.001
1.001	0.999	0.998	0.999
0.999	0.998	0.999	1.001

Find s_1^2 and s_2^2 and make the appropriate hypothesis test, using $\alpha = 0.05$.

9.6   Computer packages*

The *F*-test for comparison of two variances can be done for you by the SPSS program, T–TEST. As we have just discussed in the optional part of Section 9.5, it is sometimes appropriate to test the null hypothesis $\sigma_1^2 = \sigma_2^2$ against the alternative hypothesis $\sigma_1^2 \neq \sigma_2^2$ in order to decide which kind of *t*-test to use in a small-sample hypothesis test for the difference between two population means. In Example 18 of Chapter 8 we looked at lasting times of two waxes on test cars. The times (in days) were:

Present wax	New wax
89	94
86	91
83	88
87	92
89	87

Below is a computer printout generated by applying T–TEST to this data:

```
- - - - - - - - - - - - - - - - - - - - - - - - - T - T E S T - - - - - - - - - - - - - - - - - - - - - - - - - - -

GROUP 1 - FIRST    5 CASES
GROUP 2 - NEXT     5 CASES
                                                              * POOLED VARIANCE ESTIMATE   * SEPARATE VARIANCE ESTIMATE
                                                         *    *                            *
                     NUMBER            STANDARD  STANDARD *  F   2-TAIL *  T   DEGREES OF 2-TAIL *  T   DEGREES OF 2-TAIL
          VARIABLE  OF CASES   MEAN   DEVIATION   ERROR   * VALUE PROB. * VALUE  FREEDOM  PROB.  * VALUE  FREEDOM   PROB.
          ---------------------------------------------------------------------------------------------------------------
CSIU.                                                     $            $                         $
          GROUP 1     5      86.8000    2.490    1.114    *            *                         *
                                                          * 1.34 0.784 * -2.11     8     0.067   * -2.11    7.84    0.067
          GROUP 2     5      90.4000    2.881    1.288    *            *                         *
          ---------------------------------------------------------------------------------------------------------------
```

The comparison of variances for the two waxes can be obtained by looking at the second box in the printout, under the headings F-VALUE and 2-TAIL PROB. F-VALUE tells us that the *F*-statistic, $F = s_1^2/s_2^2$, is equal to 1.34. A look at *F*-tables will show you that such an *F*-value will not lead to rejection of the null hypothesis with $\alpha = 0.01$, 0.05, or 0.10. In fact, the 2-TAIL PROB value of 0.784 indicates that the smallest α level that could lead to rejection of the null hypothesis $\sigma_1^2 = \sigma_2^2$ would be slightly larger than $\alpha = 0.784$—not a very convincing α-level. The last two boxes in the printout contain the results of the *t*-test done in each of the two different ways given in Chapter 8—with a POOLED VARIANCE ESTIMATE and a SEPARATE VARIANCE ESTIMATE[†]. The program will not decide which kind of *t*-test to do. The computer user must make the decision from the F-VALUE as to whether $\sigma_1^2 = \sigma_2^2$, in which case the pooled estimate will be used, or $\sigma_1^2 \neq \sigma_2^2$, in which case the separate variance estimate will be used.

---

* This section is optional.

† The separate variance estimate was presented in Exercise 10, Section 8.10.

	Key Terms
population standard deviation (σ) *P*-values	
population variance (σ^2) $(1 - \alpha)$-level confidence interval	
sample standard deviation (s) *F*-statistic	
sample variance (s^2) *F*-curve	
χ^2-statistic degrees of freedom for numerator	
chi-square curve degrees of freedom for denominator	
degrees of freedom (d.f.)	

sampling distribution of the variance:

Formulas and Key Facts

If a random sample of size n is taken from a normally distributed population with variance σ^2, then the random variable

$$\chi^2 = \frac{n-1}{\sigma^2} s^2$$

has the chi-square distribution with d.f. $= n - 1$. That is, probabilities for χ^2 can be determined using areas under the chi-square curve with d.f. $= n - 1$.

formula for finding confidence intervals for σ^2 when sampled population is normally distributed:

$$\frac{(n-1)s^2}{\chi^2_{\alpha/2}} \quad \text{to} \quad \frac{(n-1)s^2}{\chi^2_{1-\alpha/2}}$$

where n = sample size, s^2 = sample variance, $(1 - \alpha)$ = confidence level, $\chi^2_{\alpha/2}$ and $\chi^2_{1-\alpha/2}$ computed for a χ^2-curve with d.f. $= n - 1$.

sampling distribution for two variances:

If two independent random samples of sizes n_1 and n_2 are taken from two normally distributed populations with variances σ_1^2 and σ_2^2, and if $\sigma_1^2 = \sigma_2^2$, then the random variable

$$F = \frac{s_1^2}{s_2^2}$$

has the F-distribution with d.f. $= (n_1 - 1, n_2 - 1)$. That is, probabilities for F can be determined using areas under the F-curve with d.f. $= (n_1 - 1, n_2 - 1)$.

In Table 9.6 we give a brief summary of the hypothesis-testing procedures covered in this chapter. For detailed procedures, see the text.

Table 9.6

Hypothesis tests for variances

Type	Assumptions	Null hypothesis	Alternative hypothesis	Test statistic	Rejection region
Single variance	Normal population	$\sigma^2 = \sigma_0^2$	$\sigma^2 > \sigma_0^2$ $\sigma^2 < \sigma_0^2$ $\sigma^2 \neq \sigma_0^2$	$\chi^2 = \dfrac{(n-1)}{\sigma_0^2} s^2$ (d.f. $= n - 1$)	$\chi^2 > \chi_\alpha^2$ $\chi^2 < \chi_{1-\alpha}^2$ $\chi^2 < \chi_{1-\alpha/2}^2 \quad$ or $\quad \chi^2 > \chi_{\alpha/2}^2$
Two variances	(1) Independent samples (2) Normal populations	$\sigma_1^2 = \sigma_2^2$	$\sigma_1^2 > \sigma_2^2$ $\sigma_1^2 < \sigma_2^2$ $\sigma_1^2 \neq \sigma_2^2$	$F = \dfrac{s_1^2}{s_2^2}$ $[\text{d.f.} = (n_1 - 1, n_2 - 1)]$	$F > F_\alpha$ $F < F_{1-\alpha}$* $F < F_{1-\alpha/2}$† $\quad$ or $\quad F > F_{\alpha/2}$

* For d.f. $= (n_1 - 1, n_2 - 1)$, $F_{1-\alpha} = 1/F_\alpha$, where F_α is computed for an F-curve with d.f. $= (n_2 - 1, n_1 - 1)$.

† For d.f. $= (n_1 - 1, n_2 - 1)$, $F_{1-\alpha/2} = 1/F_{\alpha/2}$, where $F_{\alpha/2}$ is compared for an F-curve with d.f. $= (n_2 - 1, n_1 - 1)$.

You should be able to

1 Find χ_α^2 for a chi-square curve with d.f. $= n - 1$.

2 Perform a hypothesis test of the form:

Null hypothesis: $\sigma^2 = \sigma_0^2$
Alternative hypothesis: $\sigma^2 > \sigma_0^2$ or $\sigma^2 < \sigma_0^2$ or $\sigma^2 \neq \sigma_0^2$

using data obtained from a random sample from a normal population.

3 Find a $(1 - \alpha)$-*level confidence interval for* σ^2 based on s^2, when sampling from a normal population.

4 Find F_α for an F-curve with d.f. $= (n_1 - 1, n_2 - 1)$.
Find $F_{1-\alpha}$ for an F-curve with d.f. $= (n_1 - 1, n_2 - 1)$.*

5 Perform a hypothesis test of the form:

Null hypothesis: $\sigma_1^2 = \sigma_2^2$
Alternative hypothesis: $\sigma_1^2 > \sigma_2^2$ or $\sigma_1^2 < \sigma_2^2$ or $\sigma_1^2 \neq \sigma_2^2$*

using data obtained from two independent random samples from two normal populations.

REVIEW TEST

1 For a chi-square curve with eight degrees of freedom, find
a) $\chi_{0.01}^2$ b) $\chi_{0.95}^2$ c) $\chi_{0.05}^2$

2 For a chi-square curve with ten degrees of freedom, find the χ^2-values that separate the middle 0.90 area from the two outer 0.05 areas.

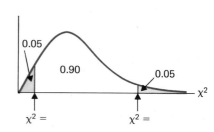

* An asterisk refers to material covered in an optional section.

3 A consumer agency suspects that the standard deviation in weights of a "12 ounce box" of crackers exceeds 0.3 ounces. They carefully weigh the contents of a random sample of $n = 21$ boxes and find $s = 0.32$ for those 21 weights.

a) Use this data to perform the hypothesis test of:

Null hypothesis: $\sigma^2 = 0.09$ $(\sigma = 0.3)$
Alternative hypothesis: $\sigma^2 > 0.09$ $(\sigma > 0.3)$,

at the 5-percent significance level.

b) Use this data to find a 95-percent confidence interval for σ^2.

4 Find:
a) $F_{0.05}$ for d.f. = (9, 6)
b) $F_{0.01}$ for d.f. = (8, 12)
c) *$F_{0.95}$ for d.f. = (6, 9)
d) *$F_{0.99}$ for d.f. = (12, 8)

5 A psychologist who works on personnel matters for two large firms suspects that one firm has more variation in employee IQ than the other. For a quick test, she takes a random sample of 21 employees from the first firm and 17 from the second, and gives each employee the same IQ test. The test score results are:

Firm 1	Firm 2
$n_1 = 21$	$n_2 = 17$
$s_1^2 = 121$	$s_2^2 = 36$

a) Perform the test:

Null hypothesis: $\sigma_1^2 = \sigma_2^2$
Alternative hypothesis: $\sigma_1^2 > \sigma_2^2$

$\alpha = 0.05$

b) What null and alternative hypotheses would the psychologist use if she suspects that variation is different in the two firms, but has no idea which is larger?

FURTHER READINGS

DIXON, W. and J. MASSEY, *Introduction to Statistical Analysis.* New York: McGraw-Hill, 1969. In addition to coverage of the material we have included here, these authors discuss the relative efficiency of alternative measures of dispersion—e.g., use of the sample range to estimate σ (pages 134–140).

NETER, J., W. WASSERMAN, and G. WHITMORE, *Applied Statistics.* Boston: Allyn and Bacon, Inc., 1978. Chapter 7 gives more detail on the chi-square and F-distributions, and states the relationship between these distributions and the t-distribution. Sections 13.2 and 13.7 cover tests concerning one and two variances, respectively.

ANSWERS TO REINFORCEMENT EXERCISES

A a) $\bar{x} \doteq 10.00$, $s^2 \doteq 0.05$, $s \doteq 0.21$

b) It would appear that the mean bolt diameter is acceptable, but that the variance of bolt diameters is too large.

c) Regarding μ:

Null hypothesis: $\mu = 10$
Alternative hypothesis: $\mu \neq 10$

Regarding σ^2:

Null hypothesis: $\sigma^2 = 0.01$
Alternative hypothesis: $\sigma^2 > 0.01$

B $\chi^2 = \dfrac{(20 - 1)}{0.01}(0.058) = 110.2$

C a)

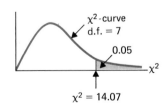

χ^2-curve
d.f. = 7
0.05
χ^2
$\chi^2 = 14.07$

---

* An asterisk refers to material covered in an optional section.

C b)

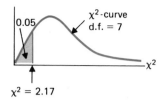

$\chi^2 = 2.17$

c)

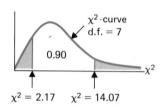

$\chi^2 = 2.17$ $\chi^2 = 14.07$

D

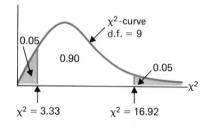

$\chi^2 = 3.33$ $\chi^2 = 16.92$

E a) $\chi^2_{0.1} = 12.02$

b) $\chi^2_{0.05} = 14.07$

c) $\chi^2_{0.01} = 18.48$

d) $\chi^2_{0.995} = 0.99$

e) $\chi^2_{0.95} = 2.17$

F *Null hypothesis:* $\sigma^2 = 0.01$
Alternative hypothesis: $\sigma^2 > 0.01$

1. $\alpha = 0.05$

2. $c = \chi^2_{0.05} = 30.14$ for d.f. $= 20 - 1 = 19$

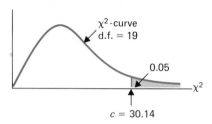

$c = 30.14$

3. $\chi^2 = \dfrac{(20-1)}{0.01}(0.008) = 15.2$

4. $\chi^2 \leqslant 30.14$, the null hypothesis cannot be rejected. The production line may be suitable.

G 1. $0.99 = 1 - 0.01$
$\alpha = 0.01$
$n = 20$
d.f. $= 20 - 1 = 19$
$\chi^2_{1-0.01/2} = \chi^2_{0.995} = 6.84$ and $\chi^2_{0.01/2} = \chi^2_{0.005} = 38.58$

2. $\dfrac{(20-1)}{38.58}(0.008)$ to $\dfrac{(20-1)}{6.84}(0.008)$
Hence, 0.004 to 0.022 is a 99-percent confidence interval for σ^2 for Production line 1.

H $F = \dfrac{34.89}{23.10} \doteq 1.51$

I a) $F_{0.05} = 3.79$ for d.f. $= (7, 7)$

b) $F_{0.05} = 19.35$ for d.f. $= (7, 2)$

c) $F_{0.025} = 2.37$ for d.f. $= (24, 21)$

d) $F_{0.025} = 2.67$ for d.f. $= (15, 18)$

e) $F_{0.01} = 7.98$ for d.f. $= (9, 6)$

f) $F_{0.01} = 5.06$ for d.f. $= (8, 10)$

J *Null hypothesis:* $\sigma_1^2 = \sigma_2^2$
Alternative hypothesis: $\sigma_1^2 > \sigma_2^2$

1. $\alpha = 0.05$
2. $F_{0.05} = 2.03$ for d.f. $= (40, 19)$

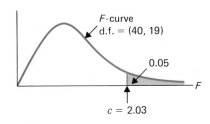

F-curve
d.f. = (40, 19)

0.05

$c = 2.03$

3. $F = \dfrac{34.89}{23.10} \doteq 1.51$
4. $F \leqslant c$, the null hypothesis cannot be rejected.

K * a) $F_{0.975}$ for d.f. $= (30, 9)$ is the reciprocal of $F_{0.025}$ for
d.f. $= (9, 30)$

$$F_{0.975} = \frac{1}{2.57} \doteq 0.39$$

---

* An asterisk refers to material covered in optional section.

b) $F_{0.95}$ for d.f. $= (12, 40)$ is the reciprocal of $F_{0.05}$ for
d.f. $= (40, 12)$

$$F_{0.95} = \frac{1}{2.43} \doteq 0.41$$

L * *Null hypothesis:* $\sigma_1^2 = \sigma_2^2$
Alternative hypothesis: $\sigma_1^2 \neq \sigma_2^2$

1. $\alpha = 0.10$
2. $F_{0.95} = 1/2.59 \doteq 0.39$ and $F_{0.05} = 3.01$ for d.f. $= (15, 9)$

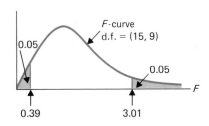

F-curve
d.f. = (15, 9)

0.05

0.05

0.39

3.01

3. $s_1^2 = 0.98$, $s_2^2 = 3.54$
$F = 0.98/3.54 \doteq 0.28$
4. Reject the null hypothesis. It appears that $\sigma_1^2 \neq \sigma_2^2$.

Chi-square tests

10.1   Introduction

In this chapter, we will look at some very widely used statistical tests based on the chi-square distribution. The first test we will discuss is called a **goodness-of-fit test.** As usual, we will begin by giving an example of a situation in which such a test might be used. In Section 10.2, we will show how such tests are actually performed.

Example **1**   ◖ In the year 1975, 36.9 percent of the people in the United States were less than 21 years of age, while 52.6 percent were in the 21–64 age group, and 10.5 percent were 65 or over. If you live in the state of Arizona (as do both authors of this book), you might guess that the many retirement communities in that state would cause it to have an unusually high percentage of people over 65. If you live in Florida, you might guess the same thing about that state. However, if you check the population figures in the *Statistical Abstract of the United States* and calculate a few percentages, you will find that the guess for Arizona in 1975 was wrong, but the guess for Florida was correct. The percentages for the entire nation and the states of Arizona and Florida are given below.

Table 10.1

Age distributions for United States, Arizona, and Florida in 1975

Area	Age		
	0–20	21–64	65 and over
United States	36.9%	52.6%	10.5%
Arizona	39.2%	50.8%	10.0%
Florida	33.2%	50.7%	16.1%

Each row of Table 10.1 can be looked at as a probability distribution. For obvious reasons, these probability distributions are also called *age distributions.* Age distributions are compiled from census data for the United States, individual states, cities, mile-square sections of cities, and even for individual blocks of large, densely populated cities. There is a good economic reason for doing this. If you think that it might be a good idea to open a children's shoe store in your local shopping center, you need to know whether there are enough children to provide a market. If the government plans to fund activities for senior citizens in a city, the government will base its spending partly on the estimated number of senior citizens in that city.

In Table 10.2 we give an age distribution for the mile-square census tract in which one of the authors now lives; it was obtained in a special census of the city of Tempe, Arizona, in 1975. The tract then had 5506 residents.

Table 10.2

Age distribution for census tract 3198 in Tempe, Arizona, in 1975

Age	0–20	21–64	65 and over
Percent	39.2%	58.4%	2.4%

Age distributions for the town of Tempe were very well tabulated in 1975. In fact, the actual sheets from the census enable a businessman in census tract 3198 to find out exactly how many ten-year-olds there were in 1975 (128), and how many people 85 and over (11). Since all this information is already available, there seems to be no need for any statistical testing, but there is. At the time of this writing, the city of Tempe was growing very rapidly. Its population increased by 30,000 between 1970 and 1975. It is still growing, and businessmen are now deciding what kind of stores to locate in new tracts that were farms in 1975.

Suppose that a planner is now interested in the age distribution for a *newly developed* tract that was not covered by the special 1975 census. If the planner feels that this new tract is similar to tract 3198, he might be tempted to use the 1975 age distribution given for tract 3198 as a working distribution for the new tract. If the new tract has 5000 residents, he will not want to take a census to check whether he is right in doing this. However, the planner could find the ages of, say, 300 residents randomly sampled from the new tract, and check whether the data from these 300 residents *fits* what he would expect from the age distribution for the old tract 3198. Table 10.3 gives *count data* from a sample of 300 residents taken in the new tract, and Fig. 10.1 illustrates the planner's checking procedure.

Age	0–20	21–64	65 and over
Number	126	165	9
Percent	42%	55%	3%

Table 10.3
Sample results, new tract

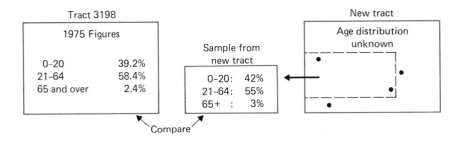

Figure 10.1

Now there are two questions to answer. First, what kind of sample would you *expect* to get if the age distribution for the new tract is the same as the age distribution for the old one? Second, how does the planner tell whether his sample data "fits" these expectations?

The first question is easy to answer. *If the new tract has the same age distribution as the old one*, 39.2 percent of its residents would be in the 0–20 age group. In a sample of size 300, the planner should expect about 39.2 percent of 300, or 117.6 individuals in the

0–20 age group. Similarly, the planner should expect about 175.2 individuals in the 21–64 age group, and about 7.2 individuals over 65. (We leave to the reader the question of what 0.2 of a person looks like; decimal parts are used for greater accuracy.) In the second and third columns of Table 10.4 we display observed sample values and expected values for the new tract. Each expected value is calculated using the formula:

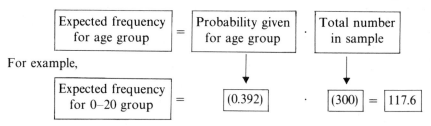

For example,

$$\boxed{\begin{array}{c}\text{Expected frequency}\\\text{for 0–20 group}\end{array}} = \boxed{(0.392)} \cdot \boxed{(300)} = \boxed{117.6}$$

Table 10.4

Age group	Observed values, O	Expected values, E	Differences, $O - E$	Squares of differences, $(O - E)^2$	$\dfrac{(O - E)^2}{E}$
0–20	126	117.6	8.4	70.56	0.60
21–64	165	175.2	−10.2	104.04	0.59
65 and over	9	7.2	1.8	3.24	0.45

Add up to 0

1.64

$$\chi^2 = \sum \frac{(O - E)^2}{E}$$

The question of whether the observed values "fit" the expected ones is harder to answer. We would like to calculate numbers that measure how close the fit is. A natural number to look at is the difference $O - E$ for each age group. We have given the values of $O - E$ in the fourth column of Table 10.4.

As you can see, the $O - E$ values don't give much information if you combine them by adding—they add up to 0. Statisticians take care of this problem by looking at $(O - E)^2$ instead of $O - E$. This situation is very similar to the one encountered when we first looked at measures of dispersion. In that situation the quantities $x - \bar{x}$ added up to 0, so we looked at their squares instead.

However, it is not enough to simply add up the terms $(O - E)^2$. A glance at the third and fifth columns of Table 10.4 will show you that bigger values of $(O - E)^2$ tend to occur for larger values of E. If we divide $(O - E)^2$ by E, differences between groups with large and small E-values are "smoothed out." In the end, to test goodness of fit, a statistician would actually calculate the statistic

$$\chi^2 = \sum \frac{(O - E)^2}{E}$$

In the last column of Table 10.4 we have calculated the $(O - E)^2/E$ values to two decimal places and added them. ▶

Exercise **A**

Suppose the planner in Example 1 had found the ages of 500 people and obtained the O-frequencies (observed frequencies) given in the second column of Table 10.5.

Table 10.5

Age group	O	E	$O - E$	$(O - E)^2$	$\dfrac{(O - E)^2}{E}$
0–20	150				
21–64	260				
65 and over	90				

Fill in the table and compute $\chi^2 = \sum[(O - E)^2/E]$.

A note on rounding

When we compute χ^2 in this chapter, we will usually round values of $(O - E)^2/E$ to two places. We do this so that answers will be consistent. If you compute χ^2 directly on a calculator, you may get slightly different answers from ours. In fact, since different calculators round off differently, two students with different calculators may get slightly different answers.

◀ We are now ready to look at the idea of a goodness-of-fit hypothesis test. We have looked at observed and expected frequencies for a sample, and introduced the χ^2-statistic, which measures goodness of fit. What we must now do is spell out carefully what kinds of hypotheses we are testing, and show how to test those hypotheses using an appropriate probability distribution. We have been trying to check out one assumption throughout this discussion—namely, that the new tract has the same age distribution as the older tract 3198 did in 1975. This assumption is the null hypothesis for our test.

Example **1** (cont.)

Null hypothesis: The age distribution for the new tract is:

Age group	Probability
0–20	0.392
21–65	0.584
65 and over	0.024

The alternative hypothesis is simply that our assumption about age distribution is not true:

Alternative hypothesis: The new tract does *not* have the age distribution given in the null hypothesis.

The population of interest to us is the *ages of the people* in the new tract. It is this population from which we have taken our sample.

To test the null hypothesis, we simply calculate the χ^2-statistic. If the null hypothesis is true, our observed frequencies (O) will generally not differ too much from the expected frequencies (E). Then the O- and E-values will match fairly well, and χ^2 will be "small." If the null hypothesis is not true, the O- and E-values will be "too far apart" and χ^2 will be "too large." As usual, we need to be more precise about the meaning of "χ^2 is too large"—that is, we need to know the probability distribution for χ^2. The exact *probability distribution* of χ^2 is not known, but a very good approximation to the distribution of χ^2 can be obtained using a familiar distribution—the *chi-square distribution*. The rule is given below.

> Suppose the sample size is at least 30, and none of the expected values (E's) is too small.* Then, if the null hypothesis is true, the random variable
>
> $$\chi^2 = \sum \frac{(O-E)^2}{E}$$
>
> has (approximately) the *chi-square distribution with $k-1$ degrees of freedom*, where k is the number of categories. In other words, probabilities for χ^2 can be found using areas under the χ^2-curve with d.f. $= k - 1$.

In the case of our tract data, the sample size is 300 and no expected value is less than 7 (see Table 10.4, page 358). The number of categories in the distribution is 3, so $k = 3$, and d.f. $= 3 - 1 = 2$. This means that we can test whether our χ^2-value is "too large" at the significance level $\alpha = 0.05$ by comparing it to $\chi^2_{0.05}$ for two degrees of freedom. Table III shows that $\chi^2_{0.05} = 5.99$ if d.f. $= 2$. Also (see Table 10.4), our calculated value of χ^2 is 1.64. We picture the comparison in Fig. 10.2.

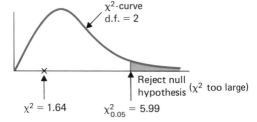

Figure 10.2
Comparison of χ^2 from tract data and $\chi^2_{0.05}$. d.f. $= 2$.

χ^2-curve d.f. $= 2$

Reject null hypothesis (χ^2 too large)

$\chi^2 = 1.64$ $\chi^2_{0.05} = 5.99$

Thus the planner would probably conclude that it was reasonable for him to apply the 1975 age distribution for census tract 3198 to the new tract. ▶

---

* We will explain the meaning of "too small" in the next section.

Only three exercises are given here; you will be given practice in hypothesis testing in Section 10.2. These exercises are designed to test your understanding.

In Exercises 1 through 3:

a) State the appropriate null hypothesis.

b) State the alternative hypothesis.

c) Calculate the expected values, E, and also calculate $O - E$ and $(O - E)^2/E$.

d) Find $\chi^2 = \sum[(O - E)^2/E]$.

e) Suppose you wish to do a test with $\alpha = 0.05$; how would you decide whether or not to reject the null hypothesis?

f) Using the given observed values and the calculated expected values, decide whether or not you would reject the null hypothesis.

1 According to *Statistical Abstract of the U.S. 1976*, households by number of persons in 1970 were as follows:

Number of persons per household	Percent of households
1	19.6
2	30.6
3	17.4
4	15.6
5 and over	16.8

A survey of 1000 households by a planning department in a midwest city yielded the following data:

Number of persons per household	Number of households observed
1	185
2	316
3	170
4	161
5 and over	168

Does this data indicate that the distribution of the number of persons per household in this city is the same as the national distribution? Do (a) through (f) above.

2 The *Statistical Abstract of the U.S. 1976*, reported the following percentages of adoptions in 1971:

By relatives: 51%

By nonrelatives: 49%

A social worker in a large southwest city found the following percentage values for adoptions from a random sample of 50 files:

By relatives: 22

By nonrelatives: 28

Do these results indicate that the southwest city has the same distribution as the rest of the nation? Do (a) through (f) above.

3 A gambler claims that a die is "loaded" (i.e., the six numbers are not equally likely). To try to "prove" his hypothesis, the gambler rolled the die 300 times and obtained the following data:

Number	Observed frequency	Number	Observed frequency
1	61	4	58
2	42	5	43
3	56	6	40

Do (a) through (f) above.

In this section we will give the general procedure for performing a **chi-square goodness-of-fit test** and apply it in some examples. Note carefully the *assumptions* that must be satisfied in order to use the chi-square goodness-of-fit test.

Chi-square goodness-of-fit test   **10.2**

Procedure for a chi-square goodness-of-fit test on a population divided into k groups or categories*

Null hypothesis: The population is grouped according to the probability distribution:

Group	Probability
1	p_1
2	p_2
.	.
.	.
.	.
k	p_k

Alternative hypothesis: The population is *not* grouped according to the probability distribution in the null hypothesis.

Assumptions: (1) *No* expected frequency (that is, no E-value) is less than 1.
(2) *At most* 20 percent of the expected frequencies (that is, E-values) are less than 5.†

1. Decide on the *significance level*, α.
2. The *critical value* is

$$c = \chi_\alpha^2, \qquad \text{d.f.} = k - 1$$

Use Table III to find χ_α^2.

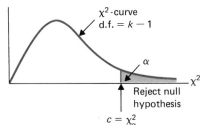

χ^2-curve
d.f. $= k - 1$

α

Reject null hypothesis

$c = \chi_\alpha^2$

3. Make a work table of observed and expected frequencies in each group and calculate

$$\chi^2 = \sum \frac{(O - E)^2}{E}$$

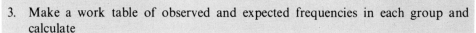

Group	O	E	$O - E$	$\dfrac{(O - E)^2}{E}$
1				
2				
.				
.				
.				
k				

$$\sum \frac{(O - E)^2}{E} = \chi^2$$

---

* This procedure, as well as all others in this chapter, is to be used only when the sample size is "relatively large"—a good rule of thumb is that $n \geqslant 30$.

† Many books give the rule that *no* expected frequency should be less than 5. The work of W. G. Cochran shows that the "rule of 5" is too restrictive.

> *Note.* The E-value for each group is $n \cdot p$, where n is the sample size and p is the probability given for the group in the null hypothesis.
> 4. If $\chi^2 > c$, *reject* the null hypothesis.

◀ As a first application of the above procedure we consider again the illustration of   Example **2**
Example 1. Recall (see page 359) that the hypothesis test the planner wants to perform is:

Null hypothesis: The age distribution for the new tract is:

Age group	Probability
0–20	0.392
21–64	0.584
65 and over	0.024

Alternative hypothesis: The age distribution for the new tract is *not* the one given above.

In this case there are three groups (the three age groups). So $k = 3$. We need to check that the assumptions (1) and (2) are satisfied. To do this, first recall that the planner took a sample of size 300 (that is, $n = 300$). Thus we get Table 10.6, showing expected frequencies (E-values).

Table 10.6

Age group	Probability, p	Expected frequency, $np = E$
0–20	0.392	300 (0.392) = 117.6
21–64	0.584	300 (0.584) = 175.2
65 and over	0.024	300 (0.024) = 7.2

Now we can check assumptions (1) and (2).

(1) No E-value is less than 1? Yes.
(2) At most 20 percent of the E-values are less than 5? Yes, no E-value is less than 5.

Since assumptions (1) and (2) are satisfied, we can now apply the above procedure to perform the hypothesis test.

1. Decide on the significance level, α. We will use $\alpha = 0.05$.
2. The critical value is $c = \chi_\alpha^2 = \chi_{0.05}^2$ which equals 5.99 for d.f. $= k - 1 = 3 - 1 = 2$. See Fig. 10.3.
3. Make a work table of observed and expected frequencies and calculate

$$\chi^2 = \sum \frac{(O - E)^2}{E}$$

(See Table 10.4, page 358, for O-values)

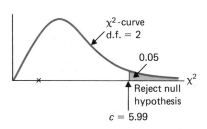

Figure 10.3

Age group	O	E	$O - E$	$\dfrac{(O - E)^2}{E}$
0–20	126	117.6	8.4	0.60
21–64	165	175.2	−10.2	0.59
65 and over	9	7.2	1.8	0.45

$$\chi^2 = 1.64$$

We have marked this value of χ^2 with a cross in Fig.10.3

4. Since $\chi^2 = 1.64$ and $c = 5.99$, we *cannot* reject the null hypothesis. In other words, based on this data, there is *insufficient evidence* for the planner to conclude that there is a difference in age distributions for the new and old tracts. ▶

Exercise **B**   The planner wishes to use the age distribution given in Table 10.2 (page 356) for a cluster of four tracts that contain a number of large mobile-home parks. He takes a random sample of 500 individuals from the four tracts. Test the null hypothesis that the age distribution is

Age	Probability
0–20	0.392
21–64	0.584
65 and over	0.024

using the sample data

Age	Observed frequency
0–20	150
21–64	260
65 and over	90

(Take $\alpha = 0.05$.)

Goodness-of-fit tests may be used to test hypotheses about any kind of percentage distribution. The example below deals with grade distributions.

Example **3**   ◀ A university science department has agreed upon the grading policy shown below for its introductory course.

Top 10% in class	A
Next 20%	B
Middle 40%	C
Next 20%	D
Bottom 10%	E

The chairman of this department is not sure that the policy was followed last semester. Since the introductory course had 2000 students in 50 sections, he does not wish to tally the grades on all those grade sheets. Instead he plans to take a random sample of 50 grades from the course grade sheets and perform the following hypothesis test to see if the policy is being followed.

Null hypothesis: The grade distribution is:

Grade	Probability
A	0.10
B	0.20
C	0.40
D	0.20
E	0.10

Alternative hypothesis: The grade distribution is *not* as given above.

The results of his sampling are given below:

Grade	Observed frequency
A	10
B	10
C	10
D	14
E	6

Perform the hypothesis test at the 0.05 significance level using the procedure given on page 362.

There are five grade groups, so $k = 5$. Next we need to check that assumptions (1) and (2) are satisfied. Since a random sample of $n = 50$ grades was taken, the table of E-values is:

Solution

Grade	Probability, p	Expected frequency, $np = E$
A	0.10	$50(0.10) = 5$
B	0.20	$50(0.20) = 10$
C	0.40	$50(0.40) = 20$
D	0.20	$50(0.20) = 10$
E	0.10	$50(0.10) = 5$

Now we can check assumptions (1) and (2).

(1) No E-value is less than 1? Yes.
(2) At most 20 percent of the E-values are less than 5? Yes, no E-value is less than 5.

We can now proceed with the hypothesis test using the procedure on page 362.

1. Decide on the significance level, α. We will use $\alpha = 0.05$.
2. The critical value is $c = \chi_\alpha^2 = \chi_{0.05}^2 = 9.49$ for d.f. $= k - 1 = 5 - 1 = 4$.
3. Work table and calculation of χ^2. (See Table 10.7.)

Table 10.7

Grade	O	E	$O - E$	$\dfrac{(O - E)^2}{E}$
A	10	5	5	5.00
B	10	10	0	0.00
C	10	20	-10	5.00
D	14	10	4	1.60
E	6	5	1	0.20

$$\chi^2 = 11.80$$

4. Since $\chi^2 = 11.80$ and $c = 9.49$, we have $\chi^2 > c$, and so we *reject* the null hypothesis.

The chairman of this science department will conclude that the departmental grading policy is probably not being followed. He may also get some ideas about further hypotheses to test, if he looks at the data in Table 10.7. He seems to have excesses of A's, D's, and E's, and far too few C's. There are further tests that he could do to see if this is the case. We cannot go through all such tests here, but the lesson is obvious. Look at your data. There is more to be learned from it than a simple χ^2-value can show. ◗

Exercise C   Perform the hypothesis test in Example 3 if the results of the sampling of 50 grades turned out to be:

Grade	Observed frequency
A	5
B	15
C	15
D	5
E	10

Use $\alpha = 0.05$.

In each of the examples we have considered, the entire percentage distribution was given and the degrees of freedom for the chi-square test was simply one less than the number of groups in the distribution, that is, d.f. $= k - 1$. There are more complex cases in which the percentages in the distribution must be calculated from the observed data. In such cases, the degrees of freedom are reduced further. We will discuss this more complex situation in the exercises. The important point to keep in mind is that we have discussed only the simplest case here, and you should not expect to be able to handle all possible goodness-of-fit problems using the techniques we have discussed so far.

EXERCISES Section **10.2**

In Exercises 1 through 6:
a) State the appropriate null hypothesis.
b) State the alternative hypothesis.
c) Use the procedure given on page 362 to perform the hypothesis test.

1 Perform a chi-square goodness-of-fit test for the family-size distribution and data given in Exercise 1 of Section 10.1. Use $\alpha = 0.05$.

2 Perform a chi-square goodness-of-fit test for the adoption distribution and data given in Exercise 2 of Section 10.1. Use $\alpha = 0.05$

3 The percentage distribution of enrollment by grade for public schools in the U.S. in 1975 is given below (Source: *Statistical Abstract of the U.S. 1976.*)

Grade	Percentage of enrollment
K–3	27.9
4–6	23.7
7–9	25.7
10–12	22.7

A random sample of 100 students in a unified school district gave the following data.

Grade	Observed frequency
K–3	15
4–6	20
7–9	35
10–12	30

Perform a chi-square goodness-of-fit test, using $\alpha = 0.05$, to test the hypothesis that the unified school district has the same distribution as does the entire U.S. (that is, it is typical).

4 In 1960, the percentages of the U.S. farm population by age groups were as follows. (Source: *Statistical Abstract of the U.S. 1976.*)

Age group	Percentage of farm population
Under 14	31.9
14–19	11.9
20–24	4.9
25–34	9.3
35–44	11.5
45–64	22.0
65 and over	8.5

In the same year, a random sample of 200 farm residents in Colorado revealed the following:

Age group	Number observed
Under 14	50
14–19	28
20–24	10
25–34	16
35–44	22
45–64	50
65 and over	24

Perform a chi-square goodness-of-fit test, using $\alpha = 0.05$, to test whether or not the age distribution for Colorado was the same as the national distribution.

5 In 1959, land was utilized in the following way. (Source: *Statistical Abstract of the U.S. 1976.*)

Major use	Percentage of land
Cropland	17.3
Grassland pasture	23.4
Woodland pasture	4.1
Woodland not pastured	3.1
Farmsteads, roads, etc.	1.6
Grazing land	14.0
Forest land not grazed	19.3
Urban, industrial, residential areas, parks, etc.	17.2

A study of 2,264 randomly selected acres in 1969 produced the following results:

Major use	Number of acres
Cropland	384
Grassland pasture	540
Woodland pasture	62
Woodland not pastured	50
Farmsteads, roads, etc.	28
Grazing land	288
Forest land not grazed	475
Urban, industrial, residential areas, parks, etc.	437

Perform a chi-square goodness-of-fit test, using $\alpha = 0.05$, to test the null hypothesis that the distribution for 1969 was the same as the distribution for 1959.

6 Perform a chi-square goodness-of-fit test for Exercise 3 of Section 10.1. [Use $\alpha = 0.05$.]

In Exercises 7 and 8 we will show you how to use a goodness-of-fit test to decide whether or not data comes from a population with a given distribution (e.g., binomial or normal).

7 A college baseball player has a batting average of .400. A fan believes that, for this player, the number of hits in four times at bat (a typical game) should follow a binomial distribution with $n = 4$, $p = 0.400$.

a) Fill out the chart below, which gives the probability of each possible number of hits in four at-bats.

Number of hits, x	0	1	2	3	4
Probability, $P(x)$					

b) The player played 40 games in each of which he had four at-bats. Use your answers from (a) to find the expected number E of games out of those 40 in which the player will get 0 hits, 1 hit, etc. Enter these in the chart in part (c).

c) The player actually had the season results listed in the third column of the table below under O:

Number of hits in a game, x	Expected number of games, E	O
0		4
1		17
2		12
3		5
4		2

Perform a goodness-of-fit test using these O- and E-values to decide whether the player's hits-per-game distribution fits the binomial distribution with $n = 4$ and $p = 0.400$. (Use $\alpha = 0.05$.)

8 A student claims that, by using a program on his calculator, he can generate numbers that are picked at random from a normal distribution with mean 0 and variance 1 (a standard normal or z-distribution).

a) What is the probability that a number from a standard normal distribution falls into each of the categories below?

	Probability	E	O
$z < -2$			2
$-2 \leqslant z < -1$			19
$-1 \leqslant z < 0$			29
$0 \leqslant z < 1$			30
$1 \leqslant z < 2$			17
$2 \leqslant z$			3

(Fill in the column headed "Probability.")

b) The student decides that he will generate 100 random numbers, using his calculator. Calculate the expected frequency E for each class above if the 100 numbers are truly picked from a standard normal distribution. (Fill in the E column.)

c) The 100 numbers that the student actually generated are given above in column O. Perform a chi-square goodness-of-fit test of the student's claim that his numbers come from a standard normal distribution. (Use $\alpha = 0.05$.)

In some attempts to fit data to a known distribution, a researcher will use the data to estimate μ, σ, or some other number associated with the population. This causes a change in the degrees of freedom, as we will show in the problem below:

9 A quality-control engineer suspects that a production line produces bolts whose diameters are normally distributed, but he has no idea of the mean and variance of that normal distribution. He takes a sample of 100 bolts from the line and measures them. The results are distributed as follows:

Diameter, d	Observed number in class
$d < 9.8$	0
$9.8 \leq d < 9.9$	12
$9.9 \leq d < 10.0$	39
$10.0 \leq d < 10.1$	32
$10.1 \leq d < 10.2$	16
$10.2 \leq d$	1

Since he has no idea of the mean and standard deviation, he estimates them from his data using $\bar{x}$ and s.

Estimate of $\mu = \bar{x} = 10.01$.
Estimate of $\sigma = s = 0.1$.

a) Use these estimates of μ and σ to find the probability that a bolt diameter is in each given class.
b) For a random sample of 100 bolts, find the expected frequency (E) for each class.
c) Use (b) and the O-values given in the table above to compute χ^2.
d) This procedure reduces the number of degrees of freedom. For six categories, we would think that d.f. $= 6 - 1 = 5$. But the degrees of freedom are reduced by one for each of the two population parameters (μ and σ) we estimated for use in finding the E-values. That is,

$$\text{d.f.} = \underbrace{6-1}_{\text{Usual}} - \underset{\underset{\substack{\text{For}\\\text{estimating}\\\mu \text{ by } \bar{x}}}{\uparrow}}{1} - \underset{\underset{\substack{\text{For}\\\text{estimating}\\\sigma \text{ by } s}}{\uparrow}}{1} = 3$$

Perform a chi-square goodness-of-fit test using the χ^2-value from (c) and d.f. $= 3$. (Take $\alpha = 0.05$.)

The most common application of the chi-square distribution is in **tests of independence**. The data for such tests is given in tables called **contingency tables**. In this section we will give examples of contingency tables, and review the concept of independence. This will prepare us to discuss tests of independence in Section 10.4.

Contingency tables— independence 10.3

Example **4**

▮ Student attitude questionnaires known as "teacher evaluations" gained widespread use in the late 1960's and early 1970's. There was (and still is) considerable debate among faculty members over the use of these forms. Some faculty felt that the highest evaluations would go to the easy classes in which students learned the least; many other faculty members maintained that this was not true—students would learn the most from a teacher with high evaluations. There were statistically analyzed studies to support both points of view.

At Arizona State University, as at many other schools, teacher evaluations were given, and some faculty criticized them. For example, all faculty members in the mathematics department were required to give a standard teacher-evaluation form to their classes. On this form the students graded their courses, using an A (most positive) to E (most negative) scale for items such as "Rate the instructor's knowledge of the subject matter" and "Give this course an overall grade." One criticism of this form was raised more frequently than all others. Many faculty felt that students would tend to give their teachers the same letter grades that the teachers were giving them: A student expecting an A in the class would tend to evaluate it with A's and B's, and a student who was failing would grade low. We decided to collect some data to see if this was true. In

December 1977, we gave all College Algebra students an evaluation form containing seven questions. Two of the questions were:

Question 2: What grade do you predict you will receive in this class?
Question 3: If you could give this *class* a grade, what would that grade be?

We were interested in whether a typical student's answer to Question 2 influenced the answer to Question 3. If the response to Question 2 did influence the response to Question 3, we could use the language of probability (Chapter 4) to say that these two responses were *dependent*. If the responses did not affect each other, we would call them *independent*. In the language of probability, we were interested in whether or not student response to Question 2 was independent of response to Question 3. This could be put in the form of a hypothesis test.

Null hypothesis: The response of students in College Algebra to Question 3 is independent of the response to Question 2.
Alternative hypothesis: The response to Question 3 is dependent upon the response to Question 2.

It is very important to discuss what population is being considered here. Professors and students who discuss teacher evaluations are not simply concerned about what has happened in a single semester. The Fall of 1977 is gone, and we are all concerned about the future. We really wish our data to tell us something about similar students in similar courses at Arizona State University (and possibly other universities) in the future. Our population could be described as follows:

> *Population*:
> All possible evaluations
> from similar College
> Algebra classes

If we had been interested only in the population of student ratings from College Algebra at Arizona State University in December 1977, there would be no need to use inferential statistics to test a hypothesis. We collected a questionnaire from every College Algebra student in December 1977. Thus we had data for the *entire* population of Fall, 1977 College Algebra evaluations, and could use that data to answer questions about the population directly. Hypothesis tests are used on *samples* from populations, not on entire populations. This is shown in Fig. 10.4. ▶

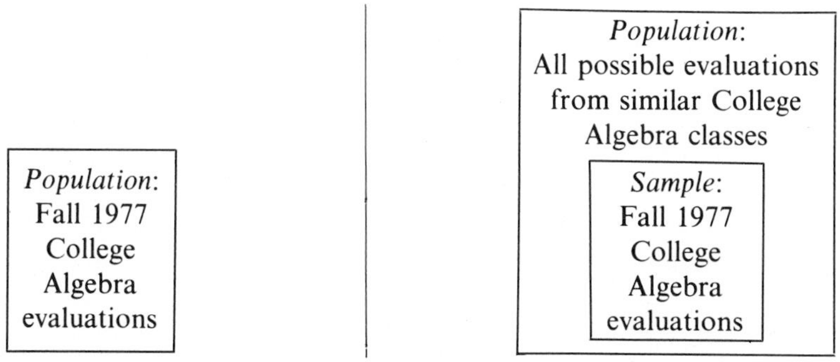

Case 1: Population completely known. No hypothesis test is necessary.

Case 2: Only sample data is known. You must use inferential statistics to test hypotheses about the population.

Figure 10.4

In December 1977, the same form was given to Intermediate Algebra students.* If you wished to test the null hypothesis of independence of response on Questions 2 and 3 for Intermediate Algebra, how would you describe your population of interest?

Exercise **D**

◖ We designed our hypothesis test *before* we collected our data. However, we will show you those data before we go through the discussion of the hypothesis-testing procedure. Our data are given in Table 10.8 below.

Example **4** (cont.)

Question 2:
Grade predicted by student

Table 10.8
Data for responses to Questions 2 and 3 on evaluation questionnaire

	A	B	C	Row total
A	42	14	1	57
B	33	34	7	74
C	9	24	7	40
D or E	2	12	4	18
Column total	86	84	19	189

Question 3: Evaluation grade (for course)

Total number of A evaluations

Total number of predicted A grades

Total number of evaluation forms received

---

* Intermediate Algebra is the prerequisite course for College Algebra at Arizona State University.

The number 42 in the upper lefthand corner of Table 10.8 tells us that 42 students predicted they would earn a grade of A in the class, and also gave the class a grade of A. The 33 below the upper left corner tells us that 33 other students who predicted they would earn A grades gave course evaluation grades of B. The column total underneath the predicted grade of A shows that a total of $42 + 33 + 9 + 2 = 86$ students expected to earn grades of A in the class. The row total to the right of the evaluation grade of A shows that a total of 57 students gave the class an evaluation grade of A. The number 189 in the lower righthand corner is the total number of evaluation forms received. This total number can be found by adding up *either* the row or column totals.

Table 10.8 is an example of a **contingency table**. The boxes inside the heavy lines in the table give the actual response data. These boxes are called the **cells** of the table. Table 10.8 has twelve cells. ◗

Exercise **E** Table 10.9 is a *contingency table* containing response data for Intermediate Algebra students in the Fall of 1977.

Table 10.9
Intermediate Algebra evaluation data, December 1977

Question 2:
Predicted grade

		A	B	C	D or E	Total
Question 3: Evaluation grade	A	96	42	8	1	
	B	81	141	53	4	
	C	30	57	52	3	
	D or E	8	18	28	5	
	Total					

a) How many cells does this table have?
b) Calculate the row and column totals for Table 10.9.
c) How many Intermediate Algebra students filled out this questionnaire? (Fill in the number in the lower righthand corner of Table 10.9.)
d) How many Intermediate Algebra students thought they would receive a grade of C, and awarded their class a grade of B?

Independence We have already used the term *independence* in a non-mathematical way. In words, two events E and F are independent if the fact that F has occurred does not affect the probability that E will occur. In the previous example, we considered events E and F such as:

> E: A student predicts he or she will get an A in College Algebra,
> F: A student gives an evaluation grade of A to College Algebra.

In words, independence means that expecting to get an A in College Algebra will not change the probability of giving an A evaluation to the class. If we wish to use mathematical symbols, there are two different formulas in Chapter 4 that can be used to characterize independent events. They are:

i) $P(E|F) = P(E)$

ii) $P(E \ \& \ F) = P(E) \cdot P(F)$.

When analyzing contingency tables, we will use the second formula. To use this formula, we must first be able to find probabilities such as $P(E) = P(\text{A randomly chosen student predicts that he or she will earn an "A" in College Algebra})$ = proportion of all students who predict they will earn an "A" in College Algebra. We do this in Example 5 below.

◀ To *estimate* the probability that a student predicts an A grade in College Algebra, we must look at our sample data. We were not given data for the entire unknown population, and the sample is all we have to look at*. The probability we want is given approximately by (see Table 10.8):   Example **5**

$$P(\text{Predicted grade is A}) \doteq \frac{\text{Number of predicted A grades}}{\text{Total number of students}}$$

$$= \frac{\text{Total of first column}}{\text{Total number of students}} = \frac{86}{189} = 0.455$$

Using the same kind of reasoning, along with Table 10.8, we find

$$P(\text{Evaluation grade is A}) \doteq \frac{\text{Total of first row}}{\text{Total number of students}}$$

$$= \frac{57}{189} \doteq 0.302$$

We can compute the (approximate) probability of any predicted grade or evaluation grade by using the appropriate row or column total in Table 10.8, as we did above. It is customary to simply fill in these probabilities around the edges of the table, as we do in Table 10.10. ▶

---

* What we are really doing here is using the *relative frequency interpretation of probability* to estimate probabilities. That is, we are using a *sample proportion to estimate a population proportion.*

Table 10.10

College Algebra response (observed) data with probabilities

Question 2:
Predicted grade

Question 3:
Evaluation grade

	A	B	C	Row total		
A	42	14	1	57	0.302 ◄——— 57/189
B	33	34	7	74	0.392 ◄——— 74/189
C	9	24	7	40	0.212 ◄——— 40/189
D or E	2	12	4	18	0.095 ◄——— 18/189
Column total	86	84	19	189	

0.455   0.444   0.101

86/189 84/189 19/189

Exercise **F**   Fill in the (approximate) probabilities for all predicted and evaluation grades for the Intermediate Algebra response data in Table 10.9.

Example **5** (cont.)   ◖ *If the null hypothesis of independence is true*, then once the probabilities of separate events are known, we can use the formula $P(E\ \&\ F) = P(E) \cdot P(F)$ to find the probability for each cell in Table 10.10. For example, let us look again at the upper lefthand cell of the table. If predicted course grade and evaluation grade are independent, then the probability for this cell is

$$P\ (Evaluation\ grade\ is\ A\ \&\ Predicted\ grade\ is\ A)$$
$$= P\ (Evaluation\ grade\ is\ A) \cdot P\ (Predicted\ grade\ is\ A)$$
$$= \frac{57}{189} \cdot \frac{86}{189} = 0.137$$

This means that (if the null hypothesis of independence is true) we would expect about 13.7 percent of the 189 students to fall in the upper left cell.

$$\text{Expected number in upper left cell} = (0.137)\,(189) = 25.9$$

It is simpler for purposes of computation to look at this expected number in another way. The probabilities that were multiplied to get the probability for the upper left cell were the probabilities for the first row and the first column of Table 10.10. Thus, we can write

$$\text{Expected number} = \left(\frac{57}{189}\right) \cdot \left(\frac{86}{189}\right) \cdot (189)$$
$$= \left(\begin{array}{c}\text{Row}\\\text{probability}\end{array}\right) \cdot \left(\begin{array}{c}\text{Column}\\\text{probability}\end{array}\right) \cdot \left(\begin{array}{c}\text{Total}\\\text{number}\end{array}\right)$$

By doing some cancellation we can simplify this calculation.

$$\text{Expected number} = \frac{57}{189} \cdot \frac{86}{189} \cdot 189$$

$$= \frac{57 \cdot 86}{189}$$

$$= \frac{(\text{Row total})(\text{Column total})}{\text{Total number}}$$

This relationship holds for all cells if the responses to Questions 2 and 3 are independent. In Table 10.11 we compute expected numbers for all cells, using the formula

$$\frac{\text{Expected number}}{\text{in cell}} = \frac{(\text{Row total})(\text{Column total})}{\text{Total number}}$$

In this work table for expected frequencies, we put the row and column totals on the edges for ready reference. ▶

Question 2: Predicted grade

		A	B	C	Total
	A	$\frac{57(86)}{189} = 25.9$	$\frac{57(84)}{189} = 25.3$	$\frac{57(19)}{189} = 5.7$	57
	B	$\frac{74(86)}{189} = 33.7$	$\frac{74(84)}{189} = 32.9$	$\frac{74(19)}{189} = 7.4$	74
Question 3: Evaluation grade	C	$\frac{40(86)}{189} = 18.2$	$\frac{40(84)}{189} = 17.8$	$\frac{40(19)}{189} = 4.0$	40
	D or E	$\frac{18(86)}{189} = 8.2$	$\frac{18(84)}{189} = 8.0$	$\frac{18(19)}{189} = 1.8$	18
	Total	86	84	19	189

Table 10.11
Work table for expected frequencies

We have rounded all expected values to one decimal place. We will do this in all our contingency-table computations in this chapter. As before, rounding differences may mean that your calculator gives a slightly different answer from ours (or from another student's calculator).

Note on rounding

Set up a work table and compute expected numbers for the Intermediate Algebra data in Table 10.9.

Exercise **G**

Example **5** (cont.) ◖ The expected numbers computed in Table 10.11 are the numbers to be expected if the responses are independent (that is, if the null hypothesis is true). To get some idea of the overall correspondence between observed and expected values, we will make a large version of Tables 10.10 and 10.11, which will have *both* observed and expected values. This is done by making large cells with a space in the lower right corner for expected values. The upper lefthand cell by itself looks like this:

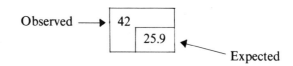

The entire table looks like Table 10.12.

Table 10.12
bserved and expected
numbers for College
Algebra data

Question 2: Predicted grade

		A	B	C	Total
Question 3: Evaluation grade	A	42 25.9	14 25.3	1 5.7	57
	B	33 33.7	34 32.9	7 7.4	74
	C	9 18.2	24 17.8	7 4.0	40
	D or E	2 8.2	12 8.0	4 1.8	18
	Total	86	84	19	189

You can see from Table 10.12 that observed and expected values do not match up very well in most cells. A good guess would be that the responses to Questions 2 and 3 are *dependent*. To do more than guess, a hypothesis test is necessary. We will show how to do this hypothesis test in the next section. ◗

Exercise **H**   Make a complete table containing both observed and expected values for the Intermediate Algebra results in Table 10.9. Would you guess that the responses to Questions 2 and 3 are dependent or independent?

In Exercises 1 through 4:

a) State the appropriate null hypothesis.

b) State the alternative hypothesis.

c) Indicate the population under discussion.

d) Construct a table for the data, including the row totals and column totals.

e) Indicate the number of cells in the table.

f) Calculate the expected value for each cell and make a table for the observed values and expected values.

g) Make a guess as to whether the observed values and expected values match well enough to support the null hypothesis.

h) Make a guess as to whether the events are dependent or independent. (You will learn how to test this hypothesis in Section 10.4.)

1 In 1975, a sociologist surveyed 450 white wage earners, 25 years and older, in an attempt to determine whether annual income is dependent on educational level. The following table displays this data.

Annual Income

Years of schooling	0–$6,999	$7,000–$14,999	$15,000–$24,999	$25,000 or over
0– 8	34	41	10	4
9–12	36	72	78	26
Over 12	10	36	58	45

2 In the Spring of 1978, a study was conducted at Arizona State University to determine whether there is an association (dependence) between the grade earned in Intermediate Algebra and hours worked on algebra per week. The following table displays the findings.

Hours

Grade earned		0–3	4–6	7–9	> 9
	A	48	51	34	13
	B	37	75	43	9
	C	22	46	32	12

3 In March 1979, a random sample of 100 college students from an undergraduate engineering school were asked two questions.

Question 1: Do you think the President is providing the needed leadership in the area of economic policy?

Question 2: Do you think the President is doing a good job overall?

Their responses were tabulated as follows:

Question 1:

		Yes	No	Total
	Yes	17	18	
Question 2:	No	5	60	
	Total			

4 A survey of 611 American males in 1974 yielded the following information on age versus smoking.

Age

	17–24	25–44	45–64	65 and over
Never smoked	60	58	39	28
Had smoked	65	162	144	55

5 The table in Exercise 3 has the form:

			Total
	a	b	$a+b$
	c	d	$c+d$
Total	$a+c$	$b+d$	

For such a table we can measure the difference between proportions in the two rows by calculating

$$d_r = \frac{a}{a+b} - \frac{c}{c+d}$$

We can also measure the difference between proportions in the columns by calculating

$$d_c = \frac{a}{a+c} - \frac{b}{b+d}$$

a) Calculate d_r for the table in Exercise 3, and describe the meaning of d_r in words.

b) Calculate d_c for the table in Exercise 3, and describe the meaning of d_c in words.

The numbers d_r and d_c are used to measure association (dependence) between rows and columns separately. We can put them together into a single measure of association by computing

$$\phi = \pm\sqrt{d_r d_c}$$

where the plus sign is used if $ad - bc \geqslant 0$, and the minus sign is used if $ad - bc < 0$.

c) Compute ϕ for the table given in Exercise 3.

The measure ϕ is such that $-1 \leqslant \phi \leqslant 1$. A rule of thumb for determining the strength of association in a 2×2 table is

$-\frac{1}{3} < \phi < \frac{1}{3}$	Weak or no association
$\frac{1}{3} \leqslant \phi \leqslant \frac{2}{3}$	Medium positive association
$-\frac{2}{3} \leqslant \phi \leqslant -\frac{1}{3}$	Medium negative association
$\frac{2}{3} < \phi \leqslant 1$	Strong positive association
$-1 \leqslant \phi < -\frac{2}{3}$	Strong negative association

d) What kind of association does the value of ϕ for the table in Exercise 3 indicate?

e) How could the values in the table be changed to give
   i) A strong positive association, or
   ii) A negative association of any kind?

6 If ϕ is as defined in Exercise 5,
a) Prove that

$$\phi = \frac{ad - bc}{\sqrt{(a+b)(c+d)(a+c)(b+d)}}$$

b) Prove that $-1 \leqslant \phi \leqslant 1$.

c) Find values for b and c that make $\phi = 1$. (This will give you a table in which there is a *perfect positive association*.)

d) Find values for a and d that make $\phi = -1$. (This will give you a table in which there is a *perfect negative association*.)

10.4 Chi-square tests of independence

In this section we will show how to use the observed and expected values from Section 10.3 to perform a hypothesis test for independence. You may have guessed what statistic we will calculate for this purpose. We will find expected values as in Section 10.3 and calculate

$$\chi^2 = \sum \frac{(O - E)^2}{E}$$

If χ^2 is "too large," we will reject the null hypothesis of independence. As with the goodness-of-fit tests in Section 10.2, we will use the chi-square probability distribution table to decide whether χ^2 is too large. To do this, we must first know how to calculate *degrees of freedom*. In Example 6 below, we will show how to find the degrees of freedom for the College Algebra data in Table 10.8 (page 371).

Example **6**   ◖ We will begin by setting up a table of empty cells for our observed data. (See the table at the top of page 379.)

There are $4 \cdot 3 = 12$ cells here, but there are *not* 12 degrees of freedom. If we did a goodness-of-fit test for Question 3 only, there would be $4 - 1 = 3$ degrees of freedom. For Question 2 alone, there would be $3 - 1 = 2$ degrees of freedom. In finding the degrees of freedom for the independence test, we remove one degree of freedom for each question, and multiply:

$$\text{d.f.} = (4-1)\,(3-1) = 3\cdot 2 = 6$$

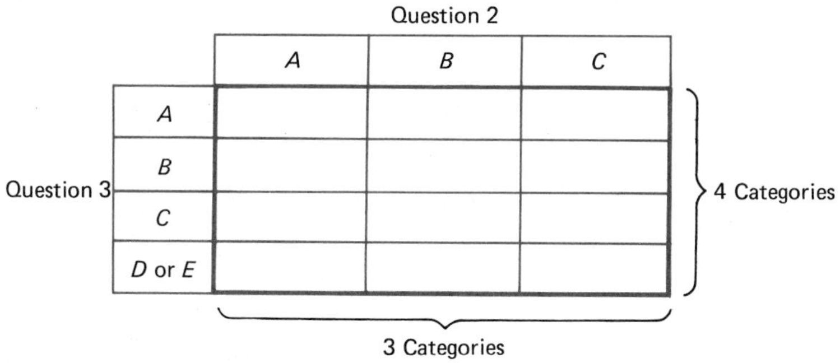

Question 2

	A	B	C
A			
B			
C			
D or E			

Question 3 · · · 4 Categories

3 Categories

In Fig. 10.5 we have shaded one row and one column of cells for the degree of freedom removed from each question. The number of degrees of freedom for the independence test is equal to the number of cells left unshaded. In this diagram, we can see the general formula used for chi-square independence tests:

$$\text{d.f.} = (\text{Number of rows} - 1)\,(\text{Number of columns} - 1)$$

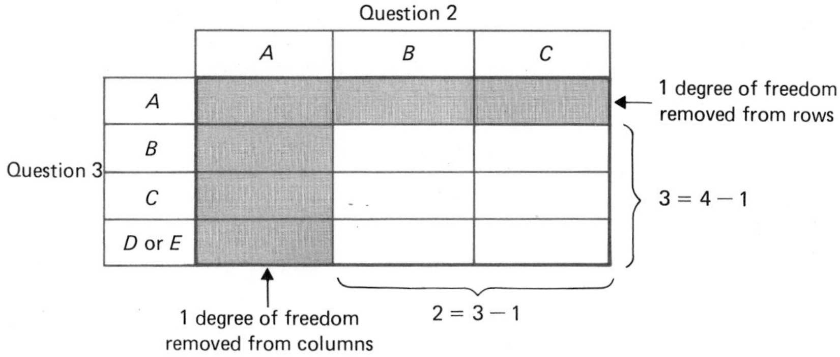

Question 2 · · A B C · Question 3 · A B C D or E · 1 degree of freedom removed from rows · $3 = 4-1$ · 1 degree of freedom removed from columns · $2 = 3-1$

Figure 10.5

Usually the following notation is used:

r = number of rows,
c = number of columns,
$\text{d.f.} = (r-1)\,(c-1).$

In the case of our College Algebra data

$$r = 4, \quad c = 3$$
$$\text{d.f.} = (4-1)\,(3-1) = 3\cdot 2 = 6$$

Exercise **I**	Find r, c, and d.f. for the Intermediate Algebra data in Table 10.9, page 372.
Minimum expected frequency	As with chi-square goodness-of-fit tests, we should not have "too many" *expected* frequencies that are "too small." Exactly how many is "too many" or how small is "too small" has been the subject of much debate. We will use the same guidelines as in Section 10.2.
Assumptions for use of chi-square independence tests	(1) No expected cell frequency is less than 1. (2) At most 20 percent of the cells have expected frequencies less than 5.

We check these assumptions for the College Algebra data in Example 7 below.

Example **7**	◖ The expected frequencies for this data set are shown in Table 10.13 (see page 375).
Table 10.13 Expected frequencies	

Question 2

		A	B	C	
	A	25.9	25.3	5.7	
Question 3	B	33.7	32.9	7.4	
	C	18.2	17.8	(4.0)	← Expected frequencies less than 5
	D or E	8.2	8.0	(1.8)	←

(1) No expected frequency is less than 1.
(2) There are 12 cells here, and two of them have expected frequencies below 5. Since $2/12 = 1/6 = 16\frac{2}{3}$ percent, the percentage of cell frequencies below 5 is at most 20 percent. ◗

Note that we had set up a combined category of "D or E" for evaluation grades, instead of separate D and E categories. We knew from prior experience that student evaluators are so generous that there would be few D's or E's. We combined those categories so that expected frequencies would not turn out to be too small for too many cells.

Exercise **J**	Look at the expected frequencies for the Intermediate Algebra study. (These were calculated in Exercise G.) a) Is any expected frequency less than 1? b) What percentage of cells have expected frequencies below 5? c) Does this satisfy our assumptions for the use of chi-square independence tests?

We can now go through the steps of the hypothesis test proposed for our College Algebra evaluation data in Section 10.3. We do this in Example 8. Following this we will state the general procedure, and then do another example.

◖ Perform the hypothesis test:

Example **8**

Null hypothesis: The responses of students in College Algebra to Questions 2 and 3 are independent.
Alternative hypothesis: Those responses are dependent.

Step 1. Decide on the significance level, α. We will use α = 0.01.

Solution

Step 2. Collect data, construct an observed data table, and calculate degrees of freedom. (See Table 10.14.)

Question 2

	A	B	C	Total
A	42	14	1	57
B	33	34	7	74
C	9	24	7	40
D or E	2	12	4	18
Total	86	84	19	189

Question 3

4 rows

3 columns

Table 10.14
Observed values (see page 371)

$$r = 4, \quad c = 3$$
$$\text{d.f.} = (r - 1)(c - 1) = 3 \cdot 2 = 6$$

Step 3. The critical value is $c = \chi^2_{0.01}$ for d.f. = 6. For d.f. = 6, $\chi^2_{0.01} = 16.81$.

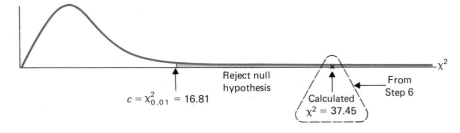

$c = \chi^2_{0.01} = 16.81$

Reject null hypothesis

Calculated $\chi^2 = 37.45$

From Step 6

Step 4. Compute expected values using the formula

$$\text{Expected number in cell} = \frac{(\text{Row total}) (\text{Column total})}{(\text{Total number in sample})}$$

and place each expected value in the lower righthand corner of the appropriate cell in the table of observed frequencies. (See Table 10.15, page 383.)

Step 5. Check that (1) no expected frequency is less than 1, and (2) the percentage of expected frequencies less than 5 is at most 20 percent. (If either (1) or (2) fails, it is inappropriate to use the chi-square test.)

   (1) There are no expected frequencies less than 1.

   (2) There are two expected frequencies less than 5: $2/12 = 16\frac{2}{3}$ percent of all cells. The number of expected values less than 5 is at most 20 percent.

Step 6. Calculate χ^2:

$$\chi^2 = \sum \frac{(O-E)^2}{E} = \frac{(42-25.9)^2}{25.9} + \frac{(14-25.3)^2}{25.3} + \frac{(1-5.7)^2}{5.7}$$

$$+ \frac{(33-33.7)^2}{33.7} + \frac{(34-32.9)^2}{32.9} + \frac{(7-7.4)^2}{7.4}$$

$$+ \frac{(9-18.2)^2}{18.2} + \frac{(24-17.8)^2}{17.8} + \frac{(7-4.0)^2}{4.0}$$

$$+ \frac{(2-8.2)^2}{8.2} + \frac{(12-8.0)^2}{8.0} + \frac{(4-1.8)^2}{1.8}$$

$$= 10.01 + 5.05 + 3.88$$
$$+ 0.01 + 0.04 + 0.02$$
$$+ 4.65 + 2.16 + 2.25$$
$$+ 4.69 + 2.00 + 2.69$$
$$= 37.45$$

Step 7. Make an inference. Since the χ^2-value is larger than 16.81, we reject the null hypothesis. Therefore, based on this data we infer that responses to Questions 2 and 3 are *dependent*.

These results should be interpreted cautiously. We were carrying out this test in order to make an inference about similar College Algebra classes. Our class had some special features. Students were tested weekly. By the end of the semester when evaluations were given, every student had a good idea of his or her grade. Our school has liberal drop procedures; a number of students who knew that they would get low grades had dropped the class right before evaluation time. You have probably noted that no students expected a grade lower than C. There are many schools similar to our

Question 2

	A		B		C		Total
A	42	25.9	14	25.3	1	5.7	57
B	33	33.7	34	32.9	7	7.4	74
C	9	18.2	24	17.8	7	4.0	40
D or E	2	8.2	12	8.0	4	1.8	18
Total	86		84		19		189

Question 3 (row label at left)

Table 10.15

own with math courses run in the same manner. This hypothesis test probably says something about evaluations in algebra at those other schools, but probably says *nothing* about evaluation of a history class with only a midterm and a final exam in a school where students who expect to fail the class cannot drop it. ▶

We now present the general procedure, illustrated above, for performing a chi-square test of independence.

Observations from a population are classified according to two different characteristics (such as response to Question 2 and response to Question 3).

The observed data is put in a table with r rows and c columns.

Procedure for performing a chi-square test of independence

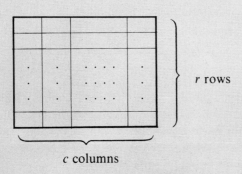

r rows

c columns

Null hypothesis: The two characteristics are independent.
Alternative hypothesis: The two characteristics are dependent.

1. Decide on the significance level, α.
2. Collect data, construct an observed data table, and calculate degrees of freedom using

$$\text{d.f.} = (r-1)(c-1).$$

3. The critical value is

$$c = \chi_\alpha^2, \qquad \text{for d.f.} = (r-1)(c-1)$$

Use Table III to find χ_α^2.

4. Compute expected values using the formula:

$$\text{Expected number in cell} = \frac{(\text{Row total})\,(\text{Column total})}{(\text{Total number in sample})}$$

and place each expected value in the lower righthand corner of the appropriate cell in the table of observed frequencies.

5. Check to see that:
   (1) No expected frequency is less than 1.
   (2) At most 20 percent of the expected frequencies are less than 5.
   (If either (1) or (2) fails, you should not use the chi-square test.)

6. Calculate χ^2 using the formula

$$\chi^2 = \sum \frac{(O-E)^2}{E}$$

7. If $\chi^2 > c$, *reject* the null hypothesis.

χ^2-curve
d.f. $= (r-1)(c-1)$

α

Reject null
hypothesis

$c = \chi_\alpha^2$

Example 9 ◀ In 1975, more than 59 million Americans suffered injuries. More males (33.6 million) were injured than females (25.6 million). Those statistics do not tell us whether males and females tend to be injured in similar circumstances. One set of categories commonly used for the circumstances of accidents is: "At work," "At home," and "Other." (The "Other" category includes car accidents.) In order to study whether accident circumstances differ by sex, a safety official in a large city took a random sample of 183 accident reports. He obtained the following data:

	Male	Female	Total
At work	40	5	45
At home	49	58	107
Other	18	13	31
Total	107	76	183

$r = 3$

$c = 2$

We will use this data to test the null hypothesis that the circumstance of an accident in that city is independent of the sex of the accident victim, at the 0.01 significance level. Specifically, we will use the procedure on page 383 to test:

Null hypothesis: The circumstance of an accident is *independent* of the sex of the accident victim.
Alternative hypothesis: Accident circumstance is *dependent* on the sex of the victim.

1. *Decide on the significance level,* α. We will use $\alpha = 0.01$.
2. *Collect data, construct an observed-data table, and calculate degrees of freedom.*

The observed data table was given at the beginning of the example.

$$\text{d.f.} = (r-1)(c-1) = (3-1)(2-1) = 2 \cdot 1 = 2$$

3. *The critical value is* $c = \chi^2_{0.01} = 9.21$ for d.f. $= 2$. (See Fig. 10.6.)

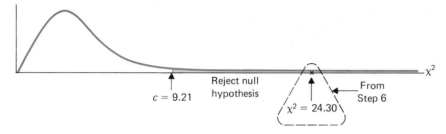

Figure 10.6

4. *Combined table:*

	Male		Female		Total
Work	40	26.3	5	18.7	45
Home	49	62.6	58	44.4	107
Other	18	18.1	13	12.9	31
Total	107		76		183

5. (1) No expected frequency is less than 1? Yes.
(2) At most 20 percent of the expected frequencies are less than 5? Yes, no expected frequency is less than 5.

6. *Calculate χ^2:*

$$\chi^2 = \sum \frac{(O-E)^2}{E} = \frac{(40-26.3)^2}{26.3} + \frac{(5-18.7)^2}{18.7}$$

$$+ \frac{(49-62.6)^2}{62.6} + \frac{(58-44.4)^2}{44.4}$$

$$+ \frac{(18-18.1)^2}{18.1} + \frac{(13-12.9)^2}{12.9}$$

$$= 7.14 + 10.04$$
$$+ 2.95 + 4.17$$
$$+ 0.001 + 0.001$$
$$= 24.30^*$$

7. Since $\chi^2 > c$, we reject the null hypothesis. In other words, on the basis of this data, we infer that the circumstance of an accident is *dependent* on the sex of the accident victim.

Exercise **K**   Use the Intermediate Algebra data in Table 10.9 (page 372) to test the null hypothesis that the responses to Questions 2 and 3 are independent. Use $\alpha = 0.01$.

One final comment regarding the use of chi-square independence tests: In Example 9, we used the chi-square independence test to make an inference about the *population* of accident victims in a city using data from a *sample* of accident victims. The data for accident victims in the entire United States population is also available to us (in *Statistical Abstract of the United States*). The actual numbers of injuries during 1974, grouped by sex and circumstance, are given below in Table 10.16.

Table 10.16

Persons injured in the U.S. in 1974, grouped by sex and circumstances. (Numbers given in millions)

	Male	Female
Work	8.0	1.3
Home	9.8	11.6
Other	17.8	12.9

---

* *Note.* For hand calculation, we usually compute expected frequencies to one decimal place and $(O-E)^2/E$ to two places. We gave the last two $(O-E)^2/E$-values to three places simply because they would have rounded off to 0.00 to two places, but were *not* exactly 0. We still rounded our final answer to two places.

It is not appropriate to do a chi-square test on the data in Table 10.16. This data represents an *entire population*—not the kind of inferential *sample* for which the chi-square test is intended. If you wish to check independence for such an entire population, just calculate observed and expected frequencies (as we have been doing) and compare them directly without bothering about chi-square tests. If O- and E-values match up, you have independence. If O- and E-values don't match up, sex and circumstance of accident victims are dependent. As you might guess, they turn out to be dependent. (See Exercise 8.)

EXERCISES Section **10.4**

In Exercises 1 through 6:
a) Give the appropriate null hypothesis.
b) Give the alternative hypothesis.
c) Use the procedure given on page 383 to perform the hypothesis test.

1 Perform a chi-square independence test on income and educational level from Exercise 1, Section 10.3, using $\alpha = 0.01$.

2 Perform a chi-square independence test on grade earned and hours worked per week from Exercise 2, Section 10.3, using $\alpha = 0.10$.

3 Perform a chi-square independence test on the opinion questionnaire from Exercise 3, Section 10.3, using $\alpha = 0.05$.

4 Perform a chi-square independence test on age and smoking from Exercise 4, Section 10.3, using $\alpha = 0.01$.

5 The following table compiled from a random sample of 490 Caucasians gives the residence of wage earners by income level.

Residence	0–$6,999	$7,000–$14,999	$15,000–$24,999	$25,000 or over
Metropolitan areas in central cities	23	42	38	18
outside central cities	26	64	75	42
Outside metropolitan areas	41	69	45	7

Formulate appropriate null and alternative hypotheses and perform a chi-square independence test, using $\alpha = 0.01$.

6 The following random sample of 208 people of all races indicates educational level in 1975 in the U.S. according to sex.

Educational level	Sex M	F	Totals
Elementary school:			
0–4	10	8	18
5–7	8	9	17
8	4	5	9
High school:			
1–3	18	24	42
4	31	42	73
College:			
1–3	13	12	25
4 and over	14	10	24
Totals	98	110	208

Formulate the appropriate null and alternative hypotheses and perform a chi-square independence test, using $\alpha = 0.01$.

7 A random sample of 100 members of a union were asked to respond to two questions:

Question 1: Are you happy with your financial situation today?

Question 2: Do you approve of the federal government's economic policies?

The responses were:

Question 1

Question 2	Yes	No	Total
Yes	22	48	70
No	12	18	30
Total	34	66	100

a) Calculate χ^2 for this table and test the null hypothesis that response to Question 1 is independent of response to Question 2 for members of this union. Use $\alpha = 0.05$.

b) Suppose that 200 members of the union (twice as many) were sampled and their responses were in exactly the same proportions.

Question 1

Question 2	Yes	No	Total
Yes	44	96	140
No	24	36	60
Total	68	132	200

Calculate χ^2 from this data, and compare your answer to the answer from part (a).

c) Suppose that 1000 members (ten times as many as in (a)) were sampled and their responses were in exactly the same proportions.

Question 1

Question 2	Yes	No	Total
Yes	220	480	700
No	120	180	300
Total	340	660	1000

Calculate χ^2 from this data, and compare your answer to your answer from (a). How would this change affect your hypothesis test?

d) What happens to χ^2 as sample proportions stay exactly the same, but sample size increases? Does this indicate that you should be cautious about the results of chi-square tests with large sample sizes?

8 Using the population data in Table 10.16, find the probabilities of the events:

a) M: The injured person is male.

b) F: The injured person is female.

c) W: The person is injured at work.

d) H: The person is injured at home.

e) A: The person is injured elsewhere.

f) Determine whether M and W are independent by checking whether $P(M \ \& \ W) = P(M) \cdot P(W)$.

g) Comparing O- and E-values for the cell in the upper lefthand corner of Table 10.16 is equivalent to doing the check in (f) above. Explain why.

h) Decide whether sex of person injured and circumstance of injury are independent.

9 In Exercises 5 and 6 of Section 10.3 we introduced a measure of association for 2×2 tables.

$$\phi = \frac{ad - bc}{\sqrt{(a+b)(c+d)(a+c)(b+d)}}$$

a) Calculate ϕ for each of the three tables in Exercise 7.

b) How strong is the association in each of those tables?

c) Is it possible that a chi-square test will lead to a conclusion of dependence between two variables when they are only weakly associated? If so, indicate how this can happen.

10 a) Suppose that you are given two 2×2 tables whose cells are in the same proportions:

Table 1	
a	b
c	d

Table 2	
Na	Nb
Nc	Nd

Show that if χ_1^2 is the χ^2-statistic for the first table and χ_2^2 is the χ^2-statistic for the second table, then

$$\chi_2^2 = N\chi_1^2$$

b) Prove that a similar result holds for any $r \times c$ table.

11 The table in Exercise 3 is of the form:

			Total
	a	b	
	c	d	
Total			

a) Fill in the totals of the rows and columns in the table.
b) Find a formula for the expected frequency in each cell of the table.
c) Prove that for such a 2×2 table,

$$\chi^2 = \frac{(a+b+c+d)(ad-bc)^2}{(a+b)(c+d)(b+d)(a+c)}$$

d) Use this formula to find the χ^2 value for the data in Exercise 3, and compare your answer here with the value you obtained in Exercise 3.

Chi-square tests of homogeneity   **10.5**

In the independence tests of the last section, we tried to determine whether two characteristics of individuals in the *same* population were independent. For example, in our teacher-evaluation questionnaire study, we considered two different responses (predicted course grade and course evaluation grade) from a single population of students. The chi-square tests of homogeneity, to be discussed in this section, look at characteristics of individuals from *different* populations, but lead to very similar data tables and use the same computational methods as independence tests. We illustrate this below in Example 10.

Example 10

A sociologist in Phoenix, Arizona, has a friend in Portland, Oregon, who is an urban planner. Census data (which is a few years old) indicates that the populations of Phoenix and Portland have somewhat similar racial makeups. The two friends decided to check this out on their own. They established three categories "White," "Black," and "Other minority." (The last category lumps smaller groups together to avoid having small cells, not to slight any particular group.) Each friend then took a random sample of 100 individuals in her city. Their results are given in Table 10.17.

Table 10.17

	White	Black	Other	Total
Phoenix	83	5	12	100
Portland	87	6	7	100
Total	170	11	19	200

As you can see, the data table looks like a data table for a test of independence, but there is an important difference. *The numbers in the two rows come from people in two entirely different cities and not from a single population.* The null and alternative hypotheses that the sociologist and the planner have in mind are also different from those in independence tests. They are:

Null hypothesis: The populations of Phoenix and Portland have the same percentage for each racial category.
Alternative hypothesis: The populations of Phoenix and Portland do not have the same percentages.

When two populations *do* have identical percentages for each category in a grouping, they are called **homogeneous** with respect to that grouping. We can rewrite the hypotheses above much more simply using this terminology.

Null hypothesis: The populations of Phoenix and Portland are homogeneous with respect to the racial categories "White," "Black," and "Other."
Alternative hypothesis: The populations are not homogeneous with respect to the given categories.

To perform this hypothesis test using a chi-square test, we must find a method for computing expected values for homogeneous populations. We will give this in Example 11, following Exercise L. ◗

Exercise L   We will give two potential hypothesis-testing situations below. Tell whether each one calls for a test of independence or a test of homogeneity. For the homogeneity test, write down the null and alternative hypotheses using the word "homogeneous."
a)   A sociologist wishes to know whether a female American's score on a test of liberalism is affected by her religion.
b)   A pollster wishes to know whether the breakdown of votes into Democrats, Republicans, and "others" is the same in New York and New Jersey.

Example 11   ◖ We will now compute an expected value for the upper left cell in Table 10.17, the cell for whites in Phoenix. The idea behind this computation is this: If the two populations really are homogeneous, then the information obtained by combining the Phoenix and Portland samples should be better than the information taken from either sample separately. So we would *estimate:*

$$\begin{array}{l} \text{Total percentage of} \\ \text{whites for two cities} \end{array} \doteq \dfrac{\begin{array}{c}\text{Total number of whites}\\\text{from two cities}\end{array}}{\begin{array}{c}\text{Total number}\\\text{sampled}\end{array}} = \dfrac{170}{200} = 0.85$$

This means that, if the null hypothesis that the two cities are racially homogeneous is true, then about 85 percent of the people in either city are white. In a sample of 100 Phoenix residents, we would expect 85 percent to be white.

$$\begin{array}{l}\text{Expected number}\\\text{of whites in}\\\text{100 Phoenicians}\end{array} = 100(0.85) = 100 \cdot \dfrac{170}{200} = 85$$

We wrote this expected number as $(100)(170/200)$ to point out an important relationship:

$$\text{Expected number} = 100 \cdot \frac{170}{200} = \frac{(100)(170)}{200} = \frac{(\text{Row total})(\text{Column total})}{\text{Total number}}$$

This is the computational rule that is used for the expected number in each cell (it is the *same* rule that we used for independence tests). Using this formula, the expected number for the lower right cell would be:

$$\frac{\text{Expected number}}{(\text{Second row, third column})} = \frac{(\text{Row total})(\text{Column total})}{\text{Total number}}$$

$$= \frac{(100)(19)}{200} = 9.5 \quad \blacktriangleright$$

Compute the expected values for all the remaining cells in Table 10.17.   Exercise **M**

Once we have computed expected values for each cell, we can perform a hypothesis test for homogeneity using the statistic

$$\chi^2 = \sum \frac{(O - E)^2}{E}$$

exactly as we did for the tests for independence in Section 10.4. (See page 383 for the procedure.)

◀ Perform the hypothesis test described in Example 10.   Example **12**

Null hypothesis: The populations of Phoenix and Portland are homogeneous with respect to the racial categories "White," "Black," and "Other."
Alternative hypothesis: The populations are not homogeneous with respect to the given categories.

1. *Decide on the significance level,* α. We will use $\alpha = 0.05$.
2. *Collect data, construct an observed-data table, and calculate degrees of freedom.*

	White	Black	Other	Total
Phoenix	83	5	12	100
Portland	87	6	7	100
Total	170	11	19	200

$$r = 2, \quad c = 3$$
$$\text{d.f.} = (r - 1)(c - 1) = (2 - 1)(3 - 1) = 1 \cdot 2 = 2$$

3. *The critical value is $c = \chi^2_{0.05} = 5.99$ for d.f. = 2.*

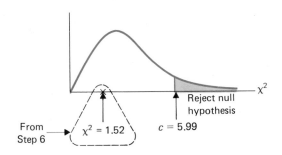

4. *Combined table:*

	White		Black		Other		Total
Phoenix	83	85	5	5.5	12	9.5	100
Portland	87	85	6	5.5	7	9.5	100
Total	170		11		19		200

5. *Check the assumptions.*
   (1) No expected frequency is less than 1? Yes.
   (2) At most 20 percent of the expected frequencies are less than 5? Yes, no expected frequency is less than 5.

6. *Calculate χ^2:*

$$\chi^2 = \sum \frac{(O-E)^2}{E}$$

$$= \frac{(83-85)^2}{85} + \frac{(5-5.5)^2}{5.5} + \frac{(12-9.5)^2}{9.5}$$

$$+ \frac{(87-85)^2}{85} + \frac{(6-5.5)^2}{5.5} + \frac{(7-9.5)^2}{9.5}$$

$$= 0.05 + 0.05 + 0.66 + 0.05 + 0.05 + 0.66$$

$$= 1.52$$

7. Since $\chi^2 = 1.52$ and $c = 5.99$, we cannot reject the null hypothesis. That is, on the basis of this data, we cannot reject the hypothesis that Phoenix and Portland are homogeneous with respect to the racial categories "White," "Black," and "Other." ▶

A political pollster is collecting data to measure voter opinion on an upcoming referendum. He samples 50 voters from each of two key districts. His data are

Exercise **N**

	Will vote yes	Undecided	Will vote no	Total
District 11	20	9	21	50
District 15	26	3	21	50
Total	46	12	42	100

Test the null hypothesis that the voter-opinion distributions in the two districts are homogeneous. Use $\alpha = 0.05$.

EXERCISES Section **10.5**

In Exercises 1 through 5:
a) State the appropriate null hypothesis.
b) State the alternative hypothesis.
c) Use the procedure given on page 383 to perform the hypothesis test.

1 Just prior to the 1976 election, a pollster took random samples of the presidential preferences of 517 registered voters in the New England States and 1098 registered voters in the South Atlantic States. The results are reported in the table below:

	New England	South Atlantic
Democratic	274	612
Republican	243	486

Test whether the two geographical areas were homogeneous in their presidential voting preference, using $\alpha = 0.05$.

2 Below is the result of a survey of 515 people in four different regions in the U.S. in 1975, concerning annual income and geographical area.

Annual income

Region	$6,999 and under	$7,000–$14,999	$15,000–$24,999	$25,000 and over
Northeast	22	44	40	20
North Central	26	41	50	22
South	19	67	49	21
West	17	32	30	15

Using $\alpha = 0.05$, test the hypothesis that the distribution of income level is the same in the four regions.

3 Random samples of 179 divorced males and 274 divorced females were interviewed. Below is a table that groups them by age and sex.

Age group

Sex	25–34	35–44	45–54
Male	67	56	56
Female	103	87	84

Using $\alpha = 0.05$, test the hypothesis that the age distribution of divorced people is the same for males and females.

4 A survey of 463 physicians licensed in four different years revealed the following information about newly licensed physicians and their training location.

Year of physician's licensing in U.S.

Origin of training	1960	1965	1970	1973
U.S. and Canada	74	76	80	92
Foreign medical schools	14	23	30	74

Using $\alpha = 0.05$, test the hypothesis that the populations of physicians first licensed in the years 1960, 1965, 1970 and 1973 are homogenous with respect to the origin of training.

5 On February 11, 1979, the television show *Roots: The Next Generation* was shown on seven consecutive nights. On the first night, *Roots* was watched by 41 percent of the viewing audience, while the remainder of the audience was split between two competing films, *American Graffiti* and *Marathon Man*. Samples of size 100 in two cities showed the following breakdown:

	Roots	Graffiti	Marathon
City 1	43	28	29
City 2	39	33	28

Using $\alpha = 0.1$, perform a chi-square homogeneity test to decide whether the viewing audiences in the two cities were distributed among the three shows in the same way.

10.6 Computer packages*

In this chapter we have shown you how to perform a chi-square test of independence given a data table such as Table 10.8 on course evaluation. Computer packages will do this for you. They will also do something even more valuable: They will *compile the table* for you from raw data such as questionnaire response sheets. We will illustrate this by discussing the interpretation of a questionnaire that we used in 1977. The questionnaire is reproduced on page 395. We gave it to all algebra students at Arizona State University in the fall of 1977.

As you can see from the instructions on the questionnaire, the students actually filled in their responses on a machine-readable multiple-choice answer sheet. A reading machine "read" these sheets and placed each student's responses on a separate computer card. The data on these cards was then processed using the SPSS program CROSSTABS. This program will compile the data table for any two questions and then perform a chi-square test of independence (and a number of other tests).

To compile the data table from questionnaires by hand without SPSS is a very tedious job. For example, 188 students in College Algebra responded to Questions 2 and 5 on this questionnaire. To obtain a data table by hand, we would have had to construct a blank work table of the form:

Question 5

		A	B	C
	A			
	B			
Question 2	C			
	D			
	E			

---

* This section is optional.

PSI Time Study

DO NOT WRITE ON THIS SHEET. IT IS TO BE REUSED.

All information requested on this form is confidential, and will be used only for statistical studies to help us improve our classes.

Instructions: Please make a check mark in the space on the IBM Sheet corresponding to the most appropriate response.

1. Which class are you in?

    (A) MAT 106   (B) MAT 115   (C) MAT 117

2. What grade do you expect to receive in this class?

    (A) A   (B) B   (C) C   (D) D   (E) E

3. How many hours per week did you spend on this course? (Include time spent in class.)

    (A) 0–3   (B) 4–6   (C) 7–9   (D) 10–12   (E) More than 12

4. How much time per week did you spend with our tutors?
    (A) None   (B) 30 minutes   (C) 1 hour   (D) 2 hours   (E) 3 or more hours

5. Which of the following best describes your learning in this class?

    (A) I went through the book carefully in most chapters. (CAREFUL)

    (B) I used the book in some chapters, but didn't need to read the book carefully in others. (MIXED)

    (C) I could usually get ready for a test by looking over the practice test, and rarely used the book. (SUPERFICIAL)

6. Which of the following best describes your background?

    (A) I had this class before and I knew most of it.

    (B) I had this class before, but forgot a lot of it.

    (C) I have never before seen most of the work in this class.

7. If you could give this class a grade, what would it be?
    (A) A   (B) B   (C) C   (D) D   (E) E

We would then have had to go through 188 questionnaires by hand, and put a tally mark in the table for each questionnaire. Instead, we had the computer read our cards, and knew within a few seconds that the joint responses were:

Question 5

		A	B	C
	A	56	23	6
Question 2	B	58	22	4
	C	13	4	2

At the same time, we knew that a chi-square test of independence indicated that the responses to the questions were independent. The actual printout from this computer run is reproduced on page 397.

As you can see, the printout contains a great deal of information—probably too much, in this case. The computer prints out an observed number for each cell, a total for each row and column, and a probability for each row and column. The number under the observed number in each cell is a percentage for the particular row in which the cell lies. For example, the 65.9 in the upper left cell tells you that, of the students expecting an "A" in the class, 65.9 percent felt that they worked carefully.

The first line below the table gives the results of the chi-square test. The number 0.8783 listed after SIGNIFICANCE = is actually a P-value. It tells us that the area under the chi-square curve (see Fig. 10.7) to the right of the calculated χ^2-value is 0.8783. Thus we could not reject a null hypothesis of independence at any reasonable α-level.

The statistics below the CHI-SQUARE line were computed only to show you that there are many other statistical methods available from SPSS for analyzing such tables. A word of caution is in order here: Computer output often tempts researchers to report

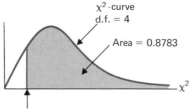

Figure 10.7 $\chi^2 = 1.19855$

measures that they do not understand but that seem to say the right words (such as CRSGRD DEPENDENT). The computer can actually be programmed so that the only statistics to be reported are the ones you know you can interpret. We recommend this procedure.

Finally, at the bottom of the printout is the NUMBER OF MISSING OBSERVATIONS. The number here is 3; that is, three students did not answer both Questions 2 and 5 and could not be tallied. Missing data is a constant problem in questionnaire work. Many people refuse to answer certain questions or simply forget them.

```
PSI TIME AND EVALUATION STUDY. DEC 77                          22 MAR 78      PAGE 16

FILE   PSIRSIM  (CREATION DATE = 22 MAR 78)   PSI TIME AND EVALUATION DATA. DEC 77
* * * * * * * * * * * * * * * * * * * C R O S S T A B U L A T I O N   O F  * * * * * * * * * * * * * * * * *
   CRSGRD   EXPECTED GRADE IN COURSE                  BY  STDMETH   STUDY METHOD
CONTROLLING FOR..
   COURSE      COURSE NUMBER                      VALUE =        3. MAT 117
* * * * * * * * * * * * * * * * * * * * * * * * * * * * * * * * * * * * * * * * * * PAGE   1 OF  1

                    STDMETH
            COUNT I
            ROW PCT I CAREFUL  MIXED     SUPERFIC  ROW
            I                            IAL       TOTAL
            I         1.I       2.I       3.I
CRSGRD     --------I---------I---------I--------I
           1. I    56  I   23  I    6  I    85         Observed number (O)
           A  I  65.9  I 27.1  I  7.1  I  45.2
              I---------I---------I--------I          Row total
           2. I    58  I   22  I    4  I    84
           B  I  69.0  I 26.2  I  4.8  I  44.7
              I---------I---------I--------I          Row percentage
           3. I    13  I    4  I    2  I    19        (of total)
           C  I  68.4  I 21.1  I 10.5  I  10.1
              I---------I---------I--------I
         COLUMN    127       49      12      188
          TOTAL   67.6      26.1     6.4    100.0

CHI SQUARE =    1.19855 WITH   4 DEGREES OF FREEDOM   SIGNIFICANCE =   .8783
CRAMER'S V =    .15646
CONTINGENCY COEFFICIENT =     .07959
LAMBDA (ASYMMETRIC) =   .01942 WITH CRSGRD   DEPENDENT.      =   .00000 WITH STDMETH  DEPENDENT.
LAMBDA (SYMMETRIC) =    .01220
UNCERTAINTY COEFFICIENT (ASYMMETRIC) =   .00324 WITH CRSGRD   DEPENDENT.      =   .00389 WITH STDMETH  DEPENDENT.
UNCERTAINTY COEFFICIENT (SYMMETRIC) =    .00354
KENDALL'S  TAU B =   -.02659  SIGNIFICANCE = .3496
KENDALL'S  TAU C =   -.02097  SIGNIFICANCE = .3496
GAMMA =      -.05050
SOMERS'S D (ASYMMETRIC) = -.02963 WITH CRSGRD   DEPENDENT.      = -.02386 WITH STDMETH  DEPENDENT.
SOMERS'S D (SYMMETRIC) = -.02644
ETA =   .03328 WITH CRSGRD   DEPENDENT.       =   .04654 WITH STDMETH  DEPENDENT.

NUMBER OF MISSING OBSERVATIONS =      3
```

CHAPTER REVIEW

Key Terms

goodness-of-fit test
chi-square distribution
observed frequencies (O)
expected frequencies (E)
test of independence
contingency table

independent
dependent
cells
homogeneous
homogeneity test

Formulas and Key Facts

distribution of χ^2 for goodness-of-fit test:

$$\chi^2 = \sum \frac{(O-E)^2}{E}$$

has (approximately) a chi-square distribution with d.f. $= k - 1$ (where $k =$ number of groups, $O =$ observed values, $E =$ expected values).

calculation of E-values for a goodness-of-fit test:

$$E = np$$

(where $n =$ sample size, $p =$ probability for group given in null hypothesis).

distribution of χ^2 for independence or homogeneity test:

$$\chi^2 = \sum \frac{(O-E)^2}{E}$$

has (approximately) a chi-square distribution with d.f. $= (r-1)(c-1)$ (where $r =$ number of rows, $c =$ number of columns).

calculation of E-values for independence or homogeneity test:

$$E = \frac{(\text{Row total})(\text{Column total})}{(\text{Total number in sample})}$$

where row totals, column totals, and total number in sample are obtained from the observed data table.

minimum expected frequency assumptions for goodness-of-fit, independence, and homogeneity tests:

(1) No expected frequency (E-value) is less than 1.
(2) At most 20 percent of the E-values are less than 5.

Goodness-of-fit test: (see page 362 for detailed procedure).

Independence test: (see page 383 for detailed procedure).

Homogeneity test: (see page 383 for detailed procedure).

You should be able to

1 Calculate *expected frequencies* (*E*-values) for a *goodness-of-fit* test.

2 Perform a *chi-square goodness-of-fit* test.

3 Calculate *expected frequencies* (*E*-values) for an *independence* or *homogeneity* test.

4 Perform a *chi-square independence test.*

5 Perform a *chi-square homogeneity test.*

REVIEW TEST

1 A percentage distribution for family income in the United States in 1974 is given below:

Under $5,000	$5,000 -9,999	$10,000 -14,999	$15,000 -24,999	$25,000 and over
13%	21%	22%	30%	14%

An economist obtained income figures for a random sample of 100 people in a large metropolitan area. They were:

Under $5,000	$5,000 -9,999	$10,000 -14,999	$15,000 -24,999	$25,000 and over
10	19	20	34	17

a) The economist wishes to test whether this metropolitan area has the same distribution as that given above for the United States. State appropriate null and alternative hypotheses.

b) For a test of these hypotheses at the level $\alpha = 0.05$,
 i) What is d.f.?
 ii) What is the critical value, c?
c) Complete the table below, and calculate χ^2.

Class	O	E	$O - E$	$\dfrac{(O - E)^2}{E}$
Under $5,000				
$5,000–9,999				
$10,000–14,999				
$15,000–24,999				
$25,000 and over				

d) Are the expected cell frequency assumptions (1) and (2), given on page 398, satisfied in this case?
e) Should the null hypothesis be rejected?

2 A random sample of 1200 voters in a city showed the following breakdown of political-party affiliation by religion:

	Catholic	Protestant	Jewish	Total
Democrat	130	360	36	526
Republican	110	540	24	674
Total	240	900	60	1200

a) State the null and alternative hypotheses for a test of independence of party affiliation and religion.

b) For a test of these hypotheses at the level $\alpha = 0.01$,
  i) What is d.f.?
  ii) What is the critical value, c?

c) Calculate expected values for each cell of the table above using the work table in the righthand column.

	Catholic	Protestant	Jewish
Democrat			
Republican			

d) Are the assumptions (1) and (2) on expected cell frequencies satisfied?

e) Complete the table below by filling in E-values in each cell.

	Catholic	Protestant	Jewish	Total
Democrat	130	360	36	526
Republican	110	540	24	674
Total	240	900	60	1200

f) Compute χ^2.

g) Should the null hypothesis be rejected?

FURTHER READINGS

ANDERSON, T. W. and S. L. SCLOVE, *An Introduction to the Statistical Analysis of Data.* Boston: Houghton Mifflin Company, 1978. Chapter 12 of this book covers the chi-square tests and also contains a careful discussion of measures of association such as ϕ, which was introduced in the exercises in this chapter.

DANIEL, W. W., *Applied Nonparametric Statistics.* Boston: Houghton Mifflin Company, 1978. Chapter 5 and Section 1 of Chapter 8 are devoted to chi-square tests. This book gives many references to recent work, but does not require a high level of sophistication in mathematics.

ANSWERS TO REINFORCEMENT EXERCISES

A

Age group	O	E	$O-E$	$(O-E)^2$	$\dfrac{(O-E)^2}{E}$
0–20	150	196	−46	2116	10.80
21–64	260	292	−32	1024	3.51
65 and over	90	12	78	6084	507.00
					521.31

$$\chi^2 = \sum \frac{(O-E)^2}{E} = 521.31.$$

B We have $k = 3$ (the three age groups). To check assumptions (1) and (2), we construct the expected frequency table.

Age group	Probability, p	Expected frequency $np = E$
0–20	0.392	$500(0.392) = 196$
21–64	0.584	$500(0.584) = 292$
65 and over	0.024	$500(0.024) = 12$

(1) No E-values less than 1? Yes.
(2) At most 20 percent of the E-values are less than 5? Yes, no E-value is less than 5.

1. *Significance level*: $\alpha = 0.05$.
2. *Critical value*: $c = \chi_\alpha^2 = \chi_{0.05}^2$ for d.f. $= k - 1 = 3 - 1$ $= 2$. By Table III, $\chi_{0.05}^2 = 5.99$ for d.f. $= 2$. So $c = 5.99$.

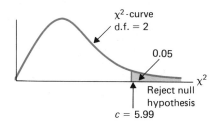

χ^2-curve
d.f. $= 2$
0.05
Reject null hypothesis
$c = 5.99$

3. *Work table for O- and E-values and computation of χ^2:*

Age group	O	E	$O - E$	$\dfrac{(O - E)^2}{E}$
0–20	150	196	−46	10.80
21–64	260	292	−32	3.51
65 and over	90	12	78	507.00
				$\chi^2 = 521.31$

4. Since $\chi^2 = 521.31$ and $c = 5.99$, we have $\chi^2 > c$. *Reject* the null hypothesis. In other words, it appears that the age distribution of the four tracts is not the one specified in the null hypothesis.

C a) We have $k = 5$ (the five grades). The expected frequency table is given on page 365. It shows that no E-value is less than 5. So assumptions (1) and (2) hold.
1. *Significance level*: $\alpha = 0.05$.
2. *Critical value*: $c = \chi_\alpha^2 = \chi_{0.05}^2 = 9.49$ for d.f. $= k - 1 = 5 - 1 = 4$.
3. *Work table for O- and E-values and computation of χ^2:*

Grade	O	E	$O - E$	$\dfrac{(O - E)^2}{E}$
A	5	5	0	0.00
B	15	10	5	2.50
C	15	20	−5	1.25
D	5	10	−5	2.50
E	10	5	5	5.00
				$\chi^2 = 11.25$

4. $\chi^2 = 11.25$, $c = 9.49$. *Reject* null hypothesis.

D *Population*: All possible evaluations (specifically, responses to Questions 2 and 3) from similar Intermediate Algebra classes.

E a) 16

b)

	A	B	C	D or E	Total
A	96	42	8	1	147
B	81	141	53	4	279
C	30	57	52	3	142
D or E	8	18	28	5	59
Total	215	258	141	13	627

c) 627
d) 53

F

	A	B	C	D or E	Total	
A	96	42	8	1	147	0.234
B	81	141	53	4	279	0.445
C	30	57	52	3	142	0.226
D or E	8	18	28	5	59	0.094
Total	215	258	141	13	627	
	0.343	0.411	0.225	0.021		

G

	A	B	C	D or E	Total
A	$\dfrac{147\,(215)}{627} = 50.4$	$\dfrac{147\,(258)}{627} = 60.5$	$\dfrac{147\,(141)}{627} = 33.1$	$\dfrac{147\,(13)}{627} = 3.0$	147
B	$\dfrac{279\,(215)}{627} = 95.7$	$\dfrac{279\,(258)}{627} = 114.8$	$\dfrac{279\,(141)}{627} = 62.7$	$\dfrac{279\,(13)}{627} = 5.8$	279
C	$\dfrac{142\,(215)}{627} = 48.7$	$\dfrac{142\,(258)}{627} = 58.4$	$\dfrac{142\,(141)}{627} = 31.9$	$\dfrac{142\,(13)}{627} = 2.9$	142
D or E	$\dfrac{59\,(215)}{627} = 20.2$	$\dfrac{59\,(258)}{627} = 24.3$	$\dfrac{59\,(141)}{627} = 13.3$	$\dfrac{59\,(13)}{627} = 1.2$	59
Total	215	258	141	13	627

H (Refer to answers to Exercises *E* and *G*.)

	A		B		C		D or E		Total
A	96		42		8		1		147
		50.4		60.5		33.1		3.0	
B	81		141		53		4		279
		95.7		114.8		62.7		5.8	
C	30		57		52		3		142
		48.7		58.4		31.9		2.9	
D or E	8		18		28		5		59
		20.2		24.3		13.3		1.2	
Total	215		258		141		13		627

I $r = 4$, $c = 4$.
d.f. $= (r - 1)(c - 1) = (4 - 1)(4 - 1) = 9$.

J (Refer to answer to Exercise *G*.)
a) No.
b) There are 16 cells here, and three of them have expected frequencies less than 5. $\frac{3}{16} = 0.1875 = 18.75\% \leqslant 20\%$.
c) Yes.

K 1. *Significance level:* $\alpha = 0.01$.
2. *Collect data, construct an observed data table, and calculate degrees of freedom:* The observed data table is given in the answer to Exercise E (page 401).

$$\text{d.f.} = (r - 1)(c - 1) = (4 - 1)(4 - 1) = 9.$$

3. *Critical value:* $c = \chi_\alpha^2 = \chi_{0.01}^2 = 21.67$ for d.f. $= 9$.

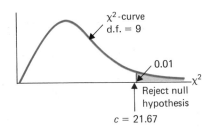

χ^2-curve
d.f. = 9

0.01

χ^2

Reject null hypothesis

$c = 21.67$

4. *Combined table containing both observed and expected values:* See answer to Exercise H (page 402).
5. In the answer to Exercise J (page 402) we saw that assumptions (1) and (2) are satisfied.

6. *Calculation of* χ^2: Referring to the table given in the answer to Exercise H (page 402), we get

$$\chi^2 = \sum \frac{(O - E)^2}{E}$$

$$= \frac{(96 - 50.4)^2}{50.4} + \frac{(42 - 60.5)^2}{60.5} + \frac{(8 - 33.1)^2}{33.1}$$

$$+ \ldots + \frac{(5 - 1.2)^2}{1.2}$$

$$= 41.26 + 5.66 + 19.03 + \ldots + 12.03.$$

7. We didn't bother adding up the last sum above to find χ^2, because $c = 21.67$ and the first value in the sum for χ^2 is 41.26, which already exceeds c. Thus, $\chi^2 > c$, and we *reject* the null hypothesis. It appears that responses to Questions 2 and 3 by Intermediate Algebra students are *dependent*.

L a) This calls for a chi-square *independence* test.
b) This calls for a chi-square test of *homogeneity*:

Null hypothesis: The voters in New York and New Jersey are homogeneous with respect to the categories "*Democrat*," "*Republican*," and "*Others*."
Alternative hypothesis: The voters in New York and New Jersey are *not* homogeneous with respect to the given categories.

M
Expected values (*E*-values)

	White	Black	Other	Total
Phoenix	85	5.5	9.5	100
Portland	85	5.5	9.5	100
Total	170	11.0	19.0	200

N 1. *Significance level:* $\alpha = 0.05$.
2. *Collect data, construct observed data table, and calculate degrees of freedom:* The observed data table is given in Exercise N (page 393).

$$\text{d.f.} = (r - 1)(c - 1) = (2 - 1)(3 - 1) = 2.$$

3. *Critical value:* $c = \chi_{0.05}^2 = 5.99$ for d.f. $= 2$.

4. *Combined table*:

	Will vote *Yes*	Undecided	Will vote *No*	Total
District 11	20 23	9 6	21 21	50
District 15	26 23	3 6	21 21	50
Total	46	12	42	100

5. Assumptions (1) and (2) hold since no expected frequency is less than 5.

6. *Calculate χ^2*:

$$\chi^2 = \sum \frac{(O-E)^2}{E}$$

$$= \frac{(20-23)^2}{23} + \frac{(9-6)^2}{6} + \frac{(21-21)^2}{21}$$

$$+ \frac{(26-23)^2}{23} + \frac{(3-6)^2}{6} + \frac{(21-21)^2}{21}$$

$$= 0.39 + 1.5 + 0 + 0.39 + 1.5 + 0 = 3.78.$$

7. Since $\chi^2 \leqslant c$, we *cannot* reject the null hypothesis.

Analysis of variance (ANOVA)

11.1 Introduction

The statistical methods referred to as analysis of variance (ANOVA) are among the most widely used methods in statistics. Surprisingly, these methods are usually not used to test hypotheses about variances; they are used to test hypotheses about means. In previous chapters we have looked at a number of examples in which *two* means are tested for equality. For example, in Chapter 8, we used tests of equality of two means to compare the effectiveness of two toothpastes and the lifetimes of two kinds of lightbulbs. Analysis of variance can be used to test hypotheses concerning more than two means. We would probably use it if we were interested in three toothpastes or five kinds of lightbulbs. To illustrate the kind of situation that calls for an analysis of variance, we will look at a simple, everyday situation: four people who weigh themselves every day for six days.

Example **1**   ◖ Since the late 1960's, Americans have become quite concerned about their health, their appearance, and consequently, their weight. The number of entries in the 1976 Boston Marathon (see Chapter 2) is just one indication of the way in which we have decided to huff and puff ourselves into shape. Many people weigh themselves every day. These people know very well that most of us do not have a single weight. Our weights may vary by a pound or two from day to day. They are affected by what we do, what we eat, what time of day we are weighed, and the accuracy of the scales we use. One of the authors weighed himself at the same time each day for six days, and "weighed" 150, 148, 149, 148, 149, 149 pounds. He has a mean weight of 148.83 for those six days, but *not* a unique weight.

In this example, we will look at four members of a health club who jog together. They appear to be the same size, and they say that they are "the same weight." Each day after they jog, they weigh themselves. Table 11.1 shows their weights for six consecutive days, and the mean for each group of weights.

Table 11.1 Daily weight results for: four club members	Steve	Harold	Gerald	Fritz
	145	148	147	147
	147	147	149	146
	145	146	146	147
	147	147	147	147
	148	147	148	147
	147	148	146	146
$\bar{x}$	146.5	147.2	147.2	146.7

The (sample) mean weights of the four joggers are reasonably close together, and we would probably be inclined to believe their claim that they are "the same weight." Before we look at the analysis-of-variance method for testing their claim, we should look at data on individuals who are probably *not* the same weight. Table 11.2 shows the weights of four more members of the same club for the same six days:

	Joe	Don	Ed	Bill	**Table 11.2**
	148	142	150	155	Daily weight results for:
	146	142	149	157	other club members
	147	143	151	156	
	148	145	152	156	
	146	144	150	156	
	146	144	149	154	
$\bar{x}$	146.8	143.3	150.2	155.7	

By simply looking at (sample) means again, we would probably agree that these four individuals have different weights.

The way we have talked about these data sets shows that we are really interested in means. We have made an assumption that each individual in Tables 11.1 and 11.2 has a true mean weight, μ, and that his daily weight fluctuates around that mean. When we say that Steve and Harold "have the same weight," we mean that their true mean weights are the same. To write this down compactly, mathematicians usually write μ_i for the true mean weight of the ith jogger, as below:

Jogger	True mean
Steve	μ_1
Harold	μ_2
Gerald	μ_3
Fritz	μ_4

If we do this, we can write the hypotheses we wish to test as:

Null hypothesis: $\mu_1 = \mu_2 = \mu_3 = \mu_4$
Alternative hypothesis: Not all of the μ_i are equal.

Note that the null hypothesis requires *all* of the means to be equal. If Steve and Harold and Fritz all had the same true mean weight but Gerald's true mean (μ_3) was different, the null hypothesis would not be true. We picture the situation described by the *null hypothesis* in Fig. 11.1 at the top of page 408. ◗

Exercise **A**

a) Following Example 1, represent the claim that the four joggers in Table 11.2 "have the same weight" as a null hypothesis with an appropriate alternative hypothesis.

b) Suppose that Joe and Don had identical true mean weights, but the true means for Ed and Bill were different. Which is true, the null hypothesis or the alternative hypothesis?

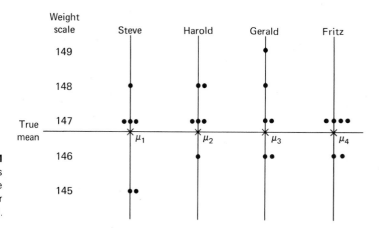

Figure 11.1
Six days of weight readings varying around the true mean weight μ for four individuals.

Example **1** (cont.)

❧ To get an idea of the manner in which analysis-of-variance methods test equality of means, we should look back at Tables 11.1 and 11.2. There are actually two kinds of variation in the data given in these tables. There is *variation within* each individual's daily weighings; for example, Steve's weight varied by three pounds, from a high of 148 to a low of 145. There is also *variation between* the means for the individuals; for example the (sample) mean weights for Steve, Harold, Gerald, and Fritz varied by 0.7 pounds, from a high of 147.2 to a low of 146.5. We picture the situation for each group of four in Figs. 11.2 and 11.3.

It should now be clear why most people would guess that Steve, Harold, Gerald, and Fritz are "the same weight" while Joe, Don, Ed, and Bill are not. The differences *between* mean weights for Joe, Don, Ed, and Bill (Fig.11.3) are much larger than the random fluctuations *within* their daily weights. The situation for Steve, Harold, Gerald

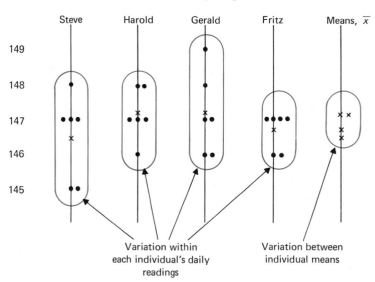

Figure 11.2
Variation within and between groups. Dots mark individual weight readings and each cross marks the sample mean $\overline{x}$ of a set of weights.

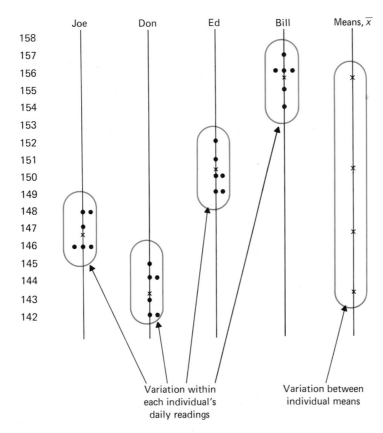

Figure 11.3
Variation within and between groups. Dots mark individual weight readings and each cross marks the sample mean $\bar{x}$ of a set of weights.

and Fritz (Fig. 11.2) is quite different. The differences *between* their mean weights are not large in comparison to the random fluctuations *within* their daily weights.

This kind of reasoning provides the basis for analysis of variance. If variation between groups (i.e., between the sample means, $\bar{x}$, of the groups) is much larger than variation within groups, we conclude that the groups do not have the same true means—as we did in Fig. 11.3. On the other hand, if variation between groups is not large relative to the variation within groups, we are forced to conclude that the groups may have the same true means, as we did in Fig. 11.2.

However, in analysis of variance we do not measure variation by simply looking at the spread of points on a graph. Instead, we calculate sums of squares and variances to *measure variation* between and within groups. The procedures that are used to evaluate variance between and within groups for the data in Tables 11.1 and 11.2 are given below in Examples 2 and 3. These procedures are lengthy and technical. It will help your understanding to keep in mind our basic goal: We are trying to define two variance measures—one to measure variance within groups and the other to measure variance between groups. We are interested in these two measures of variance because we wish to use them in testing hypotheses about equality of means. ▶

Before explaining the procedures that are used to evaluate variance between and within groups, we need to introduce some notation that is used in analysis of variance:

n = Total number of pieces of data
k = Number of groups
g = Number of pieces of data in each group

For the data in Table 11.1, $n = 24$, $k = 4$ (the four joggers), and $g = 6$ (the six weights recorded for each jogger).

Analysis of variance is one area where carrying more decimal places can make a difference. We will carry *three* decimal places in our work with analysis of variance. *

Example **2**   ◀ This example illustrates the procedure for calculating *variance within groups* using the data in Table 11.1.

Procedure for calculating variance within groups

Step 1. Compute $\bar{x}$ and $\sum(x - \bar{x})^2$ for each group of numbers. (See the four parts of Table 11.3.)

Table 11.3

Steve

x	$x - \bar{x}$	$(x - \bar{x})^2$
145	−1.500	2.250
147	0.500	0.250
145	−1.500	2.250
147	0.500	0.250
148	1.500	2.250
147	0.500	0.250
146.500 $\uparrow$ $\bar{x}$		7.500 $\sum(x \overset{\uparrow}{-} \bar{x})^2$

Harold

x	$x - \bar{x}$	$(x - \bar{x})^2$
148	0.833	0.694
147	−0.167	0.028
146	−1.167	1.362
147	−0.167	0.028
147	−0.167	0.028
148	0.833	0.694
147.167 $\uparrow$ $\bar{x}$		2.834 $\sum(x \overset{\uparrow}{-} \bar{x})^2$

Gerald

x	$x - \bar{x}$	$(x - \bar{x})^2$
147	−0.167	0.028
149	1.833	3.360
146	−1.167	1.362
147	−0.167	0.028
148	0.833	0.694
146	−1.167	1.362
147.167 $\uparrow$ $\bar{x}$		6.834 $\sum(x \overset{\uparrow}{-} \bar{x})^2$

Fritz

x	$x - \bar{x}$	$(x - \bar{x})^2$
147	0.333	0.111
146	−0.667	0.445
147	0.333	0.111
147	0.333	0.111
147	0.333	0.111
146	−0.667	0.445
146.667 $\uparrow$ $\bar{x}$		1.334 $\sum(x \overset{\uparrow}{-} \bar{x})^2$

The quantity $\sum(x - \bar{x})^2$ calculated for each group is called a **sum of squares.**

---

* The number of decimal places used in calculators varies. We have rounded to three places in order to make our displays simpler.

Step 2. Add the sums of squares from each group to obtain the **sum of squares within groups, SSW**:

$$\begin{aligned} \text{SSW} &= 7.500 + 2.834 + 6.834 + 1.334 \\ &= 18.502 \end{aligned}$$

Step 3. Calculate the **degrees of freedom for SSW** by subtracting k, the number of groups, from n, the total number of pieces of data:

$$n = \text{Total number of pieces of data} = 24$$
$$k = \text{Number of groups} = 4$$
$$\text{d.f. for SSW} = n - k = 24 - 4 = 20$$

Step 4. Calculate the **mean square within groups, MSW**, by dividing SSW by the degrees of freedom for SSW:

$$\text{MSW} = \frac{\text{SSW}}{\text{d.f. for SSW}} = \frac{\text{SSW}}{n - k}$$

We have, in this case,

$$\text{MSW} = \frac{\text{SSW}}{n - k} = \frac{18.502}{20} = 0.925. \blacktriangleright$$

What we have done here is very much like the computation of a variance, in which we find $\sum (x - \bar{x})^2$ for *one* group of n numbers, and then divide the total sum of squares by $n - 1$. Here we have found $\sum (x - \bar{x})^2$ for each of *four* groups of numbers and then divided the total sum of squares by $n - 4$. Just as the variance formula $[1/(n - 1)] \sum (x - \bar{x})^2$ measures variation for one group of numbers, the mean square within groups formula

$$\text{MSW} = \frac{\text{SSW}}{n - k}.$$

measures a kind of "average" variation within k groups of numbers.

For the data given in Table 11.2, compute (a) SSW, (b) d.f. for SSW, and (c) MSW.   Exercise **B**

◖ We now illustrate the procedure for calculating *variance between groups* using the data in Table 11.1.   Example **3**

Step 1. Compute the mean of the group means, $\bar{\bar{x}} = \dfrac{\sum \bar{x}}{k}$:   Procedure for calculating variance between groups

$$\bar{\bar{x}} = \frac{146.500 + 147.167 + 147.167 + 146.667}{4}$$

$$= \frac{587.501}{4} = 146.875$$

Step 2. Find g, the number of elements in each group. Compute $\sum(\bar{x} - \bar{\bar{x}})^2$ for the group means and multiply it by g. This quantity is the **sum of squares between groups, SSB.** (See Table 11.4.)

Table 11.4	$\bar{x}$	$\bar{x} - \bar{\bar{x}}$	$(\bar{x} - \bar{\bar{x}})^2$
	146.500	-0.375	0.141
	147.167	0.292	0.085
	147.167	0.292	0.085
	146.667	-0.208	0.043
	146.875		0.354
	$\uparrow$		$\uparrow$
	$\bar{\bar{x}}$		$\sum(\bar{x} - \bar{\bar{x}})^2$

$$\text{SSB} = g \cdot \sum(\bar{x} - \bar{\bar{x}})^2 = 6(0.354) = 2.124$$

Step 3. The **degrees of freedom for SSB** is $k-1$, where k is the number of groups. Compute the **mean square between groups, MSB,** using the formula:

$$\text{MSB} = \frac{\text{SSB}}{\text{d.f. for SSB}} = \frac{\text{SSB}}{k-1}$$

In this case,

$$\text{MSB} = \frac{\text{SSB}}{k-1} = \frac{2.124}{3} = 0.708$$

With this procedure we are simply finding the variance of the four group means $\bar{x}_1, \bar{x}_2, \bar{x}_3,$ and $\bar{x}_4,$ and multiplying it by the number of elements in a group (6). ◗

Exercise **C** For the data given in Table 11.2 compute (a) SSB, (b) d.f. for SSB, and (c) MSB.

Once we are able to measure variance within and between groups using MSW and MSB, we can easily test the hypothesis of equality of means. The idea is simply to compute

$$F^* = \frac{\text{MSB}}{\text{MSW}}$$

Here F^* is the ratio of between-group variance to within-group variance. If F^* is "too large," we will conclude that variance between groups is much larger than variance within groups, and that the true group means cannot all be equal. If F^* is not "too large," we will conclude that there is not too much variance between groups, and that we cannot reject the hypothesis that the true group means are equal. As usual, we need to have some way of deciding when F^* is too large. Since F^* is a ratio of two variance expressions, you may already have guessed how we make this decision.

If the null hypothesis of equal means is true, and if the separate groups of data are taken from normal populations with equal variances, then the random variable

$$F^* = \frac{MSB}{MSW}$$

has the F-distribution with

d.f. = (d.f. for SSB, d.f. for SSW) = $(k - 1, n - k)$

where n = total number of pieces of data, and k = number of groups. That is, probabilities for F^* can be found using areas under the F-curve with d.f. = $(k - 1, n - k)$.

Example 4 below will show how the F-distribution can be used to complete the hypothesis test.

◖ Test the hypothesis that $\mu_1 = \mu_2 = \mu_3 = \mu_4$ for the joggers in Table 11.1, using the significance level $\alpha = 0.05$.   Example **4**

We have already computed   Solution

$$MSB = 0.708, \quad d.f. = k - 1 = 4 - 1 = 3$$
$$MSW = 0.925, \quad d.f. = n - k = 24 - 4 = 20$$

To compare between- and within-group variation, we compute

$$F^* = \frac{MSB}{MSW} = \frac{0.708}{0.925} \doteq 0.77$$

Using Table VI with d.f. = (3, 20), we see that the critical value c for $\alpha = 0.05$ is 3.10. The F-curve is shown in Fig. 11.4, with c and F^*

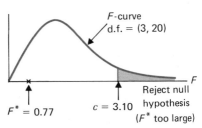

F-curve
d.f. = (3, 20)

$F^* = 0.77$ $c = 3.10$ Reject null
hypothesis
(F^* too large)   **Figure 11.4**

This means that F^* is not "too large," and we cannot reject the null hypothesis. ◗

Test the null hypothesis that $\mu_1 = \mu_2 = \mu_3 = \mu_4$ for the individuals in Table 11.2.   Exercise **D**
Use $\alpha = 0.05$.

In the next section we will look at the assumptions underlying analysis of variance more carefully, and will give some shortcut formulas. But we can sum up the simple hypothesis-testing procedure covered in this section in a few steps. In Example 5, we will do this for our weight data.

Example **5**
Hypothesis test for Table 11.1 data

◖ (Remember that μ_1, μ_2, μ_3, and μ_4 are the true mean weights for Steve, Harold, Gerald, and Fritz.)

Null hypothesis: $\mu_1 = \mu_2 = \mu_3 = \mu_4$
Alternative hypothesis: Not all the μ_i are equal.

1. *Significance level:* $\alpha = 0.05$.
2. *Degrees of freedom and critical value for F-distribution:* d.f. = (3, 20), $c = F_{0.05}$ = 3.10. See Fig. 11.5.
3. *Calculate MSB, MSW and F*:* From our previous examples,

$$MSB = 0.708$$
$$MSW = 0.925$$
$$F^* = 0.77$$

4. *Conclusion:* $F^* \leqslant c$. We cannot reject the null hypothesis. ◗

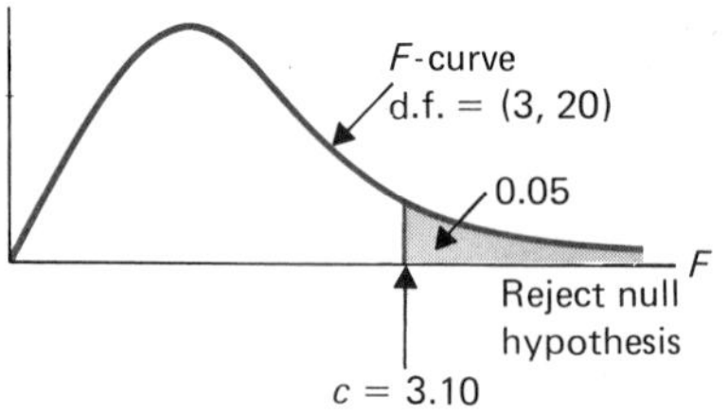

F-curve
d.f. = (3, 20)

0.05

Reject null hypothesis

$c = 3.10$

Figure 11.5

Exercise **E**

Go through the complete hypothesis-testing procedure outlined above to test the null hypothesis of equality of true mean weights for the four joggers in Table 11.2.

ANOVA tables

The entire process of finding sums of squares between and within groups, calculating mean squares, and then finding the F^*-statistic can be summarized in a single table. Such a table is commonly referred to as an *ANOVA table*. We give the results of Examples 2, 3, and 4 in Table 11.5. Remember that n = total number of pieces of data, and k = number of groups.

Table 11.5
ANOVA table for weights of Steve, Harold, Gerald, and Fritz. ($n = 24$, $k = 4$)

	Sum of squares, SS	Degrees of freedom, d.f.	Mean square, MS
Between groups	2.124 (SSB)	3 ($k-1$)	0.708 $\left(MSB = \dfrac{SSB}{k-1}\right)$
Within groups	18.502 (SSW)	20 ($n-k$)	0.925 $\left(MSW = \dfrac{SSW}{n-k}\right)$
Total	20.626 (SST)	23 ($n-1$)	$F^* = \dfrac{0.708}{0.925} \doteq 0.77 = \left(\dfrac{MSB}{MSW}\right)$ d.f. = (3, 20) (= $(k-1, n-k)$)

Table 11.5 shows at a glance how sums of squares, degrees of freedom, and mean squares break down into "between groups" and "within groups" components. This table also provides a check on your work. The degrees of freedom should always add up to $n-1$; if they do not, you have made a mistake.

It is also true that if you define a **total sum of squares**, **SST**, by the equation

$$SST = \sum (x - \bar{\bar{x}})^2$$

then

$$SSB + SSW = SST$$

Here SST represents the total amount of variation in all of the data from all of the groups. For example, in our weight example, SST would be calculated using all 24 weights (x) and the overall mean of those 24 weights ($\bar{\bar{x}}$). The relationship SSB + SSW = SST tells us that the total variation in all of the data can be broken down into two components: a component representing variation within groups and a component representing variation between groups.

Construct an ANOVA table for the hypothesis test in Exercise D (that is, for the weights in Table 11.2).

Exercise **F**

In this introduction to ANOVA, we have shown you how to test equality of means when you are given the same number of pieces of data from each group: for example, six pieces of data from each of four groups. In practice, the numbers from different groups may vary. We will discuss such situations in Exercise 6.

EXERCISES Section **11.1**

In Exercises 1 through 5,

a) state the null hypothesis,

b) state the alternative hypothesis,

c) define the population,

d) state the degrees of freedom,

e) find the critical value,

f) draw an *F*-curve with the critical value and rejection region,

g) calculate MSW,

h) calculate MSB,

i) calculate F^*, and

j) state your conclusions.

1 In Example 18 of Chapter 8, we compared the effectiveness of two waxes by looking at their lasting time. Below we give times of effectiveness for three waxes. Each wax was tested on five different cars.

Effectiveness times (in days)

Wax 1	Wax 2	Wax 3
89	94	88
86	91	87
83	88	84
87	92	85
89	87	90

Perform an appropriate ANOVA test to determine whether there is any difference between the mean effectiveness of the three waxes. Use $\alpha = 0.05$.

2 A quality-control engineer recorded the time needed by each of three workers to perform an assembly-line task. Each worker was observed five times. The times (in minutes) were:

Worker

A	B	C
8	8	10
10	9	9
9	9	10
11	8	11
10	10	9

Test the null hypothesis that the true mean times μ_i for the three workers are identical. Use $\alpha = 0.05$.

3 Three different calculator companies claim that their battery packs will work for three hours without recharging. A consumer agency tested calculators from these companies. The battery times for five calculators from each company were:

Company

1	2	3
3.16	2.91	3.27
3.26	2.15	3.14
2.64	3.10	2.76
3.27	2.82	2.77
2.81	3.13	3.01

Test the null hypothesis that the true mean lives, μ_i, are all the same. Use $\alpha = 0.05$.

4 An agricultural researcher has three new fertilizers to test for use with sugar beets. He tries each of the three new fertilizers and a commonly used older one on six different test plots. (The plots are of the same size, have similar soil, and are in the same area.) The yields are:

New fertilizers			Old fertilizer
A	B	C	
42.4	40.0	42.5	36.0
42.5	37.4	37.2	41.6
43.2	36.4	38.1	40.6
41.4	38.0	41.5	40.9
42.2	41.0	41.3	34.9
39.8	39.1	36.8	40.8

Perform a test to decide whether there is any difference in the yield that will be produced by these fertilizers. Use $\alpha = 0.05$.

5 A pig farmer wants to test three different vitamin treatments. He gives each treatment to four different pigs for two weeks, and records the weight gain (in pounds) from each pig during the second week. The weight gains are:

Treatment

1	2	3
10.5	11.3	9.8
10.9	10.9	10.5
10.7	11.8	10.4
10.1	11.8	10.2

Perform a test to determine whether there is any difference in the true mean weight gain produced by the three treatments. Use $\alpha = 0.05$

6 In some ANOVA test situations, we do not have the same sample size from each group. In such cases, the formula for SSB changes slightly. If the group j has size n_j, the formula becomes

$$\text{SSB} = \sum n_j (\bar{x}_j - \bar{\bar{x}})^2,$$

where $\bar{\bar{x}}$ is the mean of all the data given. For example, if we had four groups with sizes $n_1 = 4$, $n_2 = 6$, $n_3 = 5$, and $n_4 = 6$, the formula for SSB would be

$$\text{SSB} = 4(\bar{x}_1 - \bar{\bar{x}})^2 + 6(\bar{x}_2 - \bar{\bar{x}})^2 + 5(\bar{x}_3 - \bar{\bar{x}})^2 + 6(\bar{x}_4 - \bar{\bar{x}})^2$$

All other formulas remain the same, as is shown by Table 11.6.

Table 11.6

ANOVA table

	SS	d.f.	MS
Between groups	$\text{SSB} = \sum n_j (\bar{x}_j - \bar{\bar{x}})^2$	$k - 1$	$\text{MSB} = \dfrac{\text{SSB}}{k - 1}$
Within groups	SSW is calculated as before	$n - k$	$\text{MSW} = \dfrac{\text{SSW}}{n - k}$
Total	SST	$n - 1$	$F^* = \dfrac{\text{MSB}}{\text{MSW}}$ d.f. $= (k - 1, n - k)$

Suppose that in Table 11.1 (page 406) we had been given only four weights for the first person and five weights for the third. Then our data table would look like this:

Steve	Harold	Gerald	Fritz
145	148	147	147
147	147	149	146
145	146	146	147
147	147	147	147
	147	148	147
	148		146

Use the above formulas to do an ANOVA test with $\alpha = 0.05$ of the null hypothesis $\mu_1 = \mu_2 = \mu_3 = \mu_4$ for the true mean weights of these four men. (*Warning*: $\bar{x}$ is the mean of all 21 given pieces of data. It is *not* the mean of the four separate means for the four exercisers.)

7 In Chapter 8, we learned how to do t-tests on independent ■ samples from two populations to decide whether the two populations had the same means. This procedure could be used on our data here. For example, in Exercise 4, we could compare the fertilizers two at a time: first compare A and B, then A and C, etc. Why isn't this a desirable procedure?

8 There is a procedure for looking at the difference between any ■ two of the means in an analysis-of-variance problem. When the groups are equal in size, a 95-percent confidence interval for the difference between two group means μ_i and μ_j is given by the inequality

$$\bar{x}_i - \bar{x}_j - t_{0.025} \sqrt{\frac{2(\text{MSW})}{g}} \leq \mu_i - \mu_j \leq \bar{x}_i - \bar{x}_j$$
$$+ t_{0.025} \sqrt{\frac{2(\text{MSW})}{g}}$$

where

$\bar{x}_i$ = sample mean for group i,

g = number of observations in each group,

and the number of degrees of freedom for $t_{0.025}$ is

$$\text{d.f.} = n - k = \text{d.f. for SSW.}$$

Use this procedure on the data in Exercise 1 to find a 95-percent confidence interval for:

a) the difference between μ_1, the effectiveness time for Wax 1, and μ_3, the effectiveness time for Wax 3.

b) The difference between μ_2 and μ_1.

9 a) Derive a formula for a 99-percent confidence interval for μ_i ■ $- \mu_j$. [Refer to Exercise 8.]

b) Find a 99-percent confidence interval for the difference between μ_1 and μ_3 in Exercise 1.

10 Suppose that you have obtained 95-percent confidence intervals ■ for each of the two differences $\mu_1 - \mu_2$ and $\mu_1 - \mu_3$ in Exercise 8. Can you be "95 percent confident" of both of these results simultaneously? (In other words, can you be 95 percent confident that *both* true differences are in the intervals you find?) Explain your answer.

The kind of ANOVA test done in Section 11.1 is called a *one-way analysis of variance with equal cell sizes*. (The *cell size* of a group is just the sample size for that group. In our weight examples, each cell had size 6.) In this section we will cover shortcut formulas that will make the calculations for such a test go a bit faster. The reader who wishes only to understand the idea behind ANOVA and who will have calculations done on a computer may wish to skip the material on shortcut formulas. At the beginning of this section we will also say a bit more about the assumptions behind ANOVA. We will do all this in a very practical setting—by looking at data from the real estate sales industry.

In 1974 the national income was estimated to be 1,157.5 billion dollars. The real estate industry accounted for 88.6 billion dollars of that figure, or roughly 7.6 percent of the estimated national income. In that same year, the real estate industry employed more than 800,000 individuals. The people who sell real estate are the major force in this industry. In Example 6, we will look at a problem involving management and evaluation of these people.

ANOVA: assumptions; short cut formulas   **11.2**

Example **6** ◖ The founder and president of a real estate chain in the Southwest was interested in improving the performance of his sales people. He already had a very successful sales force, since he hired only full-time people with demonstrated professional competence, and had stated that he did not wish to retain sales people who could not eventually maintain incomes of at least $16,000 per year. However, he felt that it might still be possible to increase the incomes of his sales force. Two possibilities were suggested to him:

1. a sales contest, with bonuses to be awarded at the beginning of next year for leaders in various sales categories, and
2. a series of seminars with a speaker who specialized in sales-force motivation.

Since the company president was an objective decision maker, he decided to set up an experiment to test these methods. He chose four offices in four similar areas, each with a staff of five sales people. He then assigned the possible improvement methods to his four offices as follows:

Office	Treatment	True mean income
1	Seminars and contest	μ_1
2	Seminars only	μ_2
3	Contest only	μ_3
4	No improvement scheme	μ_4

Nothing was done in Office 4 because there is always a possibility that an improvement scheme is no better than nothing at all. We have listed the true mean income for each treatment so that we can write down the hypotheses in question here:

Null hypothesis: $\mu_1 = \mu_2 = \mu_3 = \mu_4$
Alternative hypothesis: Not all μ_i are equal.

If the null hypothesis is true, then mean income is not affected by these improvement schemes, and the company should not spend money on them. If the alternative is true, some method makes a difference and further study will be required to say what it is.

Each population under consideration here consisted of all possible one-year incomes for sales people in similar offices with the same treatment. For example, population 1 consisted of all possible incomes for similar sales people with both seminars and a sales contest.

The company president tried these four treatments, and collected income figures for his offices at the end of one year. These figures are given in Table 11.7 (rounded to the nearest $1000 for simplicity).

Office 1	Office 2	Office 3	Office 4
46	65	37	11
53	59	13	35
54	17	65	57
29	18	42	56
27	37	33	40

Table 11.7
Incomes of sales people at four branch offices (in thousand of dollars)

The president of this company was trained as an engineer and knew statistics fairly well. He wanted to do an analysis of variance to test his null hypothesis, but he knew that *ANOVA assumed* (a) *that each of his salary samples was taken from a normal population, and* (b) *that all four of these populations have the same variance.* His experience in real estate led him to believe that (a) and (b) were true, but he did not have enough data to be sure of that. He consulted a statistician friend who told him to go ahead, for the following reasons:

1. Although the theory behind ANOVA is based on the assumption that the populations sampled are normal, *in practice* ANOVA will work nearly as well for nonnormal populations, as long as their departure from normality is not extreme.
2. For a simple situation such as we have here, with the same number of individuals in each group (i.e., equal cell sizes), ANOVA will work well even if the variances of the populations involved are not identical.

The company president decided to go ahead with ANOVA. He was certain that the populations involved were nearly normal, and that their variances were not too far apart. ▶

The owner of a chain of men's-wear shops located in suburban shopping centers decided to experiment with advertising policies. For one month he tried the following approaches.

Exercise **G**

Policy 1: Advertise in local paper only (five stores).
Policy 2: Advertise in local paper, and have fliers distributed door-to-door (five stores).
Policy 3: Distribute fliers only (five stores).

a) What null and alternative hypotheses does he wish to test?
b) His sales data is given in Table 11.8.

Table 11.8	Policy 1	Policy 2	Policy 3
Total monthly sales in dollars for 15 stores (in thousands of dollars)	10	9	7
	8	11	5
	12	10	9
	9	13	8
	13	14	6

What assumptions should the owner check before he does an ANOVA test of his null hypothesis on this data?

Shortcut formulas   We will now introduce the shortcut formulas for MSB and MSW and finish off the hypothesis test for the real estate executive.

Example 7   ◖ The real estate executive wished to test his null hypothesis with $\alpha = 0.05$. He planned to do his ANOVA calculations by hand, with the help of a calculator. The shortcut formulas he used required the use of some notation:

$$n = \text{Total number of pieces of data}$$
$$g = \text{Number of pieces of data in each group}$$
$$k = \text{Number of groups}$$
$$T_i = \text{Sum of all data in group } i.$$

In Table 11.9 we repeat his data, together with the values of n, g, k, and T_i.

Table 11.9	Office 1	Office 2	Office 3	Office 4
	46	65	37	11
	53	59	13	35
	54	17	65	57
	29	18	42	56
	27	37	33	40
	$T_1 = 209$	$T_2 = 196$	$T_3 = 190$	$T_4 = 199$
	$n = 20;$	$g = 5;$	$k = 4.$	

The **shortcut formulas for SSB and SSW** are:

$$\text{SSB} = \frac{\sum T^2}{g} - \frac{(\sum T)^2}{n}$$

$$\text{SSW} = \sum x^2 - \frac{\sum T^2}{g}$$

Since the expression $(\sum T^2)/g$ appears in each formula, it is a good idea to compute it first. The computations for the real estate salaries are given below:

$$\frac{\sum T^2}{g} = \frac{209^2 + 196^2 + 190^2 + 199^2}{5}$$

$$= \frac{157798}{5} = 31559.6$$

$$\frac{(\sum T)^2}{n} = \frac{(209 + 196 + 190 + 199)^2}{20} = \frac{(794)^2}{20}$$

$$= \frac{630436}{20} = 31521.8$$

$$\sum x^2 = 46^2 + 53^2 + 54^2 + 29^2 + 27^2$$
$$+ 65^2 + 59^2 + 17^2 + 18^2 + 37^2$$
$$+ 37^2 + 13^2 + 65^2 + 42^2 + 33^2$$
$$+ 11^2 + 35^2 + 57^2 + 56^2 + 40^2$$
$$= 37046$$

$$\text{SSB} = \frac{\sum T^2}{g} - \frac{(\sum T)^2}{n}$$

$$= 31559.6 - 31521.8 = 37.8$$

$$\text{SSW} = \sum x^2 - \frac{\sum T^2}{g}$$

$$= 37046 - 31559.6 = 5486.4$$

It is now easy to complete the hypothesis test.

Null hypothesis: $\mu_1 = \mu_2 = \mu_3 = \mu_4$
Alternative hypothesis: Not all the μ_i are equal.

$$\text{d.f. for SSB} = k - 1 = 4 - 1 = 3$$

$$\text{MSB} = \frac{\text{SSB}}{k-1} = \frac{37.8}{3} = 12.6$$

$$\text{d.f. for SSW} = n - k = 20 - 4 = 16$$

$$\text{MSW} = \frac{\text{SSW}}{n-k} = \frac{5486.4}{16} = 342.9$$

$$F^* = \frac{\text{MSB}}{\text{MSW}} = \frac{12.6}{342.9} = 0.04$$

This is an extremely small value, and you might guess that it is not large enough to reject the null hypothesis. This is indeed true. For d.f. $= (3, 16)$, the critical value, $c = F_{\alpha}$, for $\alpha = 0.05$, is 3.24. This is pictured in Fig. 11.6.

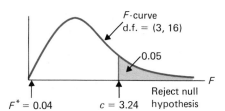

Figure 11.6 $F^* = 0.04$ $c = 3.24$ Reject null hypothesis

Thus the real estate executive should probably save his money. There is no visible effect from any of the improvement schemes he tried. ▶

The above example gives some idea of why it is important to know some statistical theory. An executive with no such knowledge might be tempted to simply compute sample means for the four offices. A person who did this would find:

Office	$\bar{x}$ (in thousands)
1	41.8
2	39.2
3	38.0
4	39.8

Without knowledge of statistics, there would be a great temptation to say that Office 1 was best and all other offices should be motivated as Office 1 was. But our ANOVA has shown us that we cannot conclude that there is any difference between the four treatments (seminars and contests, seminars only, contests only, no improvement scheme).

Exercise **H**   Test the hypothesis of equality of means for the store owner in Exercise G, using $\alpha = 0.05$ and the sales data:

Policy 1	Policy 2	Policy 3
10	9	7
8	11	5
12	10	9
9	13	8
13	14	6

A note on terminology   In situations such as the real estate and store examples, researchers often refer to such actions as motivation seminars or advertising strategies as **treatments.** They think of differences between group means as due to these treatments, and call SSB the **treatment sum of squares,** or **SSTR.** They also think of the random variations within each group as **errors** and call SSW the **error sum of squares** or **SSE.** Table 11.10 explains the relationship of the two terminologies.

Between–within terminology	Treatment–error terminology
SSB Between group sum of squares	SSTR Treatment sum of squares
MSB Mean square between groups	MSTR Treatment mean square
SSW Within group sum of squares	SSE Error sum of squares
MSW Mean square within groups	MSE Error mean square

Table 11.10
Terminology

As in our earlier discussions, we shall outline below the basic elements of the procedure we have just covered, and then illustrate with an example.

Procedure

Procedure for ANOVA tests of the form:

Null hypothesis: $\mu_1 = \mu_2 = \cdots = \mu_k$
Alternative hypothesis: Not all the μ_i are equal.

Form of data collected: Samples of size g from each of k populations:

Population 1	Population 2	$\cdots$	Population k
x_{11}	x_{21}	$\cdots$	x_{k1}
x_{12}	x_{22}	$\cdots$	x_{k2}
$\vdots$	$\vdots$	$\vdots$	$\vdots$
x_{1g}	x_{2g}	$\cdots$	x_{kg}

Data type: Metric data.

Assumptions: The k populations are normal with the same variances. (These assumptions may be violated in practice if deviations are not extreme, as long as we are in the special situation of equal cell sizes.)

1. Decide on the significance level α.

2. The critical value is

$$c = F_\alpha, \quad \text{d.f.} = (k-1, n-k)$$
$$n = \text{the total number of pieces of data} = k \cdot g$$

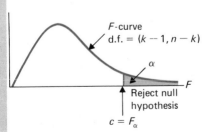

F-curve
d.f. $= (k-1, n-k)$

α

Reject null
hypothesis

$c = F_\alpha$

3. Calculate the value of

$$F^* = \frac{MSB}{MSW}$$

where

$$MSB = \frac{1}{k-1}\left[\frac{\sum T^2}{g} - \frac{(\sum T)^2}{n}\right]$$

$$MSW = \frac{1}{n-k}\left[\sum x^2 - \frac{\sum T^2}{g}\right]$$

T_i = sum of data in group i.

4. If $F^* > c$, reject the null hypothesis.

Example 8 ◖ So that you may compare the shortcut method with the original longer method, we will apply the shortcut method to Exercise 4 of Section 11.1. In that exercise, an agricultural researcher was interested in the sugar-beet yield obtained using four different fertilizers. The fertilizers and the true mean yields they give are listed below:

New brands

Fertilizer	A	B	C	Old
True mean	μ_1	μ_2	μ_3	μ_4

The researcher felt that these fertilizers gave pretty much the same yield. (If they did, farmers should simply use the cheapest one.) To decide whether his suspicions were true, the researcher set up a test with:

Null hypothesis: $\mu_1 = \mu_2 = \mu_3 = \mu_4$
Alternative hypothesis: Not all μ_i are equal.

Step 1. He decided to use $\alpha = 0.05$.

The researcher intended to try each fertilizer in six different test plots with similar soil and weather conditions. Thus he was dealing with

$$k = 4 \text{ groups,}$$
$$g = 6 \text{ observations in each group,}$$
$$n = 24 \text{ total observations.}$$

(See data table in Step 3 for actual observations.)

Step 2. With these values of k, g, and n, the test would have d.f. $= (k - 1, n - k)$ $= (3, 20)$. The critical value is $c = F_{0.05} = 3.10$.

Step 3. *Calculations*: Use the data given in Table 11.11.

	New treatments		Old treatment
A	B	C	
42.4	40.0	42.5	36.0
42.5	37.4	37.2	41.6
43.2	36.4	38.1	40.6
41.4	38.0	41.5	40.9
42.2	41.0	41.3	34.9
39.8	39.1	36.8	40.8
$T_1 = 251.5$	$T_2 = 231.9$	$T_3 = 237.4$	$T_4 = 234.8$

Table 11.11

$$\frac{\sum T^2}{g} = \frac{(251.5)^2 + (231.9)^2 + (237.4)^2 + (234.8)^2}{6}$$

$$= \frac{228519.66}{6} = 38086.610$$

$$\frac{(\sum T)^2}{n} = \frac{(955.6)^2}{24} = 38048.807$$

$$\sum x^2 = 38180.880$$

$$\text{SSB} = \frac{\sum T^2}{g} - \frac{(\sum T)^2}{n} = 38086.610 - 38048.807 = 37.803$$

$$\text{SSW} = \sum x^2 - \frac{\sum T^2}{g} = 38180.880 - 38086.610 = 94.270$$

$$\text{MSB} = \frac{\text{SSB}}{k - 1} = \frac{37.803}{3} = 12.601$$

$$\text{MSW} = \frac{\text{SSW}}{20} = \frac{94.270}{20} = 4.714$$

$$F^* = \frac{\text{MSB}}{\text{MSW}} = \frac{12.601}{4.714} \doteq 2.67$$

Step 4. *Conclusion*: Since $F^* = 2.67$ and $c = 3.10$, we have $F^* \leqslant c$. Thus, the null hypothesis *cannot* be rejected. In other words, on the basis of this data, we cannot reject the hypothesis that the four fertilizers have the same mean yields. ▶

Remarks   We have barely scratched the surface of ANOVA. Entire books and entire graduate courses are devoted to it. One problem we have not looked at has probably occurred to many readers. Suppose you find out that not all the means for a group of treatments are equal? Then one treatment is possibly superior to the others. How do you find out which treatment is the best, and how do you estimate the size of the differences? There are methods to answer this* and many other practical questions that arise in ANOVA. Although we shall not discuss those methods in this book, the interested reader might wish to look at the books listed at the end of this chapter. References 1 and 2 do not require extensive backgrounds in mathematics.

EXERCISES Section **11.2**

In Exercises 1 through 4, use the shortcut method to find SSB, SSW, and $F*$ for the indicated exercise from Section 11.1.

1 Exercise 1 of Section 11.1:

Effectiveness times

Wax 1	Wax 2	Wax 3
89	94	88
86	91	87
83	88	84
87	92	85
89	87	90

2 Exercise 2 of Section 11.1:

Task times for workers

A	B	C
8	8	10
10	9	9
9	9	10
11	8	11
10	10	9

3 Exercise 3 of Section 11.1:

Calculator battery times

Company		
1	2	3
3.16	2.91	3.27
3.26	2.15	3.14
2.64	3.10	2.76
3.27	2.82	2.77
2.81	3.13	3.01

4   Exercise 5 of Section 11.1:

Weight gains for pigs

Treatment		
1	2	3
10.5	11.3	9.8
10.9	10.9	10.5
10.7	11.8	10.4
10.1	11.8	10.2

* The methods for answering this question are called *multiple comparison methods*. They are discussed on pages 472–482 of Reference 3.

5 The manager of a small chain of three similar stores decided to try three different advertising policies in his three stores:

Store 1	Store 2	Store 3
No advertising	Advertise in neighborhood with circulars	Use circulars and advertise in local papers

He tried each of these policies for three months. The total sales for those months were (in thousands of dollars):

		Store		
		1	2	3
Month	October	20	24	25
	November	24	26	29
	December	28	29	30

Use the methods of Section 11.2 to perform an appropriate hypothesis test and decide whether the advertising policies make a difference in mean sales for the three stores. Use $\alpha = 0.05$.

Computer packages * 11.3

Most computer packages contain programs for ANOVA. The one-way analyses given in this chapter can be done by an SPSS program called **BREAKDOWN**. Below we have given a printout from this program. The data analyzed is the weight data from Example 1.

The headings SUBJECT 1, SUBJECT 2, etc., stand for the individual joggers from whom weight data were obtained. In general, the SUBJECT headings identify the

```
CRITERION VARIABLE WEIGHT
- - - - - - - - - - - - - - - - - - - - - A N A L Y S I S   O F   V A R I A N C E - - - - - - - - - - - - - - - - - - - - - -

VARIABLE          CODE    VALUE LABEL              SUM          MEAN       STD DEV    SUM OF SQ              N

SUBJECT            1.                          879.0000      146.5000      1.2247      7.5000      (       6)
SUBJECT            2.                          883.0000      147.1667      0.7528      2.8333      (       6)
SUBJECT            3.                          883.0000      147.1667      1.1690      6.8333      (       6)
SUBJECT            4.                          880.0000      146.6667      0.5164      1.3333      (       6)
                                              ---------------------------------------------------------------------
                  WITHIN GROUPS TOTAL         3525.0000      146.8750      0.9618     18.5000      (      24)

      * * * * * * * * * * * * * * * * * * * * * * * * * * * * * * * * * *
      *                                                                 *
      *               A N A L Y S I S   O F   V A R I A N C E           *
      *                                                                 *
      * * * * * * * * * * * * * * * * * * * * * * * * * * * * * * * * * *
      *                                                                 *
      *    SOURCE          SUM OF SQUARES  D.F.  MEAN SQUARE      F    SIG.  *
      *                                                                 *
      *    BETWEEN GROUPS        2.125      3       0.708     0.766  0.5266  *
      *                                                                 *
      *    WITHIN GROUPS        18.500     20       0.925              *
      *                                                                 *
      *          ETA = 0.3210 ETA SQUARED = 0.1030                     *
      *                                                                 *
      * * * * * * * * * * * * * * * * * * * * * * * * * * * * * * * * * *
```

* This section is optional.

individual data groups under consideration. BREAKDOWN first prints the sum, mean, standard deviation, and sum of squares for each group. Directly below, the program displays the sum of all 24 weights given, the mean and standard deviation of this entire data set, and the total within-groups sum of squares (SSW). Finally, within the box entitled **ANALYSIS OF VARIANCE**, BREAKDOWN gives, among other things, the mean square between groups (MSB), the mean square within groups (MSW), degrees of freedom, and the F-statistic F^*. The number 0.5266 listed under SIG is the area value, which we called a P-value in Chapter 9. The number 0.5266 is the area to the right of $F^* = 0.766$ under the F-curve with d.f. $= (3, 20)$. (See Fig. 11.7.) This tells us that we cannot reject the null hypothesis of equality of means for any value of α less than 0.5266.

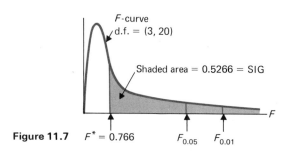

Figure 11.7 $F^* = 0.766$ $F_{0.05}$ $F_{0.01}$

BREAKDOWN contains a quantity we have not discussed in this chapter—namely, ETA SQUARED (which happens to equal SSB/SST).

CHAPTER REVIEW

Key Terms

sum of squares
sum of squares within groups (SSW)
degrees of freedom for SSW
mean square within groups (MSW)
sum of squares between groups (SSB)
degrees of freedom for SSB

mean square between groups (MSB)
total sum of squares (SST)
ANOVA table
treatment sum of squares (SSTR)
treatment mean square (MSTR)
error sum of squares (SSE)
error mean square (MSE)

In the formulas below,

Formulas and Key Facts

n = total number of pieces of data

k = number of groups

g = number of pieces of data in each group

T_i = sum of data in group i

sum of squares between groups, SSB:*

$$SSB = \frac{\sum T^2}{g} - \frac{(\sum T)^2}{n}$$

mean square between groups, MSB:

$$MSB = \frac{SSB}{k-1}$$

sum of squares within groups, SSW:*

$$SSW = \sum x^2 - \frac{\sum T^2}{g}$$

mean square within groups, MSW:

$$MSW = \frac{SSW}{n-k}$$

formula for F^*:

$$F^* = \frac{MSB}{MSW}, \qquad \text{d.f.} = (k-1, n-k)$$

You should be able to

1 Calculate *SSB, MSB, SSW, MSW, and F**.

2 Find the *degrees of freedom for F**.

3 Construct an *ANOVA table*.

4 Relate the *"between–within"* terminology to the *"treatment–error"* terminology.

5 Perform a *one-way analysis of variance test with equal cell sizes*, as described on page 423.

---

* This is the "shortcut" formula.

REVIEW TEST

1 A cereal company wished to test the possible market effect of three different artists' designs for its boxes. The company picked 12 supermarkets in similar suburban communities with similar sales volume, and stocked each design in four of the markets for a fixed test period. The sales figures for the test period were:

Test-period sales (in hundreds of boxes)

Design 1	Design 2	Design 3
58	44	59
43	59	49
57	53	50
45	46	55

Perform an appropriate hypothesis test to decide whether or not the three designs all lead to the same mean sales level in suburban supermarkets. Use $\alpha = 0.01$.

2 A consumer agency is testing a new model scale with a digital readout. To test for consistency, they buy four scales of this model and weigh the same person five times on each scale. Here are the results (we have also calculated T and T^2 for each group, and the overall sum of squares, $\sum x^2$).

	Scale 1	Scale 2	Scale 3	Scale 4
	148.2	148.4	147.6	146.8
	148.6	149.6	147.9	147.3
	147.5	149.2	148.7	147.9
	148.1	149.3	148.4	147.2
	147.3	148.6	148.2	147.5
T	739.7	745.1	740.8	736.7
T^2	547156.09	555174.01	548784.64	542726.89

$$\sum x^2 = 438{,}771.85$$

a) Use the shortcut formulas to perform the hypothesis test.

Null hypothesis: $\mu_1 = \mu_2 = \mu_3 = \mu_4$
Alternative hypothesis: Not all μ_i are equal.

Use $\alpha = 0.05$.

(*Note*: μ_i is the mean of all possible weight readings that scale i would give for the person in the test.)

b) Discuss the assumptions necessary for this test.

FURTHER READINGS

Dixon, W., and J. Massey, *Introduction to Statistical Analysis.* New York: McGraw-Hill, 1969. Chapters 10 and 15 are devoted to ANOVA.

Lapin, Lawrence, *Statistics.* New York: Harcourt Brace Jovanovich, Inc., 1975. Chapter 13 of this introductory text-book covers a broader range of topics than is covered here.

Neter, J. and W. Wasserman, *Applied Linear Statistics Models.* Homewood, Illinois: Richard D. Irwin, Inc., 1974. Pages 419–800 of this book cover a great deal of ANOVA. There are no advanced mathematical prerequisites.

ANSWERS TO REINFORCEMENT EXERCISES

A a) *Null hypothesis:* $\mu_1 = \mu_2 = \mu_3 = \mu_4$
    Alternative hypothesis: Not all the μ_i are equal.

  b) Alternative hypothesis

B a)

Joe

x	$x - \bar{x}$	$(x - \bar{x})^2$
148	1.167	1.362
146	−0.833	0.694
147	0.167	0.028
148	1.167	1.362
146	−0.833	0.694
146	−0.833	0.694
146.833		4.834
$\bar{x}$		$\sum(x-\bar{x})^2$

Don

x	$x - \bar{x}$	$(x - \bar{x})^2$
142	−1.333	1.777
142	−1.333	1.777
143	−0.333	0.111
145	1.667	2.779
144	0.667	0.445
144	0.667	0.445
143.333		7.334
$\bar{x}$		$\sum(x-\bar{x})^2$

Ed

x	$x - \bar{x}$	$(x - \bar{x})^2$
150	−0.167	0.028
149	−1.167	1.362
151	0.833	0.694
152	1.833	3.360
150	−0.167	0.028
149	1.167	1.362
150.167		6.834
$\bar{x}$		$\sum(x-\bar{x})^2$

Bill

x	$x - \bar{x}$	$(x - \bar{x})^2$
155	−0.667	0.445
157	1.333	1.777
156	0.333	0.111
156	0.333	0.111
156	0.333	0.111
154	−1.667	2.779
155.667		5.334
$\bar{x}$		$\sum(x-\bar{x})^2$

$$\text{SSW} = 4.834 + 7.334 + 6.834 + 5.334 = 24.336$$

  b) $n = 24$, $k = 4$, $n - k = 24 - 4 = 20 = $ d.f. for SSW

  c) $\text{MSW} = \dfrac{\text{SSW}}{n-k} = \dfrac{24.336}{20} = 1.217$

C

a) $\bar{\bar{x}} = \dfrac{146.833 + 143.333 + 150.167 + 155.667}{4} = \dfrac{596}{4} = 149$

$g = 6$

$\bar{x}$	$\bar{x} - \bar{\bar{x}}$	$(x - \bar{\bar{x}})^2$
146.833	-2.167	4.696
143.333	-5.667	32.115
150.167	1.167	1.362
155.667	6.667	44.449
149.00		82.622

$\uparrow$ $\bar{\bar{x}}$ $\qquad\qquad$ $\uparrow$ $\sum(\bar{x} - \bar{\bar{x}})^2$

$SSB = 6(82.622) = 495.732$

b) $k = 4$, $k - 1 = 4 - 1 = 3 = $ d.f. for SSB

c) $MSB = \dfrac{SSB}{k - 1} = 165.244$

D $MSB = 165.244,$ $\quad$ d.f. $= 3$

$MSW = 1.217,$ $\quad$ d.f. $= 20$

$F^* = \dfrac{MSB}{MSW} \doteq 135.78$

$c = 3.10$ for d.f. $= (3, 20)$

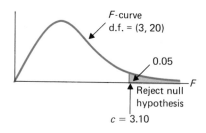

F-curve
d.f. = (3, 20)

0.05

Reject null
hypothesis

$c = 3.10$

F^* is in the rejection region. Reject the null hypothesis.

E *Null hypothesis:* $\mu_1 = \mu_2 = \mu_3 = \mu_4$

Alternative hypothesis: Not all the μ_i are equal.

1. $\alpha = 0.05$
2. $c = F_\alpha = F_{0.05} = 3.10$, for d.f. $= (3, 20)$.
3. $MSB = 165.244$
   $MSW = 1.217$
   $F^* \doteq 135.78$
4. $F^* > c$. Reject the null hypothesis.

F

ANOVA Table

	Sum of squares, SS	Degree of freedom, d.f.	Mean square, MS
Between groups	495.732 (SSB)	3 $(k-1)$	165.244 $\left(MSB = \dfrac{SSB}{k-1}\right)$
Within groups	24.336 (SSW)	20 $(n-k)$	1.217 $\left(MSW = \dfrac{SSW}{n-k}\right)$
Total	520.068 (SST)	23 $(n-1)$	$F^* = \dfrac{165.244}{1.217}$ $= 135.78$

$= \left(\dfrac{MSB}{MSW}\right)$

d.f. $= (3, 20)$

G a) *Null hypothesis:* $\mu_1 = \mu_2 = \mu_3$

Alternative hypothesis: Not all the μ_i are equal.

b) He should check to see

i) that the populations from which he took the sales data are normal or that their departure from normality is not extreme, and

ii) that all three of the populations have the same variance, or that the number of pieces of data in each group is the same.

H *Null hypothesis:* $\mu_1 = \mu_2 = \mu_3$
Alternative hypothesis: Not all μ_i are equal.
$\alpha = 0.05$

Policy 1	Policy 2	Policy 3
10	9	7
8	11	5
12	10	9
9	13	8
13	14	6
$T_1 = 52$	$T_2 = 57$	$T_3 = 35$

$n = 15, \, g = 5, \, k = 3$

$$\frac{\sum T^2}{g} = \frac{52^2 + 57^2 + 35^2}{5} = \frac{7178}{5} = 1435.6$$

$$\frac{(\sum T)^2}{n} = \frac{(52 + 57 + 35)^2}{15} = \frac{20736}{15} = 1382.4$$

$$\sum x^2 = 10^2 + \; 8^2 + 12^2 + \; 9^2 + 13^2$$
$$+ \; 9^2 + 11^2 + 10^2 + 13^2 + 14^2$$
$$+ \; 7^2 + \; 5^2 + \; 9^2 + \; 8^2 + \; 6^2$$
$$= 1480$$

$\text{SSB} = 1435.6 - 1382.4 = 53.2, \quad$ d.f. for SSB $= 3 - 1 = 2$

$$\text{MSB} = \frac{53.2}{2} = 26.6$$

$\text{SSW} = 1480 - 1435.6 = 44.4, \quad$ d.f. for SSW $= 15 - 3 = 12$

$$\text{MSW} = \frac{44.4}{12} = 3.7$$

$$F* = \frac{26.6}{3.7} = 7.19$$

For d.f. $= (2, 12)$, $c = F_{0.05} = 3.89$

$F* > c.$ Reject the null hypothesis.

CHAPTER

Regression and correlation

12.1 Straight-line equations

Most Americans are concerned about the price of gasoline, and many of us have become involved in arguments about its price: Will a rise in price lower demand and prevent shortages? People who follow the news regularly also watch the value of the dollar, and speculate as to how this value will affect the quality of life in the next year. Both of these current issues revolve around the relationship between *two variables*: between gas prices and supply, or between the value of the dollar and some index of the quality of life. Research on such relationships is done using the methods of **linear regression**, which we will study in this chapter. Since the simplest linear regression methods assume that variables are related by straight-line equations, we will begin with a review of straight lines.

Straight lines

Equations such as

$$y = 4 + 2x, \qquad y = 0.5 + 3x, \qquad \text{and} \qquad y = 18.95 + 0.23x$$

have as their graphs *straight lines*. The general form of a straight-line equation is

$$y = b_0 + b_1 x.$$

Any equation of this form has a straight-line graph, and any straight-line graph (except a vertical line) can be described by an equation of this kind. Straight-line equations are very simple and easy to use, and they are of use in a wide range of applied problems. An illustration of the use of straight lines is given in Example 1.

Example 1

On November 14, 1978, a mathematician phoned a car rental company in Colorado Springs, Colorado, to find out the rates for renting a full-size car (a Cutlass four-door). The quoted rates on that day were:

$18.95 per day plus $0.23 per mile.

The purpose of the car rental was to give a *one-day* sightseeing trip on November 21 to some visitors who would not fit in a Datsun B210. Depending on how the visitors felt on November 21, the mathematician might end up driving them 50, 100, 150, or 220 miles. To get some idea of what his costs might be, he set up a formula for costs, and did some calculations. These are given below.

Variables: $\begin{cases} x = \text{number of miles driven} \\ y = \text{cost of car rental} \end{cases}$

Formula: $y = 18.95 + 0.23x$

Calculations:
$$x = 50 \rightarrow y = 18.95 + (0.23)(50) = 18.95 + 11.50 = 30.45$$
$$x = 100 \rightarrow y = 18.95 + (0.23)(100) = 18.95 + 23.00 = 41.95$$
$$x = 150 \rightarrow y = 18.95 + (0.23)(150) = 18.95 + 34.50 = 53.45$$
$$x = 220 \rightarrow y = 18.95 + (0.23)(220) = 18.95 + 50.60 = 69.55$$

Table of values:

Miles, x	Cost, y
50	30.45
100	41.95
150	53.45
220	69.55

The basic formula for the cost of renting and using the car is a straight-line formula:

$$y = \quad 18.95 \quad + \quad 0.23x$$

$$b_0 = 18.95 \qquad b_1 = 0.23$$

If you plot the x- and y-values above, the plotted points will lie on the straight line shown in Fig. 12.1.

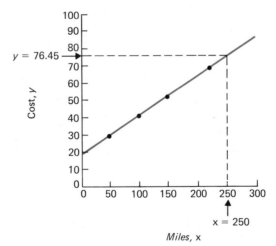

Figure 12.1

This graph can be used to (visually) estimate the cost y for any mileage x. For example, if a car renter is planning a trip of 250 miles he can estimate his cost by reading off the y-value on the graph above $x = 250$. A quick glance will show that his cost is about \$76. His exact cost for 250 miles is

$$y = 18.95 + (0.23)\,(250) = 76.45. \;▶$$

Exercise **A**   A salesman is paid $200 per week plus a 5 percent commission on the dollar value of his sales. Suppose

$$x = \text{the dollar amount of sales,}$$
$$y = \text{the salesman's earnings for the week.}$$

a)   Write an equation that gives y in terms of x.
b)   Complete the table below:

Sales, x	Earnings, y
1000	
2000	
6000	
10000	

c)   Graph the x- and y-values from the table in part (b), and connect them with a straight line.
d)   Use the graph to (visually) estimate the salesman's earnings, if his sales are $12,500. Then calculate the earnings exactly, using the equation in part (a), and compare answers.

Straight-line equations and graphs are applied in the physical sciences, biology, business, the social sciences, and many other areas. In all of these applications, the concepts of **slope** and **intercept** are very important.

Slope and intercept   For an equation of the form $y = b_0 + b_1 x$, the **slope** is b_1, and the **y-intercept** is b_0.

Example **2**   ◀ a)   For the equation $y = 18.95 + 0.23x$ the *slope* $= 0.23 = b_1$, and the *y-intercept* $= 18.95 = b_0$.
  b)   For the equation, $y = 1 - 2x$, the *slope* $= -2 = b_1$, and the *y-intercept* $= 1 = b_0$. ▶

It is very useful to know the *meaning* of the slope and y-intercept for a graph. We will review these next.

i) The **y-intercept**, b_0, gives the y-value at which the line $y = b_0 + b_1 x$ intersects the y-axis (which is found by setting $x = 0$):

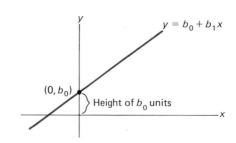

ii) **The slope**, b_1, measures how much the y-value on a straight line moves up (or down) when the x-value increases by one unit:

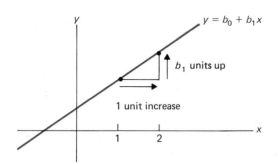

In Example 3, we will interpret the slope and intercept for the car-rental cost problem.

◀ In the car-rental cost problem, the equation is $y = 18.95 + 0.23x$.   Example **3**

i) *y-intercept*: The y-intercept is 18.95. This is the price you pay for the car if you rent it (for one day) and drive 0 miles.

$$x = 0 \rightarrow y = 18.95 + 0.23(0) = 18.95$$

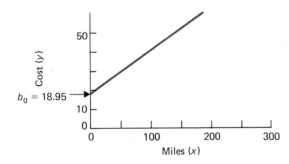

ii) *Slope*: The slope is the amount the graph (that is, the y-value) moves up when x increases by one unit. We already know that when $x = 0$, $y = 18.95$. If we increase x to $x = 1$, y becomes $y = 18.95 + 0.23(1) = 19.18$. We now have found two points on the line:

x	y
0	18.95
1	19.18

We graph the line and these two points on a magnified scale below:

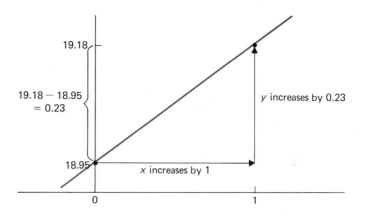

The slope here is 0.23; it is the cost per mile. This makes sense. If you increase x by one unit, you are driving one more mile. This increases your cost by \$0.23. ◗

Exercise **B** Refer back to the sales and earnings graph for the salesman in Exercise A.
a) Find the y-intercept on this graph.
b) What does this y-intercept represent?
c) Find the slope of this graph.
d) What does this slope represent?

We have just seen that, to plot the graph $y = b_0 + b_1 x$, we need only to plot two points and connect them with a straight line. The point $(0, b_0)$ is one point, and the second point may be obtained using any convenient value of x (other than $x = 0$).

Example **4** ◖ Given $b_1 = -2$ and $b_0 = 1$, graph $y = b_0 + b_1 x$. That is, graph $y = 1 - 2x$.

Solution The first step is to graph the y-intercept. See Fig. 12.2.

Figure 12.2

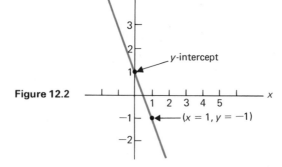

For the second step, if we choose to use $x = 1$, then $y = 1 - 2x = 1 - 2(1) = -1$. These two points are graphed in Fig. 12.2, along with the line. ▶

---

a) Given $b_1 = 3$ and $b_0 = 2$, graph $y = b_0 + b_1 x$.   Exercise **C**
b) Given $b_1 = -1$ and $b_0 = 3$, graph $y = b_0 + b_1 x$.

Note that, in Example 3, where b_1 is positive, the y-values increase as x increases. In Example 4, where b_1 is negative, the y-values decrease as x increases. This is always true. If $b_1 = 0$, the graph is just a horizontal line.

Tell whether each of the equations has a graph that is increasing, decreasing, or   Exercise **D**
horizontal.
a) $y = 1 + 2.7x$
b) $y = -4 + 2.7x$
c) $y = 2 - 5x$
d) $y = -2 - 5x$
e) $y = 4$

EXERCISES Section **12.1**

1 A crop-dusting firm charges a $50 setup fee for each spraying job, and charges an additional $5 per acre for the spraying. Let

$x =$ the number of acres for a spraying job, and
$y =$ the total cost of the job.

a) Write an equation that gives y in terms of x.
b) Construct a table for x and y where $x = 1$, 2, and 3.
c) Graph the x- and y-values from the table in part (b), and connect them with a straight line.
d) Use the graph to visually estimate the cost of a dusting job to cover 4 acres. Then calculate the cost exactly, using the equation in part (a), and compare answers.

2 A company plans to produce hair dryers. The dryers cost $5 each to produce. In addition, the company has a daily administrative and maintenance cost of $2500 for each day of production. Let

$x =$ the number of hair dryers produced in a day, and
$y =$ the total production cost for a day.

a) Write an equation that gives y in terms of x.
b) Construct a table for x- and y-values where $x = 100$, 500, and 1000.
c) Graph the x- and y-values from the table in part (b), and connect them with a straight line.
d) Use the graph to visually estimate the cost of producing 750 dryers. Then calculate the cost exactly, using the equation in part (a), and compare answers.

3 A construction firm purchased a new piece of equipment at the cost of $100,000. The company accountant depreciates the value of the equipment by $10,000 each year. Suppose

$x =$ the age of the equipment (in years), and
$y =$ the value of the equipment after x years.

a) Write an equation that gives y in terms of x.
b) Construct a table where $x = 1$, 4, and 8.
c) Graph the x- and y-values from the table in part (b), and connect them with a straight line.
d) Use the graph to visually estimate the value of the equipment in six years. Then calculate the value exactly, using the equation in part (a), and compare answers.

For the equations in Exercises 4 through 10:
a) find the slope,
b) find the y-intercept, and
c) graph the equation.

4 $y = 3 + 4x$ 5 $y = -1 + 2x$

6 $y = 6 - 7x$ 7 $y = -8 - 4x$

8 $y = \frac{1}{2}x - 2$ 9 $y = -\frac{2}{3}x + 2$

10 $y = -\frac{3}{4}x - 5$

In Exercises 11 through 16, you will be given the slope, b_1, and the y-intercept, b_0. Write the equation in the slope–intercept form, $y = b_0 + b_1 x$, and graph the equation.

11 $b_1 = 2$, $b_0 = 5$ 12 $b_1 = 4$, $b_0 = -3$

13 $b_1 = -3$, $b_0 = -2$ 14 $b_1 = \frac{2}{5}$, $b_0 = 1$

15 $b_1 = -\frac{3}{2}$, $b_0 = -1$ 16 $b_1 = 5$, $b_0 = 0$

Tell whether the straight lines in Exercises 17 through 21 are increasing, decreasing or horizontal.

17 $y = 5 + 4x$ 18 $y = 3$

19 $y = 1 - 3x$ 20 $y = -0.4 - 8x$

21 $y = 2.6x - 9$

22 Refer to the graphs for Exercises 1, 2, and 3 above.
   a) Find the y-intercept.
   b) What does this y-intercept represent?
   c) Find the slope of the straight line.
   d) What does this slope represent?

12.2 Fitting straight-line equations to data

In each example of Section 12.1, all points plotted were on the given straight line. We can also make useful observations and *predictions* if the points we plot are clustered *near* a straight line but are not exactly on it. In the next two examples, we will give some data sets that illustrate this. The data we use relate to production costs. The product should be familiar: it is graded mathematics papers.

Example **5**   ◖ The director of a large mathematics course hires upper-division science students to grade papers for him each week.* On each grading day, he records the number of papers graded, and the total amount of money paid to the graders. Table 12.1 shows his costs, and the numbers of papers graded, for 12 different grading days in the fall of 1978.

Table 12.1	Number of papers graded† (in hundreds), x	Cost of grading (in dollars), y
	22	298
	19	273
	19	265
	18	246
	18	258
	18	251
	17	250
	16	220
	16	227
	16	234
	15	223
	15	210

† We did not arrange these x-values in order. The number of students coming to class drops off through the semester.

---

* This is a common practice on state university campuses. Many freshman and sophomore classes have lecture sections containing from 100 to 300 students, for which a single instructor cannot do all the grading. An instructor with 300 students who spends 10 minutes per week grading each student's papers would grade for 50 hours per week—leaving no time for class.

The first entry in Table 12.1 shows that on the first grading day given here, 2200 papers were graded at a cost of $298. On the last grading day 1500 papers were graded at a cost of $210. These x- and y-values are plotted below in Fig. 12.3.

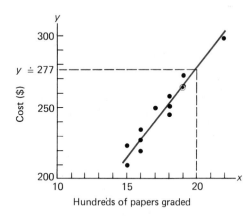

Figure 12.3
Plot of paper-grading data.

The points which we have plotted do not lie on a single straight line, but they are clustered around the line we have drawn. The director could use this line to get a rough idea of the cost of grading 2000 papers by reading up from the x-value of 20 to the line, and then reading off the y-value, as shown in Fig. 12.3. (The x-values represent 100's of papers, so $x = 20$ represents 2000 papers.)

The y-value of 277 that we have read off from the graph is only a rough approximation of the cost for 2000 papers. But that kind of rough estimate can be useful to the course director, if he is required to *predict* costs for next year's budget. ▶

Use Fig. 12.3 to give a rough approximation of the cost of grading 2100 papers.　　Exercise **E**

There is a problem in Example 5—maybe you have noticed it already. It is possible to draw many "reasonable-looking" straight lines through the cluster of points. This raises the question of deciding which straight line is "best"—and what does "best" mean in the first place? Statisticians have answers to both of these questions. Their concept of which line best fits the data is based on analysis of the errors made in using a straight line to fit the data. In simplest terms, an **error** is *the difference between the y-value of the data point and the y-value that the straight line would have predicted for it*. In Fig. 12.4, the errors for three of our data points are shown.

The idea behind finding the best line is to find the line that leads to the least "total error." To illustrate this, we will leave our real data on grading costs and look at a different, simplified example.

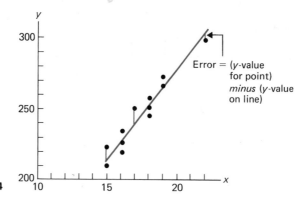

Figure 12.4

Example **6** ◖ The problem here is to fit a straight line to the four data points below, which are graphed at the right.

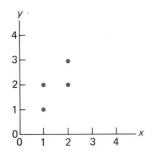

x	y
1	1
1	2
2	2
2	3

Many straight lines could be drawn near these points. Below we picture two of these many lines:

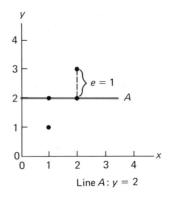

Line $A: y = 2$

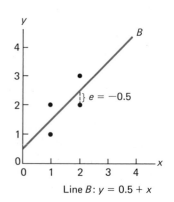

Line $B: y = 0.5 + x$ ◗

The value of y that is *predicted* for a given value of x using a straight line is denoted by $\hat{y}$. (This is pronounced "y-hat.") If we use this notation, *the error, e, can be written as* $y - \hat{y}$. In the tables below we give values of x, y, $\hat{y}$, and e for each of the two lines. We will add the e-values to attempt to measure total error.

In reading these tables keep in mind that $\hat{y}$ is computed using the given straight-line formula. For line A, $\hat{y}$, will always be 2, since the formula is $y = 2$. For line B and $x = 1$, we would use the formula $y = 0.5 + x$ to get $\hat{y} = 0.5 + 1 = 1.5$.

Line A: $y = 2$

x	y	$\hat{y}$	$y - \hat{y} = e$
1	1	2	-1
1	2	2	0
2	2	2	0
2	3	2	1
			$0 = \sum e$

Line B: $y = 0.5 + x$

x	y	$\hat{y}$	$y - \hat{y} = e$
1	1	1.5	-0.5
1	2	1.5	0.5
2	2	2.5	-0.5
2	3	2.5	0.5
			$0 = \sum e$

You can see that positive and negative errors cancel each other when we add up all the e-values. In both cases we get $\sum e = 0$. We cannot use $\sum e$ to decide which line is "better" since $\sum e$ is 0 for both lines. A solution to this problem is to *square* the e-values to make them positive. Adding values of e^2 will give a good measure of total error.

Line A: $y = 2$

x	y	$\hat{y}$	e	e^2
1	1	2	-1	1
1	2	2	0	0
2	2	2	0	0
2	3	2	1	1
			0	$2 = \sum e^2$

Line B: $y = 0.5 + x$

x	y	$\hat{y}$	e	e^2
1	1	1.5	-0.5	0.25
1	2	1.5	0.5	0.25
2	2	2.5	-0.5	0.25
2	3	2.5	0.5	0.25
			0	$1.00 = \sum e^2$

As you can see, the sum of the squared errors is smaller for line B. We say that line B fits the data better than line A, because its *total squared error* is smaller. The quantity $\sum e^2$, which we have calculated, is used in regression analysis and is usually referred to as the **error sum of squares,** written SSE:

$$\text{SSE} = \sum (y - \hat{y})^2 = \sum e^2 = \text{Error sum of squares}$$

The individual e-values are called **residuals.** ◗

Exercise **F**   Given the four data points

x	y
2	2
2	3
3	3
3	4

and the two lines

$$\text{Line } A: \quad y = 3 \quad \text{and} \quad \text{Line } B: \quad y = 0.5 + x$$

a) Graph each line together with the data points.
b) Make a table of values for x, y, $\hat{y}$, e, e^2, and find SSE for each line.
c) Decide which line fits the data better (i.e., which one has smaller total squared error).

Statisticians have shown that, for any data set, there is a **line of best fit** that makes SSE as small as possible. They have also derived formulas for the slope and y-intercept of this line, which is also called the **regression line**. We will not go through the derivations here, but will give the formulas and illustrate their use in the next example.

FORMULA   The **line of best fit (regression line)** for n data points is given by

$$\hat{y} = b_0 + b_1 x$$

where

$$b_1 = \frac{\sum xy - (\sum x)(\sum y)/n}{\sum x^2 - (\sum x)^2/n}$$

and

$$b_0 = \frac{1}{n}\left(\sum y - b_1 \sum x\right)$$

Example **7**   ◖ Find the line of best fit for the points from Example 6.

Step 1.   Make a table for calculation of x, y, xy, x^2 and their sums:

x	y	xy	x^2
1	1	1	1
1	2	2	1
2	2	4	4
2	3	6	4

Sums:

6	8	13	10
↑	↑	↑	↑
$\sum x$	$\sum y$	$\sum xy$	$\sum x^2$

Step 2. Use the sums from this table to find b_1. (Note that $n = 4$.)

$$b_1 = \frac{\sum xy - (\sum x)(\sum y)/n}{\sum x^2 - (\sum x)^2/n} = \frac{13 - (6)(8)/4}{10 - 6^2/4}$$

$$= \frac{13 - 12}{10 - 9} = 1$$

Step 3. Find b_0.

$$b_0 = \frac{1}{n}\left(\sum y - b_1 \sum x\right) = \frac{1}{4}(8 - 1 \cdot 6)$$

$$= \frac{1}{4} \cdot 2 = 0.5$$

Step 4. Write the equation:

$$\hat{y} = b_0 + b_1 x = 0.5 + 1x = 0.5 + x$$

So,

$$\hat{y} = 0.5 + x$$

is the *line of best fit (regression line)*.

The line we have found to be the line of best fit is actually the line B of Example 6. So the SSE value of 1 from that example is the smallest possible value of SSE for any line "fitted" to these four points. ▶

Find the line of best fit (regression line) for the four points in Exercise F:

Exercise **G**

x	y
2	2
2	3
3	3
3	4

For most data sets that you will encounter, determination of the line of best fit will involve much more calculation than we have had to do so far. Fortunately, most people who have large data sets also have access to computers, or to special calculators that can be programmed to do this work. In the next example, we will show how much calculation is required for our original data on paper grading.

Example **8** ◀ Find the line of best fit for the paper-grading data from Example 5.

Step 1. Table for x, y, xy, x^2: (See Table 12.2.)

Table 12.2	Number of papers, x	Cost, y	xy	x^2
	22	298	6556	484
	19	273	5187	361
	19	265	5035	361
	18	246	4428	324
	18	258	4644	324
	18	251	4518	324
	17	250	4250	289
	16	220	3520	256
	16	227	3632	256
	16	234	3744	256
	15	223	3345	225
	15	210	3150	225
Sums:	209	2955	52009	3685

Step 2. Find b_1 ($n = 12$):

$$b_1 = \frac{52009 - \dfrac{(209)(2955)}{12}}{3685 - \dfrac{(209)^2}{12}} = \frac{542.75}{44.917} = 12.083$$

Step 3. Find b_0:

$$b_0 = \frac{1}{12}(2955 - (12.083)(209)) = 35.804$$

Step 4. Write the straight-line equation:

$$\hat{y} = 35.804 + 12.083x$$

This is the regression line.

It will be worthwhile at this point to interpret the *regression line*. The y-value given is a cost in dollars, and the x-value represents the number (of hundreds) of papers graded. Thus the numbers 12.083 and 35.804 must represent dollar figures. We can see this better if we look at the equation of the regression line:

Predicted   Cost per
dollar cost   hundred papers

$$\downarrow \qquad\qquad \downarrow$$

$$\hat{y} = 35.804 \;+\; (12.083)x$$

$$\uparrow \qquad\qquad \uparrow$$

Fixed   Hundreds
cost   of papers

We refer to $12.083 as a cost per hundred papers, since, if we increase the number of papers graded by 100, the predicted cost will go up by $12.083. (If you don't see this, calculate the cost for 200 and 300 papers from the above equation using $x = 2$ and $x = 3$. The predicted cost will go up by $12.083). The figure of $35.804 represents a fixed cost: it is always there, no matter how many papers are graded. Thus the course director gets more information from his line of best fit than simple predictions of y-values; he can also get a rough estimate of his cost per hundred papers, and his fixed cost. ▶

One of the authors collected data on the heights (x) and weights (y) of 11 male college students, aged 18–19 years.   Exercise **H**

a) Complete Table. 12.3.

x	y	xy	x^2	**Table 12.3**
70	185			
65	126			
71	170			
67	140			
74	165			
71	150			
67	130			
69	160			
73	175			
69	140			
73	160			

Sums:

---

b) Find the line of best fit (regression line) for this data.

In each of our previous examples, the slope of the regression line was positive, and consequently the predicted value of y increased as the value of x did. Some regression studies yield negative slopes and decreasing predicted y-values, as the following example shows.

◀ A physician read a report that indicates that the maximum heart rate an individual   Example **9**
can reach during intensive exercise decreases with age. He decided to do his own study, and persuaded ten members of his jogging club to perform exercise tests and record

their highest pulse rate at peak effort. (There are now digital devices that an individual can wear which enable him to read out his pulse rate as he runs.) His results are shown in Table 12.4.

Table 12.4

Age x	Peak heart rate y	xy	x^2
10	210		
20	200		
20	195		
25	195		
30	190		
30	180		
30	185		
40	180		
45	170		
50	165		
Sums: 300	1870	54625	10350

We have left the spaces for individual values of xy and x^2 blank for the reader who wishes additional practice in computation, but we have calculated, that $\sum xy = 54625$ and $\sum x^2 = 10350$. Using the value $n = 10$, we can compute b_1 and b_0.

$$b_1 = \frac{54625 - (300)(1870)/10}{10350 - (300)^2/10} = \frac{-1475}{1350} = -1.093$$

$$b_0 = \frac{1}{10}(1870 - (-1.093)(300))$$

$$= \frac{1}{10}(1870 + 327.9) = \frac{1}{10}(2197.9)$$

$$= 219.79$$

This gives us the regression line equation,

$$\hat{y} = 219.79 - 1.093x$$

In Fig. 12.5 we have graphed our data points and the regression line.

The physician's study supported what he had read. The data points show a clearly decreasing trend for peak heart rate with age. The negative slope of the regression line shows the same trend: y-values on the regression line decrease as x increases. ▶

y

Peak heart rate

210

200

190

180

170

160

150

0 10 20 30 40 50 60 70 x

Age (years)

Figure 12.5

The idea behind finding a regression line (line of best fit) is based on the assumption that the data is really scattered randomly about a single straight line.*In some cases, data points may be scattered about some curve that is *not* a straight line, as shown in Fig. 12.6.

A warning on the use of regression lines

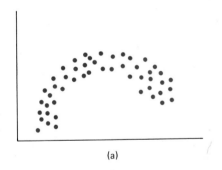

(a)

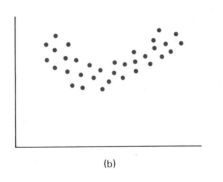

(b)

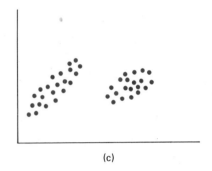

(c)

Figure 12.6

   In Fig. 12.6 (a) and (b), the data is scattered around a curve, not a straight line. In Fig. 12.6 (c), there are two separate groups of data points, each scattered about a different straight line. Unfortunately, the formulas for b_0 and b_1 will work for these data sets, and "fit" a single *inappropriate* straight line to the data. For example, the data from Fig. 12.6 (a), which really follows a curve, would be "fitted" by the straight line shown in Fig. 12.7.

---

* We shall discuss this assumption in detail in Section 12.4, and make it more precise.

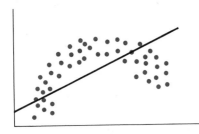

Figure 12.7
Regression line for data
that does not follow a
straight line

This procedure is misleading. It would lead you to predict that y-values will keep increasing in Fig. 12.6 (a), when they have actually begun to decrease. There are more advanced techniques which would enable you to fit *curves* to data points (as should have been done in Fig. 12.7). We will not cover these methods in this book, but any statistical consultant should be able to advise you on such curve fits. The important point is this: If you plan to find a line of best fit, you should first look at a plot of your data. If the data does not appear to be scattered randomly about a line, do not try to use the line-of-best-fit method on it. Consult with a statistician instead, since more advanced methods will probably be necessary.

Exercise I   Which of the following graphs in Fig. 12.8 represent data sets for which a regression line should be computed?

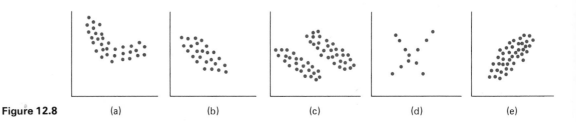

Figure 12.8   (a)   (b)   (c)   (d)   (e)

EXERCISES Section **12.2**

Note. Save your work from this section. You will need it in future sections.

In Exercises 1 and 2:

a) Graph each given line together with the given data points (as in Exercise F).
b) Make a table of values for x, y, $\hat{y}$, e, e^2, and find SSE for each line.
c) Decide which line fits the data with the least total squared error.

1 *Lines A:* $y = 1.75 + 0.25x$
    B: $y = 2$

Points

x	y
1	1
1	3
5	2
5	4

2 *Lines A:* $y = -1 - 2x$
    B: $y = -3x$

Points

x	y
1	−2
1	−4
3	−8
3	−10

3 Find the line of best fit (regression line) for the four points in Exercise 1.

4 Find the line of best fit (regression line) for the four points in Exercise 2.

5 An economist was interested in the relation between the disposable income of a family and the amount of money spent on food. He did a preliminary study on eight randomly selected middle income families. (Each family was of the same size: father, mother, and two children.) His data are given in Table 12.5 (rounded to nearest thousand or hundred for simplicity).

Table 12.5

Family disposable income (in thousands of dollars), x	Food costs (in hundreds of dollars), y
30	55
36	60
27	42
20	40
16	37
24	26
19	39
25	43

a) Find the equation of the regression line for this data.
b) Plot the data points and the line.
c) Describe the apparent relationship between food costs and family income in words.

6 The concession-sales manager for a baseball team initiated a study of the relation between attendance and total sales. He decided to do a pilot study using data from six randomly selected days in the previous season. His results are shown in Table 12.6.

Table 12.6

Attendance (in thousands), x	Total sales (in thousands of dollars), y
10	29
25	72
19	40
42	128
22	64
28	87

a) Find the regression line for this data.
b) Plot the data points and the line.
c) Describe the apparent relationship between attendance and total concession sales.

7 An instructor required each student to retake a logarithm test if the student's grade was below an announced "mastery grade." Table 12.7 gives mean scores for the first try for seven different semesters with various mastery grades.

Table 12.7

Mastery grade (required to avoid retest), x	Mean class grade on test (first try only), y
65	71
70	77
75	70
70	78
75	82
80	84
75	81

a) Find the equation of the regression line for this data.
b) Plot the data points and the line.
c) Describe the apparent relation between required grade and mean score.

8 The following ages and selling prices for Ford pickups were sampled from the classified ads of the *Denver Post* in May 1979. The prices are rounded to the nearest hundred dollars.

Age, x	Price (in hundreds of dollars), y
3	30
1	53
6	18
6	17
3	38
1	59
5	29
6	24
3	40

a) Find the regression line equation.
b) Plot the points and the regression line.
c) What price would you predict to pay for a four-year old Ford pickup?

9 A calculus instructor asked a random sample of ten students to record their study times per lesson in a beginning calculus course. He then made a table for total study times over two weeks, and a test score at the end of the two weeks. Table 12.8 shows the results.

Table 12.8

Study time (in hours), x	Grade (Percentage), y
10	92
15	81
12	84
20	74
8	85
16	80
14	84
22	80
11	91
15	79

a) Find the equation of the regression line.
b) Plot the points and the regression line.
c) Describe the apparent relation between study time and grade. (Does it surprise you?)

10 Which of the following graphs represent data sets for which a regression line should be computed?

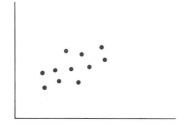

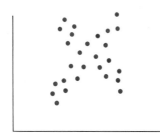

11 a) Refer to Exercise 7. Do you think that the relationship given there would hold if x (the required mastery grade) were raised to nearly 100?
b) Within what range of x-values can you safely say that the relationship found in Exercise 7 applies?
c) Refer to Exercise 8. Give an example of an x-value (age) for which the relationship found there would *not* hold.
d) Within what range of x-values can you safely say that the relationship found in Exercise 8 applies?

12 The negative relation between study time and grade in Exercise 9 has been found by many investigators, and has puzzled them. Can you think of a possible explanation for it?

13 (*Covariance*)   The *sample covariance* of a set of data points is defined by

$$\text{Cov}(x, y) = \frac{\sum (x - \bar{x})(y - \bar{y})}{n - 1}$$

a) Compute the sample covariance for the points in Exercise 1.
b) Compute the sample covariance for the points in Exercise 2.

14 (*Alternative formulas for b_0, b_1*)   Suppose that we use $s(x)$ to stand for the standard deviation of the x-values in a set of data points. The numbers b_0 and b_1 can also be found using the formulas

$$b_1 = \frac{\text{Cov}(x, y)}{[s(x)]^2} \qquad \text{(cf. Exercise 13)}$$

$$b_0 = \bar{y} - b_1 \bar{x}$$

a) Compute b_1 and b_0 for the points in Exercise 1 using these formulas.
b) Compute b_1 and b_0 for the points in Exercise 2 using these formulas.

15 a) Prove that the formulas for b_0 and b_1 given in Exercise 14 are equivalent to the formulas given in the text.
b) Show that the formula for b_1 in Exercise 14 is equivalent to

$$b_1 = \frac{\sum (x - \bar{x})(y - \bar{y})}{\sum (x - \bar{x})^2}$$

In the previous section we found straight lines that related two variables such as height and weight. Many researchers wish only to see whether such a relationship exists, and to determine the *strength* of such a relationship when it does exist. The **correlation coefficient**, r, is a single number that can be used to measure the degree of "straight-line" relationship between two variables. In addition, r determines the precision with which predictions can be made using the regression line. We will show how to calculate the correlation coefficient in Examples 10, 11 and 12. These examples will make more sense if we first discuss the *meaning* of the correlation coefficient.

1. The correlation coefficient, r, is always a number between -1 and $+1$; that is, $-1 \leqslant r \leqslant 1$.

2. If the correlation coefficient for a set of points is exactly 1, *all* of those points *lie on* a straight line with *positive* slope.*

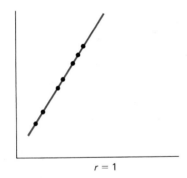

$r = 1$

3. If $0 < r < 1$, then the data points are *scattered around* a straight line of *positive* slope.* Values of r close to 1 indicate that the data points are scattered near the straight line. Values of r farther from 1 indicate that the data points are scattered farther from the line.

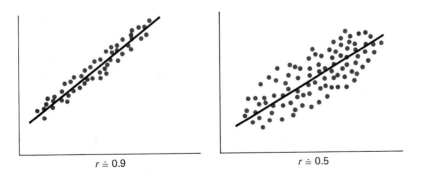

$r \doteq 0.9$ $r \doteq 0.5$

---

* This straight line is the regression line.

When $r > 0$ (i.e., the regression line has positive slope), we say that there is a *positive linear relationship in the data.*

4. If the correlation coefficient is -1, all data points *lie on* a straight line with *negative* slope.*

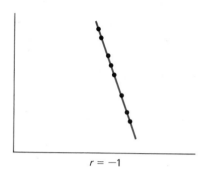

$r = -1$

5. If $-1 < r < 0$, then the data points are *scattered around* a straight line with *negative* slope.* Values of r close to -1 indicate that the data points are scattered near the straight line. Values of r farther from -1 indicate that the data points are scattered farther from the line.

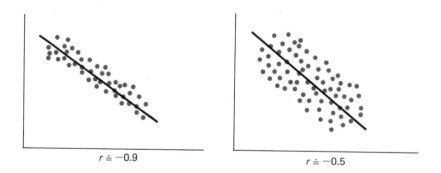

$r \doteq -0.9$ $r \doteq -0.5$

When $r < 0$ (i.e., the regression line has negative slope), we say that there is a *negative linear relationship in the data.*

6. If r is 0, then the data points are scattered around a *horizontal* line,* and therefore don't indicate either a positive or a negative linear relationship.

---

* This straight line is the regression line.

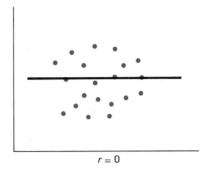

$r = 0$

Researchers are interested in finding *useful* linear relationships between variables. With a *positive linear relationship* ($r > 0$), we can predict that y will *tend to increase* as x increases—e.g., weight will tend to increase as height increases. The closer r is to 1, the more precise the prediction will be. With a *negative linear relationship* ($r < 0$), we can predict that y will *tend to decrease* as x increases—e.g., maximum pulse rate will tend to decrease as age increases. The closer r is to -1, the more precise the prediction will be. In the situation where there is neither a positive nor a negative linear relationship ($r = 0$), no such predictions can be made.

The procedure followed by researchers who are examining variables for useful linear relationships is this: Compute r. If r is 0 or near 0, conclude that there is probably no useful linear relationship. If r is close to ± 1, conclude that there is a useful linear relationship and begin further study. (The question of how "close" r must be to ± 1 to conclude that there is a useful linear relationship will be discussed in Sec. 12.6.)

We are now ready to calculate r for specific data sets. The formula used to find r is as follows:*

The **correlation coefficient**, r, for n data points $(x_1, y_1), (x_2, y_2), \ldots, (x_n, y_n)$ is given by   FORMULA

$$r = \frac{n\sum xy - (\sum x)(\sum y)}{\sqrt{n\sum x^2 - (\sum x)^2} \; \sqrt{n\sum y^2 - (\sum y)^2}}.$$

This formula uses the quantities $\sum x, \sum y, \sum xy$, and $\sum x^2$, which we used in finding regression lines. In addition, it uses the sum $\sum y^2$. The calculation of r will then be very much like the calculation of b_0 and b_1 for a regression line: set up a table, find the necessary sums, and calculate r using these sums and the formula for r. We will not derive this formula, but you will gain a good idea of its workings if you do Exercise 11 of this section.

---

* This is actually the "shortcut" formula for r. The defining formula is:

$$r = \frac{\sum(x - \bar{x})(y - \bar{y})}{\sqrt{\sum(x - \bar{x})^2 \sum(y - \bar{y})^2}}.$$

Example **10** ◀ Find r for the paper-grading data from Example 8.

Solution   We have already found $\sum x$, $\sum y$, $\sum xy$, and $\sum x^2$ in Example 8. We need only add to our work table a column for y^2. The work table is shown in Table 12.9. The reader should remember that the x-, y-, xy- and x^2-columns do not need to be recalculated. They can be copied from Example 8 (see page 448).

Table 12.9	x	y	xy	x^2	y^2
	22	298	6556	484	88804
	19	273	5187	361	74529
	19	265	5035	361	70225
	18	246	4428	324	60516
	18	258	4644	324	66564
	18	251	4518	324	63001
	17	250	4250	289	62500
	16	220	3520	256	48400
	16	227	3632	256	51529
	16	234	3744	256	54756
	15	223	3345	225	49729
	15	210	3150	225	44100
Sums:	209	2955	52009	3685	734653

Using these sums and $n = 12$, we get

$$r = \frac{n\sum xy - (\sum x)(\sum y)}{\sqrt{n\sum x^2 - (\sum x)^2}\sqrt{n\sum y^2 - (\sum y)^2}}$$

$$= \frac{12(52009) - (209)(2955)}{\sqrt{12(3685) - (209)^2}\sqrt{12(734653) - (2955)^2}}$$

$$= \frac{6513}{\sqrt{539}\sqrt{83811}} = \frac{6513}{(23.22)(289.5)} = 0.969$$

This is a high correlation. The r-value of 0.969 is very close to 1. This is exactly what you would expect if you look at the graph of this data and its regression line shown in Fig. 12.9. The regression line has positive slope, and the data points are scattered closely around the regression line. ▶

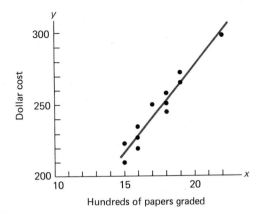

Figure 12.9

Complete Table 12.10, which contains the height and weight data from Exercise H, and   Exercise **J**
find the correlation coefficient r between height and weight.

Height, x	Weight, y	xy	x^2	y^2
70	185			
65	126			
71	170			
67	140			
74	165			
71	150			
67	130			
69	160			
73	175			
69	140			
73	160			

Table 12.10

Sums:

Example 10 and Exercise J illustrate *positive* correlations. In Example 11 below we
return to the data on peak heart rate for an example of a *negative* correlation.

◖ The physician who collected the data on peak heart rate and age wishes to find   Example **11**
the correlation between those two variables. He already knows that $\sum x = 300$,
$\sum y = 1870$, $\sum xy = 54625$, and $\sum x^2 = 10350$. To find r he needs only to find $\sum y^2$. We
repeat his data for x and y in Table 12.11, and include y^2 values.

Table 12.11

Age, x	Peak heart rate, y	y^2
10	210	44100
20	200	40000
20	195	38025
25	195	38025
30	190	36100
30	180	32400
30	185	34225
40	180	32400
45	170	28900
50	165	27225
Sums: 300	1870	351400

Thus,

$$r = \frac{n\sum xy - (\sum x)(\sum y)}{\sqrt{n\sum x^2 - (\sum x)^2}\ \sqrt{n\sum y^2 - (\sum y)^2}}$$

$$= \frac{10(54625) - (300)(1870)}{\sqrt{10(10350) - (300)^2}\ \sqrt{10(351400) - (1870)^2}}$$

$$= \frac{-14750}{\sqrt{13500}\sqrt{17100}} = \frac{-14750}{(116.19)(130.77)} = -0.971.$$

Looking back at the regression line previously calculated for this data we see that this answer makes sense. We have obtained a strong negative correlation, and the data is indeed scattered closely around the regression line of negative slope as shown in Fig. 12.5, page 451. ▶

Exercise **K**   Find the correlation coefficient for the data points given below:

x	y
1	10
2	9
2	8
3	7
4	5
5	3

Then graph the points to see whether your value of r is consistent with the data.

We have now looked at examples resulting in positive and negative values of r. Although you will rarely come up with an r-value that is exactly 0, you will often see values *near* 0. This occurs in the next example.

◀ In Exercise J we looked at the relation between height and weight for eleven students.   Example **12**
There we had good reason to suspect that a relationship would appear. We will now look at data points that we would *not* suspect to be related; height and score on a differential-equations test. (All students were in the same class, which was a top section in an ability-grouped program.) Our data are given in Table 12.12 along with all of the remaining numbers that must be computed.

Height x	Test score y	xy	x^2	y^2	Table 12.12
70	85	5950	4900	7225	
65	83	5395	4225	6889	
71	87	6177	5041	7569	
67	91	6097	4489	8281	
74	84	6216	5476	7056	
71	89	6319	5041	7921	
67	83	5561	4489	6889	
69	87	6003	4761	7569	
73	89	6497	5329	7921	
69	83	5727	4761	6889	
73	83	6059	5329	6889	
Sums: 769	944	66001	53841	81098	

Using this table, along with the fact that $n = 11$, we get

$$r = \frac{n\sum xy - (\sum x)(\sum y)}{\sqrt{n\sum x^2 - (\sum x)^2}\ \sqrt{n\sum y^2 - (\sum y)^2}}$$

$$= \frac{11(66001) - (769)(944)}{\sqrt{11(53841) - (769)^2}\ \sqrt{11(81098) - (944)^2}}$$

$$= \frac{75}{\sqrt{890}\ \sqrt{942}} = \frac{75}{(29.83)(30.69)} = 0.082$$

This correlation is not zero, but is very close to zero. The scatter of the points also indicates that there is very little positive or negative linear relationship in the data.

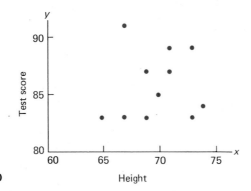

Figure 12.10

In our previous examples, the *y*-values increased or decreased *noticeably* as *x* increased. In this picture, no such thing happens. Test scores fall in roughly the same range no matter what the height values are; being tall doesn't seem to noticeably raise or lower the score range. This is more or less what we expected in advance. We did not expect height and test score in differential equations to show either a positive or negative linear relationship. ▶

We should emphasize that the correlation coefficient, *r*, is used to detect *linear* relationships. If *r* = 0, we can say that there is *no positive or negative linear* relationship in our data, but we cannot say that there is no relationship at all. Suppose, for example, that we are given the following data:

x	y	xy	x^2	y^2
-3	9	-27	9	81
-2	4	-8	4	16
-1	1	-1	1	1
0	0	0	0	0
1	1	1	1	1
2	4	8	4	16
3	9	27	9	81
0	28	0	28	196

Since $n = 7$,

$$r = \frac{7(0) - (0)28}{\sqrt{7(28) - (0)^2} \ \sqrt{7(196) - (28)^2}} = \frac{0}{\sqrt{7(28)} \ \sqrt{7(196) - (28)^2}} = 0$$

There is a great temptation to say that *x* and *y* are unrelated, since *r* = 0. However, if we graph these values, we can see that this is not true.

The points in Fig. 12.11 are related, but not by a straight line. They all lie on the *curve* we have drawn through them, which is the graph of the equation $y = x^2$.

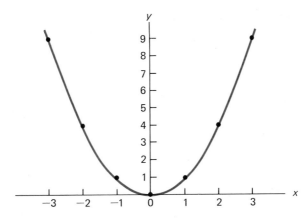

Figure 12.11

It is easy to fall into the trap of feeding data into a computer, reading out values of r and making statements about relationships between two variables *without* looking at a data plot. As the last example shows, this can lead to problems. Whenever possible, you should look at plots of your data to make sure that the straight-line assumption is not badly violated. Most computer programs that calculate r will also do such data plots for you.

The square of r, r^2, is often spoken of as the percentage of variation "explained" by a linear relationship. For example, a mathematics teacher we know designed his own diagnostic test, and correlated diagnostic scores with final-exam grades for 30 students. He came up with a correlation of $r = 0.5$. An educator friend of his looked at this result and said "If $r = 0.5$, $r^2 = 0.25$. So your regression explains only 25 percent of all the variation in your data. That is not very good." To give some idea of the real meaning of this terminology, we will look carefully at a very simple data set in Example 13.

Explained variation

◀ In Example 6, we looked at the data set

Example **13**

x	y
1	1
1	2
2	2
2	3

We found that the regression line for this data set is $\hat{y} = 0.5 + x$, and that the predicted y values, $\hat{y}$, and errors, e, are given by the following table:

x	y	$\hat{y}$	e	e^2
1	1	1.5	-0.5	0.25
1	2	1.5	0.5	0.25
2	2	2.5	-0.5	0.25
2	3	2.5	0.5	0.25
Sums: 6	8	8	0	1.00

We can measure three different kinds of variation for the y values given here:

1. We can compute the usual sum of squares that would be used in finding the variance of the *observed* y-values, $\sum(y - \bar{y})^2$. This is called the **total sum of squares**, or **SST**:

$$\text{SST} = \sum(y - \bar{y})^2$$

In this example, $\bar{y} = \sum y/n = 8/4 = 2$. Thus

$$\begin{aligned} \text{SST} &= \sum(y - \bar{y})^2 \\ &= (1 - 2)^2 + (2 - 2)^2 + (2 - 2)^2 + (3 - 2)^2 \\ &= 1^2 + 0^2 + 0^2 + 1^2 = 2 \end{aligned}$$

The measure of the total variation in our observed y-values is $\text{SST} = 2$.

2. We can measure the variation in the *predicted* y-values on the regression line in the same manner, by computing $\sum(\hat{y} - \bar{y})^2$. This is called the **sum of squares for regression, SSR**:

$$\text{SSR} = \sum(\hat{y} - \bar{y})^2$$

In our example, since $\bar{y} = 2$,

$$\begin{aligned} \text{SSR} &= \sum(\hat{y} - \bar{y})^2 \\ &= (1.5 - 2)^2 + (1.5 - 2)^2 + (2.5 - 2)^2 + (2.5 - 2)^2 \\ &= (-0.5)^2 + (-0.5)^2 + (0.5)^2 + (0.5)^2 \\ &= 0.25 + 0.25 + 0.25 + 0.25 = 1 \end{aligned}$$

The measure of variation in our predicted regression-line values is $\text{SSR} = 1$.

3. Finally, we can measure the amount of *variation due to errors* by computing $\sum e^2$ (that is, $\sum(y - \hat{y})^2$). This quantity is called the **error sum of squares, SSE**:

$$\text{SSE} = \sum(y - \hat{y})^2$$

In our example

$$SSE = \sum (y - \hat{y})^2 = \sum e^2$$
$$= 0.25 + 0.25 + 0.25 + 0.25 = 1$$

The measure of variation due to errors is SSE = 1.

The reader may already have noted that in our example SSE + SSR = 1 + 1 = 2 = SST. This is always the case. That is,

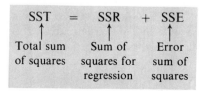

Many people speak of SSR as that part of the total sum of squares (total variation) that is "explained by regression," and SSE as the part that is "unexplained," because it is due to errors:

$$\begin{array}{ccccc} SST & = & SSR & + & SSE \\ \uparrow & & \uparrow & & \uparrow \\ \text{Total sum} & & \text{Part "explained"} & & \text{"Unexplained"} \\ \text{of squares} & & \text{by regression} & & \text{part} \end{array}$$

In our example, we would have

$$\begin{array}{ccccc} 2 & = & 1 & + & 1 \\ \uparrow & & \uparrow & & \uparrow \\ \text{Total sum} & & \text{Explained} & & \text{Unexplained} \end{array}$$

(*Note.* It is simply an accident that SSR and SSE turned out to have the same value in this example. In most cases they will not be equal.)

If we divide through the equation SST = SSR + SSE by SST, we get the equation

$$1 = \frac{SSR}{SST} + \frac{SSE}{SST}$$

or

$$1 - \frac{SSE}{SST} = \frac{SSR}{SST}$$

Portion of variation not explained by regression   Portion of variation explained by regression

In our example, this simply means that

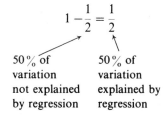

$$1 - \frac{1}{2} = \frac{1}{2}$$

50% of
variation
not explained
by regression

50% of
variation
explained by
regression

The quantity SSR/SST, which represents the portion of total variation explained by regression, is called the **coefficient of determination.** It turns out that the coefficient of determination, SSR/SST, is equal to the *square of the correlation coefficient, r.* That is,

$$r^2 = \frac{SSR}{SST} = \text{Portion of variation explained by regression}$$

In our example, $r^2 = \frac{1}{2}$.

If you take the square root of r^2, of course, you get r as an answer. In our example,

$$r = \sqrt{r^2} = \sqrt{0.5} \doteq 0.707.^*$$

If you compute r using the original formula given on page 457, you will get the same answer; we have just shown you another way to compute r. However, the latter method is not an efficient way to compute r. ◗

Exercise **L**   Given the data:

x	y
1	1
1	3
5	2
5	4

The regression line is $\hat{y} = 1.75 + 0.25x$. This is given; you do not have to find it.
a) Compute SST, SSR, and SSE
b) Verify that SST = SSR + SSE
c) Find the percent of SST "explained" by regression.
d) Find the percent of SST "not explained" by regression.
e) Find the coefficient of determination, r^2.
f) Find r in two different ways.

---

* Actually $r = \pm \sqrt{r^2}$, where the $+$ sign is taken if r is positive, and the $-$ sign is taken if r is negative. The sign of r is always the same as the sign of the slope, b_1, of the regression line. In this case, $b_1 = 1$ (that is, $+1$), so $r = +\sqrt{r^2} = +\sqrt{0.5} = +0.707 = 0.707$.

We can now explain more clearly what was meant when the educator said to the mathematician "If $r = 0.5$, $r^2 = 0.25$. You have explained only 25 percent of the variation." He meant that, if you look at the equation

$$1 = \frac{SSR}{SST} + \frac{SSE}{SST}$$

then, since $SSR/SST = r^2 = 0.25$, we must have $SSE/SST = 0.75$. Thus the *error sum of squares* represents 75 percent of the total variation. Only 25 percent of the total variation is "explained" by the regression line.

EXERCISES Section **12.3**

Note: Save your work from this section. You will need it in future sections.

In Exercises 1 and 2:

a) Find r for the given data set, using the computational formula given on page 457.

b) Compute SSR, SST, and SSE. (You will be given the regression line equation necessary for these calculations.)

c) Compute the portion of variation "explained" by the regression line and the portion that is "unexplained."

d) Compute r using the formula $\pm\sqrt{SSR/SST}$ and compare your results to the answer in (a).

1

x	y
1	2
1	4
5	1
5	3

Regression line: $\hat{y} = 3.25 - 0.25x$

2

x	y
1	-2
1	-4
3	-8
3	-10

Regression line: $\hat{y} = -3x$

In Exercises 3 through 7, find r for the given data, using the computational formula given on page 457.

3 The food cost and family disposable income data (from Exercise 5 of Section 12.2):

Family income, x	Food cost, y
30	55
36	60
27	42
20	40
16	37
24	26
19	39
25	43

4 The sales and attendance data (from Exercise 6 of Section 12.2):

Attendance, x	Sales, y
10	29
25	72
19	40
42	128
22	64
28	87

5 The grade data (from Exercise 7 of Section 12.2):

Mastery grade, x	Mean class grade, y
65	71
70	77
75	70
70	78
75	82
80	84
75	81

6 The car age and price data (from Exercise 8 of Section 12.2):

Age, x	Price, y
3	30
1	53
6	18
6	17
3	38
1	59
5	29
6	24
3	40

7 The study time and grade data (from Exercise 9 of Section 12.2):

Study time, x	Grade, y
10	92
15	81
12	84
20	74
8	85
16	80
14	84
22	80
11	91
15	79

8 Determine whether each of the following sets of points has positive r, negative r, or r equal to zero.

9 a) Find r for the data set given below:

x	y
0	0
1	5
2	8
3	9
4	8
5	5
6	0

b) Is there a positive or negative linear relationship between x and y? (Give a reason for your answer.)

c) Plot the data. Does there appear to be a relationship between x and y?

10 In Exercise 13 of Section 12.2, we defined

$$\text{Cov}(x, y) = \frac{\sum (x - \bar{x})(y - \bar{y})}{n - 1}$$

It can be shown that

$$r = \frac{\text{Cov}(x, y)}{s(x)\,s(y)}$$

a) Use this formula to find r for the data set in Exercise 1, and compare answers.

b) Use this formula to find r for the data set in Exercise 2, and compare answers.

11 (*Interpretation of covariance*) In this problem we will use the following data set:

x	y
1	2.1
2	2.9
3	4.0
4	4.8
5	6.1
6	7.1

a) For this data, $\bar{x} = 3.5$ and $\bar{y} = 4.5$. Below we have given an x, y-coordinate system graph with a second set of axes running through the point (3.5, 4.5). Graph this data on that coordinate system.

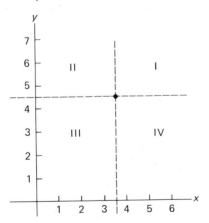

b) The dashed lines above divide the x, y-plane into four regions, which we have labelled I, II, III, and IV. In region I, $x - \bar{x}$ and $y - \bar{y}$ are both positive, so $(x - \bar{x})(y - \bar{y})$ is positive. Fill in the remainder of the table below using similar reasoning.

Region	Sign of $(x - \bar{x})(y - \bar{y})$
I	+
II	
III	
IV	

c) Use the information in (a) and (b) to decide whether $\sum (x - \bar{x})(y - \bar{y})$ will be positive or negative, without calculating that sum.

d) Use a similar graphing procedure to decide whether the covariance will be positive or negative for the following data set:

x	y
1	7.1
2	6.1
3	4.8
4	4.0
5	2.9
6	2.1

e) Below are two data sets, one scattered about a straight line of positive slope and the other scattered about a straight line of negative slope.

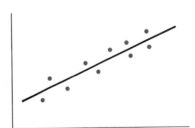

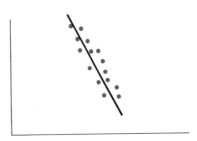

Complete the statements below.

i) For data scattered about a straight line of positive slope, the covariance will be _____.

ii) For data scattered about a straight line of negative slope, the covariance will be _____.

f) Use the relation $r = \mathrm{Cov}(x, y)/s(x)s(y)$ and the reasoning above, to describe in your own words why r is positive for data scattered around a line of positive slope and negative for data scattered around a line of negative slope.

12 Show that the two formulas for r are equivalent, by deriving the algebraic identity

$$\frac{\mathrm{Cov}(x, y)}{s(x)s(y)} = \frac{n\sum xy - (\sum x)(\sum y)}{\sqrt{n\sum x^2 - (\sum x)^2} \ \sqrt{n\sum y^2 - (\sum y)^2}}$$

12.4 The regression model; mean square error

Up to this point we have given only *descriptive statistics* for regression and correlation. We have shown how to find a regression line for a *sample* of heights and weights without making any inferences about the "true" line that relates height and weight for an entire *population*. We have also shown how to find the correlation coefficient for a *sample* of heights and weights without making any statements about the "true" correlation for the *population* of all heights and weights. In other words, up to this point we have seen how to find a *sample regression line* and how to compute a *sample correlation coefficient*.

In the following sections, we will cover the *inferential theory* for regression and correlation. Among other things, we will show how to make inferences concerning the slope of the *population regression line*, and how to test hypotheses about the *population correlation coefficient*.

The inferential methods for regression are based on some assumptions about the entire population. We will go through these assumptions in the next example.

Example 14 ◀ In Example 8 we found that the line of best fit (*sample regression line*) for a data set (sample) of paper grading costs was

$$\hat{y} = 35.804 + 12.083x$$

(see page 448). The data and regression line were shown in Fig. 12.9 (page 459).

We pointed out that the course director could predict future costs for work loads, such as 2000 papers, using this line. But as you can see from the data, you cannot expect such predictions to be completely accurate. For example, for the grading of 1800 papers there are three different costs given: $246, $258, and $251.

This variation should be expected: on different days, different things happen on the same job. Two fast workers (who save you money when they work) are sick; an answer key disappears and people spend time (and money) looking for it; and so on. A large number of random factors make costs vary for jobs of the same size.

Thus, there is a population of *all possible costs* for each job size. In other words, for each x-value (job size), there is a corresponding population of y-values (costs). The first assumption made for the inferential theory is that, for each x-value, the *mean* of the corresponding population of y-values lies on a straight line,

$$y = \beta_0 + \beta_1 x$$

We will refer to this straight line as the **population regression line.**

For example, for $x = 17$ (1700 papers), the mean of the corresponding population of y-values (i.e., the mean cost for grading 1700 papers) is $y = \beta_0 + \beta_1(17)$. For $x = 20$ (2000 papers), the mean cost is $y = \beta_0 + \beta_1(20)$. This is illustrated in Fig. 12.12

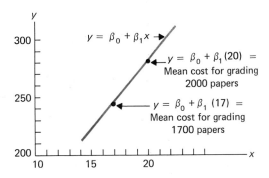

Figure 12.12
Population regression line.

In general, the population regression line is *not* known; and, in fact, one of the main reasons for finding the line of best fit (sample regression line) for the data is to *estimate* the population regression line. Of course, the sample regression line $\hat{y} = b_0 + b_1 x$ (in this case, $\hat{y} = 35.804 + 12.083x$) will ordinarily not be the same as the population regression line $y = \beta_0 + \beta_1 x$. (This is similar to the fact that a sample mean, $\bar{x}$, will usually not equal the population mean, μ.) We picture the situation in Fig. 12.13.

The dotted line in Fig. 12.13 is the *population* regression line. The solid line is the best approximation we can make to the population regression line using our sample data.

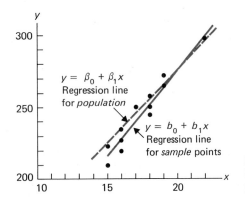

Figure 12.13
Regression lines for population and sample.

The next assumptions for the inferential theory concern the *distribution* of the population of y-values for each x-value. For example, consider the population of y-values (costs) for grading 1700 papers (that is $x = 17$). Since these costs vary because of a large number of random factors, *statisticians assume that they are distributed according to a normal distribution*. The mean of this normal distribution will be the regression-line value, $y = \beta_0 + \beta_1(17)$. The distribution is pictured in Fig. 12.14, which magnifies a small portion of the complete graph.

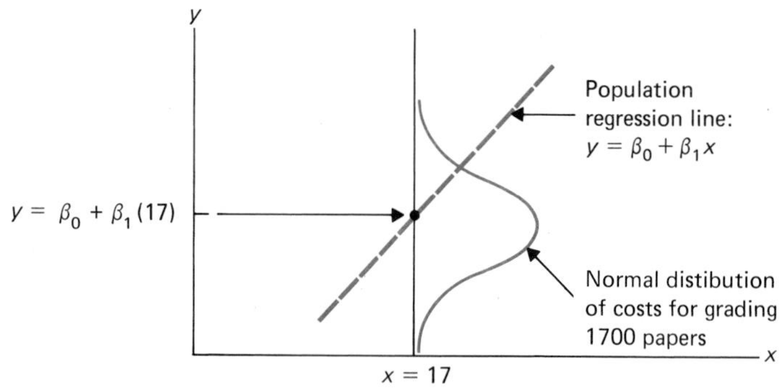

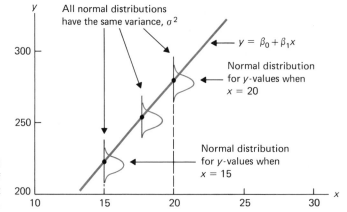

Figure 12.14
Normal distribution of
possible costs for
grading 1700 papers.

(It would be nice to be able to give a numerical value for the mean cost, $y = \beta_0 + \beta_1(17)$, but we don't know β_1 and β_0.)

The second and third major assumptions for our regression inferences are that the kind of normal distribution occurring at $x = 17$ will also appear at every other value of x. That is, for each x-value, the distribution of the population of y-values is a normal distribution, *and* the variance σ^2 of that normal distribution is the same for every x. This is shown in Fig. 12.15.

Figure 12.15
Normal distributions of
population y-values about
regression line.

You should recall that the variance, σ^2, indicates the shape of the normal distribution curve. We have shown identical normal curves in Fig. 12.15 because these curves all have the same variance—and so the *same shape.* ▶

The three assumptions we have considered above are the basic ones for the inferential methods we need.

Assumptions for
regression inferences

I. *Regression line*: For each x-value, the *mean* of the corresponding population of y-values is $\beta_0 + \beta_1 x$.
II. *Normality*: For each x-value, the distribution of the corresponding population of y-values is normal.
III. *Equal variances*: The variances, σ^2, of these normal distributions are identical.

In other words, Assumptions I–III state that, for each x-value, the corresponding population of y-values is normally distributed with mean $\beta_0 + \beta_1 x$, and variance σ^2.

These three assumptions can be rephrased if we consider error terms. For any y-value in the population, the vertical distance between the y-value on the (population) regression line and the population y-value is called an *error term* (written using the Greek letter ε). (See Fig. 12.16.)

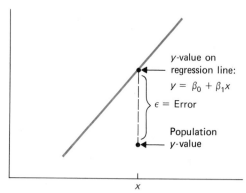

y-value on regression line:
$y = \beta_0 + \beta_1 x$

$\varepsilon = $ Error

Population y-value

Figure 12.16
Magnified view of an error term from the paper-grading data.

Saying that, for each x-value, the corresponding population of y-values is normally distributed is the same as saying that the error terms, ε, are normally distributed. The relationship between the population y-values and the population regression line can be written as:

$$y \;\; = \;\; \beta_0 + \beta_1 x \;\; + \;\; \varepsilon$$

population y-value | population regression line value | error term

Using this equation, referred to as the **regression model**, we can rephrase Assumptions I–III simply by saying that the *error terms, ε, are normally distributed with mean 0 and the same variance σ^2.*

◀ We have referred to the paper-grading data in this discussion. Below we summarize the application of the basic regression concepts and assumptions to the paper-grading data problem.

Example **15**

a) *Population:* All *possible* pairs of values (x, y), where x is a number of papers to be graded (in hundreds) and y is the cost of grading the papers.

b) *Assumptions:*

  I. *Regression line*: For each x-value (number of hundreds of papers), the mean cost for grading is $\beta_0 + \beta_1 x$.

  II. *Normality*: For any given x-value (number of hundreds of papers), the corresponding population of y-values (costs) is normally distributed.

  III. *Equal variances*: These normal distributions are identical in shape for every x (that is, they all have the same variance, σ^2).

c) *Picture illustrating these assumptions*: Fig. 12.15.

d) *Regression model:* The number of papers, x, and cost, y, are related by the equation:

$$y = \beta_0 + \beta_1 x + \varepsilon$$

where ε represents a normally distributed error term with mean 0 and variance σ^2.

Exercise **M** In Example 9 (page 449) we gave data on the age and peak heart rate of ten joggers.
a) Describe the population from which these sample data were taken.
b) What basic assumptions must be made about this population if we are to use standard inferential methods on our sample data?
c) Draw a picture (similar to Fig. 12.15) to illustrate those assumptions.
d) State the *regression model* for this problem.

There are methods for checking the regression-model assumptions. At the level of this book, it is difficult to check Assumption II (normality). However, "rough" checks can be made on Assumptions I and III by simply looking at data plots to see if there are any obvious departures from the model. We have already noted in Section 12.2 that you should look to see if the data is scattered around a single straight line. This check should take care of the regression-line assumption (Assumption I, page 472). The most obvious violation of the equal-variance assumption (Assumption III, page 472) would occur when the data is scattered, as in Fig. 12.17(a). There are great deviations in the width of the scatter, and you can see from Fig. 12.17(b) that the normal distributions for scatter (error) at different x-values have *different* variances.

Figure 12.17
Data for which the scatter distributions do not have the same variance (violation of Assumption III)

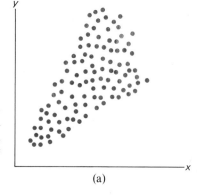

(a)

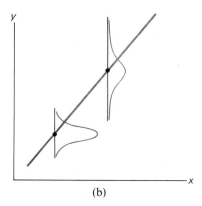

(b)

Exercise **N** Which of the data sets in Fig. 12.18 appear to come from populations that satisfy all of the assumptions of the regression model?

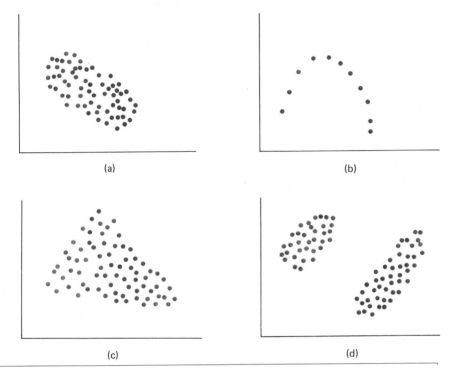

(a)   (b)

(c)   (d)

Figure 12.18

There are methods that can be used to deal with data sets that do not satisfy the Assumptions I–III. That is, data can still be analyzed when the assumptions are not satisfied, but you need to apply more sophisticated methods than those covered in this book. The data sets we have used in this chapter show no obvious violation of the Assumptions I–III. This will be true of many data sets that you consider, but *it always pays to look at a plot of your data and check that you are not about to do something inappropriate.*

In the paper-grading illustration, we mentioned that, for each x-value, the population of y-values has the same variance, σ^2 (Assumption III). Although we cannot know this variance exactly without knowing the entire population, we can estimate it from sample data. (This is similar to the situation for β_0 and β_1. We cannot know them exactly, but we can estimate them by finding b_0 and b_1.) To estimate σ^2, we find the error sum of squares SSE and divide by $n-2$. The quantity obtained is called the **mean square error, MSE**.

Mean square error

$$\text{MSE} = \frac{\text{SSE}}{n-2} = \frac{\sum(y - \hat{y})^2}{n-2} = \frac{\sum e^2}{n-2} *$$

---

* The shortcut formula for computing MSE will be discussed in the exercises.

Example **16** ◀ Estimate σ^2 for the paper-grading cost problem.

Solution To estimate σ^2, we must find values of $\hat{y}$ and then of $(y - \hat{y})^2$. We have already found the (sample) regression line for our paper-grading data; it is $\hat{y} = 35.804 + 12.083x$. For any x-value, we will find $\hat{y}$ by substituting x into this equation; for example, if $x = 18$, we find that $\hat{y} = 35.804 + 12.083(18) = 253.30$. The original x, y-values and the other numbers we need are displayed in Table 12.13.*

Table 12.13

x	y	$\hat{y}$	$y - \hat{y} = e$	e^2
22	298	301.63	-3.63	13.18
19	273	265.38	7.62	58.06
19	265	265.38	-0.38	0.14
18	246	253.30	-7.30	53.29
18	258	253.30	4.70	22.09
18	251	253.30	-2.30	5.29
17	250	241.22	8.78	77.09
16	220	229.13	-9.13	83.36
16	227	229.13	-2.13	4.54
16	234	229.13	4.87	23.72
15	223	217.05	5.95	35.40
15	210	217.05	-7.05	49.70
			0	425.86 = SSE

Since $n = 12$, we find that

$$\text{MSE} = \frac{\text{SSE}}{n-2} = \frac{425.86}{12-2} = \frac{425.86}{10} = 42.586$$

This means that our estimate of σ^2 is 42.586, and our estimate of σ is $\sqrt{42.586} = 6.526$.

We have calculated $\sum e$ (sum of the fourth column) in this example only as a check. The theory says that $\sum e = 0$. Rounding errors may give values for $\sum e$ like ± 0.01 or ± 0.02 for some data sets; such values are close enough to 0. But if you get $\sum e = 6$, or $\sum e = -20$, check your work. Such values are too far away from 0. ◗

The estimates of σ^2 and σ, obtained from MSE, can give us some idea of how the population is spread around its own regression line. Even more important, the quantity MSE is used in each of the regression inference formulas that we will give in the next section.

---

* In the calculations, we used the three-place numbers 12.083 and 35.804 to start, but rounded each answer to two-places. If you check this on a calculator and do not round, your values will be slightly different.

In Table 12.14 we give the (sample) regression-line equation and data for age and peak   Exercise **O**
heart rate. Complete the table, find MSE, and use it to estimate σ^2 and σ.

Table 12.14

Age, x	Heart rate, y	$\hat{y}$	e	e^2
10	210			
20	200			
20	195			
25	195			
30	190			
30	180			
30	185			
40	180			
45	170			
50	165			

Sample regression line equation: $\hat{y} = 219.79 - 1.093x$

EXERCISES Section **12.4**

Note: Save your work from this section. You will need it in future
sections.

In Exercises 1 through 4:

a) Describe the population from which the given sample data was
   taken.

b) State the basic assumptions that must be made about this
   population if standard inferential methods are to be used.

c) Draw a picture to illustrate these assumptions.

d) State the *regression model* for this problem.

1 The data on family disposable income and food costs in
   Exercise 5 of Section 12.2.

2 The data on attendance and concession sales in Exercise 6 of
   Section 12.2.

3 The data on age and selling price for Ford pickups in Exercise 8
   of Section 12.2.

4 The data on study time and grade in Exercise 9 of Section 12.2.

5 Which of the following data sets appear to be from populations
   that satisfy all of the assumptions of the regression model?

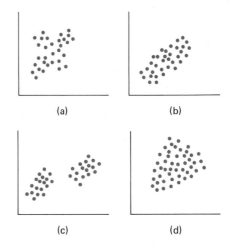

(a)   (b)

(c)   (d)

In Exercises 6 through 10 below:
a) Find MSE for the given data. (The sample regression line is given.)
b) Estimate the variance around the regression line, σ^2, for the population from which the data was sampled.

6

x	y
1	1
1	3
5	2
5	4

Regression line $\hat{y} = 1.75 + 0.25x$

7

x	y
1	-2
1	-4
3	-8
3	-10

Regression line $\hat{y} = -3x$

8 (*Data from Exercise 5, Section 12.2*) $\hat{y} = 12.86 + 1.21x$

Family disposable income, x	Food cost, y	$\hat{y}$	e	e^2
30	55			
36	60			
27	42			
20	40			
16	37			
24	26			
19	39			
25	43			

9 (*Data from Exercise 6, Section 12.2*) $\hat{y} = -9.63 + 3.27x$

Attendance, x	Sales, y	$\hat{y}$	e	e^2
10	29			
25	72			
19	40			
42	128			
22	64			
28	87			

10 (*Data from Exercise 8, Section 12.2*) $\hat{y} = 59.95 - 6.81x$

Age x	Price y
3	30
1	53
6	18
6	17
3	38
1	59
5	29
6	24
3	40

11 (*Shortcut formula for SSE*) You may also compute SSE using the shortcut formula

$$SSE = \sum y^2 - b_0 \left(\sum y\right) - b_1 \left(\sum xy\right)$$

Use this formula to find

a) MSE for the data in Exercise 7;

b) MSE for the data in Exercise 9.

12 (*Analysis of residuals*) One method that is often used in checking the appropriateness of the regression model is to graph the error terms e with their x-values. This is referred to as a *residual plot* since the error terms e are called residuals.

a) Using your work from Exercise 10, plot the residuals e with their associated x-values on a graph labeled like the one below:

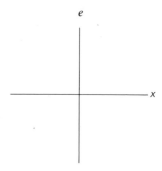

b) Compute the residuals e for the data given below. The regression line is $\hat{y} = 4$.

x	y	$\hat{y}$	e
1	9		
2	4		
3	1		
4	0		
5	1		
6	4		
7	9		

c) Using axes similar to those in (a), plot these residuals with their associated x-values.

d) You now have two data sets. Which one looks as if it came from a population satisfying the regression model? Which one doesn't?

e) What is the difference between the residual plots for the two data sets?

f) Explain in your own words how you would use residual plots to decide whether or not the regression-model assumptions are satisfied.

In this section we will learn how to do hypothesis tests concerning the slope, β_1, of the (population) regression line. Specifically, we will learn how to test the null hypothesis $\beta_1 = 0$ against the alternative $\beta_1 \neq 0$. We will also learn how to find intervals for predictions by using the (sample) regression line. There are other inferential procedures, but these two are the most important. The test of $\beta_1 = 0$ enables us to see if there is a positive or negative linear relationship between our variables, and the prediction intervals enable us to make reasonable predictions if such a relationship exists.

As you can see from Fig. 12.19, it is only when $\beta_1 \neq 0$ that the regression line is of any special value in making predictions.

Inferences in regression   **12.5**

Testing the hypothesis $\beta_1 = 0$

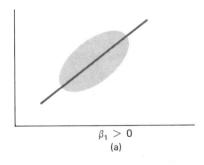

$\beta_1 > 0$
(a)

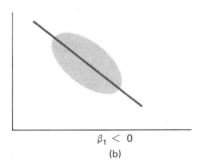

$\beta_1 < 0$
(b)

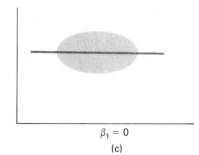

$\beta_1 = 0$
(c)

Figure 12.19
Populations with positive, negative, and zero β_1 values

When $\beta_1 > 0$, we can predict increases in y, and when $\beta_1 < 0$ we can predict decreases. When $\beta_1 = 0$, there is *no* increase or decrease to predict. In order to test the hypothesis $\beta_1 = 0$ against the alternative $\beta_1 \neq 0$, we will follow the usual pattern. Compute the value of a statistic and reject the null hypothesis if that value is "too large" or "too small." The statistic used here is:

$$t = \frac{b_1}{s(b_1)}$$

You already know how to compute b_1 from sample data (see page 446). The quantity $s(b_1)$, the sample standard deviation of b_1, is new, but it can be found using quantities already calculated:

$$s(b_1) = \sqrt{\frac{\text{MSE}}{\sum (x - \bar{x})^2}} = \sqrt{\frac{\text{MSE}}{\sum x^2 - (\sum x)^2/n}}$$

where $\text{MSE} = [\sum (y - \hat{y})^2]/(n - 2)$; see page 475.

The procedure will be to calculate b_1 and $s(b_1)$, and find t. To see whether t is "too large" or "too small," we use the following fact:

Let β_1 be the slope of the *population* regression line, $y = \beta_0 + \beta_1 x$, and let b_1 be the slope of the *sample* regression line, $y = b_0 + b_1 x$, based on n data pairs. Then the random variable

$$t = \frac{b_1 - \beta_1}{s(b_1)}$$

has the *t-distribution with* d.f. $= n - 2$.

In particular, then, *if the null hypothesis $\beta_1 = 0$ is true*, the random variable

$$t = \frac{b_1}{s(b_1)}$$

has the *t-distribution with* d.f. $= n - 2$. This test is very much like the t-test done in Chapter 8. Looking at $(b_1 - \beta_1)/[s(b_1)]$ is similar to looking at $(\bar{x} - \mu)/[s/\sqrt{n}]$ in a t-test for a population mean.

Example **17** ◀ In Example 13 we looked at the data set

x	y
1	1
1	2
2	2
2	3

Suppose we are asked to perform the hypothesis test:

Null hypothesis: $\beta_1 = 0$
Alternative hypothesis: $\beta_1 \neq 0$

We found in Example 13 (page 463) that

$$n = 4, \quad \text{SSE} = 1, \quad \sum x = 6$$

Also $\sum x^2 = 1^2 + 1^2 + 2^2 + 2^2 = 10$. Using these values we find that

$$\text{MSE} = \frac{\text{SSE}}{n-2} = \frac{1}{2} = 0.5.$$

$$\sum(x-\bar{x})^2 = \sum x^2 - \frac{(\sum x)^2}{n} = 10 - \frac{36}{4} = 1$$

Consequently,

$$s(b_1) = \sqrt{\frac{\text{MSE}}{\sum(x-\bar{x})^2}} = \sqrt{\frac{0.5}{1}} = \sqrt{0.5} = 0.707$$

In Example 7 we found that $b_1 = 1$ for this data, so

$$t = \frac{b_1}{s(b_1)} = \frac{1}{0.707} \doteq 1.41$$

If we wish to perform a test at the level of $\alpha = 0.05$, we need to look at the t-distribution for $n-2 = 4-2 = 2$ degrees of freedom. Our alternative $\beta_1 \neq 0$ gives us a two-tailed test, so with $\alpha = 0.05$ we need to find the t-values that leave areas of 0.025 on either end of the t-curve for d.f. $= 2$. Table II shows that these values are $t_{0.025} = 4.30$ and $-t_{0.025} = -4.30$. This is pictured in Fig. 12.20.

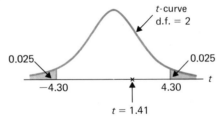

Figure 12.20

The t-value we computed is $t = 1.41$, which is in the acceptance region. We cannot reject the null hypothesis $\beta_1 = 0$. ▶

Procedure for performing a hypothesis test of the form:

Procedure

Null hypothesis: $\beta_1 = 0$
Alternative hypothesis: $\beta_1 \neq 0$

Assumptions: Assumptions I–III (page 472).

1. Decide on the significance level, α.
2. Find the degrees of freedom, d.f. $= n - 2$, where n is the number of data pairs (x, y) in the sample.
3. Using the t-distribution with d.f. $= n - 2$, the critical values are

$$-c = -t_{\alpha/2} \quad \text{and} \quad c = t_{\alpha/2}.$$

Use Table II to find $t_{\alpha/2}$.

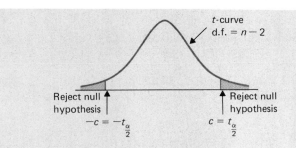

4. Compute

$$t = \frac{b_1}{s(b_1)},$$

where

$$s(b_1) = \sqrt{\frac{\text{MSE}}{\sum(x - \bar{x})^2}} \quad \text{and} \quad \text{MSE} = \frac{\sum(y - \hat{y})^2}{n - 2}$$

5. If either $t < -c$ or $t > c$, *reject* the null hypothesis.

Example **18**   ◖ Perform a test of the hypotheses:

Null hypothesis: $\beta_1 = 0$
Alternative hypothesis: $\beta_1 \neq 0$

on the paper-grading cost data given in Example 5. Use the results of previous computations whenever possible. Take $\alpha = 0.01$.

Solution   *Step 1.* Significance level: given as $\alpha = 0.01$.

Step 2. For that data set, $n = 12$. So d.f. $= 12 - 2 = 10$.

Step 3. The critical values are

$$-t_{0.01/2} = -t_{0.005} = -3.17 \quad \text{and} \quad t_{0.01/2} = t_{0.005} = 3.17.$$

Step 4. We must compute

$$t = \frac{b_1}{s(b_1)}$$

In Example 16 (page 476) we found

$$\text{MSE} = 42.586.$$

In Example 8 (page 448) we found

$$\sum x = 209, \qquad \sum x^2 = 3685.$$

Therefore

$$\sum(x - \bar{x})^2 = \sum x^2 - \frac{(\sum x)^2}{n} = 3685 - \frac{(209)^2}{12} = 44.917,$$

and

$$s(b_1) = \sqrt{\frac{MSE}{\sum(x - \bar{x})^2}} = \sqrt{\frac{42.586}{44.917}} = \sqrt{0.948} = 0.974.$$

In Example 8 we also found

$$b_1 = 12.083.$$

Consequently

$$t = \frac{b_1}{s(b_1)} = \frac{12.083}{0.974} \doteq 12.41.$$

Step 5. Since $12.41 > 3.17$, we reject the null hypothesis and conclude $\beta_1 \neq 0$. ◗

This example seemed quite short, but it relied on many computations previously done in this chapter. *When doing regression analyses, save your work sheets.* Many values you compute will be used again.

Perform a test with

Null hypothesis: $\beta_1 = 0$
Alternative hypothesis: $\beta_1 \neq 0$

on the age and heart-rate data given in Exercise O. The quantities you will need to use have already been calculated in Exercise O and Example 9. They are $n = 10$, $\sum x = 300$, $\sum x^2 = 10350$, $b_1 = -1.093$, and MSE $= 12.3$ (Use $\alpha = 0.05$.)

Exercise **P**

We introduced the (sample) regression line as a useful tool for making predictions. For example, we pointed out that the course director who was reviewing the paper-grading cost data might like to use the (sample) regression line to predict the cost of grading 2000 papers. We have seen that the cost of grading 2000 papers varies (i.e., it is a random variable). Thus, it makes more sense to establish a prediction interval for the cost of grading 2000 papers, than to give a single predicted value.* We can find such a prediction interval by using numbers most of which have already been computed. In our computations we will need to specify

Prediction intervals

$$x_0 = \text{the } x\text{-value for which a prediction is required.}$$

---

* Prediction intervals are similar to confidence intervals. The term "confidence" is usually reserved for intervals involving population parameters such as μ and σ. The term "prediction" is used when considering intervals for the value of a random variable (in this case, the cost of grading 2000 papers).

For the paper-grading supervisor, $x_0 = 20$. (Remember that his x-values are measured in hundreds of papers, and he needs a cost prediction for $2000 = 20(100)$ papers.)

Procedure

Procedure for finding prediction intervals for the y-values when $x = x_0$:

1. Find $n - 2$. (This will be the degrees of freedom for a t-distribution.)
2. If the desired prediction level is $1 - \alpha$, use Table II and d.f. $= n - 2$ to find $t_{\alpha/2}$.
3. Find the predicted value

$$\hat{y} = b_0 + b_1 x_0$$

4. The desired prediction interval extends from

$$\hat{y} - t_{\alpha/2} \sqrt{\text{MSE}} \sqrt{1 + \frac{1}{n} + \frac{(x_0 - \bar{x})^2}{\sum x^2 - (\sum x)^2/n}}$$

to

$$\hat{y} + t_{\alpha/2} \sqrt{\text{MSE}} \sqrt{1 + \frac{1}{n} + \frac{(x_0 - \bar{x})^2}{\sum x^2 - (\sum x)^2/n}}$$

Example **19** ◖ Find a 95-percent prediction interval for the cost of grading 2000 papers, using the data given in Example 5 (page 442) and the results of previous examples.

Solution *Step 1.* $n - 2 = 12 - 2 = 10$

Step 2. The desired prediction level is $1 - 0.05$, so $\alpha = 0.05$.

For d.f. $= 10$, $t_{\alpha/2} = t_{0.025} = 2.23$.

Step 3. $x_0 = 20$, since we want the cost of 2000 papers. The regression line equation, derived in Example 8 (page 448), is $\hat{y} = 35.804 + 12.083x$. The predicted y-value is therefore

$$\hat{y} = 35.804 + (12.083)(20) = 277.464.$$

Step 4. To use our prediction-interval formula, we need

$$\text{MSE} = 42.586 \qquad \text{(from Example 16)}$$

$$\sum x^2 - \frac{(\sum x)^2}{n} = 44.917 \qquad \text{(from Example 18)}$$

$$\bar{x} = \frac{\sum x}{n} = \frac{209}{12} = 17.417 \qquad \text{(from Example 5)}$$

$$(x_0 - \bar{x})^2 = (20 - 17.417)^2 = (2.583)^2 = 6.672$$

Thus, our prediction interval extends from

$$\hat{y} - t_{\alpha/2} \sqrt{\text{MSE}} \sqrt{1 + \frac{1}{n} + \frac{(x_0 - \bar{x})^2}{\sum x^2 - (\sum x)^2/n}}$$

$$= 277.464 - 2.23 \sqrt{42.586} \sqrt{1 + \frac{1}{12} + \frac{6.672}{44.917}}$$

$$= 277.464 - (2.23)(6.526)\sqrt{1.232}$$

$$= 277.464 - (2.23)(6.526)(1.110)$$

$$= 261.31$$

to

$$\hat{y} + t_{\alpha/2} \sqrt{\text{MSE}} \sqrt{1 + \frac{1}{n} + \frac{(x_0 - \bar{x})^2}{\sum x^2 - (\sum x)^2/n}}$$

$$= 277.464 + (2.23)(6.526)(1.110)$$

$$= 293.62.$$

(Note that, in the second computation, we did not rework the term of the formula with the radical, since we had already calculated it.)

The 95-percent prediction interval for the cost of grading 2000 papers extends from $261.31 to $293.62. That is, we can be 95 percent certain that the cost of grading 2000 papers will be somewhere between $261.31 and $293.62. ▶

Find 95-percent prediction intervals for the costs of grading   Exercise **Q**
a) 1700 papers
b) 1500 papers

If you look at the lengths of the prediction intervals in Example 19 and Exercise Q, you will see that the intervals are narrower for $x_0 = 1700$ papers than for 1500 or 2000 papers. It is always the case that the prediction intervals are narrowest nearest to the center of the point cluster, and widen out as x_0 is taken farther away from the center. More precisely, you will get the smallest prediction interval when $x_0 = \bar{x}$. The prediction intervals will get wider as you move away from $\bar{x}$ in either direction. This pattern is illustrated in Fig. 12.21.

There are also confidence-interval formulas for the mean value of y when $x = x_0$. These are given in Exercise 12 on page 487.

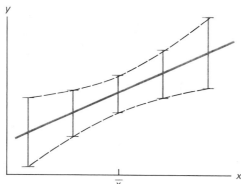

Figure 12.21
Pattern of increase in size of
prediction intervals for predicted
values of *y*. Prediction invervals
are shown by vertical line segments.

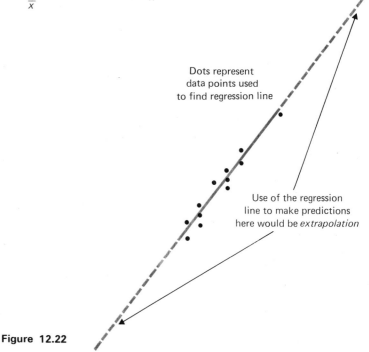

Dots represent
data points used
to find regression line

Use of the regression
line to make predictions
here would be *extrapolation*

Figure 12.22

Extrapolation   It is important to warn here against the danger of **extrapolation** in regression. Extrapolation means that we use the regression line to make predictions for new *x*-values that are far removed from our original data. For example, the data we used to find a regression line for paper-grading costs involved only numbers of papers between 1500 and 2200. It would be *extrapolation* to use the regression line we found to estimate the costs of grading as few as 100 papers, or as many as 100,000 papers. Each of these new tasks would require changes in staffing and management. These tasks are too different from the jobs used to provide data for the regression line. The regression line should *not* be used to make predictions for very large or very small jobs. We illustrate this in Fig. 12.22.

In Exercise 1 through 5, use the data indicated to test the null hypothesis "$\beta_1 = 0$" against the alternative hypothesis $\beta_1 \neq 0$ at the indicated significance level.

1 (*Data from Exercise* 1 *of Section* 12.2)

x	y	
1	1	Regression line: $\hat{y} = 1.75 + 0.25x$
1	3	
5	2	
5	4	

Use $\alpha = 0.05$.

2 Use the family disposable income and food data from Exercise 5 of Section 12.2. (MSE was calculated in Exercise 8 of Section 12.4.) Use $\alpha = 0.05$.

3 Use the attendance and sales data from Exercise 6 of Section 12.2. (MSE was calculated in Exercise 9 of Section 12.4.) Use $\alpha = 0.01$.

4 Use the age and price data from Exercise 8 of Section 12.2. (MSE was calculated in Exercise 10 of Section 12.4.) Use $\alpha = 0.01$.

5

x	y
1	2.4
2	2.7
3	4.0
4	4.9
5	6.2
6	7.7
7	8.1
8	8.9

Use $\alpha = 0.05$.

6 Find a 95-percent prediction interval for the food costs of a family with a disposable income of $22,000, using the data given in Exercise 5 of Section 12.2 and the value of MSE calculated in Exercise 8 of Section 12.4.

7 Find a 99-percent prediction interval for the sales at a baseball game with attendance of 35,000, using the data given in Exercise 6 of Section 12.2 and the value of MSE calculated in Exercise 9 of Section 12.4.

8 Find a 99-percent prediction interval for the price of a four-year-old Ford pickup, using the data given in Exercise 8 of Section 12.2 and the value of MSE calculated in Exercise 10 of Section 12.4.

9 Use the data from Exercise 5 to find a 95-percent prediction interval for y, given that $x = 3$.

10 Find a 95-percent prediction interval for the cost of grading 1900 papers, using the data given in Example 5 of Section 12.2. (Most of the numbers you need will be found in Example 19 of this section.)

11 a) Use the data from Exercise 5 to find 95-percent prediction intervals for y for each of the x-values 1, 2, 3, 4, 5, 6, 7, 8. (The interval for $x = 3$ was found in Exercise 9.)

b) Graph these prediction intervals around the regression line to obtain a figure similar to Fig. 12.21).

c) Calculate and graph 99-percent prediction intervals for each of the x-values given in (a).

d) What is the effect of increasing the prediction level to 99-percent?

12 (*Confidence intervals for means*). In this section we gave a formula for a prediction interval for an individual y-value given that $x = x_0$. This would enable us to estimate with 95-percent confidence the food costs of an individual family with a disposable income of $22,000, or the price of an individual Ford pickup that was four years old. One can also estimate the mean value of y, given x: What is a 95-percent confidence interval for the mean food cost of a family with income of $22,000, or for the mean price of a four-year-old Ford pickup? For a given value of $x = x_0$, a $(1 - \alpha)$-level confidence interval extends from

$$\hat{y} - t_{\alpha/2} \sqrt{\text{MSE}} \sqrt{\frac{1}{n} + \frac{(x_0 - \bar{x})^2}{\Sigma x^2 - (\Sigma x)^2/n}}$$

to

$$\hat{y} + t_{\alpha/2} \sqrt{\text{MSE}} \sqrt{\frac{1}{n} + \frac{(x_0 - \bar{x})^2}{\Sigma x^2 - (\Sigma x)^2/n}}$$

a) Find a 95-percent confidence interval for the mean food expenditure of a family with a disposable income of $22,000. Use the results of Exercise 6.

b) Find a 99-percent confidence interval for the mean price of a four-year-old Ford pickup. Use the results of Exercise 8.

c) Explain in your own words the difference between the intervals found here and those found in Exercises 6 and 8.

12.6 Inferences in correlation analysis

We have pointed out that some researchers will not go through a complete regression analysis. They will compute the *sample correlation coefficient*, r, and use it to determine whether there is a positive or negative linear relationship between two variables. As in the case of regression inferences, they are actually trying to make inferences about the correlation coefficient for an entire *population* by using data from a sample of that population. The **population correlation coefficient** is denoted by the Greek letter, ρ.

Researchers who are looking to see if there is *a positive or negative linear relationship* between their variables will take as their null hypothesis the statement $\rho = 0$ (there is no such linear relationship). The alternative hypothesis will be $\rho \neq 0$ which would indicate that there is a positive or negative linear relationship between the variables. In the latter case, we will say that there is a *significant* linear relationship between the variables. The procedure used in this type of hypothesis test should not surprise you:

Calculate r. If r is "too big" (too near $+1$) or "too small" (too near -1), reject the null hypothesis. Otherwise, accept it.

The decision as to whether r is "too big" or "too small" is a very simple one. It can be made directly from Table VII. A portion of this table is reproduced here as Table 12.15.

Table 12.15

d.f.	α 0.10	0.05	0.02	0.01
1	0.988	0.997	0.9995	0.9999
2	0.900	0.950	0.980	0.990
3	0.805	0.878	0.934	0.959
:	:	:	:	:
:	:	:	:	:
9	0.521	0.602	0.685	0.735
10	0.497	(0.576)	0.658	0.708
11	0.476	0.553	0.634	0.684
12	0.458	0.532	0.612	0.661
:	:	:	:	:
:	:	:	:	:

For tests of $\rho = 0$, d.f. $= n - 2$, where n is the number of data pairs in the sample. Suppose that you wish to do a test of $\rho = 0$ with d.f. $= 10$ (that is, a sample of size 12) and $\alpha = 0.05$. The value in the table under $\alpha = 0.05$ for d.f. $= 10$ is the circled value of 0.576. This means that, if $r > 0.576$ or $r < -0.576$, you reject the null hypothesis.

Example **20**   The supervisor who collected the paper grading data wanted to test whether there is a significant linear relationship between number of papers graded and cost, using a test with $\alpha = 0.05$.

Step 1. We will use $\alpha = 0.05$ Solution

Step 2. Since $n = 12$, d.f. $= 10$.

Step 3. We have just found that for $\alpha = 0.05$ and d.f. $= 10$, the value in Table VII is 0.576. The acceptance–rejection regions are:

Step 4. $r = 0.969$ from Example 10 (see page 458).

Step 5. This value falls in the righthand part of the rejection region. The supervisor can reject the null hypothesis, and conclude that there is a significant linear relationship between the variables. ▶

The procedure used above on the paper-grading cost problem is summarized below.

Procedure

Procedure for performing a hypothesis test of the form:

Null hypothesis: $\rho = 0$
Alternative hypothesis: $\rho \neq 0$

Assumptions: Regression-model assumptions (see page 472).

1.   Decide on the significance level, α.
2.   Find d.f. $= n - 2$
3.   Use Table VII to set up the rejection region.

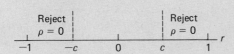

$c =$ value from Table VII.

4.   Calculate r from your sample data.
5.   If $r < -c$ or $r > c$, reject the null hypothesis.

◀ A physician is interested in determining whether or not there is a significant linear   Example **21**
relationship between age and the maximum heart rate an individual can reach during intensive exercise.

Data from a random sample of ten individuals are given in Table 12.16.

Table 12.16

Age x	Peak heart rate y
10	210
20	200
20	195
25	195
30	190
30	180
30	185
40	180
45	170
50	165

Use these data to perform the hypothesis test:

Null hypothesis: $\rho = 0$
Alternative hypothesis: $\rho \neq 0$

at the 0.01 significance level ($\alpha = 0.01$).

Solution   We apply the step-by-step procedure given above.

Step 1. We will use $\alpha = 0.01$.

Step 2. Since $n = 10$ and d.f. $= n - 2$, we have d.f. $= 8$.

Step 3. The value in Table VII for $\alpha = 0.01$ and d.f. $= 8$ is 0.765:

Step 4. Using the data and the formula

$$r = \frac{n\sum xy - (\sum x)(\sum y)}{\sqrt{n\sum x^2 - (\sum x)^2}\sqrt{n\sum y^2 - (\sum y)^2}}$$

we find r (see Example 11, page 460) to be -0.971.

Step 5. Since $r < -0.765$, we reject the null hypothesis and conclude $\rho \neq 0$. That is, there appears to be a significant linear relationship between age and peak heart rate during intensive exercise. ▶

Exercise **R**   In Example 12, we found a sample correlation coefficient of $r = 0.082$ between height and test score in a differential equations class. The sample size was $n = 11$. Test the null hypothesis $\rho = 0$ against the alternative hypothesis $\rho \neq 0$, using $\alpha = 0.05$.

Some readers may have noted that we now have two tests of whether there is a significant linear relation between two variables. One is the test of $\rho = 0$, which we have just done; the other is the test of $\beta_1 = 0$ done in Section 12.5. These two tests are equivalent: For any data set, they will either both say "accept" or both say "reject." This means that, if you are doing a regression analysis and have already tested $\beta_1 = 0$, you do not need to do any correlation analysis to test $\rho = 0$ since you already know what the answer is.

A special word on the assumptions for correlation tests is necessary here. The test of $\rho = 0$ is the simplest test that can be done for correlations. There are other tests and techniques for studying correlations; for example, there are methods for finding *confidence intervals* for ρ. These methods are derived using the assumption that our data pairs come from a special distribution called the *bivariate normal distribution*. The distribution requires not only that y-values be normally distributed for any fixed x-value, but also that x-values be normally distributed for any fixed y-value. Some of the populations we have looked at in this chapter are bivariate normal; for example, the population of heights and weights for college students. However, others are not. The possible number of papers and costs for paper grading form a population that is probably *not* bivariate normal.

Although we will not pursue more advanced correlation methods here, the reader should be aware that most correlation analyses are based on more demanding assumptions than is simple regression. A few of the data sets we have looked at do not satisfy those assumptions.

Two variables may be correlated without having any *causal* relationship. Here, for example, are some data we gathered on attendance in public colleges and total parimutuel turnover at racetracks (money bet on the horses).

Correlation is not causation

Year	Total parimutuel turnover (in millions of dollars), x	Public college population (in thousands), y
1970	5977	6428
1971	6350	6804
1972	6401	7070
1973	7027	7420
1974	7513	7988

The correlation between x and y is 0.9617, and is significant at the 0.01 level. But this does not mean that, when people go to race-tracks, they are somehow inspired to go to a public college. It simply means that the two variables x and y happened to be increasing at the same time. You can have a great deal of fun looking through *Statistical Abstract* or an almanac, picking out unrelated quantities that happen to increase at the same time, and proving that they are correlated. That is how we got the data above. We also showed, using another data set, that the size of the public college population is significantly associated with the size of the prison population.

It may also happen that two variables are correlated because they are both associated with a *third* variable that wasn't considered. For example, a study was once done to show that teachers' salaries were positively correlated with the dollar amount of liquor sales. This does not mean that heavy drinking makes individuals into successful teachers. It just means that both of these variables are tied to other variables (such as the rate of inflation), which pull them along together.

EXERCISES Section **12.6**

1 A psychologist gave 25 students a questionnaire that measures self-image in relation to mathematics. She then gave the students a mathematics test. The correlation between self-image scores and mathematics scores was $r = 0.31$. Is this significant at the level $\alpha = 0.05$ (that is, can the null hypothesis "$\rho = 0$" be rejected at the 5% significance level)?

2 The correlation between family income and food costs for the eight families in Exercise 5 of Section 12.2 is $r = 0.74$. Is this significant at the level $\alpha = 0.01$ (that is, can the null hypothesis "$\rho = 0$" be rejected at the 1% significance level)?

3 The correlation between attendance and sales for the six days considered in Exercise 6 of Section 12.2 is $r = 0.98$. Is this significant at the level $\alpha = 0.05$?

4 For a random sample of 30 college freshmen, the correlation between grade in Calculus I and grade in Calculus II is $r = 0.7$. Is this significant at the level $\alpha = 0.01$?

5 The correlation between age and price for the nine Ford pickups in Exercise 8 of Section 12.2 is $r = -0.95$. Is this significant at the level $\alpha = 0.10$?

6 An industrial psychologist measured the times (in seconds) for two assembly-line tasks for ten workers. His data were:

Task 1, x	Task 2, y	Task 1, x	Task 2, y
15	31	16	31
17	36	13	32
18	35	14	30
14	33	20	40
21	39	12	29

a) Find the correlation coefficient r.
b) Is r significant at the level $\alpha = 0.05$?

7 a) Find r for each of the following data sets:

i) x	y	ii) x	y	iii) x	y
2	3	2	30	20	30
4	5	4	50	40	50
6	4	6	40	60	40
8	7	8	70	80	70
10	11	10	110	100	110

b) In many exercises in this chapter, we gave data in hundreds or thousands; for example, an income of $22,000 might be represented by the number 22. Does such scaling change the value of r?

8 Suppose a and b are positive numbers. Show that the following two (general) data sets have the same correlation coefficient.

Data set I		Data set II	
x_1	y_1	ax_1	by_1
x_2	y_2	ax_2	by_2
.	.	.	.
.	.	.	.
.	.	.	.
x_n	y_n	ax_n	by_n

(This will justify the answer to 7(b): scaling changes do not change r.)

The regression-analysis programs in packages such as SPSS will plot data in addition to providing b_0, b_1, r, and other appropriate statistics. Below we have given the printout from the SPSS program SCATTERGRAM for the paper-grading cost data that was used throughout this chapter.

Computer packages*

12.7

```
GRADES
FILE   FILE₁   (CREATION DATE = 05/25/79)
SCATTERGRAM OF  (DOWN) COST                          (ACROSS) NUMBER
             15.35    16.05    16.75    17.45    18.15    18.85    19.55    20.25    20.95    21.65
         .+----+----+----+----+----+----+----+----+----+----+----+----+----+----+----+----+----+.
  298.00 +                        I                        I                            *+ 298.00
         I                        I                        I                             I
         I                        I                        I                             I
         I                        I                        I                             I
         I                        I                        I                             I
  289.20 +                        I                        I                             + 289.20
         I                        I                        I                             I
         I                        I                        I                             I
         I                        I                        I                             I
         I                        I                        I                             I
  280.40 +                        I                        I                             + 280.40
         I                        I                        I                             I
         I                        I                        I                             I
         I                        I                        I                             I
         I                        I              *         I                             I
  271.60 +                        I                        I                             + 271.60
         I                        I                        I                             I
         I- - - - - - - - - - - - I- - - - - - - - - - - - I- - - - - - - - - - - - - - -I
         I                        I                        I                             I
         I                        I              *         I                             I
  262.80 +                        I                        I                             + 262.80
         I                        I                        I                             I
         I                        I                        I                             I
         I                        I    *                  .I                             I
         I                        I                        I                             I
  254.00 +                        I                        I                             + 254.00
         I                        I                        I                             I
         I                  *     I    *                   I                             I
         I                        I                        I                             I
         I                        I                        I                             I
  245.20 +                        I    *                   I                             + 245.20
         I                        I                        I                             I
         I                        I                        I                             I
         I- - - - - - - - - - - - I- - - - - - - - - - - - I- - - - - - - - - - - - - - -I
         I                        I                        I                             I
  236.40 +                        I                        I                             + 236.40
         I          *             I                        I                             I
         I                        I                        I                             I
         I                        I                        I                             I
         I                        I                        I                             I
  227.60 +          *             I                        I                             + 227.60
         I                        I                        I                             I
         I                        I                        I                             I
         I.*                      I                        I                             I
         I          *             I                        I                             I
  218.80 +                        I                        I'                            + 218.80
         I                        I                        I                             I
         I                        I                        I                             I
         I                        I                        I                             I
         I                        I                        I                             I
  210.00 +.*                      I                        I                             + 210.00
         .+----+----+----+----+----+----+----+----+----+----+----+----+----+----+----+----+.
         15.00    15.70    16.40    17.10    17.80    18.50    19.20    19.90    20.60    21.30    22.00
```

STATISTICS..

CORRELATION (R)—	0.96903	R SQUARED —	0.93901	SIGNIFICANCE —	0.00001
STD ERR OF EST —	6.52638	INTERCEPT (A) —	35.79592	SLOPE (B) —	12.08349
PLOTTED VALUES —	12	EXCLUDED VALUES—	0	MISSING VALUES —	0

'********' IS PRINTED IF A COEFFICIENT CANNOT BE COMPUTED.

---

* This section is optional.

As you can see from this display, the printout begins with a plot of the data points, which are displayed as asterisks. It is necessary to look carefully: Two asterisks are nearly hidden on the left of the plot, while one more is on the extreme upper right. Immediately below the data plot are the calculated values for slope, intercept, and correlation coefficient, along with the usual extra statistics. The number printed after SIGNIFICANCE is the P-value for r; the r calculated for this data set would be significant for α as small as 0.00001.

Note. The computer printout uses different letters from those found in the text for the (sample) correlation coefficient, the slope of the (sample) regression line, and the y-intercept of the (sample) regression line. The following table shows the correspondence.

Text	Computer printout
r	R
b_1	B
b_0	A

Finally, the quantity STD ERR OF EST (standard error of the estimate), whose value is given to be 6.52638 on the computer printout, is $\sqrt{\text{MSE}}$. In other words STD ERR OF EST is the sample estimate for σ.

CHAPTER REVIEW

Key Terms

linear regression
straight lines
slope
y-intercept
error
predicted value
error sum of squares (SSE)
residuals
line of best fit
b_0, b_1
(sample) regression line
(sample) correlation
   coefficient (r)

explained variation
total sum of squares (SST)
sum of squares for regression (SSR)
coefficient of determination
population regression line
β_0, β_1
regression model
mean square error (MSE)
$s(b_1)$
prediction interval
extrapolation
population correlation
   coefficient (ρ)

Formulas and Key Facts

equation of a straight line (slope–intercept form):

$$y = b_0 + b_1 x$$

$[b_1 = \text{slope}, b_0 = y\text{-intercept}]$

error sum of squares, SSE:

$$\text{SSE} = \sum (y - \hat{y})^2 = \sum e^2$$

line of best fit (sample regression line):

$$\hat{y} = b_0 + b_1 x$$

where

$$b_1 = \frac{\sum xy - (\sum x)(\sum y)/n}{\sum x^2 - (\sum x)^2/n}$$

and

$$b_0 = \frac{1}{n}(\sum y - b_1 \sum x)$$

(where n = number of data points).

sample correlation coefficient, r:

$$r = \frac{n\sum xy - (\sum x)(\sum y)}{\sqrt{n\sum x^2 - (\sum x)^2}\ \sqrt{n\sum y^2 - (\sum y)^2}}$$

total sum of squares, SST:

$$\text{SST} = \sum (y - \bar{y})^2$$

sum of squares for regression, SSR:

$$\text{SSR} = \sum (\hat{y} - \bar{y})^2$$

coefficient of determination:

$$r^2 = \frac{\text{SSR}}{\text{SST}}$$

regression model:

$$y = \beta_0 + \beta_1 x + \varepsilon$$

(ε is normally distributed with mean 0 and variance σ^2.)

mean square error, MSE:

$$\text{MSE} = \frac{\text{SSE}}{n-2} = \frac{\sum(y-\hat{y})^2}{n-2}$$

(used to estimate the variance, σ^2, in the regression model).

sampling distribution of b_1:

If β_1 is the slope of the population regression line, and b_1 is the slope of the sample regression line based on n data pairs, then the random variable

$$t = \frac{b_1 - \beta_1}{s(b_1)}$$

has the *t-distribution with d.f. $= n - 2$*. Here,

$$s(b_1) = \sqrt{\frac{\text{MSE}}{\sum(x-\bar{x})^2}}.$$

formula for finding prediction intervals for the y-value when $x = x_0$:

$$\hat{y} - t_{\alpha/2}\sqrt{\text{MSE}}\sqrt{1 + \frac{1}{n} + \frac{(x_0 - \bar{x})^2}{\sum x^2 - (\sum x)^2/n}}$$

to

$$\hat{y} + t_{\alpha/2}\sqrt{\text{MSE}}\sqrt{1 + \frac{1}{n} + \frac{(x_0 - \bar{x})^2}{\sum x^2 - (\sum x)^2/n}}$$

($n =$ number of data pairs, $\hat{y} = b_0 + b_1 x_0$, $1 - \alpha =$ prediction level, and $t_{\alpha/2}$ computed for a *t*-curve with d.f. $= n - 2$).

In Table 12.17 we give a brief summary of the hypothesis testing procedures covered in this chapter. For detailed procedures, see the text.

Type	Assumptions	Null hypothesis	Alternative hypothesis	Test statistic	Rejection region
Slope of regression line	Regression model	$\beta_1 = 0$	$\beta_1 \neq 0$	$t = \dfrac{b_1}{s(b_1)}$ (d.f. $= n - 2$)	$t < -t_{\alpha/2}$ or $t > t_{\alpha/2}$
Correlation coefficient	Regression model	$\rho = 0$	$\rho \neq 0$	r	$r < -c$ or $r > c$ $[c =$ value from Table VII, d.f. $= n - 2]$

Table 12.17
Hypothesis tests for regression and correlation

Note. We have discussed only two-tailed tests, but one-tailed tests can also be performed using the above procedure with the usual modifications. For example, to test $\beta_1 = 0$ vs. $\beta_1 > 0$, the rejection region is $t > t_\alpha$.

You should be able to

1 Interpret the meaning of the *slope* and *y-intercept* of a straight line.

2 *Graph a straight line*, given its slope and *y*-intercept.

3 Determine whether a straight line is *increasing, decreasing*, or *horizontal*, knowing its slope.

4 Find the *line of best fit (sample regression line)* for *n* data points.

5 Decide whether or not it is appropriate to fit a regression line to a data set.

6 Compute and explain the meaning of the *sample correlation coefficient.*

7 Interpret the sample correlation coefficient in terms of *explained variation.*

8 State the *assumptions* required for the inferential methods in regression and correlation.

9 Describe the assumptions required for the inferential methods in regression and correlation in terms of the *regression model.*

10 Calculate the *mean square error, MSE,* for *n* data points, and explain its uses.

11 Compute $s(b_1)$ for *n* data points.

12 Perform a hypothesis test of the form:

Null hypothesis: $\beta_1 = 0$
Alternative hypothesis: $\beta_1 \neq 0$

13 Find a *prediction interval* for the *y*-value when $x = x_0$.

14 Perform a hypothesis test of the form

Null hypothesis: $\rho = 0$
Alternative hypothesis: $\rho \neq 0$

REVIEW TEST

The data set below will be used in problems 1, 2 and 4.

Data set 1

x	y
1	5
2	7
3	12
4	14
5	18

1 a) Find the equation of the regression line for the Data set 1.
b) What value, $\hat{y}$, would this regression line predict if $x = 3.5$?

2 Find r for Data set 1.

3 Classify each of the following data sets as having $r > 0, r < 0,$ or $r \doteq 0$.

a)

b)

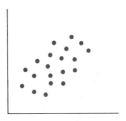

c)

4 a) Calculate SSE for Data set 1.

b) Calculate SST for Data set 1.

c) Use the results of (a) and (b) to find SSR.

d) What percent of the total variation in Data set 1 is explained by the regression?

e) Use the result of part (d) to find r.

5 A study of the relation between height x (in inches) and weight y (in pounds) for 23 subjects gave the following results: $\Sigma x = 1612$, $\Sigma y = 3574$, $\Sigma x^2 = 113106.25$, $b_0 = -128.46$, $b_1 = 4.05$, and MSE $= 166.74$.

a) Perform the hypothesis test:

Null hypothesis: $\beta_1 = 0$
Alternative hypothesis: $\beta_1 \neq 0$

at the 0.05 significance level.

b) Find a 95-percent prediction interval for the weight, y, of an individual whose height is $x_0 = 70$ inches.

c) The value of r calculated from the $n = 23$ data pairs in this study was $r = 0.61$. Is this significant at the level $\alpha = 0.05$?

d) State the regression-model assumptions used in such an analysis as the above.

FURTHER READINGS

FREEDMAN, D., R. PISANI, and R. PURVIS, *Statistics*. New York: W.W. Norton & Co., 1978. Chapters 7–12 deal with regression and correlation. The authors work on a very intuitive level, with extremely interesting examples.

NETER, J., and W. WASSERMAN, *Applied Linear Statistical Models*. Homewood, Illinois: Richard D. Irwin, Inc., 1974. This text can be read by users of statistics who have not taken calculus, even though it covers a wide range of advanced applications (Chapters 1–12 deal with regression). It is an excellent text or reference for researchers who need to use statistics, but do not have an extensive mathematical background.

ANSWERS TO REINFORCEMENT EXERCISES

A a) $y = 200 + 0.05x$

   b) 250, 300, 500, 700

   c)

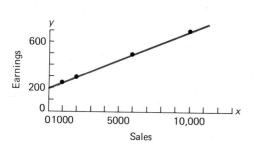

   d) 825

B a) $y = 200$ b) Base salary

   c) Slope = 0.05   d) Commission rate

C a) $y = 2 + 3x$ b) $y = 3 - x$

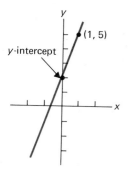   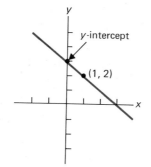

D a) Increasing   b) Increasing   c) Decreasing

   d) Decreasing   e) Horizontal

E Between $280 and $290

F a)

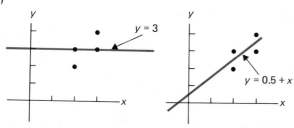

b)

x	y	$\hat{y}$	e	e^2
2	2	3	-1	1
2	3	3	0	0
3	3	3	0	0
3	4	3	1	1
			0	$2 = \Sigma e^2$

SSE = 2

x	y	$\hat{y}$	e	e^2
2	2	2.5	-0.5	0.25
2	3	2.5	0.5	0.25
3	3	3.5	-0.5	0.25
3	4	3.5	0.5	0.25
			0	$1 = \Sigma e^2$

SSE = 1

c) $y = 0.5 + x$

G

x	y	xy	x^2
2	2	4	4
2	3	6	4
3	3	9	9
3	4	12	9
10	12	31	26

$$b_1 = \frac{31 - 10(12)/4}{26 - (10^2/4)} = \frac{1}{1} = 1$$

$$b_0 = \frac{1}{4}(12 - 1(10)) = 0.5$$

$$\hat{y} = 0.5 + x$$

H a)

x	y	xy	x^2
70	185	12950	4900
65	126	8190	4225
71	170	12070	5041
67	140	9380	4489
74	165	12210	5476
71	150	10650	5041
67	130	8710	4489
69	160	11040	4761
73	175	12775	5329
69	140	9660	4761
73	160	11680	5329
769	1701	119315	53841

b) $b_1 = \dfrac{119315 - (769)(1701)/11}{53841 - (769)^2/11}$

$= \dfrac{399.633}{80.909} = 4.939$

$b_0 = \dfrac{1}{11}(1701 - 4.939(769)) = -190.645$

$\hat{y} = -190.645 + 4.939x$

I (b) and (e)

J

x	y	xy	x^2	y^2
70	185	12950	4900	34225
65	126	8190	4225	15876
71	170	12070	5041	28900
67	140	9380	4489	19600
74	165	12210	5476	27225
71	150	10650	5041	22500
67	130	8710	4489	16900
69	160	11040	4761	25600
73	175	12775	5329	30625
69	140	9660	4761	19600
73	160	11680	5329	25600
769	1701	119315	53841	266651

$r = \dfrac{11(119315) - 769(1701)}{\sqrt{11(53841) - (769)^2}\ \sqrt{11(266651) - (1701)^2}}$

$= \dfrac{4396}{29.833(199.399)} = 0.739$

K

x	y	xy	x^2	y^2
1	10	10	1	100
2	9	18	4	81
2	8	16	4	64
3	7	21	9	49
4	5	20	16	25
5	3	15	25	9
17	42	100	59	328

$r = \dfrac{6(100) - 17(42)}{\sqrt{6(59) - 17^2}\ \sqrt{6(328) - 42^2}}$

$= \dfrac{-114}{8.062(14.283)} = -0.990$

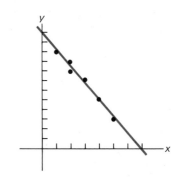

L

x	y	$\hat{y}$	e	e^2
1	1	2	-1	1
1	3	2	1	1
5	2	3	-1	1
5	4	3	1	1
	10			4

$\bar{y} = \dfrac{10}{4} = 2.5$

$\text{SST} = (1 - 2.5)^2 + (3 - 2.5)^2 + (2 - 2.5)^2 + (4 - 2.5)^2 = 5$

$\text{SSR} = (2 - 2.5)^2 + (2 - 2.5)^2 + (3 - 2.5)^2 + (3 - 2.5)^2 = 1$

$\text{SSE} = (1 - 2)^2 + (3 - 2)^2 + (2 - 3)^2 + (4 - 3)^2 = 4$

b) $\text{SSR} + \text{SSE} = 1 + 4 = 5 = \text{SST}$

c) $\dfrac{\text{SSR}}{\text{SST}} = \dfrac{1}{5} = 0.2$

20 % of SST is "explained" by regression.

d) $\dfrac{\text{SSE}}{\text{SST}} = \dfrac{4}{5} = 0.8$

80 % of SST is "not explained" by regression.

e) $r^2 = \dfrac{\text{SSR}}{\text{SST}} = 0.2$

f) i) $r = \sqrt{\dfrac{\text{SSR}}{\text{SST}}} = \sqrt{0.2} = 0.447$

ii) $r = \dfrac{n\Sigma xy - (\Sigma x)(\Sigma y)}{\sqrt{n(\Sigma x^2) - (\Sigma x)^2}\ \sqrt{n\Sigma y^2 - (\Sigma y)^2}}$

$= \dfrac{4(34) - (12)(10)}{\sqrt{4(52) - (12)^2}\ \sqrt{4(30) - (10)^2}} = 0.447$

M a) Age and heart-rate for joggers in jogging clubs.
   b) Assumptions I–III on page 472.
   c)

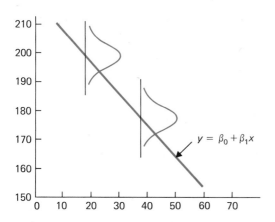

$y = \beta_0 + \beta_1 x$

d) $y = \beta_0 + \beta_1 x + \varepsilon$, where ε is normally distributed with mean 0 and variance σ^2.

N (a)

O

x	y	$\hat{y}$	e	e^2
10	210	208.86	1.14	1.30
20	200	197.93	2.07	4.28
20	195	197.93	−2.93	8.58
25	195	192.47	2.54	6.43
30	190	187.00	3.00	9.00
30	180	187.00	−7.00	49.00
30	185	187.00	−2.00	4.00
40	180	176.07	3.93	15.44
45	170	170.61	−0.61	0.37
50	165	165.14	−0.14	0.02
			0	98.42 = SSE

$$\text{MSE} = \frac{98.42}{10-2} \doteq 12.30$$

Estimate for $\sigma^2 = 12.30$
Estimate for $\sigma = 3.51$

P 1. $\alpha = 0.05$

2. d.f. $= n - 2 = 10 - 2 = 8$

3. $t_{\alpha/2} = t_{0.025} = 2.31$ for d.f. $= 8$

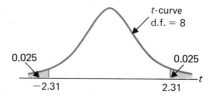

t-curve
d.f. $= 8$

0.025   0.025

−2.31   2.31

4. MSE $= 12.30$ (from Exercise O)

$$\Sigma(x - \bar{x})^2 = \Sigma x^2 - \frac{(\Sigma x)^2}{n} = 10350 - \frac{(300)^2}{10} = 1350$$

$$s(b_1) = \sqrt{\frac{\text{MSE}}{\Sigma(x - \bar{x})^2}} = \sqrt{\frac{12.30}{1350}} \doteq 0.095$$

$$b_1 = -1.093$$

$$t = \frac{b_1}{s(b_1)} = \frac{-1.093}{0.095} = -11.51$$

5. Reject the null hypothesis.

Q a) $n = 12$, d.f. $= 12 - 2 = 10$, $t_{0.025} = 2.23$, $x_0 = 17$, MSE $= 42.586$; $\sum x^2 - \frac{(\sum x)^2}{n} = 44.917$, $\bar{x} = 17.417$

$\hat{y} = 35.804 + 12.083(17) = 241.215$
$(x_0 - \bar{x})^2 = (17 - 17.417)^2 = 0.174$

$$241.215 \pm (2.23)\sqrt{42.586}\sqrt{1 + \frac{1}{12} + \frac{0.174}{44.917}}$$

$241.215 \pm (2.23)(6.526)(1.043)$
241.215 ± 15.179
\$226.04   to   \$256.40

b) $x_0 = 15$
$\hat{y} = 35.804 + 12.083(15) = 217.049$
$(x_0 - \bar{x})^2 = (15 - 17.417)^2 = 5.842$

$$217.049 \pm 2.23 \cdot \sqrt{42.586}\sqrt{1 + \frac{1}{12} + \frac{5.842}{44.917}}$$

$217.049 \pm 2.23(6.526)(1.102)$
217.049 ± 16.037
\$201.01   to   \$233.09

R $n = 11$, d.f. $= 11 - 2 = 9$, $r = 0.082$.

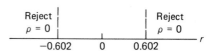

Reject   Reject
$\rho = 0$ $\rho = 0$

−0.602   0   0.602

The null hypothesis cannot be rejected.

Nonparametric statistics

13.1 Introduction

In the previous chapters, we looked at statistical tests derived for use on data from normal populations. Assumptions of normal populations were made for *t*-tests, analysis of variance, and regression analysis. In practice, most users of these statistical methods do not worry too much about working with data taken from populations that are not exactly normal, as long as they are nearly so. These methods still work fairly well if their assumptions are only *slightly* violated, and for this reason they are called **robust**.

However, there is still a serious question to be answered: What would you do if the assumptions are very badly violated? Populations that are very different from normal populations do occur. For example, one of the authors found in the early 1970's that the test-grade distributions in his calculus classes consistently looked like Fig. 13.1.

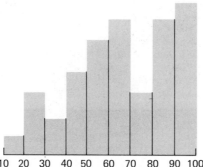

Figure 13.1   10   20   30   40   50   60   70   80   90   100

The population depicted in Fig. 13.1 is very far from being normally distributed. To use a test such as a *t*-test on data from this population would be inappropriate.

To handle data from nonnormal populations, we need new tests that do not rely on normality assumptions. Such tests are called **distribution-free** because you can apply them to populations with almost any distribution. In this chapter we will look at a few of the most important distribution-free tests.

Some of our distribution-free tests will also be referred to as **nonparametric tests.** You might recall that numbers such as μ and σ^2, which describe important features of a population, are called *parameters*. Many tests, such as the *t*-test, deal with the true values of parameters like μ; for this reason, they are called **parametric tests.** Other tests, such as the chi-square test for independence, do not deal with parameters. The chi-square test of independence applies only to the concept of independence, and is not used to test the true value of μ or any other parameter. The tests that do not deal with parameters are called *nonparametric*.

Many distribution-free tests are actually parametric, since they deal with such population parameters as the median. However, it has become common practice to refer to most of the tests that can be used without checking assumptions about normality as *nonparametric*. Books and surveys entitled "nonparametric statistics" typically cover both nonparametric and distribution-free tests.

Nonparametric tests have other advantages. They often require fewer (or simpler) calculations than parametric tests, and some of them can be used on ordinal or qualitative data, for which the parametric tests of our previous chapters are inappropriate.

By this time the reader is probably asking, "Why not use nonparametric tests all the time? They are simpler and based on fewer assumptions." The answer is that some things are lost when you use a nonparametric test. Nonparametric tests often do not make use of all the information contained in a data set. If you are fairly sure that your data comes from a normal population, it is probably better to use a parametric test.

In this chapter, we will consider some examples that illustrate the differences between parametric and nonparametric tests. In Section 13.7, we will also discuss the problem of deciding whether to use a parametric or nonparametric test in various situations. However, you should remember that any rule of thumb we give in this introductory book is not a final answer to every problem. It may not be obvious at first glance, even to professional statisticians, whether a parametric or nonparametric test is best for a specific research project. Frequently quite complex and technical concepts must be considered in order to decide which type of test is more appropriate. It is fairly easy to see the basic differences between parametric and nonparametric tests, but there is no simple rule that will always tell you which one is best for a given study.

The first nonparametric test we will look at, called the **sign test,** depends on the binomial distribution, which was introduced in Section 4.8. In this section we will prepare for our study of the sign test by reviewing the binomial distribution, and introducing tables for it. The basic facts about the binomial distribution, given in Chapter 4, are as follows:

The binomial distribution **13.2**

> Suppose n independent success–failure experiments are performed, with the probability of success being p. If x represents the number of successes, then
>
> $$P(x = k) = \binom{n}{k} p^k (1 - p)^{n-k}$$
>
> where
>
> $$\binom{n}{k} = \frac{n!}{k!(n - k)!}$$

In Example 1 below, we review Example 42 of Chapter 4, which gave a typical situation in which the binomial distribution could be applied.

◀ In an ESP experiment, a person in one room has cards numbered 1 through 10. This person selects a card at random, and a second person in another room is supposed to guess the number. This experiment is repeated three times. If the person guessing has no ESP and is just guessing at random, what is the probability that he guesses correctly exactly twice?

Example **1**

Solution In this problem the three different guessing experiments are the independent trials; that is, $n = 3$.

For each individual trial, since there are 10 cards, the guesser has only one chance in ten of guessing correctly:

$$p = \text{Probability of success} = P(\text{guess correctly}) = \frac{1}{10}$$

$$1 - p = \text{Probability of failure} = P(\text{guess incorrectly}) = \frac{9}{10}$$

The probability of guessing correctly exactly twice is

$$P(x = 2) = \binom{3}{2}\left(\frac{1}{10}\right)^2\left(\frac{9}{10}\right)^1 = 3(0.1)^2(0.9) = 0.027$$ ◗

Exercise **A** Find the probability that the guesser in Example 1 makes exactly one correct guess in three tries.

Tables have been compiled to save you the work of doing these calculations, and are given in the back of the book in Table VIII. In Table 13.1 we give a portion of one of these tables, where

n = number of trials,

p = probability of success,

k = number of successes.

Table 13.1

		p				
n	k	.1	.2	. . .	.5	. . .
3	0	.7290	.5120	. . .	.1250	. . .
	1	.2430	.3840	. . .	.3750	. . .
	2	.0270	.0960	. . .	.3750	. . .
	3	.0010	.0080	. . .	.1250	. . .

To find the desired probability, we simply go to the part of the table with the correct number of trials ($n = 3$) and read the entry for the given values of p and k. For $n = 3$, $p = 0.1$, and $k = 2$, the table gives a value of 0.0270. This is exactly what we found in Example 1.

The table also makes it easy to find more complicated probabilities. For example, to find the probability of fewer than two successes ($x < 2$) when $n = 3$ and $p = 0.1$, we simply add up the first two entries in the table under .1:

$$P(x < 2) = 0.7290 + 0.2430 = 0.9720$$

Similarly,

$$P(x \geq 2) = 0.0270 + 0.0010 = 0.0280$$

a) Use the table to find the probability of *exactly one* success in three trials if   Exercise **B**
    $p = 0.1$.
b) Also use the table to find the probability of *at least* one success.

In each of Exercises 1 through 5, find the required probability using the formula for $P(x = k)$, and then check your answer using Table VIII (if that is possible for the given values of n and p).

1   The probability of the birth of a female child is approximately 0.5 (the actual value is closer to 0.49). Use $p = 0.5$ to find the probability that exactly three out of ten babies born in a big-city hospital today will be female.

2   The probability that a randomly selected household in a southern city has a room air-conditioner is $p = 0.4$. A telephone sales operation plans to phone ten randomly selected households in that city in the next hour. What is the probability that exactly five of these households will already have room air-conditioners?

3   a) In 1977, the probability that a United States resident was under five years of age was $p = 0.07$. If eight U.S. residents were selected at random in that year, what is the probability that none of them was under five years of age?
    b) In that same year, the probability that a resident of Utah was under five was $p = 0.12$. What is the probability that none of eight randomly selected Utah residents was under 5?

4   The probability that a resident of a midwestern state has ever had a chest x-ray is $p = 0.8$. Find the probability that

exactly four out of seven randomly selected residents have ever had a chest x-ray.

5   The probability that a U.S. citizen had an upper respiratory infection between 1970 and 1976 is approximately $p = 0.6$. Twelve U.S. citizens were randomly selected for a medical experiment in 1977. Find the probability that exactly four of these twelve citizens had suffered upper respiratory infections between 1970 and 1976.

In Exercises 6 through 9, use Table VIII to find the indicated probability.

6   Find the probability that at least five out of ten babies born at the hospital in Exercise 1 are female ($p = 0.5$).

7   Find the probability that at most three of the five households phoned in Exercise 2 will have room air-conditioners ($p = 0.4$).

8   Find the probability that more than four of the seven people selected in Exercise 4 have ever had a chest x-ray ($p = 0.8$).

9   Find the probability that at least six of the twelve citizens selected in Exercise 5 had suffered upper respiratory infections between 1970 and 1976 ($p = 0.6$).

The sign test for medians   **13.3**

We are beginning with the **sign test** because the reasoning behind it is fairly simple. By looking at the sign test, you can get an idea of how nonparametric tests actually work. In Example 2 below, we will illustrate both the reasoning behind the sign test and the steps in its application.

◀ An economic analyst in a southwestern city was given an estimate that her   Example **2**
city's *median* family income was $14,000. (The estimated median family income

for the entire United States in 1975 was $13,719.) This city was actually a high-income suburb of a large city, and the economist thought that the figure of $14,000 might be low. She decided to do a quick test, and took a random sample of 12 family incomes from the city and arranged them in order from the largest to the smallest. They were:

$$
\begin{array}{l}
\left.\begin{array}{l}
\$60,000 \\
25,700 \\
22,400 \\
20,100 \\
17,300 \\
16,100 \\
15,300 \\
14,800 \\
14,300 \\
14,100
\end{array}\right\} \text{10 above \$14,000} \\[2pt]
\left.\begin{array}{l}
10,400 \\
6,200
\end{array}\right\} \text{2 below \$14,000}
\end{array}
$$

The economist seemed to be right. Ten families were above $14,000, and only two were below. (A population with a median of $14,000 must have half of the population values above $14,000 and half below.) However, the economist had only *sampled* the population. It is possible that the true median was $14,000, and that she had sampled ten of the higher incomes by chance. To convince her supervisor that the median income was larger than $14,000, the analyst needed a statistical test to show that such a chance was highly unlikely. The test she chose was the *sign test.** She began by setting up null and alternative hypotheses, using the abbreviation Md for median.

Null hypothesis: Md = 14,000
Alternative hypothesis: Md > 14,000

To perform the hypothesis test, she reasoned as follows: In a large population with Md = 14,000, you would expect about half of the family incomes sampled to be above $14,000 and about half to be below. Suppose that you are picking just one family income from this population. Then the probability that this one income will be greater than $14,000 is 0.5, and the probability that it will be less than $14,000 is also 0.5. This situation is pictured in Fig. 13.2.

The economist did not sample just one income, however; she sampled 12, and found ten incomes over $14,000. We can calculate the probability of picking such a sample from a population with median $14,000 by looking at the sampling process as a binomial experiment.

---

* Actually, if the population of incomes was *normally distributed,* a *t*-test of the form:

Null hypothesis: μ = 14,000
Alternative hypothesis: μ > 14,000

could (and should) be used, because the mean and median are the same for a normally distributed population and because for such a population a *t*-test can be used. However, the economist did not know whether or not the population of incomes was normally distributed.

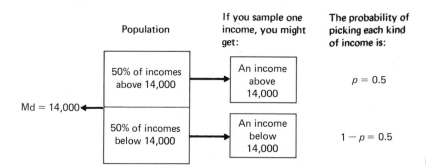

Md = 14,000

Figure 13.2

Each single selection of a sample income is one trial in the experiment, and the total number of trials is $n = 12$. If we call each selection of a single income over \$14,000 a success, then

$$p = \text{Probability of success} = 0.5.$$

The economist's sample had ten incomes over \$14,000, or ten successes. So the probability of selecting her sample is the probability of ten successes in 12 trials with $p = 0.5$; it is:

$$\binom{12}{10}(0.5)^{10}(1 - 0.5)^2 \doteq 0.0161.$$

(This probability can be obtained even more simply by use of Table VIII).

What we have just seen is that we can find the probability of any particular number of incomes over \$14,000 in a sample by using the binomial distribution formula. To perform the hypothesis test, we need to look at the entire binomial distribution with $n = 12$ and $p = 0.5$. Table 13.2 is a table for that distribution; we shall show how it is used in setting up the hypothesis test.

Number of incomes over 14,000, k	Probability, $P(x = k)$
0	0.0002
1	0.0029
2	0.0161
3	0.0537
4	0.1208
5	0.1934
6	0.2256
7	0.1934
8	0.1208
9	0.0537
10	0.0161
11	0.0029
12	0.0002

Table 13.2
Binomial distribution for $n = 12$ and $p = 0.5$

10 or more incomes over 14,000 { 10, 11, 12 }

From this table we can see that:

$$P(10 \text{ or more incomes over } 14{,}000)$$
$$= P(10 \text{ or } 11 \text{ or } 12 \text{ incomes over } 14{,}000)$$
$$= P(x = 10) + P(x = 11) + P(x = 12)$$
$$= 0.0161 + 0.0029 + 0.0002$$
$$= 0.0192.$$

Since the probabilities in the table add up to 1,

$$P(9 \text{ or fewer incomes over } 14{,}000) = 1 - 0.0192 = 0.9808.$$

We picture these probabilities as areas in a binomial-distribution histogram in Fig. 13.3. The two areas are separated by a dashed line.

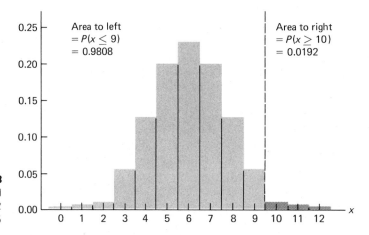

Figure 13.3
Histogram of binomial distribution with $n = 12$ and $p = 0.05$

Figure 13.3 should remind you of the acceptance-rejection pictures we drew with normal distribution curves in Chapter 8. It tells you that if the null hypothesis is true, samples with ten or more incomes over $14,000 have a probability of 0.0192; they will occur less than 2% of the time. Such samples are so unlikely when Md = 14,000 that they give us evidence to conclude that Md > 14,000. In other words, the probability histogram enables us to set up the rejection region for a test of Md = 14,000 against the alternative Md > 14,000. (See Fig. 13.4.)

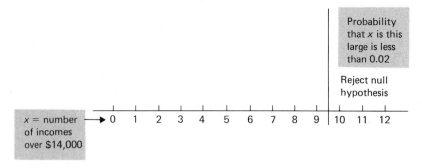

Figure 13.4

The economist could use the diagram in Fig. 13.4 to support her claim that the true median was higher than $14,000. The rejection region consists of 10, 11, or 12 incomes over $14,000. Since she found 10, she can reject her null hypothesis. The significance level for this test is found exactly as for our normal tests: It is the rejection probability for the null hypothesis and is also equal to the histogram area above the rejection region. For this test, $\alpha \doteq 0.02$ is the significance level. ▶

This test was a one-tailed test with alternative Md > 14,000. We went through it more to illustrate basic reasoning patterns than to illustrate the procedure. In the next example, we will go through the same problem, illustrating the step-by-step procedure for the sign test.

◀ (Continuation of Example 2) The steps the economist must perform to test whether the median is greater than $14,000 are: Example **3**

Step 1. Identify null and alternative hypotheses.

> *Null hypothesis:*   Md = $14,000
> *Alternative hypothesis:*   Md > $14,000

Step 2. Decide on n, the number of pieces of data you will collect. Here, $n = 12$.

Step 3. Using Table VIII, make a table and a histogram for the binomial distribution probabilities with $n = 12$ and $p = 0.5$. Your table should look like Table 13.3.

Successes, k	Probability, $P(x = k)$
0	0.0002
1	0.0029
2	0.0161
3	0.0537
4	0.1208
5	0.1934
6	0.2256
7	0.1934
8	0.1208
9	0.0537
10	0.0161
11	0.0029
12	0.0002

Table 13.3

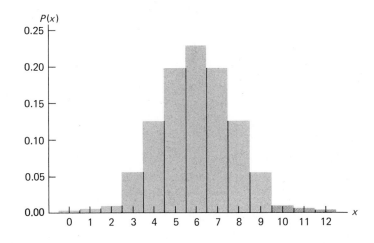

Figure 13.5

Step 4. Use the table and histogram from Step 3 to help decide on a significance level for the test, and to find the rejection region.

For this one-tailed test, we would like to reject the null hypothesis if the number of incomes over $14,000 is large. So our rejection region should be on the right end of the histogram. The areas under the rectangles on the right end of the histogram are:

Rectangle group	Area
Last rectangle ($k = 12$)	0.0002
Last 2 rectangles ($k = 11$ or 12)	0.0031
Last 3 rectangles ($k = 10, 11$ or 12)	0.0192
Last 4 rectangles ($k = 9, 10, 11$ or 12)	0.0729

As you can see, you cannot cut off exactly 5 percent of the area on the right of the histogram without breaking up one of the rectangles.

The economist looked at the area values above to determine which one would make a useful and convincing significance level for her test; $\alpha = 0.0002$ and $\alpha = 0.0031$ looked too small and demanding, and $\alpha = 0.0729$ looked too large. The value $\alpha = 0.0192$ was closer to the commonly used significance levels, which range from 0.05 to 0.01. So she chose a significance level of 0.0192, which she rounded to $\alpha = 0.02$ for simplicity. The rejection region that goes with this choice is shown in Fig. 13.6.

Figure 13.6

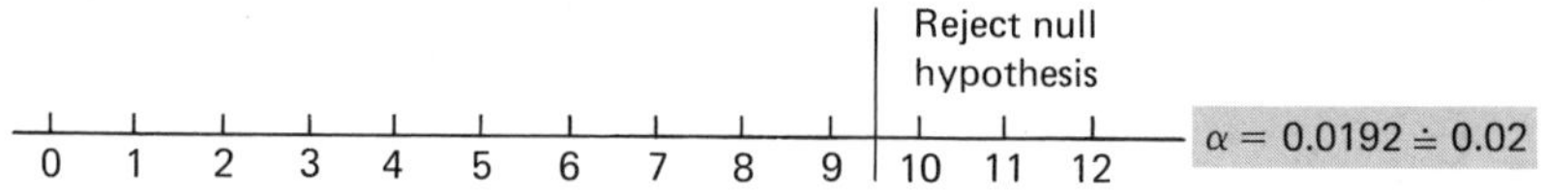

Step 5. Collect data. Mark each data value above the value of Md in the null hypothesis with a "+", and each value below it with a "−".

Income	Sign
$60,000	+
25,700	+
22,400	+
20,100	+
17,300	+
16,100	+
15,300	+
14,800	+
14,300	+
14,100	+
10,400	−
6,200	−

Step 6. Count the number, x, of "+" signs. Reject the null hypothesis if this number is in the rejection region.

The number of "+" signs is ten, so $x = 10$. This is in the rejection region. The economist rejected the null hypothesis and concluded Md > \$14,000 at the significance level $\alpha = 0.02$. ▶

The selection of a significance level for a sign test is somewhat different from the selection procedures for the t-test and other normal tests. For those tests it is always possible to pick an exact significance level, α, in advance, and then to find a rejection region that gives an area under the curve exactly equal to α. For the sign test you cannot always get an exact significance level like $\alpha = 0.05$ or $\alpha = 0.01$. You need to look at the histogram and the table to determine a reasonable significance level. (Another possible method is to pick a significance level in advance, and then get as close to that value as you can.)

A commuter in New York City was told that the median time for commuting to work from his neighborhood was 47 minutes. His own time was 62 minutes. He felt that he could prove the median was greater, and decided to obtain times from a random sample of ten commuters in his area and perform a test. Set up his test for him. Use a value of α as close to 0.01 as possible. His data set was

Exercise **C**

Times	
23	56
34	61
39	63
43	70
52	81

(He did not use his own commuting time in the data set.) Complete the hypothesis test using this data.

The next example gives a one-tailed test with rejection region on the left instead of the right.*

◀ A college athlete was trying to impress a coed at a party. He told her that he could shot-put an average (median) of 50 feet. She was skeptical, and challenged him to prove his claim by making eight throws. She then planned to use a sign test on the data.

Example **4**

The coed began by setting up the test (steps 1 through 4 of our procedure).

Step 1. The null hypothesis was the athlete's boast.

Null hypothesis: Md = 50

Since the coed did not believe him, her alternative hypothesis was:

Alternative hypothesis: Md < 50

---

* Examples involving two-tailed sign tests will be given in the exercises.

Step 2. There will be eight throws: $n = 8$.

Step 3. (Table and histogram for binomial distribution with $n = 8$ and $p = 0.5$.) See Table 13.4 and Fig. 13.7.

Table 13.4

Successes, k	Probability, $P(x = k)$
0	0.0039
1	0.0312
2	0.1094
3	0.2188
4	0.2734
5	0.2188
6	0.1094
7	0.0312
8	0.0039

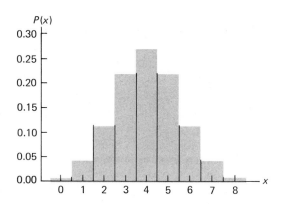

Figure 13.7

Step 4. (Decide on significance level and rejection region.) In this problem the alternative hypothesis is Md < 50. If there are too few throws over 50 feet, then the null hypothesis (Md $= 50$) will be rejected in favor of the alternative hypothesis (Md < 50). Thus our rejection region will be on the left of the histogram where the smaller numbers are. As before, we begin by listing areas on the left to find a reasonable significance level.

Rectangles	Area
First rectangle ($k = 0$)	0.0039
First two rectangles ($k = 0, 1$)	0.0351 ←$\boxed{0.0039 + 0.0312}$
First three rectangles ($k = 0, 1, 2$)	0.1445

The skeptical coed had wanted to do a test with $\alpha = 0.05$. The choice of $\alpha = 0.0351$ was the best she could do here. Her rejection region is shown in Fig. 13.8.

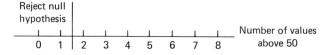

Figure 13.8

Step 5. (Collect data and attach signs.) The next day the athlete, the coed, and a number of other curious people went out to the shot-put area at the track. The athlete made his eight throws, and they were measured (in feet). We have put them in decreasing order, and given them in the following table.

Throw	Sign
50.2	+
50.1	+
49.6	−
49.5	−
49.2	−
49.0	−
48.4	−
47.0	−

Step 6. (Conclusion) There are two plus signs. The number two is *not* in the rejection region, so the null hypothesis cannot be rejected. (The coed was still skeptical, but agreed to stick by her test.) ▶

Suppose that the athlete in the last example had made four more throws under **Exercise D** 50 feet for a total of two throws above 50 feet and ten below. Set up the same hypothesis test with a significance level as near to $\alpha = 0.02$ as possible, and use this data to complete the test.

In all of our examples and exercises, no data value has been equal to the median given in the null hypothesis. Some readers may already have noted that we did not give any instructions for the sign of a data value *equal* to the median. There is a reason for this. A basic assumption of the sign test is that the data used in it is continuous data. For such data, there is almost no chance that a data value will be exactly equal to the given median value. (How many times would the athlete put the shot *exactly* 50 feet?) When a data value appears that is equal to the given median, it is customary to remove it from the data set and reduce the sample size by 1. This assumption means that you should not use the sign test for data such as ratings of 1, 2, 3, 4, 5 on a preference scale. Such data is not continuous, and could lead to too many values equal to the median.

In the next example we will indicate some of the possible differences in the results of parametric and nonparametric tests.

◀ The supervisor of a youth baseball program in a large city gave out a press **Example 5** release in which he described his program. The release stated that the "average" age of his coaches was 47. One coach read this release and decided that this estimate was a bit high. (The supervisor was known to make up statistics to save time.) The coach took a random sample of eight other coaches. Their ages (to the nearest tenth of a year) are shown at the right.

The skeptical coach now has to choose a method of data analysis. The press release (like most newspaper copy) simply refers to an "average" without stating whether it is a median or a mean. The sign test could be used on the null hypothesis Md = 47, or a *t*-test could be done to test the null hypothesis $\mu = 47$. We will give the results of each of these tests below, without going through the details. (The purpose of this example is not to practice calculations, but to compare test results.)

28.5 ⎫
33.8 ⎪
36.3 ⎬ 5 below 47
24.6 ⎪
43.2 ⎭
47.4 ⎫
47.1 ⎬ 3 above 47
50.8 ⎭

a) *Sign test*

Null hypothesis: Md = 47
Alternative hypothesis: Md < 47

We have already done a one-tailed test of this kind in Example 4. We chose a significance level of $\alpha = 0.0351$. As in Example 4, we reject the null hypothesis if the number of values above 47 is 0 or 1.

Since three of the given ages are above 47, *we cannot reject the null hypothesis* Md = 47 *using the sign test.*

b) *t-test*

Null hypothesis: $\mu = 47$
Alternative hypothesis: $\mu < 47$

Here we chose a significance level from our tables as close to 0.0351 as possible; we chose $\alpha = 0.025$. The rejection region for this test (with d.f. = 7) is given by:

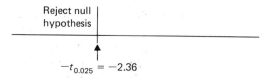

If you calculate the *t*-statistic from the given data set,* you will get:

$$t = \frac{\bar{x} - \mu}{s/\sqrt{n}} = \frac{38.9625 - 47}{9.5989/\sqrt{8}} \doteq -2.368.$$

Thus we reject the null hypothesis $\mu = 47$ *at the level* $\alpha = 0.025$.

The two tests seem to be in conflict. The first says to accept the null hypothesis Md = 47; the second says to reject the null hypothesis $\mu = 47$. If you simply look at the data, you will probably be more inclined to believe the second test because 47 does not seem to be a "typical" value.

We are inclined to believe the *t*-test, since we have studied the population from which this data was actually sampled; it is a normal population with $\mu = 38$ and $\sigma = 10$. For such a population, the mean and the median are the same, so Md = 38, also. It really *was* appropriate to use a *t*-test for this population, and less appropriate to use the sign test. The reason that the sign test did not reject the false null hypothesis Md = 47 is primarily that the sign test did not use all of the information given. The sign-test calculations were based only upon whether values were above or below 47; the actual values were not used. For example,

---

* We calculated this value using all the decimal places that our calculator would carry. If each quantity were rounded to two decimal places, we would get $t = -2.37$.

the sign test counts a value of 44 in exactly the same way as a value of 28.6; both are merely tagged as "below 47." On the other hand, the *t*-test uses the actual age values without throwing any information away. ▶

Example 5 was intended to show that different statistical tests may lead you to make different conclusions *using the same data*. In this example, the *t*-test was actually better; it correctly rejected the null hypothesis $\mu = 47$ while the sign test incorrectly accepted the equivalent null hypothesis Md = 47. When one test outperforms another in this way, statisticians say that it is *more powerful*.

We will discuss the concept of "power" for tests in more detail in Chapter 14. For the time being, it is simplest to think of a more powerful test as one that gives you a better chance of rejecting a false null hypothesis, and thus a better chance of being correct in a statistical inference. In the next section, we will look at the Wilcoxon signed-rank test. This test is also nonparametric, but is more powerful than the sign test in certain situations.

EXERCISES Section **13.3**

1 In 1977, farmers obtained an average of 12.1 cents per pound for chickens. A market analyst who felt that prices had gone up in 1979 did a spot check on a random sample of sales figures during a two-week period. She found costs per pound of:

12.0	11.9	12.3	12.4
12.2	11.7	12.4	12.5

Use this data to test the null hypothesis Md = 12.1 against the alternative hypothesis Md > 12.1. Take α as close to 0.05 as possible.

2 A transportation expert was told that the median load carried by merchant vessels in 1977 was 43 tons. He suspected that this figure was too small, so he checked the load records of a random sample of 11 vessels. Their tonnages were:

44	42	39	44
49	41	35	37
46	47	51	

Use this data to test the null hypothesis Md = 43 against the alternative hypothesis Md > 43. Take α as close to 0.05 as possible.

3 In 1977, the median income of women working full time in the United States was $8618. A random sample of seven working women in a California city gave the incomes:

8610	11428	6092	9385
9346	17347	7817	

Test the null hypothesis Md = 8618 (for women in the city in question) against the alternative hypothesis Md < 8618. Take α as near to 0.01 as possible.

4 A personnel officer for a company is attempting to design a qualifying test in mathematics for prospective employees. She would like to have Md = 70 for all possible test scores, but suspects that the test is too hard. To decide this, she gave the test to a random sample of 20 prospective employees.

a) State the appropriate null and alternative hypotheses.

b) Of the 20 subjects, 12 scored below 70 and eight above. Use this information to perform the hypothesis test from (a) at a level α as close to 0.05 as possible.

5 An employers' association in an Eastern city announces that the median hourly wage for secretaries in that city is $5.00. A secretary who is active in a secretaries' association suspects that this figure is too high.

a) State the appropriate null and alternative hypotheses for the concerned secretary.

b) The secretary took a random sample of 12 other secretaries. Ten of them had hourly wages below $5 and two had hourly wages above $5. Use this information to perform the hypothesis test from (a) at a level α as close to 0.05 as possible.

6 A sports-car enthusiast was told (in early 1979) that he can expect to pay an "average" of $12,000 for a 1978 Corvette. He thought that this might be too low, and decided to perform a test of Md = 12000 against Md > 12000. He took a random sample of prices quoted in the classified ads. They were:

13135	11800	12500	12100
11650	11950	11900	11750

Use this data to perform the hypothesis test at a level α as near to 0.05 as possible.

7 (*Two-tailed sign tests*) If we wish to test the null hypothesis "Md = M" against the alternative hypothesis "Md ≠ M," we will reject the null hypothesis if the total number of + signs is either too large or too small. The following problem illustrates this:

A student is told that the "average" cost of a paperback book is $1.75. She is not sure whether this is correct, but has no idea as to whether it is high or low. She decides to perform the test:

Null hypothesis: Md = 1.75
Alternative hypothesis: Md ≠ 1.75

She intends to take a sample of nine paperbacks and to take α near 0.05.
a) Make a table and a histogram for the binomial distribution with $n = 9$ and $p = 0.5$.
b) Find the values of k (the number of + signs) that give areas on each end of the histogram as close to 0.025 as possible.
c) Use your results from (b) to set up the rejection region.
d) Complete the test using the sample data:

2.95	1.50	2.50
2.50	3.25	2.50
1.95	1.95	2.25

8 The recommended retail price for a certain calculator is $45. A manufacturer's representative wishes to determine whether the median price at retail outlets is $45. He checks a random sample of 14 retail outlets in order to perform a *two-tailed* hypothesis test.

a) Give the null hypothesis, alternative hypothesis, and rejection region if α is to be near 0.05.
b) The sample data showed 11 retail prices below 45 and three above. Use this data to complete the hypothesis test.

9 (*Normal approximation for large n*) For *large* sample sizes, we may use the normal approximation to the binomial distribution to complete a sign test. The following problem illustrates this approach.

A university recommends that its students study two hours outside of class for each hour in class. A professor suspects that most students do not study this much and takes a confidential survey of 400 students.
a) State the null and alternative hypotheses for the professor.
b) Use the normal approximation to the binomial distribution to find a normal curve that approximates the binomial distribution with $n = 400$ and $p = 0.5$.
c) Use the normal approximation from (b) to find the rejection region for the test given in (a). (Use $\alpha = 0.05$.)
d) The professor found that 50 students studied more than two hours for each class hour and 350 students studied less. Use this information to complete the hypothesis test.

10 (*Confidence intervals for the median*) Reasoning similar to that used for the sign test can be used to find confidence intervals for the median. Suppose, for example, that nine numbers are sampled from a population with median M, and are arranged in order from smallest to largest: $x_1, x_2, x_3, x_4, x_5, x_6, x_7, x_8, x_9$.
a) Use the binomial distribution with $p = 0.5$ and $n = 9$ to find $P(M < x_3)$. (*Hint.* $M < x_3$ if and only if at most 2 of the x's fall below M.)
b) Find $P(M > x_7)$.
c) From (a) and (b), find $P(x_3 \leq M \leq x_7)$.
d) Suppose the nine numbers

2.7, 3.1, 3.6, 4.0, 4.4, 5.1, 6.1, 7.2, 8.4

are sampled from a population with median M. Use your result from (c) to find a confidence interval for M.

13.4   The Wilcoxon signed-rank test

In the last section we pointed out that the sign test was inferior to the *t*-test for a study of ages drawn from a normal population. It is more likely that the sign test will accept a false null hypothesis, because it allows too much information to be "thrown away." In this section we will look at the Wilcoxon signed-rank test. This nonparametric test does not "throw away" as much information as the sign test, and is less likely to accept a false null hypothesis. We will use it in Example 6 below to re-analyze the shot-put data from Example 4.

◀ The problem in Example 4 was to test the

Example **6**

Null hypothesis: Md = 50

against the

Alternative hypothesis: Md < 50

for a population consisting of all possible shot-put throws by a college athlete. A coed (who suspected that the alternative hypothesis was true) arranged for a sample of eight throws by that athlete. The distances were

50.2, 50.1, 49.6, 49.5, 49.2, 49.0, 48.4, 47.0

The idea behind the signed-rank test is to calculate the difference between each of these throws and 50, in order to see which throws are farthest from 50. We do this in Table 13.5.

Table 13.5

Throw (in feet), x	Difference, $D = x - 50$	Positive difference, $|D|$	Rank of $|D|$	Signed rank, R
50.2	+0.2	0.2	2	+2
50.1	+0.1	0.1	1	+1
49.6	−0.4	0.4	3	−3
49.5	−0.5	0.5	4	−4
49.2	−0.8	0.8	5	−5
49.0	−1.0	1.0	6	−6
48.4	−1.6	1.6	7	−7
47.0	−3.0	3.0	8	−8

Step 1. Subtract 50 from x

Step 2. Make each difference D positive by taking its absolute value $|D|$

Step 3. Rank the positive differences in order from smallest (1) to largest (8)

Step 4. Give each rank a + sign or − sign, which is the same as the sign in the column for D

Before showing you how to perform a Wilcoxon signed-rank test, we should discuss what the numbers in Table 13.5 mean. The positive differences $|D|$ actually tell you how far away from 50 each x-value is. The ranks R tell you which x-values are closer to 50 and which are farther away, and the signs on these ranks

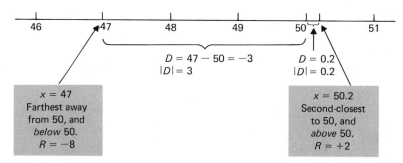

Figure 13.9

tell you whether an x-value is above 50 ($+$ sign) or below 50 ($-$ sign). We picture this situation in Fig. 13.9 for the throws $x = 47.0$ and $x = 50.2$.

The reasoning behind the Wilcoxon test is this: If the null hypothesis "Md $= 50$" is true, then we would expect the sum of the positive ranks and the sum of the negative ranks to have *about* the same *magnitude*—in this case, *about* 18.* If the sum of the positive ranks is "too much smaller" than 18, then we reject the null hypothesis "Md $= 50$" in favor of the alternative hypothesis "Md < 50".

Table 13.6
Critical values and significance levels for a one-tailed Wilcoxon signed-rank test.

Sample size, n	Critical value, d	Significance level, α
⋮	⋮	⋮
7	0	.008
	1	.016
	2	.023
	3	.039
	4	.055
	6	.109
8	1	.008
	2	.012
	3	.020
	5	.039
	⑥	.055
	8	.098
9	3	.010
	5	.020
	8	.049
	11	.102
⋮	⋮	⋮

* The sum of the magnitudes of all the ranks must, in this example, always equal $1 + 2 + 3 + 4 + 5 + 6 + 7 + 8 = 36$, and half of 36 is 18.

Now, for our data we have (see Table 13.5) that the sum of the positive ranks is equal to $2 + 1 = 3$. This seems to indicate that the median is truly below 50, but we really need a table of critical values such as Table IX to tell us when a small positive-rank sum is "too small." Table 13.6 shows a portion of such a table.

The circled numbers in the table tell us that $d = 6$ is the critical value for the level $\alpha = 0.055$ when the sample size is 8. This means that, if our positive-rank sum is *less than or equal to* 6, we can reject the hypothesis $\text{Md} = 50$ and conclude that $\text{Md} < 50$ at the significance level $\alpha = 0.055$.

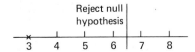

Since our positive-rank sum is actually 3, we can now reject the null hypothesis and conclude that $\text{Md} < 50$, at the level $\alpha = 0.055$.

In Example 4 we performed a sign test on the same data and were unable to reject the null hypothesis. The level we used there was $\alpha = 0.0351$, while the level used here for the Wilcoxon test was $\alpha = 0.055$. But even if we used a significance level as small as $\alpha = 0.02$ in the Wilcoxon test, we could still reject the null hypothesis.*

The population from which this data was sampled actually has a median of 49.3; that is, 49.3 is the median of all shot-puts made by the athlete during the year of this test. Thus you can see that, in this example, the Wilcoxon test is in some sense a better test than the sign test. It rejected a null hypothesis that was false, while the sign test did not reject that same null hypothesis with the same data. ▶

We will now give the step-by-step procedure for a Wilcoxon test, and apply that procedure to the shot-put data.

Procedure for performing a Wilcoxon signed-rank test of the form: Procedure
Null hypothesis: $\text{Md} = M$
Alternative hypothesis: $\text{Md} < M$ (or $\text{Md} > M$)

Step 1. Decide on a sample size n. Use Table IX to find an appropriate significance level α and critical value d, for this sample size.

Step 2. Take a sample of size n, and make a work table of the form:

Sample values, x	Differences, $D = x - M$	Positive differences, $\lvert D \rvert$	Rank of $\lvert D \rvert$	Signed rank, R
⋮	⋮	⋮	⋮	⋮

---
* For a Wilcoxon test with $n = 8$ and $\alpha = 0.020$, the table gives the critical value $d = 3$. Since the positive rank sum was 3, we would *reject* the null hypothesis at the level $\alpha = 0.02$.

Step 3. Compute:

For alternative $Md < M$	For alternative $Md > M$
$R^+ =$ sum of positive ranks	$R^- =$ sum of negative ranks
	$|R^-| = R^-$ made positive

Step 4. Conclusion:

For alternative $Md < M$	For alternative $Md > M$
Reject null hypothesis if $R^+ \le d$.	Reject null hypothesis if $|R^-| \le d$.

Example **7** ◖ Application of step-by-step procedure to Example 6:

Null hypothesis: $Md = 50$
Alternative hypothesis: $Md < 50$

Step 1. We intend to allow eight throws. Thus $n = 8$. Looking at Table IX for a value of α as close to 0.05 as possible, we see $\alpha = 0.055$ for $n = 8$ with a critical value of $d = 6$. We choose:

$$\alpha = 0.055, \qquad d = 6$$

Step 2. Take a sample of size $n = 8$, and construct a work table as shown in Table 13.5 on page 519.

Step 3. (Sums of positive ranks)

Positive ranks
2
$\underline{1}$
$3 = R^+$

Step 4. (Conclusion)

$$R^+ = 3 \text{ and } d = 6, \text{ so } R^+ \le d.$$

We reject the null hypothesis, and conclude $Md < 50$. ◗

Exercise **E** Suppose that the athlete had made one additional throw of 51.1. Then his throws would be:

$$51.1, \; 50.2, \; 50.1, \; 49.6, \; 49.5, \; 49.2, \; 49.0, \; 48.4, \; 47.0$$

Perform a Wilcoxon test of $Md = 50$ against $Md < 50$, using the data and a significance level α as near to 0.05 as possible.

In Example 8 below, we will perform a Wilcoxon test with an alternative hypothesis of the form Md $> M$.* In this test we will look at the sum of the negative ranks, rather than the sum of the positive ranks:

◀ In the autumn of 1977, the Bureau of Labor Statistics estimated that a family   Example **8** of four with an intermediate budget would spend approximately \$4100 for food for one year. A consumer researcher had arranged for twelve randomly chosen, four-member middle-class families in Tempe, Arizona, to make careful records of their food expenditures for 1977. He suspected that prices in that town required *more than* \$4100 per year in food expenditures for such families. He wished to perform a test of the form

Null hypothesis: Md $= 4100$
Alternative hypothesis: Md > 4100

Step 1. Since he had selected 12 families, $n = 12$. He decided to use a significance level as near to $\alpha = 0.02$ as possible. Looking at Table IX, he chose $\alpha = 0.021$ and $d = 13$.

Step 2. (Data and work table) In January, 1978, the researcher collected the food expenditure totals and arranged them in order. They were (to the nearest dollar):

5197	4723	4501	3963
4982	4695	4320	3872
4817	4632	4111	3604

Using this data, the researcher made a work table (Table 13.7).

Income, x	Difference, $D = x - 4100$	$|D|$	Rank	Signed rank, R
5197	1097	1097	12	$+12$
4982	882	882	11	$+11$
4817	717	717	10	$+10$
4723	623	623	9	$+ 9$
4695	595	595	8	$+ 8$
4632	532	532	7	$+ 7$
4501	401	401	5	$+ 5$
4320	220	220	3	$+ 3$
4111	11	11	1	$+ 1$
3963	-137	137	2	$- 2$
3872	-228	228	4	$- 4$
3604	-496	496	6	$- 6$

Table 13.7

---

* Two-tailed Wilcoxon tests will be covered in the exercises.

Step 3. (Signed rank sum) Since the alternative hypothesis is Md > 4100, we need only the negative signed-rank sum here:

$$
\begin{array}{r}
-2 \\
-4 \\
-6 \\
\hline
-12 = R^- \\
12 = |R^-|
\end{array}
$$

Step 4. (Conclusion) Since $d = 13$ and $|R^-| = 12$, $|R^-| \leqslant d$. Thus we can reject the null hypothesis and conclude that Md > 4100. ❯

Exercise **F**   Suppose that the researcher had not been able to get the food expenditures from the first family (whose budget was 5197). His numbers would then be:

4982	4695	4320	3872
4817	4632	4111	3604
4723	4501	3963	

Use this data to perform the same hypothesis test: Md = 4100 against Md > 4100, with a significance level as close to $\alpha = 0.02$ as possible.

Assumptions   At this point the reader may be wondering why there is a sign test at all; the Wilcoxon test often seems better. One reason is that the Wilcoxon test has more restrictive assumptions than the sign test. The basic assumptions of the Wilcoxon test are:

A1. Sample data is continuous
A2. The sampled population is symmetric.

The new assumption, which was not required for the sign test, is A2. To say that a population is **symmetric** simply means that its probability distribution curve can be cut into two halves that are *mirror images* of each other.

The Assumption A2 is hard to check, but there may be times when you are quite sure that it does *not* hold. In such cases, you would use a sign test rather than a Wilcoxon test.

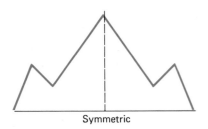
Symmetric

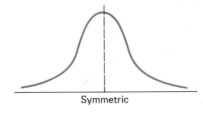
Symmetric

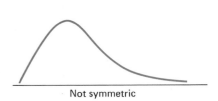
Not symmetric

This discussion allows us to make an important point about nonparametric tests. They are not free of all assumptions; they have their own assumptions that must be checked. Many people who have taken one or two statistics courses seem to believe that, since nonparametric tests do not assume normality, these tests can be used without checking anything at all. That is not true.

Occasionally two differences will have the same rank. If two differences are tied for second place, we give them each rank 2.5, and award rank 4 to the next value (which is really fourth). If there is a tie for third, we give two ranks of 3.5, award rank 5 to the next value, and so on. We illustrate this below in Table 13.8, which is a work table for application of a Wilcoxon test to the age data of Example 5. (In that example, the null hypothesis was that the median equals 47.)

Ties

Table 13.8

Age, x	$x - 47 = D$	$\|D\|$	Rank	Signed rank, R
28.5	− 18.5	18.5	7	−7
33.8	− 13.2	13.2	6	−6
36.3	− 10.7	10.7	5	−5
47.4	+ 0.4	0.4	2	+2
24.6	− 22.4	22.4	8	−8
47.1	+ 0.1	0.1	1	+1
43.2	− 3.8	3.8	3.5	−3.5 ← ┐
50.8	+ 3.8	3.8	3.5	+3.5 ← ┘ Tied

Exercise G

Use Table 13.8 to perform a hypothesis test of the form:

Null hypothesis: Md = 47
Alternative hypothesis: Md < 47

for the population of ages of youth baseball coaches. Use the significance level $\alpha = 0.020$.

Comparison of the Wilcoxon signed-rank test and the t-test for a single mean

For data from normal populations, there is a slight chance that a t-test will lead to rejection of a false null hypothesis while a Wilcoxon test leads to the misleading conclusion that the null hypothesis cannot be rejected (that is, the t-test is slightly more powerful). This will not happen very often. On the other hand, if data is taken from a symmetric, nonnormal population, the Wilcoxon signed-rank test will be as good as the t-test or better in the accuracy of its conclusions.

If you have reason to believe your underlying population is symmetric but cannot be sure that it is normal, use the Wilcoxon test instead of the t-test.

EXERCISES Section **13.4**

1 Suppose that the athlete in Example 6 had made throws of

$$50.5, 50.2, 50.1, 49.4, 49.3, 48.9, 48.1, 46.5$$

Use this data and the Wilcoxon test to test the null hypothesis Md = 50 against the alternative hypothesis Md < 50. Use an α-level as close to 0.05 as possible.

2 Use the Wilcoxon signed-rank test to perform the hypothesis test on shipping tonnage given in Exercise 2 of Section 13.3:
Null hypothesis: Md = 43
Alternative hypothesis: Md > 43
$\alpha \doteq 0.05$
Data: 44, 49, 46, 42, 41, 47, 39, 35, 51, 44, 37

3 Use the Wilcoxon test to perform the test in Exercise 3 of Section 13.3.
Null hypothesis: Md = 8618
Alternative hypothesis: Md < 8618
$\alpha \doteq 0.01$
Data: 8610, 9346, 11428, 17347, 6092, 7817, 9385

4 A psychologist is told that the median IQ of an assembly-line worker at a large plant is 97. She feels that this is too low, and gives IQ tests to a random sample of 12 workers. The scores are:

95	98	105	115
93	87	110	88
121	90	99	104

Perform an appropriate hypothesis test with $\alpha \doteq 0.01$.

5 In 1977, the median age of a person arrested in the United States was approximately 23. A district attorney believed that the median age of persons arrested for gambling was higher than 23. A random sample of arrest records showed the following ages for ten gambling arrests:

35	28	31	53	35
60	22	18	42	47

Use this data and the Wilcoxon signed-rank test to perform the appropriate hypothesis test. Take α as near to 0.05 as possible.

6 In 1977, the median age of a United States citizen was 29.4 years. A random sample of 14 trailer-park residents in Phoenix, Arizona, gave ages of:

68	14	18	47	6
72	59	79	63	64
38	25	83	65	

Use the Wilcoxon signed-rank test to perform the test:
Null hypothesis: Md = 29.4
Alternative hypothesis: Md > 29.4
$\alpha = 0.01$

7 (*Two-tailed tests*) The Wilcoxon signed-rank test was used only for one-tailed tests in this section. It may also be used for two-tailed tests of the form:

Null hypothesis: Md = M
Alternative hypothesis: Md ≠ M

The idea in such a test is to reject the null hypothesis if either R^+ or $|R^-|$ is too small. This is illustrated in the following exercise:

The owner of a shopping mall will ask the ages of a random sample of 20 shoppers in order to perform the test:

Null hypothesis: Md = 29.4
Alternative hypothesis: Md ≠ 29.4

a) Use Table IX to find the critical value d for a one-tailed test with $\alpha = 0.020$.

b) A two-tailed test with $\alpha = 0.04$ can be performed by using the value of d from (a) and the rule "Reject the null hypothesis if *either* the positive rank sum, R^+, *or* the absolute value of the negative rank sum, $|R^-|$, is less than or equal to d." Perform this test using the data:

35	28	14	68	43	17	25
62	31	54	47	24	25	18
17	22	12	21	19	36	

c) Use the reasoning in (b) to determine the critical value, d, that would be used for a two-tailed test with $\alpha = 0.02$.

8 In this problem we will find the probability distribution of R^+ for $n = 3$ and $n = 4$. This will enable you to see how the critical values for the Wilcoxon test are derived.

a) In Table 13.9 the rows give all possible signs for the signed ranks in a Wilcoxon test with $n = 3$. For example, row 1 covers the possibility that all three data values are above M and thus have the sign +. There is an empty column for values of R^+. Fill it in. (*Hint.* The first entry is 6 and the last is 0.)

Table 13.9

Rank			
1	2	3	R^+
+	+	+	
+	−	+	
+	+	−	
+	−	−	
−	+	+	
−	−	+	
−	+	−	
−	−	−	

b) If the null hypothesis for a Wilcoxon signed-rank test is true, what is the probability that a sample will fit any single row in the table? (The answer is the same for each row.)

c) Use the answer from (b) to find the probability of each possible value of R^+ and fill in the table below.

R^+	Probability
0	
1	
2	
3	
4	
5	
6	

d) Draw a histogram of the probability distribution of R^+.

e) What critical value would you use for R^+ if you wished to test Md $= M$ against Md $< M$ at the level $\alpha = 0.125$? (Read your answer from the histogram.)

f) Compare your critical value with the critical value given for $n = 3$ in Table IX.

g) Go through the above procedure for a Wilcoxon test with $n = 4$, and compare the results in your histogram to the critical values in Table IX.

9 (*Normal approximation for large samples*) For large sample sizes, the statistics R^+ and $|R^-|$ considered in this section are approximately normally distributed with mean and variance

$$\mu = \frac{n(n+1)}{4}, \qquad \sigma^2 = \frac{n(2n+1)(n+1)}{24}$$

The use of this approximation can be illustrated using the situation in Exercise 6, since $n = 14$ is in the sample size range considered "sufficiently large."

a) Compute μ and σ above using $n = 14$.

b) Use the normal distribution with μ and σ as above, to find the rejection region for a one-tailed (lower-tailed) test with $\alpha = 0.01$.

c) Compare your answer to that found in Exercise 6. Would the use of this normal approximation have changed your results in the hypothesis test?

The Mann–Whitney test   13.5

In Sections 13.3 and 13.4 we showed how to test a hypothesis about the median of a single population using data from one sample. In this section we will discuss the most widely used distribution-free test for comparison of the medians (or means) of *two* populations using two independent samples. This test involves ranks, and should remind you of the Wilcoxon test.

◖ A nationwide shipping firm purchased a new computer system to keep track of the present status of all its current shipments, pickups, and deliveries. This system was linked to computer terminals in all regional offices, where office personnel could type in requests for information on the location of shipments, and get answers immediately on display screens. The company had to set up a training program for use of the computer terminals, and decided to hire a technical writer to write a short self-study manual for this purpose. This manual was designed so that a person could read it and be ready to use the computer terminal

Example **9**

in two hours. The last page of the manual contained some information requests for a person to submit to the computer. If a person had learned to use the terminal correctly, the request he typed in would be answered correctly. If not, the request would be answered with the reply "STUDY SOME MORE", and the person would go back to the manual.

The company found that, in practice, the manual took very little time for some people and quite a bit of time for others. Someone suggested that this could have happened because some employees had previous experience with computers and others did not. To test this suggestion, the company decided to time a sample of 15 employees with similar educational levels but different computer backgrounds; eight had computer experience of some kind, and seven did not. The times required for them to learn to use the computer were:

With experience (I)	No experience (II)
2.33 hours	2.31 hours
1.81	1.96
2.17	2.73
1.78	2.51
1.74	3.04
1.46	2.34
1.58	2.24
1.92	

The most obvious way to compare the two populations seemed to be a *t*-test for comparison of means, but the company statistician had already observed that the times seemed far from normally distributed for either population. Thus the statistician decided to perform a *Mann–Whitney test*. In other words, her null and alternative hypotheses were:

Null hypothesis: There is no difference in median learning times for the populations of experienced and inexperienced workers.
Alternative hypothesis: The median learning time for experienced workers is smaller.

The idea behind this test is a simple one: Rank all of the times from the two samples as in Table 13.10.

Group I (with computer experience)		Group II (without computer experience)		Table 13.10
Time	Overall rank	Time	Overall rank	
2.33	11	2.31	10	
1.81	5	1.96	7	
2.17	8	2.73	14	
1.78	4	2.51	13	
1.74	3	3.04	15	
1.46	1	2.34	12	
1.58	2	2.24	9	
1.92	6			

If experienced workers really do have a smaller median learning time, then the ranks for Group I should be lower. A natural first step in this test would be to compute the sums of the ranks for each of the two groups and compare the results.

Rank sums for groups

I	II
11	10
5	7
8	14
4	13
3	15
1	12
2	9
6	$\overline{80} = S_{II}$
$\overline{40} = S_I$	

The fact that the rank sum for Group I is smaller seems to indicate that experienced workers have a smaller median learning time. But we have not discussed any statistical test that shows this with reasonable probability. We will not do this directly with rank sums. Instead we will find critical values for a test statistic, T, computed from the rank sums. To define T, we need some notation. We will let:

n = the size of the first sample,

k = the size of the second sample.

For our problem here,

$$n = 8, \qquad k = 7.$$

The formula for T is

$$T = S_1 - \frac{n(n + 1)}{2}.$$

For our problem here,

$$T = 40 - \frac{8 \cdot (8 + 1)}{2} = 40 - 36 = 4.$$

(The statistic T is very closely related to S_1. The number $n(n + 1)/2$ is the smallest value that S_1 could possibly have, and T measures how close the actual value of S_1 is to its smallest possible value.)

We have already pointed out that small values of S_1 indicate that experienced workers learn faster. Small values of T indicate the same thing. Table 13.11 shows the critical values that tell you how small T needs to be in order to reject the null hypothesis at the significance level $\alpha = 0.05$. To read the table, recall that n is the size of the first sample and k is the size of the second sample. For our current problem, $n = 8$ and $k = 7$. The critical value we want is circled in the table.

Table 13.11

One-tailed critical values of the Mann–Whitney test statistic T for $\alpha = 0.05$

n \ k	2	3	4	5	6	7	8	9	10	11	12	13	14	15
2	*	*	*	0	0	0	1	1	1	1	2	2	3	3
3	*	0	0	1	2	2	3	4	4	5	5	6	7	7
4	*	0	1	2	3	4	5	6	7	8	9	10	11	12
5	0	1	2	4	5	6	8	9	11	12	13	15	16	18
6	0	2	3	5	7	8	10	12	14	16	17	19	21	23
7	0	2	4	6	8	11	13	15	17	19	21	24	26	28
8	1	3	5	8	10	(13)	15	18	20	23	26	28	31	33
9	1	4	6	9	12	15	18	21	24	27	30	33	36	39
10	1	4	7	11	14	17	20	24	27	31	34	37	41	44
11	1	5	8	12	16	19	23	27	31	34	38	42	46	50
12	2	5	9	13	17	21	26	30	34	38	42	47	51	55
13	2	6	10	15	19	24	28	33	37	42	47	51	56	61
14	3	7	11	16	21	26	31	36	41	46	51	56	61	66
15	3	7	12	18	23	28	33	39	44	50	55	61	66	72

* No test can be done for these sample sizes with $\alpha \leq 0.05$.

Here the critical value is 13. For rejection of the null hypothesis, the T-value must be *less than or equal to* 13. Since, in this case, our T-value calculated from the data is 4, we reject the null hypothesis, and conclude that computer-experienced workers have shorter learning times on the average. ▶

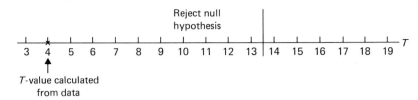

T-value calculated
from data

We will summarize the procedure we have just discussed, and then show how it would actually be applied, step-by-step, to the previous example. We use Md_1 to stand for the median of Population I, and Md_2 for the median of Population II.

Procedure for a Mann–Whitney test on independent samples from two populations*
Procedure

Null hypothesis: The two populations have the same median. That is, $Md_1 = Md_2$.

Alternative hypothesis: The median of Population I is smaller than the median of Population II. That is, $Md_1 < Md_2$.

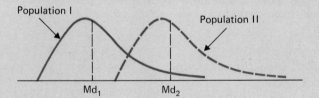

Step 1. Decide on sample sizes:

$$n = \text{sample size from I.}$$
$$k = \text{sample size from II.}$$

Step 2. Use Table 13.11 to find the critical value c for the level $\alpha = 0.05$. (We will restrict our attention to $\alpha = 0.05$ to make this section simpler. Books on nonparametric statistics contain tables for other values of α.)

Step 3. Take the sample data, assign ranks, and compute $S_I = $ sum of the ranks from I.

Step 4. Calculate $T = S_I - \dfrac{n(n + 1)}{2}$

Step 5. Reject the null hypothesis if $T \leq c$.

---

* This test is based on the assumption that the two populations have distribution curves of the same "shape," as pictured above. In many texts, the null hypothesis is "the two populations have the same distribution." We have stated the hypotheses in terms of medians to maintain consistency with the previous section.

Example **10** ◖ The two populations here are

   I: learning times for all workers with computer experience;
  II: learning times for all workers without such experience.

The hypotheses to be tested are:

Null hypothesis: $Md_1 = Md_2$. That is, there is no difference in median learning times for experienced and inexperienced workers.

Alternative hypothesis: $Md_1 < Md_2$. That is, experienced workers have a smaller median learning time.

Step 1. Decide on sample sizes:

$$n = 8 \quad \text{from I.}$$
$$k = 7 \quad \text{from II.}$$

Step 2. From Table 13.11, the critical value for $\alpha = 0.05$, $n = 8$, and $k = 7$ is $c = 13$.

Step 3. a) Collect and rank sample data (see page 529). We repeat Table 13.10 as Table 13.12.

      b) Compute the rank sum for I: $S_1 = 11 + 5 + 8 + 4 + 3 + 1 + 2 + 6 = 40$.

Table 13.12

Group I (with computer experience)		Group II (without computer experience)	
Time	Overall rank	Time	Overall rank
2.33	11	2.31	10
1.81	5	1.96	7
2.17	8	2.73	14
1.78	4	2.51	13
1.74	3	3.04	15
1.46	1	2.34	12
1.58	2	2.24	9
1.92	6		

Step 4. Calculate T:

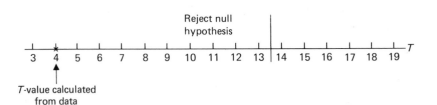

Step 5. $T = 4$ and $c = 13$. Thus, $T \leq c$, and so we reject the null hypothesis, and conclude that workers with computer experience have a smaller median learning time. ▶

Exercise **H**

An algebra teacher always gives a placement test with 30 questions to her students at the beginning of the semester. She tells students with scores of 8 or less that they are unprepared, and students with scores of 20 or more that they are well-prepared. (She is not sure about students between 8 and 20.) She decides to perform a hypothesis test to determine whether students who score high (≥ 20) on the placement test actually perform better on a final exam than students who score low (≤ 8) on the placement test. The two populations are:

I: Final-exam scores for students with low placement scores;
II: Final-exam scores for students with high placement scores.

a) State appropriate null and alternative hypotheses for this test.
b) The teacher randomly selects five "lows" and six "highs." She gives these 11 students a comprehensive 1000-point final exam. The results are shown in Table 13.13.

Low placement (I)	High placement (II)
427	532
583	671
658	735
722	892
842	927
	958

Table 13.13
Final-exam scores

Use these data to perform the hypothesis test in (a) at the level $\alpha = 0.05$.

Ranks are awarded to tied values by the method used in the Wilcoxon test. We illustrate this with an example:

Ties

◀ The company in Example 9 tested two *new* samples of workers on the learning package for computer-terminal use. The learning times were:

Example **11**

Workers with computer experience (I)	Workers with no computer experience (II)
1.81	1.97
2.16	2.30
2.23	2.47
2.30	2.51

We rank them as shown in Table 13.14. ▶

Table 13.14

	I (Experience)				II (No experience)			
Time	1.81	2.16	2.23	2.30	1.97	2.30	2.47	2.51
Rank	1	3	4	5.5	2	5.5	7	8

↖ Tied for ↗
rank 5

Exercise I   Use the data in Table 13.14 to test the hypothesis that workers with computer experience learn terminal use, on the average, faster than workers without computer experience (that is, experienced workers have a smaller median learning time).

Assumptions for the Mann–Whitney test   The Mann–Whitney test is based on the assumptions that data is continuous, and that we have independent samples. It is also based on the assumption that the two populations compared have distributions of the same "shape."

Other tests for comparison of two groups   In this section we showed how to do comparisons based on two *independent* samples. There are also nonparametric tests for comparisons based on *paired data* from two *dependent* samples. These tests actually are applications of the sign test and the Wilcoxon test covered in Sections 13.3 and 13.4. We will show how to do such tests in the exercises.

EXERCISES Section **13.5**

1   A consumer organization collected information on the lifetimes (up to major breakdown) of power lawnmowers from two manufacturers. The sampled lifetime data (in years) on five randomly selected mowers from each manufacturer were:

Manufacturer I	Manufacturer II
2.3	1.9
3.7	3.8
5.9	6.4
6.8	5.6
3.5	4.9

Use a Mann–Whitney test with $\alpha = 0.05$ to test the null hypothesis of no difference in median lifetime between mowers from the two manufacturers, against the alternative hypothesis that mowers from Manufacturer II have a larger median lifetime.

2   An appliance dealer did a study on the time necessary for experienced and inexperienced workers to install a major appliance in a new tract home. The times (in hours) for nine randomly selected jobs involving the same appliance and home model were:

Experienced	Inexperienced
32	45
38	42
41	51
33	48
35	

Use a Mann–Whitney test with $\alpha = 0.05$ to test the null hypothesis of no difference in median time between experienced and inexperienced workers, against the alternative hypothesis that experienced workers have a smaller median time for installation.

3 A college chemistry teacher was concerned about the possible bad effects of poor math background on his students. He randomly selected ten students for a study, and divided them according to mathematics background. Their semester averages are shown in Table 13.15.

Table 13.15

Group I (Less than two years of high school algebra)	Group II (Two or more years of high school algebra)
81	75
62	91
53	76
58	83
	64
	49

Use a Mann–Whitney test with $\alpha = 0.05$ to test the null hypothesis of no difference in median performance of students with the two backgrounds against the alternative hypothesis that students with less than two years of high school algebra have a lower median score in chemistry.

4 A business researcher priced a specific model of color TV at a random sample of discount stores and neighborhood TV stores. The prices were:

Discount stores	TV stores
600	550
530	590
540	610
570	630
580	620
	595

Use a Mann–Whitney test with $\alpha = 0.05$ to test the null hypothesis of no median price difference against the alternative hypothesis that discount houses have a lower median price. (Such a test does not take into account possible service benefits from a neighborhood-store purchase.)

5 A testing laboratory subjected samples of 50-pound test cord from two manufacturers to a stress test. Below are the forces that the sample cords could actually stand:

Manufacturer I	Manufacturer II
52	58
55	57
49	56
53	53
52	55

Use a Mann–Whitney test with $\alpha = 0.05$ to test the null hypothesis of no difference between cords from the two manufacturers, against the alternative hypothesis that Manufacturer I makes cord with less strength. (Note that there are *ties* in this data set.) Be sure to give precise statements of the hypotheses.

6 (*Distribution of the statistic T*) Suppose that the sample sizes from the populations are $n = 3$ members from Population I and $k = 2$ members from Population II. We can display all possible arrangements of ranks in our data by making a table like Table 13.16, in which the letter A stands for a member of Population I, and the letter B stands for a member of Population II. [The column headed "U" is explained in part (d).]

Table 13.16

Rank							
1	2	3	4	5	S_1	T	U
A	A	A	B	B	6	0	6
A	A	B	A	B			
A	A	B	B	A			
A	B	A	A	B			
A	B	A	B	A			
A	B	B	A	A			
B	A	A	A	B			
B	A	A	B	A			
B	A	B	A	A			
B	B	A	A	A			

a) Compute the statistics S_1 and T for each row of the table, and enter those values in the appropriate columns.

b) Use your results from (a) to fill out the following table, which gives the probability distribution of T. (You may assume that each row of the table is equally likely.)

T	Probability
0	
1	
2	
3	
4	
5	
6	

c) Use your results from (b) to explain why we cannot use $\alpha = 0.05$ as a significance level for a Mann–Whitney test with $n = 3$ and $k = 2$. Give the smallest α-level that is possible for such a test.

d) Often another statistic, U, is defined for use with Mann–Whitney tests, where U is simply the number of data pairs (a, b) for which a is from Population I, b is from Population II, and $a < b$. In row 1 of Table 13.16 in this problem, $U = 6$ because each of the two elements from Population II is larger than all three elements from Population I. Compute U for each subsequent row of the table, and state the relationship between U and T that is apparent.

7 (*Normal approximation for large sample sizes*) The distribution of the T-statistic used here may be approximated for large sample sizes by a normal distribution with

$$\mu = \frac{nk}{2}, \qquad \sigma = \sqrt{\frac{n \cdot k \cdot (n + k + 1)}{12}}$$

a) Find μ and σ above for $n = 15$ and $k = 10$.

b) Use the normal distribution with μ and σ as given in (a) to find the rejection region for a lower-tailed test with $\alpha = 0.05$. Compare the critical value obtained here with that given in Table 13.11 for T under $n = 15$ and $k = 10$.

c) Do a similar comparison of critical values for $n = 15$, $k = 15$.

8 (*Dependent samples*) The samples used in the Mann–Whitney test are independent samples from two populations. A different test is used for paired dependent samples, the Wilcoxon test. We illustrate this below.

A remedial-mathematics teacher gave a standardized test to ten remedial students on the first and last days of class. The scores are shown in Table 13.17.

Table 13.17

Student	First-day score, x	Last-day score, y	Difference $d = y - x$
1	10	8	
2	12	20	
3	8	12	
4	14	12	
5	9	21	
6	6	10	
7	20	18	
8	6	8	
9	19	27	
10	10	14	

a) The differences $d = y - x$ indicate whether a student improved or showed no improvement. Calculate these d-values.

b) Let Md stand for the median of the d-values. The teacher plans to do a test with:

Null hypothesis: Md = 0.

He wishes to use an alternative hypothesis that would show that he brought about some improvement. What should it be?

c) Use the d-values given above to perform this hypothesis test, using the Wilcoxon signed-rank procedure with $\alpha = 0.05$.

13.6 The runs test

We have used the word "random" many times in this book, but we have never given a statistical test for randomness. The **runs test** is such a test. We illustrate its use directly in Example 12.

Example 12 ◀ Twenty students in a university political-science class were assigned a project; each student was to make a *random* selection of ten other university students to fill out a voter-preference survey form. The survey form that was filled out for the class did not have a space for the respondent's name, but the respondent was asked to check off boxes that indicated age, income level, and sex. Each political-science student who administered the survey was asked to number his ten forms, so that the teacher could see the order in which they were given out. When the teacher was given the forms, he became worried about the samples taken by two students, whom we shall call A and B. The teacher was worried because it appeared, from the sexes of the respondents, that the forms had not been handed out randomly: The order in which the forms had been handed out to males (M) and females (F) was:

Student A: MMMMMFFFFF
Student B: MFMFMFMFMF

The teacher called in Students A and B, and asked how they had handed out their forms. Their answers showed that their samples were *not* random. Student A had given out five forms at his fraternity house, then decided that his sample should be half female, and gave out five forms at his girl friend's sorority house. Student B had gone to the cafeteria and asked passersby to fill out his forms. However, he also had decided that his sample should be half male and half female, so he alternated asking men and women to fill out his form. Neither sample was random, nor was it representative of the university population. The university actually had a sex distribution of 57 percent males and 43 percent females. The teacher had noticed this departure from randomness, because he had looked at the number of *runs* in the sexes on the questionnaires. A **run** is a sequence in which the same letter is repeated one or more times with a different letter (or a blank space) at each end. Here are the runs for Students A and B, and the total number of runs, R, for each student:

A: M̲M̲M̲M̲M̲ F̲F̲F̲F̲F̲ $R = 2 =$ total number of runs

   Run 1   Run 2

B: M̲ F̲M̲ F̲M̲ F̲M̲ F̲M̲ F̲ $R = 10$

Run: 1 2 3 4 5 6 7 8 9 10

As you can see, a run can be several letters long or only one letter long. The fact that Student A had only a few (long) runs indicated that he was consistently sampling the same kind of person many times in a row. The fact that Student B had a large number of (short) runs indicated that he was systematically switching from M to F and back. This is the general idea behind the runs test: If you keep track of the order in which a sample was made and count runs, too *many* runs or too *few* runs will indicate that the sample was not randomly selected. We have actually been talking about two separate runs tests here: one for Student A, who seems to have too few runs, and one for Student B, who seems to have too many. Table X shows which values of R (the number of runs) are "too large" or "too small." For these tests we have reproduced part of that table as Table 13.18.

Table 13.18
Critical values for a two-tailed
runs test with $\alpha = 0.05$.

Number of times
one letter appears
↓

	5	6	7	8
2	*	*	*	*
3	*	2 8	2 8	2 8
4	2 9	2 9	2 10	3 10
5	(2 10)	3 10	3 11	3 11
6	3 10	3 11	3 12	3 12

Number of times →
other letter appears

* Boxes with an * represent cases for which no
$\alpha = 0.05$ critical values can be found.

The numbers at the top and the left of the table are the numbers of times the two letters appear. In our example with Students A and B, each letter appears a total of five times, so the entry to look at inside the table is the one we have circled. The two numbers inside the table tell us which numbers of runs are too large or too small. When we see the table entry $\boxed{\tfrac{2}{10}}$ we read it as

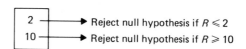

Since the R-value for Student A was $R = 2$, the teacher could conclude that Student A's sample was not random at the level $\alpha = 0.05$. The same conclusion could be made for Student B (where $R = 10$) at the same level. We will summarize this test below and apply it to another example. ▶

Procedure

Procedure for a runs test of the form:
Null hypothesis: Two symbols (such as M and F) occur in random order in a sampling sequence.
Alternative hypothesis: The symbols do not occur in random order (i.e., the sample was not random).

Significance level: We will perform this test only at the level $\alpha = 0.05$.

Step 1. Write down the letter sequence and mark off each run in it.

Step 2. Find

$$R = \text{the number of runs.}$$

Step 3. Count the number of times each letter appears in the sequence. This will give two numbers. Enter Table X under these two numbers and obtain two critical values.

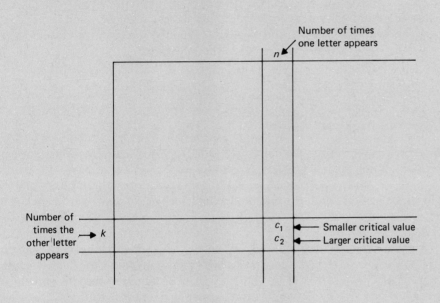

Step 4. Reject the null hypothesis if $R \leqslant c_1$ or if $R \geqslant c_2$.

◖ The teacher of the political science class in Example 12 sent out Student C to give the voter-preference form to 11 students. The teacher decided that he would do a runs test on sex to decide whether or not the sample taken by Student C was random (according to sex). The teacher intended to do a hypothesis test of the following form: Example **13**

Null hypothesis: In Student C's sample, the letters M and F occur in random order.

Alternative hypothesis: The letters M and F do not occur in random order (the sample is not random).

Student C's sampling sequence was:

MMMMFFMMMMF

The steps in the runs test are:

Step 1. (Identify runs in letter sequence)

MMMM	FF	MMMM	F
Run 1	2	3	4

Step 2. (Number of runs)

$$R = 4$$

Step 3. (Find critical values in Table X.) There are eight M's and three F's, so we will use the table entry under 8 (from the top of the table) and 3 (from the side). The critical values in the table are:

Step 4. We cannot reject the null hypothesis. $R = 4$ is neither less than or equal to 2, nor greater than or equal to 8. ▶

Exercise **J**   Perform the same hypothesis test for a sample of ten students with the sequence

MFMMFFMFMF

We should discuss Example 13 a bit further since it illustrates how statistical tests may not be powerful enough to reject false hypotheses. The teacher in charge of this project talked to Student C, and C admitted that his sample was not randomized. He had given his forms only to his personal friends, who were predominantly male and all upper-division political-science majors. He had made no attempt to get a truly random or representative sample of the university community. However, the runs test was not powerful enough to establish this. A statistician who had done a runs test and then talked to the student would ignore the results of the runs test, and say that the sample taken by Student C was not random.

EXERCISES Section **13.6**

1 A chamber of commerce representative was assigned to interview people at a beach to determine why they had come to this particular beach. The sexes of the people he interviewed (in order of interview) were:

FFFMMMFFFFMFFMMF.

Perform a runs test to determine whether this interviewer had selected people at random with respect to sex (as he was instructed to do).

2 A young employee in a department store was assigned to stop randomly chosen shoppers to ask whether they had noticed a display at the store entrance. Here is the sequence of her interviews, classified as to over thirty (O) and under thirty (U):

UOUOUOUUOUOU.

(If the shopper refused to reveal his or her age, that interview was not counted.) Perform a runs test to determine whether the new employee had selected people at random with respect to age.

3 The daily high temperature in a Colorado town was recorded each day for part of a winter month in 1979. Each day was classified as above normal (A) or below normal (B). Here is the sequence of classifications.

AABBBBBBBBBBBBAAAABBAABBB.

Determine whether the daily high temperatures were following a random pattern.

4 (*Probability distribution for number of runs*)
a) Suppose that you are to pick a sample of three males and two females. Below we have listed one sequence in which they might be selected. There are nine more sequences; find them.

Sequence 1:   MMMFF

b) Find the number of runs for each sequence.
c) Use your result in (b) to find the probability for each possible number of runs with $n = 3$ and $k = 2$. (You may assume that each sequence has an equal probability.)
d) Explain why no runs test with $\alpha \leq 0.05$ can be done for $n = 3$ and $k = 2$.

You will often have to decide whether to use a parametric or a nonparametric test in a particular situation. As you have seen in the last few sections, a parametric test that requires that the population be normally distributed will usually correspond to a nonparametric test that does nearly the same job for data from nonnormal populations. We list a few of these corresponding tests below.

Comparison of parametric and nonparametric tests 13.7

Parametric	Nonparametric
t-test for $\mu = \mu_0$	Sign test for Md $= M$
	Wilcoxon test for Md $= M$
t-test for $\mu_1 = \mu_2$	Mann–Whitney test for $Md_1 = Md_2$

There are many more nonparametric tests (ones we have not covered here). There are nonparametric correlation measures and nonparametric methods for analysis of variance. We have already seen how several of these parametric and nonparametric tests compare. For data from *normal* populations, the parametric tests are more powerful.*

However, the nonparametric tests are more powerful for data from quite *nonnormal* populations, and are often "nearly as good" as the parametric tests for normal population data. For this reason, some statisticians recommend the use of nonparametric tests in a wide variety of situations. Entire books have been written on nonparametric statistics, and a great variety of nonparametric tests are available. For the general reader of this book, we recommend the following rule of thumb for choosing the appropriate kind of test.

1. If you are sure that your data is taken from an approximately *normal population*, use a *parametric test*.

---

* Recall that "more powerful" means "more likely to reject a false null hypothesis." In simpler terms, a more powerful test is less likely to lead to a false conclusion.

2. If you have reason to suspect that your data is taken from a population that is *not normal,* use a *nonparametric test.* (For some nonparametric tests, you will have to look beyond this book. We have included references to some readable books on nonparametric statistics at the end of this section.)

3. If your data is *ordinal* or *qualitative* (i.e., not metric), look for an appropriate *nonparametric test.* Parametric tests are designed for use on metric data, and generally should not be used on other kinds of data.

In the past it was common to recommend nonparametric tests because their calculations were simpler. The computer has changed that. Most researchers have all of their calculations done by computers; the time it takes to do a calculation by hand (and head) is no longer an issue.

Finally, some authors state that nonparametric methods are better for beginning students because the logic behind them is simpler and easier to understand. We will leave it up to the reader to decide that issue individually.

The reader has probably noticed that, although we are quite favorable toward nonparametric tests, we have spent most of this book on the development of parametric tests. There is a reason for that: The normal distribution and the tests based upon it played a major part in the development of inferential statistics. Despite the fact that nonparametric tests are quite useful, you will probably see far more parametric tests in any reading you do. A major reason for studying statistics is to be able to read and understand reports that contain statistical analysis. Many of our readers will never set up a real statistical experiment, but most of them will read studies that use a z-score, a correlation coefficient, or a confidence interval for a mean. A knowledge of parametric statistics is required for this purpose.

CHAPTER REVIEW

Key Terms		
	robust	Wilcoxon signed-rank test
	normally distributed population	rank
	distribution-free tests	signed rank
	nonparametric tests	$R^+, R^-, |R^-|$
	parametric tests	symmetric
	binomial distribution	Mann–Whitney test
	sign test	rank sums
	median (Md)	T-statistic
	t-test	runs test
	power	run

binomial distribution:

If n independent success–failure experiments are performed with success probability p, and if x is the random variable representing the number of successes, then x has the *binomial distribution* with parameters n and p. That is,

$$P(x = k) = \binom{n}{k} p^k (1 - p)^{n-k}$$

where

$$\binom{n}{k} = \frac{n!}{k!(n-k)!}$$

test statistic for the sign test:

x = number of + signs = number of data values above the value of the median given in the null hypothesis.

test statistics for the Wilcoxon signed-rank test:

i) For a test of the form:

   Null hypothesis: Md $= M$
   Alternative hypothesis: Md $< M$

   use R^+ = sum of positive ranks.

ii) For a test of the form:

   Null hypothesis: Md $= M$
   Alternative hypothesis: Md $> M$

   use $|R^-|$, where R^- = sum of negative ranks.

T-statistic for the Mann–Whitney test:

$$T = S_1 - \frac{n(n+1)}{2}$$

(S_1 = rank sum for Group I, n = sample size for Group I.)

test statistic for the runs test:

$$R = \text{number of runs}$$

You should be able to

1 Calculate *binomial probabilities*.

2 Perform a *sign test* of the form:

   Null hypothesis: Md = M
   Alternative hypothesis: Md < M or Md > M

3 Perform a *Wilcoxon signed-rank test* of the form:

   Null hypothesis: Md = M
   Alternative hypothesis: Md < M or Md > M

4 Perform a *Mann–Whitney test* of the form:

   Null hypothesis: $Md_1 = Md_2$.
   Alternative hypothesis: $Md_1 < Md_2$.

5 Perform a *runs test* of the form:

   Null hypothesis: Two symbols occur in random order in a sampling sequence.
   Alternative hypothesis: The symbols do not occur in random order.

REVIEW TEST

1 The suggested retail price of a color TV is $599. A random sample of stores showed the set offered for:

   $530, $540, $570, $580, $600, $610, $595, $620

   Use this data and a sign test to perform the test:

   Null hypothesis: Md = 599
   Alternative hypothesis: Md < 599

   $\alpha \doteq 0.05$

2 Use the Wilcoxon signed-rank test and the chicken-price data:

   12.0, 12.2, 11.9, 11.7, 12.3, 12.4, 12.5, 12.45

   to perform the test:

   Null hypothesis: Md = 12.1
   Alternative hypothesis: Md > 12.1

   $\alpha \doteq 0.04$

3 A manufacturer of battery packs developed a new pack that is supposed to last longer than the present model. A preliminary test on five randomly selected packs of each type gave the following results:

Lifetime (in hours)

Present model	New model
2.93	3.07
2.74	3.12
2.81	2.85
3.02	2.94
2.98	3.05

Use a Mann–Whitney test with $\alpha = 0.05$ to test the null hypothesis of no difference in median lifetime between the two packs against the alternative hypothesis that the present model has a shorter median lifetime.

4 A political worker was assigned to walk through a shopping mall, stop randomly selected passersby, and determine their feelings on a proposed bond issue. The people he stopped were classified as O (over thirty) and U (under thirty). The sequence of interviews was

   O O O O U U O O O O O U U U O O O O O

Use a runs test with $\alpha = 0.05$ to determine whether the worker was sampling randomly (with regard to being over or under thirty) from all passersby.

FURTHER READINGS

DANIEL, WAYNE W., *Applied Nonparametric Statistics*. Boston: Houghton Mifflin Company, 1978. This book is written in a concrete and easy-to-read fashion, but is quite sophisticated in the wide range of topics it covers and in the professional tone of its examples (most of which are taken from published research papers).

MOSTELLER, F., AND R. ROURKE, *Sturdy Statistics*. Reading, Mass.: Addison-Wesley Publishing Company, 1973. As with the work above, this text is at the same time understandable and comprehensive, with an abundance of realistic examples.

NOETHER, G. E., *Introduction to Statistics: A Nonparametric Approach*. Boston: Houghton Mifflin Company, 1976. This book is intended as an introductory text for a first course in statistics, and covers many of the same topics that you studied here in Chapter 13. However, it relies more on nonparametric methods to illustrate the basic principles and ideas of statistics.

ANSWERS TO REINFORCEMENT EXERCISES

A $P(x = 1) = \binom{3}{1} \cdot (0.1)^1 \cdot (0.9)^2 = 0.243$

B a) 0.2430
   b) 0.2430 + 0.0270 + 0.0010 = 0.2710

C *Step 1.*

   Null hypothesis: Md = 47
   Alternative hypothesis: Md > 47

   Step 2. $n = 10$

   Step 3. Table and histogram for binomial distribution with $n = 10$ and $p = 0.5$.

Successes, k	Probability, $P(x = k)$
0	0.0010
1	0.0098
2	0.0439
3	0.1172
4	0.2051
5	0.2461
6	0.2051
7	0.1172
8	0.0439
9	0.0098
10	0.0010

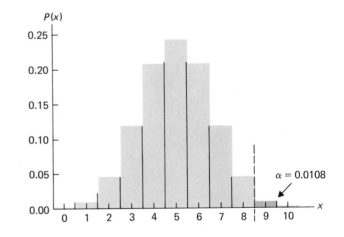

Step 4. Significance level and rejection region.
   $\alpha = 0.0108$

Step 5. Collect data and mark signs.

Time	Sign
23	−
34	−
39	−
43	−
52	+
56	+
61	+
63	+
70	+
81	+

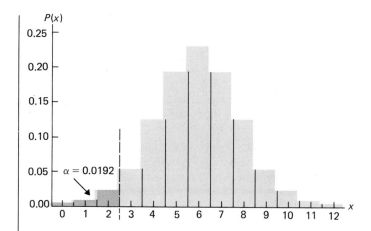

Step 6. $x = 6$. The null hypothesis of Md = 47 cannot be rejected.

D   *Step 1.*

Null hypothesis: Md = 50
Alternative hypothesis: Md < 50

Step 2. $n = 12$

Step 3. Table and histogram for binomial distribution with $n = 12$ and $p = 0.5$.

Successes, k	Probability, $P(x = k)$
0	0.0002
1	0.0029
2	0.0161
3	0.0537
4	0.1208
5	0.1934
6	0.2256
7	0.1934
8	0.1208
9	0.0537
10	0.0161
11	0.0029
12	0.0002

Step 4. Significance level and rejection region.
     $\alpha = 0.0192$

Step 5. Collect data and mark signs. We are given that there are two throws above 50 and ten below 50.

Step 6. $x = 2$. Reject the null hypothesis, and conclude that Md < 50.

E   *Null hypothesis:*   Md = 50
Alternative hypothesis: Md < 50

Step 1. $n = 9$, $\alpha = 0.049$, $d = 8$

Step 2. Work table.

x	$D = x - M$	$|D|$	*Rank*	*R*
51.1	+1.1	1.1	7	+7
50.2	+0.2	0.2	2	+2
50.1	+0.1	0.1	1	+1
49.6	−0.4	0.4	3	−3
49.5	−0.5	0.5	4	−4
49.2	−0.8	0.8	5	−5
49.0	−1.0	1.0	6	−6
48.4	−1.6	1.6	8	−8
47.0	−3.0	3.0	9	−9

Step 3. Positive-rank sum:

$$
\begin{array}{r}
7 \\
2 \\
1 \\
\hline
10 = R^+
\end{array}
$$

Step 4. $R^+ = 10$ and $d = 8$. The null hypothesis *cannot* be rejected.

Reject null hypothesis

F   *Null hypothesis:*   Md = 4100
Alternative hypothesis: Md > 4100

Step 1. $n = 11$, $\alpha = 0.021$, $d = 10$

Step 2. Work table.

x	D	$|D|$	Rank	R
4982	+882	882	11	+11
4817	+717	717	10	+10
4723	+623	623	9	+ 9
4695	+595	595	8	+ 8
4632	+532	532	7	+ 7
4501	+401	401	5	+ 5
4320	+220	220	3	+ 3
4111	+ 11	11	1	+ 1
3963	−137	137	2	− 2
3872	−228	228	4	− 4
3604	−496	496	6	− 6

Step 3. Negative-rank sum:

$$
\begin{array}{r}
-2 \\
-4 \\
-6 \\
\hline
-12 = R^- \\
12 = |R^-|
\end{array}
$$

Step 4. $|R^-| = 12$ and $d = 10$. The null hypothesis cannot be rejected.

Reject null hypothesis

G   *Null hypothesis:*   Md = 47
Alternative hypothesis: Md < 47

Step 1. $n = 8$, $\alpha = 0.020$, $d = 3$

Step 2. Work table (see page 525).

Step 3. Positive-rank sum:

$$
\begin{array}{r}
2 \\
1 \\
3.5 \\
\hline
6.5 = R^+
\end{array}
$$

Step 4. $R^+ = 6.5$, $d = 3$. The null hypothesis cannot be rejected.

H   a) *Null hypothesis:* $Md_1 = Md_2$. That is, there is no difference in median final-exam scores for students with low placement scores and those with high placement scores.

Alternative hypothesis: $Md_1 < Md_2$. That is, students with high placement scores have a higher median final-exam score than those with low placement scores.

b) *Step 1.* Sample sizes: $n = 5$ from I, $k = 6$ from II.

Step 2. $c = 5$.

Step 3.

Low placement (I)		High placement (II)	
Score	Rank	Score	Rank
427	1	532	2
583	3	671	5
658	4	735	7
722	6	892	9
842	8	927	10
		958	11

$$S_I = 1 + 3 + 4 + 6 + 8 = 22$$

Step 4. Calculate T.

$$T = S_1 - \frac{n(n+1)}{2} = 22 - \frac{5 \cdot 6}{2} = 7$$

Step 5. $T = 7$, $c = 5$. The null hypothesis cannot be rejected.

I   *Null hypothesis:* $Md_1 = Md_2$.
Alternative hypothesis: $Md_1 < Md_2$.

Step 1. Sample sizes: $n = 4$, $k = 4$

Step 2. $c = 1$

Step 3. (See page 534.)

$$S_1 = 1 + 3 + 4 + 5.5 = 13.5$$

Step 4.

$$T = S_1 - \frac{n(n+1)}{2} = 13.5 - \frac{4 \cdot 5}{2} = 3.5$$

Step 5. $T = 3.5$, $c = 1$. The null hypothesis cannot be rejected.

J   *Null hypothesis:*   The letters M and F occur in random order.
Alternative hypothesis: The letters M and F do not occur in random order.

Step 1. Identify runs.

	M	F	MM	FF	M	F	M	F
Run	1	2	3	4	5	6	7	8

Step 2. Number of runs: $R = 8$

Step 3. Critical values: $c_1 = 2$, $c_2 = 10$

Step 4. $R = 8$, $c_1 = 2$, $c_2 = 10$. The null hypothesis cannot be rejected.

Planning a study

14.1   Introduction

Since Chapter 4, we have been concentrating on the mathematics necessary for inferential statistics. We have shown how to calculate confidence intervals and perform various kinds of hypothesis tests using data from random samples. In most of our examples the random sample data were simply given to you; you did not have to think about how to take a random sample. Furthermore, most of our examples were designed to illustrate a specific kind of statistical test. If a problem appeared in Chapter 10, which dealt with chi-square tests, you knew in advance that you would probably have to use a chi-square test to solve it. It is necessary to keep things this simple for a student who is learning computational methods for the first time, but life is not that simple. When a researcher plans a real study, he must think about how he will obtain his sample data, and what statistical test he will use.

It is not enough to simply know how to do tests such as *t*-tests, *z*-tests, ANOVA, and regression. Once you know all of these tests, you should be able to choose from among them the test that is best for a particular study, and to go through the *entire process* of collecting sample data, performing the test, and drawing a conclusion.

In this chapter we will look at some of the key aspects of planning a complete study. We will begin by discussing a more important, preliminary question: Should you be doing a statistical study in the first place?

14.2   Is a study necessary?

Throughout this book we have seen examples of people performing their own studies. A consumer group wants to know about gas mileage, so it gathers data on a random sample of cars of a particular model. A teacher wants to know about the comparative merits of two teaching methods, so she tests these methods in two randomly selected classes. This reflects a healthy attitude: If you don't know about something, collect data and find out about it. However, it also requires time, effort, and money from the person doing the testing, and this *could* be time, effort, and money wasted: *It is always possible that someone else has already done the study or gathered the information that you need.* Before you plan a complicated statistical research effort, you should check to see whether some other researcher has already done your study for you. This does not require going through all the books in the library. There are information-collection agencies that specialize in finding studies on specific topics in specific areas. For example, the *Educational Resources Information Center* (ERIC) collects studies in the field of education. There are publications entitled *Chemical Abstracts* and *Psychological Abstracts,* which collect the results of studies in chemistry and psychology. The *National Library of Medicine* compiles lists of medical studies and makes them available on computer tapes that are sold to research centers and universities. Searching through these resources has become a complex technical job, but most libraries, universities, and research centers have information specialists who will search for you, either as a service or for a small fee. A specialist at a computer terminal can find you 20 references to a subject of interest within 15 minutes.

A great deal of information about Americans is also available from the *Census Bureau*. You can find out about income, age, number of appliances owned, and hundreds of other variables, for areas ranging from the entire United States to a single city block. The Bureau of the Census has published a special guide to its resources, titled *A Student's Workbook on the 1970 Census*.* This guide explains the use of census data in some detail.

It is not the purpose of this book to show you how to search through journal articles or census data. In most cities and/or most campuses there are services that will do this for you. The important principle is this: *You can often avoid the effort and expense of a study if someone else has already done that study and published the results.*

Sampling   14.3

Once a researcher has decided that the information he needs is not already available, he must decide how to obtain sample data for his own study. In practice, researchers use many different sampling methods. However, all of the statistical formulas given in this book are designed for use with one kind of sample: the **simple random sample,** for which each possible sample of the population of interest has an equal chance of being the one selected. In Example 1 below, we will show a way of obtaining a *simple random sample*.

Example 1

◀ We have already discussed student evaluations of courses and teachers in Chapter 10. There we looked at data obtained from questionnaires given to all students in a course immediately after the final examination. Student evaluation of teaching is not usually done at final-exam time, however. It is common practice to hand out evaluation forms during a class period a few weeks before the final, and to have the teacher leave the classroom while forms are filled out. (This is done to avoid intimidation of students by the teacher.) There are some problems with this practice. On some class days, only about 60 percent of the students registered for a class are in attendance. In addition, many of the attending students have other classes to prepare for, and will fill out their forms in a hurry in order to leave class early.

In such circumstances, it may well be better to select a random sample of students from a class, and then to interview those selected students thoroughly and carefully. This is exactly the kind of situation in which a simple random sample must be found.

In the fall of 1980, Prof. H wanted to sample the attitudes of the students in College Algebra at his school. This course had 728 students, and Prof. H wished to interview 15 of them. He had a registration list of all 728 students, numbered from 1 to 728. To select a random sample of 15 students from this list, Prof. H really needed only to select 15 numbers *at random* between 1 and 728. To do this he used a **table of random numbers.** Such tables are constructed by statisticians so that all of the digits in them are randomly selected. The table of 1000 random digits used by Prof. H is given on page 552 as Table 14.1.

---

* For sale by the Superintendent of Documents, U.S. Government Printing Office, Washington, D.C. 20402; GPO Stock No. 003–024–1642–7.

Table 14.1

Table of 1000 random digits

Line number	Column number									
	00–09		10–19		20–29		30–39		40–49	
00	15544	80712	97742	21500	97081	42451	50623	56071	28882	28739
01	01011	21285	04729	39986	73150	31548	30168	76189	56996	19210
02	47435	53308	40718	29050	74858	64517	93573	51058	68501	42723
03	91312	75137	86274	59834	69844	19853	06917	17413	44474	86530
04	12775	08768	80791	16298	22934	09630	98862	39746	64623	32768
05	31466	43761	94872	92230	52367	13205	38634	55882	77518	36252
06	09300	43847	40881	51243	97810	18903	53914	31688	06220	40422
07	73582	13810	57784	72454	68997	72229	30340	08844	53924	89630
08	11092	81392	58189	22697	41063	09451	09789	00637	06450	85990
09	93322	98567	00116	35605	66790	52965	62877	21740	56476	49296
10	80134	12484	67089	08674	70753	90959	45842	59844	45214	36505
11	97888	31797	95037	84400	76041	96668	75920	68482	56855	97417
12	92612	27082	59459	69380	98654	20407	88151	56263	27126	63797
13	72744	45586	43279	44218	83638	05422	00995	70217	78925	39097
14	96256	70653	45285	26293	78305	80252	03625	40159	68760	84716
15	07851	47452	66742	83331	54701	06573	98169	37499	67756	68301
16	25594	41552	96475	56151	02089	33748	65289	89956	89559	33687
17	65358	15155	59374	80940	03411	94656	69440	47156	77115	99463
18	09402	31008	53424	21928	02198	61201	02457	87214	59750	51330
19	97424	90765	01634	37328	41243	33564	17884	94747	93650	77668

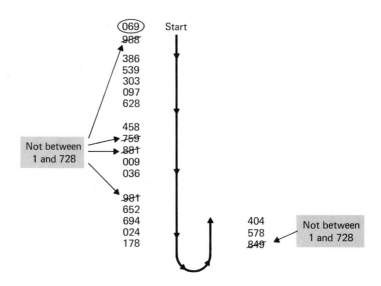

Figure 14.1

To obtain 15 random numbers between 1 and 728, all you need to do is pick a random starting point: close your eyes and put your finger down on the table. Starting from the three digits under your finger, read off three-digit numbers, reading down until you reach the bottom of the table. Since you want only numbers between 1 and 728, throw out numbers between 729 and 999.*

If you have not found enough numbers by the time you reach the bottom of the table, move over to the next column of three digits and read up. Prof. H did this and obtained as a starting point the numer 069, which is circled in the table. Reading down and then up, he decided to interview students numbered as shown in Fig. 14.1.

The final list of students to interview is:

069	303	458	652	178
386	097	009	694	578
539	628	036	024	404 ▶

Pick a random starting point in the table and make your own list of 15 randomly selected students for the interview.

Exercise **A**

Many calculators and most computers now furnish **random-number generators.** These enable you to pick one number at random and then obtain as many random numbers as you like within a given range. Tables such as the one above were used much more frequently before calculators with random-number generators were available at low prices. If your calculator has a random-number generator, it will be simpler to use that than to use tables.

In some cases, it is not possible to obtain simple random samples using random-number tables, but samples that appear to be random can be obtained. We illustrate such a situation in Example 2.

◀ A teacher at a state university wished to compare two methods of teaching college algebra by using these two methods on two randomly selected groups of students. The university had 10,000 students who had never taken college algebra, so the ideal solution would be to select two groups at random from those 10,000 students. Unfortunately, the university did not require anybody to take college algebra, and most of those 10,000 students would not. As a matter of fact, only 728 students registered for college algebra.

Example **2**

The next obvious solution would be to pick two randomly selected groups from these 728 students to be taught by the two different methods. Unfortunately, this would require that students in each of the two selected groups be put in the same classes. For example, students selected for one of the methods might be told that they were in a class at 8:40. Students at large universities cannot be scheduled like this. Some work, and some are in car pools. They tell the university when they can take a class; the university does not tell them.

There was one further problem for the teacher who wanted to do a comparison study: the other teachers. Six other teachers were scheduled to teach college

---

* Also, throw out any number that has occurred previously.

algebra, and none of them wanted to do a comparison study. The teacher who wished to do a comparison was scheduled for only one class, at 9:40.

This leaves only one group to sample from: the students who signed up for class at 9:40. What the teacher did was the only thing possible: She assumed that the registration process for this large course gave her a random sample of all students at 9:40. She then took *every other* student on her roll, divided her class into two groups, and found a teacher from another course who was willing to volunteer to teach the second group, using an experimental method. We picture the whole process below in Fig. 14.2.

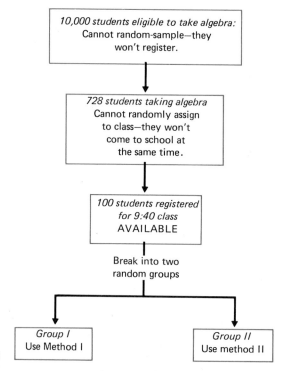

Figure 14.2

Samples of this kind are called **samples of convenience.** They are available, they seem to be random, and there is no other way to get a random sample. When samples of convenience are used, it is a good idea to try some common-sense checks to be sure that they are not biased. In this particular case, the teacher gave each of Groups I and II a standard placement test. There was no significant test-score difference between Group I and Group II, and their results were similar to those of other algebra classes. The teacher decided that it was reasonable to go ahead with her comparative study of teaching methods using Groups I and II as samples for comparison. ▶

There is a danger in samples of convenience. You can never be sure if they are truly random. For example, at one school we know of, it has been observed that classes scheduled at 9:40 in remedial algebra always have better average grades than classes scheduled at 1:40. In such a situation, 9:40 students are a *biased* sample of all remedial algebra students; a sample of convenience at 9:40 might not be representative of all remedial algebra students. Samples of convenience should be looked at very carefully.

Describe another situation in which a sample of convenience would be used.　　　Exercise **B**

　　There are occasions when a simple random sample is not possible, but we can do somewhat better than a sample of convenience. We illustrate this below.

◀ A southwestern city of 100,000 was under pressure from citizens' groups to install bike paths. The members of the city council wanted to be sure that they had the support of a majority of taxpayers, and decided to poll the homeowners in the city. Their first attempt at surveying opinion was a questionnaire, which was mailed out with the city's 18,000 homeowner water bills. This mailing did not work very well. Only 3,500 questionnaires were returned, and a large proportion of these had written comments that indicated that they had come from avid bicyclists (pro), or people who strongly resented bicyclists (con). The questionnaire had generally not been returned by the average voter, and the city council realized this.

　　Example **3**

　　The city had an employee in the planning department who had some survey-sample experience. The council called him in and told him to do a survey. He was given two helpers to help him interview a representative selection of voters. He was also told to report back in ten days.

　　The planner thought about taking a simple random sample of 300 city voters: 100 interviews for each helper, and 100 for himself. However, obtaining a simple random sample created some time problems. The city was spread out. An interviewer with a list of 100 names randomly scattered around the city would have to drive an average of 18 minutes from one interview to the next. This would require nearly 30 hours of driving time for each worker, and might delay completion of the report. Obviously the random-sample method would not do. ▶

To save time, the planner decided to use a method called **cluster sampling.** The residential portion of the city was divided into 347 blocks, each containing about 20 houses. See Fig. 14.3.

　　Cluster sampling

A Typical Block of Houses

Figure 14.3

Example **3**
(cont.)

◀ The planner numbered the blocks on the city map from 1 to 347, and then used a table of random numbers to select 15 of those blocks at random. Each interviewer then took 5 of those 15 blocks as his interviewing territory. This gave each interviewer approximately 100 households to visit, but saved a great deal of travel time. An interviewer could work on a block for nearly a full day without having to drive to another neighborhood. The report was finished on time. ▶

This method is called *cluster-sampling,* because the interviewer concentrates on *clusters* of voters; a block is a cluster of houses. It saves time and money, but can have its drawbacks.

For example, let us look at a simplified small town (Fig. 14.4) in which there are only ten clusters. The town council is considering building a town swimming pool.

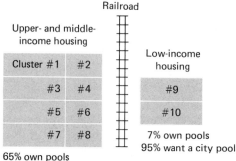

Figure 14.4

Suppose that a planner in this town wishes to sample voter sentiment on using public funds to build the town swimming pool. Upper- and middle-income homeowners will probably say "No"; they own pools, or can use a neighbor's. Low-income voters will say "Yes"; they do not have access to pools. If the planner decides to interview the residents of three randomly selected clusters, there is a good chance that he will randomly select *no* low-income clusters.* Suppose, for example, that he selects clusters 3, 5, and 8 to interview. His survey will show that only 30 percent of the voters want a pool, but this is not true. More than 40 percent of the voters actually want a pool; the planner has left out the clusters that most strongly support it. In this hypothetical example, the town is so small that common sense would show the cluster sample to be nonrepresentative. In situations with hundreds of clusters, the problems are more difficult to see.

Stratified sampling

There is another sampling method, known as **stratified sampling,** which is often more reliable and precise than cluster sampling. With a stratified-sampling procedure, you divide your population into groups at the same level, called **strata,** and then sample at each level. For example, in the town swimming pool example,

---

* This will happen 46.6% of the time; 46.6% of all "three-cluster" random samples will contain neither cluster 9 nor cluster 10.

you would divide voters into three strata: *high income, middle income,* and *low income.* You would then take a *simple random sample* from *each* of the three groups. This procedure would ensure that no income group is missed, and would improve the precision of your estimates. (Such sampling also enables you to determine the opinions of the separate groups; e.g., you can find out the opinion of the high-income voters, because they are in a separate sample.)

The groups are often sampled in *proportion* to their actual percentages. If your entire city is 20 percent low-income, 70 percent middle-income, and 10 percent high-income, and you want to sample 100 individuals, you would sample 20 low-income residents, 70 middle-income residents, and 10 high-income residents. The mathematics required for statistical tests based on stratified samples is somewhat different from what you have learned in this book. For example, slightly different formulas are used to calculate the sample mean and variance from your sample data.

The methods and formulas you have learned in this book are designed for data obtained from *simple random samples,* and also may occasionally be used on *samples of convenience.* Although it is important to know about stratified and cluster samples, you should *not* apply the methods you have learned here to data obtained from stratified and cluster samples.

The whole question of how to sample and how to interview is a complex one, and entire books are devoted to it. A highly readable introduction that requires very little mathematics is *An Introduction to Survey Research and Data Analysis,* by Herbert Weisberg and Bruce Bowen (San Francisco: W. H. Freeman & Co., 1977).

EXERCISES Section **14.3**

1 The owner of a large business with 685 employees wishes to select 25 of them at random for extensive interviewing. Make a list of 25 random numbers between 1 and 685 to be used in obtaining this sample. (Use Table 14.1)

2 The players on a football team wear the numbers from 1 to 65. Make a list of five random numbers to select five of these players at random. (Use Table 14.1)

3 The cars registered with a university have parking stickers numbered 1 through 8493. Make a list of 30 random numbers to obtain a random sample of size 30 from this list. (Use Table 14.1)

4 Define ''sample of convenience'' in your own words, and give an example of such a sample.

5 Students in the dormitories of a university in the state of New York live in clusters of four double rooms called *suites*. There are 48 such suites, each with eight occupants.
a) Describe a cluster-sampling procedure for finding 32 dormitory residents to interview.
b) Students typically choose friends in their own classes to share suites with. Is cluster sampling a good procedure for obtaining a random sample of students?
c) The university housing office has separate lists of dormitory residents by year:

Freshmen	128
Sophomores	112
Juniors	96
Seniors	48

Design a sampling procedure to use this list to obtain a stratified random sample of 48 students.

14.4 Type I and Type II errors

Statistical tests of hypotheses do not always yield correct conclusions; they have built-in margins of error. Another big part of planning a study is to take an advance look at the types of errors that could be made, and to determine the possible effects of these errors on the conclusions to be made at the end of the study. In the following examples, we will illustrate the two main types of error that must be looked at in planning a study.

Example 4
Type I error

◀ In Example 2 of Chapter 8, we considered a hypothesis test concerning gas mileages. A car manufacturer claimed that one of his car models had an average gas mileage of 35 mpg, and a consumer group thought that the average gas mileage was less than 35 mpg. If μ stands for the true mean gas mileage of this model, we can write down a hypothesis test for the consumer agency to perform:

Null hypothesis: $\mu = 35$ (manufacturer's claim)
Alternative hypothesis: $\mu < 35$ (consumer group's suspicion)

In Chapter 8, we showed how to test this hypothesis at the level $\alpha = 0.05$ using mileage date from a sample of $n = 30$ cars. The approach used in Chapter 8 was based on the use of z-statistics and the standard normal curve. In this chapter we will look at that hypothesis test in a slightly different fashion. The hypothesis test was based on the following analysis of the sample mean x: If the null hypothesis is true, then the possible sample means $\bar{x}$ for gas mileages of 30 cars are normally distributed with mean $\mu_{\bar{x}} = \mu = 35$. For simplicity we will assume that the standard deviation of gas mileages is known to be $\sigma \doteq 1.45$ (say, from a previous study). Then $\sigma_{\bar{x}} \doteq \sigma/\sqrt{n} = 1.45/\sqrt{30} = 0.265$. Consequently, if the mean gas mileage is truly $\mu = 35$, the normal curve for all possible values of $\bar{x}$ looks like Fig. 14.5.*

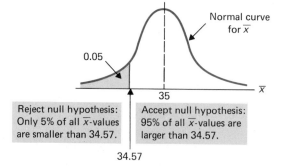

Figure 14.5
Normal curve for $\bar{x}$, if $\mu = 35$ (null hypothesis is true).

Also shown in Fig. 14.5 are the acceptance and rejection regions for a 0.05-level test.

---

* We have calculated and displayed some curve areas and probabilities necessary for the examples of this section. Although you have learned to calculate them, it is not necessary for you to do the calculations here.

Recall that the reasoning behind the definition of acceptance and rejection regions is this: If $\mu = 35$ is true, the sample mean $\bar{x}$ should be smaller than 34.57 only 5 percent of the time. Thus we will consider $\bar{x}$ values less than 34.57 as "too small," and use such values as a justification to reject the null hypothesis.

There is a problem with using this reasoning. If $\mu = 35$ is true there is still a 5 percent chance that a random sample of size 30 would give an $\bar{x}$ value smaller than 34.57, and lead you to the *false* conclusion that $\mu < 35$. Such false conclusions occur 5 percent of the time. If 100 different consumer groups performed this experiment independently, it is likely that about five of them would mistakenly reject the null hypothesis. This incorrect rejection of the null hypothesis is called a *Type I error.* ◗

Type I error: Rejection of the null hypothesis when it is in fact true.   DEFINITION

Every hypothesis test has the possibility of Type I error built into it. *The probability of a Type I error is always equal to α, the significance level of the test.* If you use $\alpha = 0.05$ in your test, you have a 5 percent chance of a Type I error. If you use $\alpha = 0.01$, you reduce the chance of a Type I error to 1 percent.

To see how Type I error affects the planning of an experiment, we will look at the experiment in Example 4 from two points of view:

a)   *The manufacturer's point of view:*   The car manufacturer would like people to believe his claim that $\mu = 35$. He would not like to see a Type I error, which would cause the consumer group to announce that $\mu < 35$ (when in fact $\mu = 35$). He would have liked to see the experiment planned with a level of $\alpha = 0.01$. This would have reduced the chance of bad publicity for his car.

b)   *The consumer group's point of view:*   The consumer group is not as worried as the manufacturer about Type I error. They are more concerned about protecting the public from misleading claims. If they use a level of $\alpha = 0.01$, their rejection region will be smaller, and false claims harder to detect. They would like to keep the level at $\alpha = 0.05$. (We have calculated the rejection regions for both values of α. They are pictured below:)

Acceptance and rejection regions for $\alpha = 0.05$ and $\alpha = 0.01$.
a) $\alpha = 0.05$; *probability of Type I error* $= 0.05$

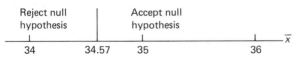

b) $\alpha = 0.01$; *probability of Type I error* $= 0.01$.

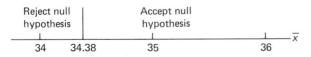

21   Exercise **C**   Suppose the situation in Example 4 were reversed. The consumer group claims that the mean gas mileage is $\mu = 35$, and the manufacturer claims that it is even better ($\mu > 35$). In this case, the manufacturer wants to sponsor the hypothesis test

Null hypothesis: $\mu = 35$
Alternative hypothesis: $\mu > 35$

with $\alpha = 0.05$.

The normal curve for sample mean gas mileages still looks the same, but we break up the areas under it differently for this test. Figure 14.6 below shows that curve broken up into acceptance and rejection regions.

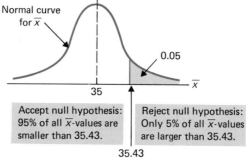

Figure 14.6

Normal curve for $\overline{x}$

0.05

35

$\overline{x}$

Accept null hypothesis: 95% of all $\overline{x}$-values are smaller than 35.43.

Reject null hypothesis: Only 5% of all $\overline{x}$-values are larger than 35.43.

35.43

a)   Describe in your own words how Type I error would occur. (Assume $\mu = 35$ is true.)
b)   What is the probability of a Type I error?
c)   Discuss the desirability of Type I error for (i) the manufacturer and (ii) the consumer group.

So far we have only discussed the chance of mistakenly rejecting the null hypothesis when it is *true*. It is also possible to accept the null hypothesis when it is *false*. This is a Type II error:

DEFINITION   **Type II error:**   Acceptance of the null hypothesis when it is in fact false.

We illustrate Type II errors in Example 5 below.

Example **5**   ◖   We will return to the gas mileage test of Example 4. We already know that this
Type II error   test has the form:

Null hypothesis: $\mu = 35$
Alternative hypothesis: $\mu < 35$

with $\alpha = 0.05$.

Regions:

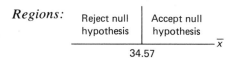

Reject null hypothesis | Accept null hypothesis

34.57

We will now look at the situation in which the null hypothesis is actually false. Suppose that the model we are considering has, in fact, a true mean gas mileage of $\mu = 34.3$ miles per gallon. Then the sample mean gas mileages ($\bar{x}$) will be normally distributed with mean $\mu_{\bar{x}} = \mu = 34.3$. We have drawn this new curve in Fig. 14.7.

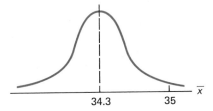

34.3   35

Figure 14.7
Normal curve for $\bar{x}$
if $\mu = 34.3$

The people performing the test do not know this true mean, and will accept the null hypothesis ($\mu = 35$) if $\bar{x} \geq 34.57$. We have calculated the curve area that tells you $P(\bar{x} \geq 34.57)$ when $\mu = 34.3$. We show this, and the acceptance and rejection regions for the original test, in Fig. 14.8.

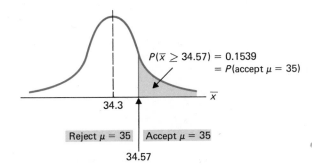

$P(\bar{x} \geq 34.57) = 0.1539$
$= P(\text{accept } \mu = 35)$

34.3

Reject $\mu = 35$ | Accept $\mu = 35$

34.57

Figure 14.8

Remember that the normal curve we have drawn in Fig. 14.8 is not the curve based on the (false) null hypothesis. The curve we have drawn is based on the true value of μ, which in this case is $\mu = 34.3$. This new curve tells us the probability of accepting $\mu = 35$ even though it is really true that $\mu = 34.3$. That probability is equal to the *shaded area*!

$$\left(\begin{array}{c} \text{Probability of accepting} \\ \mu = 35 \text{ when } \mu = 34.3 \end{array} \right) = P(\bar{x} \geq 34.57) = 0.1539$$

This means that if, in fact, $\mu = 34.3$, then about 15 percent of the time you will accept the false null hypothesis $\mu = 35$. As we said before, the error of accepting a false null hypothesis is called a Type II error. Type II errors also affect the planning of experiments. We will look at Type II errors from two points of view:

a) *The manufacturer:* He is not too worried about a Type II error. If the true gas mileage is less than 35, a Type II error would still permit him to advertise that his car gets 35 mpg.

b) *The consumer group:* They would like to avoid Type II error. If the true gas mileage is less than 35 mpg, they would like to detect that fact. ▶

Exercise **D** As in Exercise C, we will reverse the situation again. The manufacturer would like to sponsor the test

Null hypothesis: $\mu = 35$ (consumer claim)
Alternative hypothesis: $\mu > 35$ (manufacturer's claim)

with $\alpha = 0.05$.

Regions (see Exercise C).

Accept $\mu = 35$	Reject $\mu = 35$

35.43　$\bar{x}$

Now suppose that the true value of the mean is $\mu = 35.3$. In Fig. 14.9 we have sketched the appropriate normal curve for $\bar{x}$ when $\mu = 35.3$, and have shaded in areas corresponding to acceptance and rejection for the test of $\mu = 35$ versus $\mu > 35$.

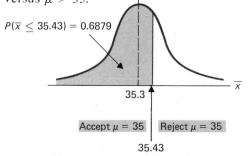

$P(\bar{x} \leq 35.43) = 0.6879$

35.3

Accept $\mu = 35$ │ Reject $\mu = 35$
35.43

Figure 14.9
Normal curve for
$\bar{x}$, if $\mu = 35.3$

a) Describe in your own words how a Type II error would occur.
b) What is the probability of a Type II error?
c) How will a Type II error affect (i) the manufacturer, and (ii) the consumer group?

We have not yet discussed how to lower the probability of a Type I or Type II error. In most cases, this is done by increasing sample size. A statistician who is planning an experiment will usually look at the possible Type I and Type II errors for various sample sizes when he is deciding how large a sample to use. We illustrate this in Example 6.

◀ A statistician suggests to the consumer group and the manufacturer that they   Example **6**
test 100 cars for gas mileage, instead of 30. He also suggests that they use the
same acceptance and rejection regions used in Examples 4 and 5. The test would
look like this:

Null hypothesis: $\mu = 35$
Alternative hypothesis: $\mu < 35$

Regions:

| Reject null hypothesis | | Accept null hypothesis |

$$\underset{34.57 \qquad\qquad 35}{\underline{\hspace{6cm}}}\ \bar{x}$$

We have not given a value of α because changing the sample size (from 30
to 100), while keeping the same acceptance–rejection regions, changes α. To find
the new value of α, we look (in Fig. 14.10) at the normal curve for $\bar{x}$ based on the
null hypothesis $\mu = 35$ and the new sample size of $n = 100$. (The new sample size
changes $\sigma_{\bar{x}}$ to $\sigma/\sqrt{n} = 1.45/\sqrt{100} = 0.145$.)

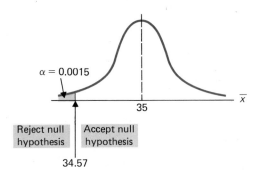

$\alpha = 0.0015$

| Reject null hypothesis | Accept null hypothesis |

34.57

Figure 14.10
Normal curve for
$\bar{x}$ if $\mu = 35$ and $n = 100$.

The shaded area on the left in Fig. 14.9 is the probability of a Type I error,
and is equal to the significance level, α, of the test for 100 cars. It is very low:

$$\alpha = 0.0015 = \text{Probability of Type I error.}$$

This should make the manufacturer happy. If his car truly has $\mu = 35$, there
is little chance of falsely rejecting his claim.

Changing the sample size will also change the probability of a Type II error.
Let us look at the situation we previously considered in Example 5, where

Null hypothesis: $\mu = 35$
Alternative hypothesis: $\mu < 35$,

and the true value of μ is 34.3. The normal curve for $\bar{x}$-values with $\mu = 34.3$ and
$n = 100$ is shown in Fig. 14.11 at the top of page 564.

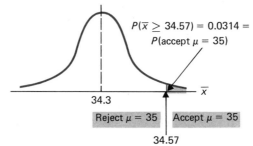

Figure 14.11
Normal curve for
$\bar{x}$ if $\mu = 34.3$ and
$n = 100$.

The shaded area in Fig. 14.11 gives the probability of a Type II error; it is 0.0314. This is a smaller probability of Type II error than occurred in Example 5 (where 30 cars, instead of 100, were sampled). The consumer group should be happier with 100 cars. There is less chance of failing to detect a false null hypothesis. We summarize the results from Examples 4, 5, and 6 in Table 14.2.

Null hypothesis: $\mu = 35$
Alternative hypothesis: $\mu < 35$

Regions:

Reject null hypothesis		Accept null hypothesis
	34.57	$\bar{x}$

Table 14.2

		True value of μ	
		$\mu = 35$	$\mu = 34.3$
Sample size		Probability of Type I error (falsely rejecting $\mu = 35$)	Probability of Type II error (falsely accepting $\mu = 35$)
30		0.05	0.1539
100		0.0015	0.0314

It would seem better all around to test 100 cars, but there is still a decision to be made: A test on 100 cars would cost far more money. More cars, more drivers, more gas will all be required, and somebody must pay for these. The planner of an experiment must balance all of these factors. There is no final right answer here. You must look at how much money you have, and balance this against the possible cost of errors. ▶

The analyses done by statisticians are more complex than those in Example 6. They usually look at a wide range of possible Type II errors for different possible values of μ; say,

$$\mu = 34.9, 34.8, 34.7, 34.6, 34.5, \text{etc.}$$

There are also charts and tables for determining Type II errors and finding the sample sizes necessary to reduce Type II errors to desired levels. We will not go through such analyses in this book, but will leave you with a simple principle.

Increasing sample size for a test without changing the acceptance and rejection regions will lower the probability of both Type I and Type II errors.

For the hypothesis test (and regions) described on page 564, we have also calculated the probabilities of Type I and Type II errors using $n = 64$ cars. Table 14.3 is an error probability table for sample sizes 30, 64, and 100.

Exercise **E**

Table 14.3

	True value of μ	
	$\mu = 35$	$\mu = 34.3$
Sample size	Type I error probability	Type II error probability
30	0.0500	0.1539
64	0.0089	0.0681
100	0.0015	0.0314

Suppose the costs for these tests were:

Sample size	Cost
30	$35,000
64	$50,000
100	$63,000

Which sample size would you use if you were:
a) The manufacturer?
b) The consumer group?
(There are no "right" answers. Your reasoning is the important thing here.)

Power   The probability of a Type II error is the probability of making a certain kind of mistake. We can subtract that probability from 1 to find the probability of *not* making that mistake. For example, in our test of the hypothesis $\mu = 35$ with 30 cars,

$$P(\underbrace{\text{accepting } \mu = 35 \text{ when } \mu = 34.3}_{\text{Type II error}}) = 0.1539$$

$$P(\overbrace{\text{accepting } \mu < 35 \text{ when } \mu = 34.3}^{\text{No Type II error}}) = 1 - P(\text{accepting } \mu = 35 \text{ when } \mu = 34.3)$$
$$= 1 - 0.1539 = 0.8461$$

The probability of avoiding Type II errors is called the **power** of a test.

DEFINITION   **Power of a test:**   The probability of avoiding Type II error.

Power $= 1 - P(\text{Type II error})$

If $\mu = 34.3$, the power of our test on 30 cars is 0.8461. Below we give a chart of power for various sample sizes for our gas mileage test in case the true value of $\mu = 34.3$.

Sample size	Type II error probability	Power
30	0.1539	0.8461
64	0.0681	
100	0.0314	0.9686

Exercise **F**   Fill in the power for $n = 64$ in the table above.

As you can see, increasing sample size increases the power of a test. Since higher power means greater probability of avoiding a Type II error, people look for higher power in their tests. Statisticians also use the concept of power in comparing different kinds of tests. Here we have shown only that increasing sample size improves power for the *same test*. In practice, a statistician might use power to decide between different tests. For example, a statistician might be forced to use a sample of size 15, and decide to use a *t*-test instead of a sign test because the *t*-test will be more powerful than the sign test for that sample size.

Exercise **G**   You are going to do a study on the average final examination scores of 22 psychology students. You see a consulting statistician, and ask whether you should use a sign test (for the median) or a *t*-test (for the mean). He advises you to use the *t*-test because it is more powerful. Explain in your own words what he means.

1 A counselor is going to take a random sample of IQ scores from children in his county in order to perform the hypothesis test:

Null hypothesis: $\mu = 100$
Alternative hypothesis: $\mu > 100$

with $\alpha = 0.05$.
a) How would he make a Type I error?
b) What is the probability of a Type I error?
c) How would he make a Type II error?
d) Suppose that he had decided on his critical value for $\bar{x}$. Would increasing his sample size increase or decrease the probability of a Type II error?

2 A consumer agency tests ten watches of a model that is advertised to lose at most one second a day. Their hypothesis test is:

Null hypothesis: $\mu = 1$
Alternative hypothesis: $\mu > 1$

with $\alpha = 0.01$.
a) How could a Type I error be made?
b) Who would benefit from a Type I error?
c) How could a Type II error be made?
d) Who would benefit from a Type II error?

3 In Example 6 of Chapter 8, we considered a situation in which a soft-drink manufacturer sold "1-liter" bottles of soda, and a regulatory agency suspected that these bottles contained less than 1 liter. The regulatory agency decided to take a random sample of 100 bottles, and perform the test:

Null hypothesis: $\mu = 1$
Alternative hypothesis: $\mu < 1$

with $\alpha = 0.05$.
The company obtained an estimate of $s = 0.021$ liters for σ.
a) Estimate $\sigma_{\bar{x}}$ using the s given above, and $n = 100$.
b) For a lower-tailed test with $\alpha = 0.05$, the critical value of z is $c = -1.64$. Find the corresponding critical value

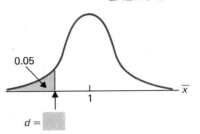

0.05

$\bar{x}$

1

$d =$

d for $\bar{x}$ by solving the equation $-1.64 = (d - 1)/\sigma_{\bar{x}}$, and write this value of d in the figure above.
c) The regulatory agency will accept the null hypothesis if $\bar{x} \geq d$ and reject the null hypothesis if $\bar{x} < d$. What is the probability of a Type I error if this is done?
d) Explain how a Type I error would occur, and describe how the soft-drink manufacturer would regard a Type I error.
d) Explain how a Type II error would occur in this experiment, and describe how the regulatory agency might feel about a Type II error.
f) Find the probability of a Type II error if the true value of μ is $\mu = 0.995$ liters.
g) Suppose that the critical value, d, from part (b) is used, but the sample size is changed. Complete Table 14.4, using $s = 0.021$ as your estimate of σ.

Table 14.4

	True value of μ	
	$\mu = 1$	$\mu = 0.995$
Sample size	Probability of Type I error	Probability of Type II error
64		
81		
100	0.05	
144		

There are many cases in which simple hypothesis tests about a single mean or comparison of two groups are not enough. In such cases, you must use some ingenuity to decide how to arrange your groups to account for all the variables of interest. This will be illustrated in Example 7.

Design of experiments

14.5

Example **7**   A large corporate farming company grows corn on three farms—two in Indiana and one in Iowa. The company has been using the same fertilizer for the last three years, but is considering two new fertilizers for this year. This means that the company needs to do a test to see which fertilizer produces the best yields. (All three fertilizers cost nearly the same amount, so cost is not a factor.) The company has set aside one large field on each farm for testing purposes. Since there are three fertilizers and three farms, one possible method of approach is to test one fertilizer on each farm:

Farm A	Farm B	Farm C
Fertilizer 1	Fertilizer 2	Fertilizer 3

This is not a very good method. The farms are hundreds of miles apart. One farm may have better soil to begin with; another may have better weather. In other words, we are looking *not* at one variable (fertilizer), but at two (fertilizer and farm). A testing method that would make more sense would be to use each fertilizer on each farm. A statistician arranged for the test field on each farm to be divided into three plots. He then made a random assignment of fertilizers to plots:

Farm A	Farm B	Farm C
Fertilizer 3	Fertilizer 1	Fertilizer 2
Fertilizer 1	Fertilizer 2	Fertilizer 3
Fertilizer 2	Fertilizer 3	Fertilizer 1

Now the fertilizers can be compared without any bias due to difference in farms. ▶

The mathematical methods used to make the comparison in Example 7 are simply an extension of the analysis-of-variance methods covered in Chapter 11. These methods are covered in more advanced books on analysis of variance and experimental design. The important point here is that researchers must often think about what other variables influence the variable of interest, and choose an experimental design that takes those other variables into consideration.

Randomized block design   The design we have just shown is called a **randomized block design.** The three farm test fields are called *blocks;* they are blocks of land. The terminology of blocks and plots comes directly from agriculture, but is used in other kinds of problems as well. Much of the theory of experimental design and analysis of variance was developed by Sir Ronald Fisher when he was working at an agricultural experiment station on problems such as Example 7.

A teacher wants to compare two teaching methods for advanced calculus. This subject is taught at 8:40, 10:40, and 1:40 in his college. There is some evidence that the time a class is offered affects student performance. Set up an experimental design that would allow the two methods to be compared without having time differences introduce bias into the experiment.

Exercise **H**

The simpler studies we have done already in this book are based upon extremely simple experimental designs. The gas mileage study is a good example. If there are no other important variables affecting your results, it is a simple matter to test whether mean gas mileage is $\mu = 35$. Your design is to take a simple random sample and perform a z-test. The reason that more complex experimental designs are necessary is that in real life other variables do interfere.

Name some variables that we did not take into account in our gas mileage study, but that might affect the results of our experiment.

Exercise **I**

1   A restaurant chain has three restaurants in Phoenix, Arizona, three in El Paso, Texas, and six in Los Angeles, California. The chain wishes to try out three different promotional schemes for stimulating business from people who live within two miles of a particular restaurant. How should the chain design its experiment with these three promotional schemes?

2   Identify the two most important variables affecting the gas mileage study from among those you listed in Exercise I. Indicate how you would set up an experiment to determine the effects of these variables on gas mileage.

3   Suppose that you are assigned to study the effect of educational level on income. Identify another important variable affecting income, and indicate how you would set up a study of the effects of both variables on income.

We have covered a few key aspects of planning a study in this chapter, but we have covered them one by one. It is now time to look at the entire process. We will do this for a real example from the field of education.

◀ In the fall of 1977, the mathematics department at the United States Air Force Academy sponsored a study for comparison of traditional instruction in calculus and a new individualized instructional method. The study was designed and supervised by Major Samuel Thompson of the Academy (Major Thompson has a doctorate in Mathematics Education). We can only touch briefly on the details of this study; the final report by Major Thompson took more than 160 pages. A brief survey should be enough to give you some idea of the planning that goes into a well-designed study.

1. *Literature search.* Major Thompson surveyed writing on individualized instruction, and found a number* of studies comparing traditional and individualized instruction. He analyzed these studies and found that they did not answer the question the Air Force Academy was asking: "Is this new method better for *our* calculus course?"

The study: putting it all together   **14.6**

Example **8**

---

* His bibliography contains 55 references to books and articles.

2. *Other preparation.* To ensure that the comparison was not unfair to the new method, the Academy ran a pilot study in which the individualized method was tried out for a semester before the experiment. This enabled the experimenters to gain some experience with their new method, and to solve the most obvious basic instructional problems before the experiment began.

Personnel from the Academy also went to conferences on educational innovation in order to learn more about individualized instruction and ensure that the new method would be implemented correctly in the experiment.

3. *Population and sample.* The population of interest was all students who would ever take calculus at the Air Force Academy. The Academy used as its sample the 800 students who took calculus in the fall of 1977.

4. *Experimental design.* A number of variables can affect student performance in a comparison. Differences can arise because of student aptitude, instructor abilities, time of day for the class, and a number of other variables. The experimenters thought about this in advance and set up an elaborate design that had students of all ability levels taking calculus by both methods at all class hours. Even the teaching assignments were balanced between the two methods: Each method was carefully given the same mix of experienced and inexperienced instructors.

5. *Variables for comparison.* A number of variables are important to students and faculty alike. The new instructional method would be compared to the old one using:

a) A common final examination to be given to students taught by the two methods.

b) Records of student study time, to determine whether one method produced time savings.

c) Measurements of instructor work time in the two methods.

d) A comparison of failure rates under the two methods.

e) A follow-up study of the success of the two groups of students in their next mathematics course.

f) A study of student and faculty attitudes toward the new method and the old one.

These are not the only variables that were studied, but they are the major ones. The experimenter considered in advance how he would collect data for each of these variables, and planned his mathematical analyses, considering such important things as the power of various tests he could use.

6. *Analysis.* As supervisor of the experiment, Major Thompson did not teach a section of the course. He spent the entire semester of the experiment observing, collecting data, and ensuring that the experiment was run correctly. Once the experiment was finished, Major Thompson spent an entire semester doing an intensive analysis of his data according to the plans that were made previously. Other statisticians on the Academy faculty also did independent analyses of the data.

7. *Results.* The new method and the old produced no significant difference on the

common final examination. However, the students in the new method were found to have spent less total study time in calculus, and to have higher grades in other classes. The failure rate for the new method was less than half that of the old method. However, students and faculty alike indicated that they did not like the new method. It was decided not to force students or faculty to use a method they did not like. The Academy decided to continue using its old method of calculus instruction. ❿

The entire process—from asking the question "Is this new method better?" down to the decision *not* to implement it—took more than two years, hundreds of hours of careful planning, and additional hundreds of hours of analysis. Not every study needs to be this elaborate, but this example shows something about real-life uses of statistics. They are usually part of a decision-making process that takes some time from start to finish. Real uses of statistics are never as simple as the ten-minute problems in which you take given data, calculate a predetermined statistic, and draw a conclusion.

Beyond this book   14.7

The material covered in this book should enable you to do fairly simple studies, read reports in which statistics is used, or work intelligently on a larger project designed by a professional statistician. However, there is much more to statistics than the basics presented in this book. We have already pointed out that there are entire separate advanced books (and courses) dealing with topics that we have introduced in a chapter or less: regression analysis, analysis of variance, nonparametric statistics, sampling theory, and design of experiments. There is one other extremely important advanced area that we should mention here, since we have hardly covered it at all in this book: the area of **multivariate statistics.**

Multivariate statistics

In *most* of the applications in this book, we have looked at only one variable of interest. We are doing *univariate* (one-variable) statistics when we studied such things as the gas mileage for cars, the tar content for cigarettes, or the mathematics aptitude scores of college students. In multivariate statistics we are interested in *more* than one variable for each individual in our population.

For example, if we are studying characteristics of entering freshmen at a school, we may wish to know verbal aptitude and high school record as well as math aptitude. We illustrate this below:

Univariate statistics: Analyzes one variable at a time.
Data: One number per individual.

Example:

Individual	Data
John Smith	500
Harold Jones	525

↑
Mathematics
aptitude scores

Multivariate statistics: Analyzes many variables at one time.

Data: More than one number per individual.

Example:	Individual	Data		
	John Smith	(500,	610,	3.17)
	Harold Jones	(525,	450,	2.71)
		↑	↑	↑
		Math aptitude	Verbal aptitude	High school grade-point average

There are multivariate methods similar to most of the univariate methods in this book. You can test hypotheses concerning multivariate means, do a multivariate analysis of variance, or even find a nonparametric multivariate test. We would recommend study of multivariate methods to any reader who plans to do research in the life or social sciences.

The reader who wishes to do further work in statistics will discover that more advanced methods are covered in two kinds of courses at most schools. There are mathematical statistics courses, which develop mathematical theory in detail, and require a great deal of prerequisite mathematical coursework. The average user of statistics in applied areas such as education, psychology, or biology probably will not have the background for such theoretical courses.

For such applied users of statistics, there are special courses that concentrate on the practical use of statistics, and do not require prerequisite advanced courses in mathematics. (Most of these applied courses are accessible to researchers who have never learned calculus.) Those of our readers who wish to learn more statistics can easily find a number of practical advanced courses designed for nonmathematicians. Applied courses are available on almost every university campus in such departments as mathematics, statistics, psychology, business, and zoology. A brief conversation with the teacher is usually enough to determine whether a particular statistics course fits your own needs and mathematical level.

CHAPTER REVIEW

Key Terms

simple random sample
table of random numbers
random number generator
samples of convenience
cluster sampling
stratified sampling
Type I error

Type II error
probability of Type I error
probability of Type II error
power
randomized block design
multivariate statistics

sampling distribution of the mean: Formulas and Key Facts

If a random sample of size $n \geq 30$ is taken from a population with mean μ and standard deviation σ, then the random variable

$$z = \frac{\bar{x} - \mu}{\sigma/\sqrt{n}}$$

has (approximately) the standard normal distribution.

power of a test:

$$\text{Power} = 1 - P(\text{Type II error})$$

You should be able to

1 Obtain a *simple random sample* using a *table of random numbers*.

2 Distinguish between the various types of *sampling procedures* (e.g., simple random sampling, stratified sampling).

3 Calculate the *probability of a Type I error and a Type II error*, as done in Section 14.4.

4 Calculate the *power of a test*, as done in Section 14.4.

5 State the key aspects in *planning a study* as explained in Section 14.6.

REVIEW TEST

An industrial psychologist would like a company to try a motivational procedure to improve the output of assembly-line workers. The company is not convinced that the motivational procedure is worth the expense involved, and would like to be shown proof that it would increase output over present personnel management procedures. The workers work in three different eight-hour shifts, and company records show that there are differnces in output, absentee records, and morale across these shifts. Outline the steps the psychologist should go through in gathering information to make his case for the new motivational procedure. (These steps should cover the essential points mentioned in each section of this chapter. No mathematical calculations are necessary.)

FURTHER READINGS

For the reader who wishes to use more advanced statistical methods, but has had either little or no calculus, we recommend the following texts as useful references.

Analysis of variance and regression analysis

NETER, J., and W. WASSERMAN, *Applied Linear Statistical Models*. Homewood, Ill.: Richard D. Irwin, Inc., 1974.

Nonparametric statistics

DANIEL, W., *Applied Nonparametric Statistics*. Boston: Houghton Mifflin Co., 1978.

Multivariate statistics

HARRIS, R., *A Primer of Multivariate Statistics*. New York: Academic Press, 1975.

General reference

For the reader who would like a single book that covers a wide selection of the more advanced methods in the references above, we recommend:

DIXON, W., and F. MASSEY, *Introduction to Statistical Analysis* (Third Edition). New York: McGraw-Hill, 1969.

Statistics with calculus

For the reader who has had some calculus and would like to pursue some of the theoretical background of statistics without dealing with an excessively high level of mathe-

matical abstraction, we recommend:

WALPOLE, R., and R. MYERS, *Probability and Statistics for Engineers and Scientists*. New York: The Macmillan Company, 1972.

Computer packages

For the reader who is interested in learning to use the SPSS computer packages we have discussed in this book, we recommend:

KLECKA, W., N. NIE, and C. HULL, *SPSS Primer*. New York: McGraw-Hill, 1975.

ANSWERS TO REINFORCEMENT EXERCISES

A   Answers will vary.

B   Answers will vary.

C   a) A Type I error would occur if the sample mean gas mileage, $\bar{x}$, of the cars tested turned out to be greater than 35.43 mpg.
b) $P(\bar{x} > 35.43) = 0.05$.
c) i) The manufacturer is not too worried about a Type I error, since this would allow him to announce $\mu > 35$, even though, in fact, $\mu = 35$.
ii) The consumer group would *not* like to see a Type I error, since this would be misleading to the public; the manufacturer could claim $\mu > 35$, when, in fact, that is not true.

D   a) A Type II error would occur if the sample mean gas mileage of the cars tested out to be less than or equal to 35.43.
b) $P(\bar{x} \leq 35.43) = 0.6879$
c) i) The manufacturer would *not* like to see a Type II error, for then he would not be able to advertise $\mu > 35$, when, in fact, that is the case.
ii) The consumer group is not so concerned about a Type II error.

E   Answers will vary.

F   For $n = 64$, Power $= 1 - 0.0681 = 0.9319$

G   He means that by using a t-test there is a higher probability of avoiding a Type II error than by using a sign test.

H   Answers may vary, but here is a possibility: Divide each class into two groups, and teach one group by one method and the other group by the other method.

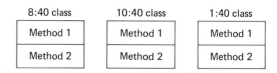

I   Answers will vary, but here are some variables that we did not take into account in the gas mileage study:
1. The drivers.
2. Where the cars are driven (e.g., geographical location, road conditions, city vs. highway driving).
3. Type of gasoline used.
4. Speed at which cars are driven.

Tables

Table I

Areas under the standard normal curve (Areas to the left)

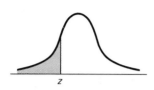

z	0	1	2	3	4	5	6	7	8	9
−3.0*	.0013	.0013	.0013	.0012	.0012	.0011	.0011	.0011	.0010	.0010
−2.9	.0019	.0018	.0017	.0017	.0016	.0016	.0015	.0015	.0014	.0014
−2.8	.0026	.0025	.0024	.0023	.0023	.0022	.0021	.0021	.0020	.0019
−2.7	.0035	.0034	.0033	.0032	.0031	.0030	.0029	.0028	.0027	.0026
−2.6	.0047	.0045	.0044	.0043	.0041	.0040	.0039	.0038	.0037	.0036
−2.5	.0062	.0060	.0059	.0057	.0055	.0054	.0052	.0051	.0049	.0048
−2.4	.0082	.0080	.0078	.0075	.0073	.0071	.0069	.0068	.0066	.0064
−2.3	.0107	.0104	.0102	.0099	.0096	.0094	.0091	.0089	.0087	.0084
−2.2	.0139	.0136	.0132	.0129	.0125	.0122	.0119	.0116	.0113	.0110
−2.1	.0179	.0174	.0170	.0166	.0162	.0158	.0154	.0150	.0146	.0143
−2.0	.0228	.0222	.0217	.0212	.0207	.0202	.0197	.0192	.0188	.0183
−1.9	.0287	.0281	.0274	.0268	.0262	.0256	.0250	.0244	.0239	.0233
−1.8	.0359	.0351	.0344	.0336	.0329	.0322	.0314	.0307	.0301	.0294
−1.7	.0446	.0436	.0427	.0418	.0409	.0401	.0392	.0384	.0375	.0367
−1.6	.0548	.0537	.0526	.0516	.0505	.0495	.0485	.0475	.0465	.0455
−1.5	.0668	.0655	.0643	.0630	.0618	.0606	.0594	.0582	.0571	.0559
−1.4	.0808	.0793	.0778	.0764	.0749	.0735	.0721	.0708	.0694	.0681
−1.3	.0968	.0951	.0934	.0918	.0901	.0885	.0869	.0853	.0838	.0823
−1.2	.1151	.1131	.1112	.1093	.1075	.1056	.1038	.1020	.1003	.0985
−1.1	.1357	.1335	.1314	.1292	.1271	.1251	.1230	.1210	.1190	.1170
−1.0	.1587	.1562	.1539	.1515	.1492	.1469	.1446	.1423	.1401	.1379
− .9	.1841	.1814	.1788	.1762	.1736	.1711	.1685	.1660	.1635	.1611
− .8	.2119	.2090	.2061	.2033	.2005	.1977	.1949	.1922	.1894	.1867
− .7	.2420	.2389	.2358	.2327	.2296	.2266	.2236	.2206	.2177	.2148
− .6	.2743	.2709	.2676	.2643	.2611	.2578	.2546	.2514	.2483	.2451
− .5	.3085	.3050	.3015	.2981	.2946	.2912	.2877	.2843	.2810	.2776
− .4	.3446	.3409	.3372	.3336	.3300	.3264	.3228	.3192	.3516	.3121
− .3	.3821	.3783	.3745	.3707	.3669	.3632	.3594	.3557	.3520	.3483
− .2	.4207	.4168	.4129	.4090	.4052	.4013	.3974	.3936	.3897	.3859
− .1	.4602	.4562	.4522	.4483	.4443	.4404	.4364	.4325	.4286	.4247
− .0	.5000	.4960	.4920	.4880	.4840	.4801	.4761	.4721	.4681	.4641

* For $z \leq -4$ the areas are 0 to four decimal places.

TABLE I   **577**

Table I (Cont.)

z	0	1	2	3	4	5	6	7	8	9
.0	.5000	.5040	.5080	.5120	.5160	.5199	.5239	.5279	.5319	.5359
.1	.5398	.5438	.5478	.5517	.5557	.5596	.5636	.5675	.5714	.5753
.2	.5793	.5832	.5871	.5910	.5948	.5987	.6026	.6064	.6103	.6141
.3	.6179	.6217	.6255	.6293	.6331	.6368	.6406	.6443	.6480	.6517
.4	.6554	.6591	.6628	.6664	.6700	.6736	.6772	.6808	.6844	.6879
.5	.6915	.6950	.6985	.7019	.7054	.7088	.7123	.7157	.7190	.7224
.6	.7257	.7291	.7324	.7357	.7389	.7422	.7454	.7486	.7517	.7549
.7	.7580	.7611	.7642	.7673	.7704	.7734	.7764	.7794	.7823	.7852
.8	.7881	.7910	.7939	.7967	.7995	.8023	.8051	.8078	.8106	.8133
.9	.8159	.8186	.8212	.8238	.8264	.8289	.8315	.8340	.8365	.8389
1.0	.8413	.8438	.8461	.8485	.8508	.8531	.8554	.8577	.8599	.8621
1.1	.8643	.8665	.8686	.8708	.8729	.8749	.8770	.8790	.8810	.8830
1.2	.8849	.8869	.8888	.8907	.8925	.8944	.8962	.8980	.8997	.9015
1.3	.9032	.9049	.9066	.9082	.9099	.9115	.9131	.9147	.9162	.9177
1.4	.9192	.9207	.9222	.9236	.9251	.9265	.9279	.9292	.9306	.9319
1.5	.9332	.9345	.9357	.9370	.9382	.9394	.9406	.9418	.9429	.9441
1.6	.9452	.9463	.9474	.9484	.9495	.9505	.9515	.9525	.9535	.9545
1.7	.9554	.9564	.9573	.9582	.9591	.9599	.9608	.9616	.9625	.9633
1.8	.9641	.9649	.9656	.9664	.9671	.9678	.9686	.9693	.9699	.9706
1.9	.9713	.9719	.9726	.9732	.9738	.9744	.9750	.9756	.9761	.9767
2.0	.9772	.9778	.9783	.9788	.9793	.9798	.9803	.9808	.9812	.9817
2.1	.9821	.9826	.9830	.9834	.9838	.9842	.9846	.9850	.9854	.9857
2.2	.9861	.9864	.9868	.9871	.9875	.9878	.9881	.9884	.9887	.9890
2.3	.9893	.9896	.9898	.9901	.9904	.9906	.9909	.9911	.9913	.9916
2.4	.9918	.9920	.9922	.9925	.9927	.9929	.9931	.9932	.9934	.9936
2.5	.9938	.9940	.9941	.9943	.9945	.9946	.9948	.9949	.9951	.9952
2.6	.9953	.9955	.9956	.9957	.9959	.9960	.9961	.9962	.9963	.9964
2.7	.9965	.9966	.9967	.9968	.9969	.9970	.9971	.9972	.9973	.9974
2.8	.9974	.9975	.9976	.9977	.9977	.9978	.9979	.9979	.9980	.9981
2.9	.9981	.9982	.9982	.9983	.9984	.9984	.9985	.9985	.9986	.9986
3.0†	.9987	.9987	.9987	.9988	.9988	.9989	.9989	.9989	.9990	.9990

† For $z \geq 4$ the areas are 1 to four decimal places.

Adapted from *Probability with Statistical Applications,* second edition, by F. Mosteller, R. E. K. Rourke, and G. B. Thomas, Jr. Reading, Mass.: Addison-Wesley, 1970, p. 473.

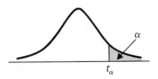

d.f.	$t_{.10}$	$t_{.05}$	$t_{.025}$	$t_{.01}$	$t_{.005}$
1	3.08	6.31	12.71	31.82	63.66
2	1.89	2.92	4.30	6.96	9.92
3	1.64	2.35	3.18	4.54	5.84
4	1.53	2.13	2.78	3.75	4.60
5	1.48	2.02	2.57	3.36	4.03
6	1.44	1.94	2.45	3.14	3.71
7	1.42	1.89	2.36	3.00	3.50
8	1.40	1.86	2.31	2.90	3.36
9	1.38	1.83	2.26	2.82	3.25
10	1.37	1.81	2.23	2.76	3.17
11	1.36	1.80	2.20	2.72	3.11
12	1.36	1.78	2.18	2.68	3.05
13	1.35	1.77	2.16	2.65	3.01
14	1.35	1.76	2.14	2.62	2.98
15	1.34	1.75	2.13	2.60	2.95
16	1.34	1.75	2.12	2.58	2.92
17	1.33	1.74	2.11	2.57	2.90
18	1.33	1.73	2.10	2.55	2.88
19	1.33	1.73	2.09	2.54	2.86
20	1.33	1.72	2.09	2.53	2.85
21	1.32	1.72	2.08	2.52	2.83
22	1.32	1.72	2.07	2.51	2.82
23	1.32	1.71	2.07	2.50	2.81
24	1.32	1.71	2.06	2.49	2.80
25	1.32	1.71	2.06	2.49	2.79
26	1.32	1.71	2.06	2.48	2.78
27	1.31	1.70	2.05	2.47	2.77
28	1.31	1.70	2.05	2.47	2.76
29	1.31	1.70	2.05	2.46	2.76
∞	1.28	1.64	1.96	2.33	2.58

Adapted from D. B. Owen, *Handbook of Statistical Tables*. Courtesy of the Atomic Energy Commission. Reading, Mass., Addison-Wesley, 1962

Note: The last row of the table (d.f. = ∞) gives values for z_α. For example, the table shows that $z_{.10} = 1.28$ and $z_{.05} = 1.64$.

TABLE III **579**

Table III
Chi-square distribution
(Values of χ_α^2)

	5	.95	.90	.10	.05	.025	.01	.005
		.00	.02	2.71	3.84	5.02	6.63	7.88
		.10	.21	4.61	5.99	7.38	9.21	10.60
		.35	.58	6.25	7.81	9.35	11.34	12.84
		.71	1.06	7.78	9.49	11.14	13.28	14.86
		1.15	1.61	9.24	11.07	12.83	15.09	16.75
		1.64	2.20	10.64	12.59	14.45	16.81	18.55
		2.17	2.83	12.02	14.07	16.01	18.48	20.28
		2.73	3.49	13.36	15.51	17.54	20.09	21.96
		3.33	4.17	14.68	16.92	19.02	21.67	23.59
		3.94	4.87	15.99	18.31	20.48	23.21	25.19
		4.57	5.58	17.28	19.68	21.92	24.72	26.76
		5.23	6.30	18.55	21.03	23.34	26.22	28.30
		5.89	7.04	19.81	22.36	24.74	27.69	29.82
		6.57	7.79	21.06	23.68	26.12	29.14	31.32
		7.26	8.55	22.31	25.00	27.49	30.58	32.80
		7.96	9.31	23.54	26.30	28.85	32.00	34.27
		8.67	10.09	24.77	27.59	30.19	33.41	35.72
		9.39	10.86	25.99	28.87	31.53	34.81	37.16
		10.12	11.65	27.20	30.14	32.85	36.19	38.58
20	7.43	8.26	9.59	10.85	12.44	28.41	31.41	34.17	37.57	40.00
21	8.03	8.90	10.28	11.59	13.24	29.62	32.67	35.48	38.93	41.40
22	8.64	9.54	10.98	12.34	14.04	30.81	33.92	36.78	40.29	42.80
23	9.26	10.20	11.69	13.09	14.85	32.01	35.17	38.08	41.64	44.18
24	9.89	10.86	12.40	13.85	15.66	33.20	36.42	39.36	42.98	45.56
25	10.52	11.52	13.12	14.61	16.47	34.38	37.65	40.65	44.31	46.93
26	11.16	12.20	13.84	15.38	17.29	35.56	38.89	41.92	45.64	48.29
27	11.81	12.88	14.57	16.15	18.11	36.74	40.11	43.19	46.96	49.65
28	12.46	13.56	15.31	16.93	18.94	37.92	41.34	44.46	48.28	50.99
29	13.12	14.26	16.05	17.71	19.77	39.09	42.56	45.72	49.59	52.34
30	13.79	14.95	16.79	18.49	20.60	40.26	43.77	46.98	50.89	53.67
50	27.99	29.71	32.36	34.76	37.69	63.17	67.50	71.42	76.15	79.49
100	67.33	70.06	74.22	77.93	82.36	118.5	124.3	129.6	135.8	140.2
500	422.3	429.4	439.9	449.1	459.9	540.9	553.1	563.9	576.5	585.2
1000	888.6	898.8	914.3	927.6	943.1	1058	1075	1090	1107	1119

Adapted from D. B. Owen, *Handbook of Statistical Tables*. Courtesy of the Atomic Energy Commission. Reading, Mass.: Addison-Wesley, 1962.

Table IV
Values of $F_{0.01}$

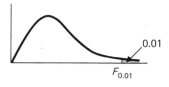

0.01

$F_{0.01}$

d. f. for numerator

	1	2	3	4	5	6	7	8	9
1	4052	4999.5	5403	5625	5764	5859	5928	5981	6022
2	98.50	99.00	99.17	99.25	99.30	99.33	99.36	99.37	99.39
3	34.12	30.82	29.46	28.71	28.24	27.91	27.67	27.49	27.35
4	21.20	18.00	16.69	15.98	15.52	15.21	14.98	14.80	14.66
5	16.26	13.27	12.06	11.39	10.97	10.67	10.46	10.29	10.16
6	13.75	10.92	9.78	9.15	8.75	8.47	8.26	8.10	7.98
7	12.25	9.55	8.45	7.85	7.46	7.19	6.99	6.84	6.72
8	11.26	8.65	7.59	7.01	6.63	6.37	6.18	6.03	5.91
9	10.56	8.02	6.99	6.42	6.06	5.80	5.61	5.47	5.35
10	10.04	7.56	6.55	5.99	5.64	5.39	5.20	5.06	4.94
11	9.65	7.21	6.22	5.67	5.32	5.07	4.89	4.74	4.63
12	9.33	6.93	5.95	5.41	5.06	4.82	4.64	4.50	4.39
13	9.07	6.70	5.74	5.21	4.86	4.62	4.44	4.30	4.19
14	8.86	6.51	5.56	5.04	4.69	4.46	4.28	4.14	4.03
15	8.68	6.36	5.42	4.89	4.56	4.32	4.14	4.00	3.89
16	8.53	6.23	5.29	4.77	4.44	4.20	4.03	3.89	3.78
17	8.40	6.11	5.18	4.67	4.34	4.10	3.93	3.79	3.68
18	8.29	6.01	5.09	4.58	4.25	4.01	3.84	3.71	3.60
19	8.18	5.93	5.01	4.50	4.17	3.94	3.77	3.63	3.52
20	8.10	5.85	4.94	4.43	4.10	3.87	3.70	3.56	3.46
21	8.02	5.78	4.87	4.37	4.04	3.81	3.64	3.51	3.40
22	7.95	5.72	4.82	4.31	3.99	3.76	3.59	3.45	3.35
23	7.88	5.66	4.76	4.26	3.94	3.71	3.54	3.41	3.30
24	7.82	5.61	4.72	4.22	3.90	3.67	3.50	3.36	3.26
25	7.77	5.57	4.68	4.18	3.85	3.63	3.46	3.32	3.22
26	7.72	5.53	4.64	4.14	3.82	3.59	3.42	3.29	3.18
27	7.68	5.49	4.60	4.11	3.78	3.56	3.39	3.26	3.15
28	7.64	5.45	4.57	4.07	3.75	3.53	3.36	3.23	3.12
29	7.60	5.42	4.54	4.04	3.73	3.50	3.33	3.20	3.09
30	7.56	5.39	4.51	4.02	3.70	3.47	3.30	3.17	3.07
40	7.31	5.18	4.31	3.83	3.51	3.29	3.12	2.99	2.89
60	7.08	4.98	4.13	3.65	3.34	3.12	2.95	2.82	2.72
120	6.85	4.79	3.95	3.48	3.17	2.96	2.79	2.66	2.56
∞	6.63	4.61	3.78	3.32	3.02	2.80	2.64	2.51	2.41

d.f. for denominator

Adapted from D. B. Owen, *Handbook of Statistical Tables*. Courtesy of the Atomic Energy Commission. Reading, Mass.: Addison-Wesley, 1962.

TABLE IV **581**

Table IV (Cont.)

d.f. for numerator

10	12	15	20	24	30	40	60	120	∞
6056	6106	6157	6209	6235	6261	6287	6313	6339	6366
99.40	99.42	99.43	99.45	99.46	99.47	99.47	99.48	99.49	99.50
27.23	27.05	26.87	26.69	26.60	26.50	26.41	26.32	26.22	26.13
14.55	14.37	14.20	14.02	13.93	13.84	13.75	13.65	13.56	13.46
10.05	9.89	9.72	9.55	9.47	9.38	9.29	9.20	9.11	9.02
7.87	7.72	7.56	7.40	7.31	7.23	7.14	7.06	6.97	6.88
6.62	6.47	6.31	6.16	6.07	5.99	5.91	5.82	5.74	5.65
5.81	5.67	5.52	5.36	5.28	5.20	5.12	5.03	4.95	4.86
5.26	5.11	4.96	4.81	4.73	4.65	4.57	4.48	4.40	4.31
4.85	4.71	4.56	4.41	4.33	4.25	4.17	4.08	4.00	3.91
4.54	4.40	4.25	4.10	4.02	3.94	3.86	3.78	3.69	3.60
4.30	4.16	4.01	3.86	3.78	3.70	3.62	3.54	3.45	3.36
4.10	3.96	3.82	3.66	3.59	3.51	3.43	3.34	3.25	3.17
3.94	3.80	3.66	3.51	3.43	3.35	3.27	3.18	3.09	3.00
3.80	3.67	3.52	3.37	3.29	3.21	3.13	3.05	2.96	2.87
3.69	3.55	3.41	3.26	3.18	3.10	3.02	2.93	2.84	2.75
3.59	3.46	3.31	3.16	3.08	3.00	2.92	2.83	2.75	2.65
3.51	3.37	3.23	3.08	3.00	2.92	2.84	2.75	2.66	2.57
3.43	3.30	3.15	3.00	2.92	2.84	2.76	2.67	2.58	2.49
3.37	3.23	3.09	2.94	2.86	2.78	2.69	2.61	2.52	2.42
3.31	3.17	3.03	2.88	2.80	2.72	2.64	2.55	2.46	2.36
3.26	3.12	2.98	2.83	2.75	2.67	2.58	2.50	2.40	2.31
3.21	3.07	2.93	2.78	2.70	2.62	2.54	2.45	2.35	2.26
3.17	3.03	2.89	2.74	2.66	2.58	2.49	2.40	2.31	2.21
3.13	2.99	2.85	2.70	2.62	2.54	2.45	2.36	2.27	2.17
3.09	2.96	2.81	2.66	2.58	2.50	2.42	2.33	2.23	2.13
3.06	2.93	2.78	2.63	2.55	2.47	2.38	2.29	2.20	2.10
3.03	2.90	2.75	2.60	2.52	2.44	2.35	2.26	2.17	2.06
3.00	2.87	2.73	2.57	2.49	2.41	2.33	2.23	2.14	2.03
2.98	2.84	2.70	2.55	2.47	2.39	2.30	2.21	2.11	2.01
2.80	2.66	2.52	2.37	2.29	2.20	2.11	2.02	1.92	1.80
2.63	2.50	2.35	2.20	2.12	2.03	1.94	1.84	1.73	1.60
2.47	2.34	2.19	2.03	1.95	1.86	1.76	1.66	1.53	1.38
2.32	2.18	2.04	1.88	1.79	1.70	1.59	1.47	1.32	1.00

Table V
Values of $F_{0.025}$

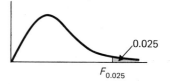

0.025

$F_{0.025}$

d.f. for numerator

	1	2	3	4	5	6	7	8	9
1	647.79	799.50	864.16	899.58	921.85	937.11	948.22	956.66	963.28
2	38.51	39.00	39.17	39.25	39.30	39.33	39.36	39.37	39.39
3	17.44	16.04	15.44	15.10	14.89	14.74	14.62	14.54	14.47
4	12.22	10.65	9.98	9.60	9.36	9.20	9.07	8.98	8.90
5	10.00	8.43	7.76	7.39	7.15	6.98	6.85	6.76	6.68
6	8.81	7.26	6.60	6.23	5.99	5.82	5.70	5.60	5.52
7	8.07	6.54	5.89	5.52	5.29	5.12	4.99	4.90	4.82
8	7.57	6.06	5.42	5.05	4.82	4.65	4.53	4.43	4.36
9	7.21	5.71	5.08	4.72	4.48	4.32	4.20	4.10	4.03
10	6.94	5.46	4.83	4.47	4.24	4.07	3.95	3.85	3.78
11	6.72	5.26	4.63	4.28	4.04	3.88	3.76	3.66	3.59
12	6.55	5.10	4.47	4.12	3.89	3.73	3.61	3.51	3.44
13	6.41	4.97	4.35	4.00	3.77	3.60	3.48	3.39	3.31
14	6.30	4.86	4.24	3.89	3.66	3.50	3.38	3.29	3.21
15	6.20	4.77	4.15	3.80	3.58	3.41	3.29	3.20	3.12
16	6.12	4.69	4.08	3.73	3.50	3.34	3.22	3.12	3.05
17	6.04	4.62	4.01	3.66	3.44	3.28	3.16	3.06	2.98
18	5.98	4.56	3.95	3.61	3.38	3.22	3.10	3.01	2.93
19	5.92	4.51	3.90	3.56	3.33	3.17	3.05	2.96	2.88
20	5.87	4.46	3.86	3.51	3.29	3.13	3.01	2.91	2.84
21	5.83	4.42	3.82	3.48	3.25	3.09	2.97	2.87	2.80
22	5.79	4.38	3.78	3.44	3.22	3.05	2.93	2.84	2.76
23	5.75	4.35	3.75	3.41	3.18	3.02	2.90	2.81	2.73
24	5.72	4.32	3.72	3.38	3.15	2.99	2.87	2.78	2.70
25	5.69	4.29	3.69	3.35	3.13	2.97	2.85	2.75	2.68
26	5.66	4.27	3.67	3.33	3.10	2.94	2.82	2.73	2.65
27	5.63	4.24	3.65	3.31	3.08	2.92	2.80	2.71	2.63
28	5.61	4.22	3.63	3.29	3.06	2.90	2.78	2.69	2.61
29	5.59	4.20	3.61	3.27	3.04	2.88	2.76	2.67	2.59
30	5.57	4.18	3.59	3.25	3.03	2.87	2.75	2.65	2.57
40	5.42	4.05	3.46	3.13	2.90	2.74	2.62	2.53	2.45
60	5.29	3.93	3.34	3.01	2.79	2.63	2.51	2.41	2.33
120	5.15	3.80	3.23	2.89	2.67	2.52	2.39	2.30	2.22
∞	5.02	3.69	3.12	2.79	2.57	2.41	2.29	2.19	2.11

d.f. for denominator

Adapted from D. B. Owen, *Handbook of Statistical Tables*. Courtesy of the Atomic Energy Commission. Reading, Mass.: Addison-Wesley, 1962.

TABLE V **583**

Table V (Cont.)

d.f. for numerator

10	12	15	20	24	30	40	60	120	∞
968.63	976.71	984.87	993.10	997.25	1001.4	1005.6	1009.8	1014.0	1018.3
39.40	39.42	39.43	39.45	39.46	39.47	39.47	39.48	39.49	39.50
14.42	14.34	14.25	14.17	14.12	14.08	14.04	13.99	13.95	13.90
8.84	8.75	8.66	8.56	8.51	8.46	8.41	8.36	8.31	8.26
6.62	6.52	6.43	6.33	6.28	6.23	6.18	6.12	6.07	6.02
5.46	5.37	5.27	5.17	5.12	5.07	5.01	4.96	4.90	4.85
4.76	4.67	4.57	4.47	4.42	4.36	4.31	4.25	4.20	4.14
4.30	4.20	4.10	4.00	3.95	3.89	3.84	3.78	3.73	3.67
3.96	3.87	3.77	3.67	3.61	3.56	3.51	3.45	3.39	3.33
3.72	3.62	3.52	3.42	3.37	3.31	3.26	3.20	3.14	3.08
3.53	3.43	3.33	3.23	3.17	3.12	3.06	3.00	2.94	2.88
3.37	3.28	3.18	3.07	3.02	2.96	2.91	2.85	2.79	2.72
3.25	3.15	3.05	2.95	2.89	2.84	2.78	2.72	2.66	2.60
3.15	3.05	2.95	2.84	2.79	2.73	2.67	2.61	2.55	2.49
3.06	2.96	2.86	2.76	2.70	2.64	2.59	2.52	2.46	2.40
2.99	2.89	2.79	2.68	2.63	2.57	2.51	2.45	2.38	2.32
2.92	2.82	2.72	2.62	2.56	2.50	2.44	2.38	2.32	2.25
2.87	2.77	2.67	2.56	2.50	2.44	2.38	2.32	2.26	2.19
2.82	2.72	2.62	2.51	2.45	2.39	2.33	2.27	2.20	2.13
2.77	2.68	2.57	2.46	2.41	2.35	2.29	2.22	2.16	2.09
2.73	2.64	2.53	2.42	2.37	2.31	2.25	2.18	2.11	2.04
2.70	2.60	2.50	2.39	2.33	2.27	2.21	2.14	2.08	2.00
2.67	2.57	2.47	2.36	2.30	2.24	2.18	2.11	2.04	1.97
2.64	2.54	2.44	2.33	2.27	2.21	2.15	2.08	2.01	1.94
2.61	2.51	2.41	2.30	2.24	2.18	2.12	2.05	1.98	1.91
2.59	2.49	2.39	2.28	2.22	2.16	2.09	2.03	1.95	1.88
2.57	2.47	2.36	2.25	2.19	2.13	2.07	2.00	1.93	1.85
2.55	2.45	2.34	2.23	2.17	2.11	2.05	1.98	1.91	1.83
2.53	2.43	2.32	2.21	2.15	2.09	2.03	1.96	1.89	1.81
2.51	2.41	2.31	2.20	2.14	2.07	2.01	1.94	1.87	1.79
2.39	2.29	2.18	2.07	2.01	1.94	1.88	1.80	1.72	1.64
2.27	2.17	2.06	1.94	1.88	1.82	1.74	1.67	1.58	1.48
2.16	2.05	1.95	1.82	1.76	1.69	1.61	1.53	1.43	1.31
2.05	1.94	1.83	1.71	1.64	1.57	1.48	1.39	1.27	1.00

Table VI
Values of $F_{0.05}$

	d.f. for numerator								
	1	2	3	4	5	6	7	8	9
1	161.4	199.5	215.7	224.6	230.2	234.0	236.8	238.9	240.5
2	18.51	19.00	19.16	19.25	19.30	19.33	19.35	19.37	19.38
3	10.13	9.55	9.28	9.12	9.01	8.94	8.89	8.85	8.81
4	7.71	6.94	6.59	6.39	6.26	6.16	6.09	6.04	6.00
5	6.61	5.79	5.41	5.19	5.05	4.95	4.88	4.82	4.77
6	5.99	5.14	4.76	4.53	4.39	4.28	4.21	4.15	4.10
7	5.59	4.74	4.35	4.12	3.97	3.87	3.79	3.73	3.68
8	5.32	4.46	4.07	3.84	3.69	3.58	3.50	3.44	3.39
9	5.12	4.26	3.86	3.63	3.48	3.37	3.29	3.23	3.18
10	4.96	4.10	3.71	3.48	3.33	3.22	3.14	3.07	3.02
11	4.84	3.98	3.59	3.36	3.20	3.09	3.01	2.95	2.90
12	4.75	3.89	3.49	3.26	3.11	3.00	2.91	2.85	2.80
13	4.67	3.81	3.41	3.18	3.03	2.92	2.83	2.77	2.71
14	4.60	3.74	3.34	3.11	2.96	2.85	2.76	2.70	2.65
15	4.54	3.68	3.29	3.06	2.90	2.79	2.71	2.64	2.59
16	4.49	3.63	3.24	3.01	2.85	2.74	2.66	2.59	2.54
17	4.45	3.59	3.20	2.96	2.81	2.70	2.61	2.55	2.49
18	4.41	3.55	3.16	2.93	2.77	2.66	2.58	2.51	2.46
19	4.38	3.52	3.13	2.90	2.74	2.63	2.54	2.48	2.42
20	4.35	3.49	3.10	2.87	2.71	2.60	2.51	2.45	2.39
21	4.32	3.47	3.07	2.84	2.68	2.57	2.49	2.42	2.37
22	4.30	3.44	3.05	2.82	2.66	2.55	2.46	2.40	2.34
23	4.28	3.42	3.03	2.80	2.64	2.53	2.44	2.37	2.32
24	4.26	3.40	3.01	2.78	2.62	2.51	2.42	2.36	2.30
25	4.24	3.39	2.99	2.76	2.60	2.49	2.40	2.34	2.28
26	4.23	3.37	2.98	2.74	2.59	2.47	2.39	2.32	2.27
27	4.21	3.35	2.96	2.73	2.57	2.46	2.37	2.31	2.25
28	4.20	3.34	2.95	2.71	2.56	2.45	2.36	2.29	2.24
29	4.18	3.33	2.93	2.70	2.55	2.43	2.35	2.28	2.22
30	4.17	3.32	2.92	2.69	2.53	2.42	2.33	2.27	2.21
40	4.08	3.23	2.84	2.61	2.45	2.34	2.25	2.18	2.12
60	4.00	3.15	2.76	2.53	2.37	2.25	2.17	2.10	2.04
120	3.92	3.07	2.68	2.45	2.29	2.17	2.09	2.02	1.96
∞	3.84	3.00	2.60	2.37	2.21	2.10	2.01	1.94	1.88

d.f. for denominator

Adapted from D. B. Owen, *Handbook of Statistical Tables*. Courtesy of the Atomic Energy Commission. Reading, Mass.: Addison-Wesley, 1962.

TABLE VI **585**

Table VI (Cont.)

d.f. for numerator

10	12	15	20	24	30	40	60	120	∞
241.9	243.9	245.9	248.0	249.1	250.1	251.1	252.2	253.3	254.3
19.40	19.41	19.43	19.45	19.45	19.46	19.47	19.48	19.49	19.50
8.79	8.74	8.70	8.66	8.64	8.62	8.59	8.57	8.55	8.53
5.96	5.91	5.86	5.80	5.77	5.75	5.72	5.69	5.66	5.63
4.74	4.68	4.62	4.56	4.53	4.50	4.46	4.43	4.40	4.36
4.06	4.00	3.94	3.87	3.84	3.81	3.77	3.74	3.70	3.67
3.64	3.57	3.51	3.41	3.41	3.38	3.34	3.30	3.27	3.23
3.35	3.28	3.22	3.15	3.12	3.08	3.04	3.01	2.97	2.93
3.14	3.07	3.01	2.94	2.90	2.86	2.83	2.79	2.75	2.71
2.98	2.91	2.85	2.77	2.74	2.70	2.66	2.62	2.58	2.54
2.85	2.79	2.72	2.65	2.61	2.57	2.53	2.49	2.45	2.40
2.75	2.69	2.62	2.54	2.51	2.47	2.43	2.38	2.34	2.30
2.67	2.60	2.53	2.46	2.42	2.38	2.34	2.30	2.25	2.21
2.60	2.53	2.46	2.39	2.35	2.31	2.27	2.22	2.18	2.13
2.54	2.48	2.40	2.33	2.29	2.25	2.20	2.16	2.11	2.07
2.49	2.42	2.35	2.28	2.24	2.19	2.15	2.11	2.06	2.01
2.45	2.38	2.31	2.23	2.19	2.15	2.10	2.06	2.01	1.96
2.41	2.34	2.27	2.19	2.15	2.11	2.06	2.02	1.97	1.92
2.38	2.31	2.23	2.16	2.11	2.07	2.03	1.98	1.93	1.88
2.35	2.28	2.20	2.12	2.08	2.04	1.99	1.95	1.90	1.84
2.32	2.25	2.18	2.10	2.05	2.01	1.96	1.92	1.87	1.81
2.30	2.23	2.15	2.07	2.03	1.98	1.94	1.89	1.84	1.78
2.27	2.20	2.13	2.05	2.01	1.96	1.91	1.86	1.81	1.76
2.25	2.18	2.11	2.03	1.98	1.94	1.89	1.84	1.79	1.73
2.24	2.16	2.09	2.01	1.96	1.92	1.87	1.82	1.77	1.71
2.22	2.15	2.07	1.99	1.95	1.90	1.85	1.80	1.75	1.69
2.20	2.13	2.06	1.97	1.93	1.88	1.84	1.79	1.73	1.67
2.19	2.12	2.04	1.96	1.91	1.87	1.82	1.77	1.71	1.65
2.18	2.10	2.03	1.94	1.90	1.85	1.81	1.75	1.70	1.64
2.16	2.09	2.01	1.93	1.89	1.84	1.79	1.74	1.68	1.62
2.08	2.00	1.92	1.84	1.79	1.74	1.69	1.64	1.58	1.51
1.99	1.92	1.84	1.75	1.70	1.65	1.59	1.53	1.47	1.39
1.91	1.83	1.75	1.66	1.61	1.55	1.50	1.43	1.35	1.25
1.83	1.75	1.67	1.57	1.52	1.46	1.39	1.32	1.22	1.00

d.f. \\ α	.10	.05	.02	.01
1	.988	.997	.9995	.9999
2	.900	.950	.980	.990
3	.805	.878	.934	.959
4	.729	.811	.882	.917
5	.669	.754	.833	.874
6	.622	.707	.789	.834
7	.582	.666	.750	.798
8	.549	.632	.716	.765
9	.521	.602	.685	.735
10	.497	.576	.658	.708
11	.476	.553	.634	.684
12	.458	.532	.612	.661
13	.441	.514	.592	.641
14	.426	.497	.574	.623
15	.412	.482	.558	.606
16	.400	.468	.543	.590
17	.389	.456	.528	.575
18	.378	.444	.516	.561
19	.369	.433	.503	.549
20	.360	.423	.492	.537
21	.352	.413	.482	.526
22	.344	.404	.472	.515
23	.337	.396	.462	.505
24	.330	.388	.453	.496
25	.323	.381	.445	.487
26	.317	.374	.437	.479
27	.311	.367	.430	.471
28	.306	.361	.423	.463
29	.301	.355	.416	.456
30	.296	.349	.409	.449
40	.257	.304	.358	.393
50	.231	.273	.322	.354
60	.211	.250	.295	.325
70	.195	.232	.274	.302
80	.183	.217	.257	.283
90	.173	.205	.242	.267
100	.164	.195	.230	.254

Entries for d.f. = 1 to 23 adapted with permission from D. B. Owen, *Handbook of Statistical Tables*, Addison-Wesley, 1962.
Other entries derived using the relation $t_{d.f.} = r\sqrt{d.f./(1 - r^2)}$.
Note: A value given in the table is the *right-hand* critical value for a *two-tailed* test at the significance level indicated. The left-hand critical value is just the negative of the right-hand critical value.

TABLE VIII **587**

Table VIII
Binomial probabilities

							p					
n	k	.1	.2	.25	.3	.4	.5	.6	.7	.75	.8	.9
1	0	.9000	.8000	.7500	.7000	.6000	.5000	.4000	.3000	.2500	.2000	.1000
	1	.1000	.2000	.2500	.3000	.4000	.5000	.6000	.7000	.7500	.8000	.9000
2	0	.8100	.6400	.5625	.4900	.3600	.2500	.1600	.0900	.0625	.0400	.0100
	1	.1800	.3200	.3750	.4200	.4800	.5000	.4800	.4200	.3750	.3200	.1800
	2	.0100	.0400	.0625	.0900	.1600	.2500	.3600	.4900	.5625	.6400	.8100
3	0	.7290	.5120	.4219	.3430	.2160	.1250	.0640	.0270	.0156	.0080	.0010
	1	.2430	.3840	.4219	.4410	.4320	.3750	.2880	.1890	.1406	.0960	.0270
	2	.0270	.0960	.1406	.1890	.2880	.3750	.4320	.4410	.4219	.3840	.2430
	3	.0010	.0080	.0156	.0270	.0640	.1250	.2160	.3430	.4219	.5120	.7290
4	0	.6561	.4096	.3164	.2401	.1296	.0625	.0256	.0081	.0039	.0016	.0001
	1	.2916	.4096	.4219	.4116	.3456	.2500	.1536	.0756	.0469	.0256	.0036
	2	.0486	.1536	.2109	.2646	.3456	.3750	.3456	.2646	.2109	.1536	.0486
	3	.0036	.0256	.0469	.0756	.1536	.2500	.3456	.4116	.4219	.4096	.2916
	4	.0001	.0016	.0039	.0081	.0256	.0625	.1296	.2401	.3164	.4096	.6561
5	0	.5905	.3277	.2373	.1681	.0778	.0313	.0102	.0024	.0010	.0003	.0000
	1	.3281	.4096	.3955	.3602	.2592	.1563	.0768	.0284	.0146	.0064	.0005
	2	.0729	.2048	.2637	.3087	.3456	.3125	.2304	.1323	.0879	.0512	.0081
	3	.0081	.0512	.0879	.1323	.2304	.3125	.3456	.3087	.2637	.2048	.0729
	4	.0004	.0064	.0146	.0283	.0768	.1563	.2592	.3602	.3955	.4096	.3281
	5	.0000	.0003	.0010	.0024	.0102	.0313	.0778	.1681	.2373	.3277	.5905
6	0	.5314	.2621	.1780	.1176	.0467	.0156	.0041	.0007	.0002	.0001	.0000
	1	.3543	.3932	.3560	.3025	.1866	.0938	.0369	.0102	.0044	.0015	.0001
	2	.0984	.2458	.2966	.3241	.3110	.2344	.1382	.0595	.0330	.0154	.0012
	3	.0146	.0819	.1318	.1852	.2765	.3125	.2765	.1852	.1318	.0819	.0146
	4	.0012	.0154	.0330	.0595	.1382	.2344	.3110	.3241	.2966	.2458	.0984
	5	.0001	.0015	.0044	.0102	.0369	.0938	.1866	.3025	.3560	.3932	.3543
	6	.0000	.0001	.0002	.0007	.0041	.0156	.0467	.1176	.1780	.2621	.5314
7	0	.4783	.2097	.1335	.0824	.0280	.0078	.0016	.0002	.0001	.0000	.0000
	1	.3720	.3670	.3115	.2471	.1306	.0547	.0172	.0036	.0013	.0004	.0000
	2	.1240	.2753	.3115	.3177	.2613	.1641	.0774	.0250	.0115	.0043	.0002
	3	.0230	.1147	.1730	.2269	.2903	.2734	.1935	.0972	.0577	.0287	.0026
	4	.0026	.0287	.0577	.0972	.1935	.2734	.2903	.2269	.1730	.1147	.0230
	5	.0002	.0043	.0115	.0250	.0774	.1641	.2613	.3177	.3115	.2753	.1240
	6	.0000	.0004	.0013	.0036	.0172	.0547	.1306	.2471	.3115	.3670	.3720
	7	.0000	.0000	.0001	.0002	.0016	.0078	.0280	.0824	.1335	.2097	.4783

							p					

Table VIII (Cont.)

n	k	.1	.2	.25	.3	.4	.5	.6	.7	.75	.8	.9
8	0	.4305	.1678	.1001	.0576	.0168	.0039	.0007	.0001	.0000	.0000	.0000
	1	.3826	.3355	.2670	.1977	.0896	.0312	.0079	.0012	.0004	.0001	.0000
	2	.1488	.2936	.3115	.2965	.2090	.1094	.0413	.0100	.0038	.0011	.0000
	3	.0331	.1468	.2076	.2541	.2787	.2188	.1239	.0467	.0231	.0092	.0004
	4	.0046	.0459	.0865	.1361	.2322	.2734	.2322	.1361	.0865	.0459	.0046
	5	.0004	.0092	.0231	.0467	.1239	.2188	.2787	.2541	.2076	.1468	.0331
	6	.0000	.0011	.0038	.0100	.0413	.1094	.2090	.2965	.3115	.2936	.1488
	7	.0000	.0001	.0004	.0012	.0079	.0312	.0896	.1977	.2670	.3355	.3826
	8	.0000	.0000	.0000	.0001	.0007	.0039	.0168	.0576	.1001	.1678	.4305
9	0	.3874	.1342	.0751	.0404	.0101	.0020	.0003	.0000	.0000	.0000	.0000
	1	.3874	.3020	.2253	.1556	.0605	.0176	.0035	.0004	.0001	.0000	.0000
	2	.1722	.3020	.3003	.2668	.1612	.0703	.0212	.0039	.0012	.0003	.0000
	3	.0446	.1762	.2336	.2668	.2508	.1641	.0743	.0210	.0087	.0028	.0001
	4	.0074	.0661	.1168	.1715	.2508	.2461	.1672	.0735	.0389	.0165	.0008
	5	.0008	.0165	.0389	.0735	.1672	.2461	.2508	.1715	.1168	.0661	.0074
	6	.0001	.0028	.0087	.0210	.0743	.1641	.2508	.2668	.2336	.1762	.0446
	7	.0000	.0003	.0012	.0039	.0212	.0703	.1612	.2668	.3003	.3020	.1722
	8	.0000	.0000	.0001	.0004	.0035	.0176	.0605	.1556	.2253	.3020	.3874
	9	.0000	.0000	.0000	.0000	.0003	.0020	.0101	.0404	.0751	.1342	.3874
10	0	.3487	.1074	.0563	.0282	.0060	.0010	.0001	.0000	.0000	.0000	.0000
	1	.3874	.2684	.1877	.1211	.0403	.0098	.0016	.0001	.0000	.0000	.0000
	2	.1937	.3020	.2816	.2335	.1209	.0439	.0106	.0014	.0004	.0001	.0000
	3	.0574	.2013	.2503	.2668	.2150	.1172	.0425	.0090	.0031	.0008	.0000
	4	.0112	.0881	.1460	.2001	.2508	.2051	.1115	.0368	.0162	.0055	.0001
	5	.0015	.0264	.0584	.1029	.2007	.2461	.2007	.1029	.0584	.0264	.0015
	6	.0001	.0055	.0162	.0368	.1115	.2051	.2508	.2001	.1460	.0881	.0112
	7	.0000	.0008	.0031	.0090	.0425	.1172	.2150	.2668	.2503	.2013	.0574
	8	.0000	.0001	.0004	.0014	.0106	.0439	.1209	.2335	.2816	.3020	.1937
	9	.0000	.0000	.0000	.0001	.0016	.0098	.0403	.1211	.1877	.2684	.3874
	10	.0000	.0000	.0000	.0000	.0001	.0010	.0060	.0282	.0563	.1074	.3487
11	0	.3138	.0859	.0422	.0198	.0036	.0005	.0000	.0000	.0000	.0000	.0000
	1	.3835	.2362	.1549	.0932	.0266	.0054	.0007	.0000	.0000	.0000	.0000
	2	.2131	.2953	.2581	.1998	.0887	.0269	.0052	.0005	.0001	.0000	.0000
	3	.0710	.2215	.2581	.2568	.1774	.0806	.0234	.0037	.0011	.0002	.0000
	4	.0158	.1107	.1721	.2201	.2365	.1611	.0701	.0173	.0064	.0017	.0000
	5	.0025	.0388	.0803	.1321	.2207	.2256	.1471	.0566	.0268	.0097	.0003
	6	.0003	.0097	.0268	.0566	.1471	.2256	.2207	.1321	.0803	.0388	.0025
	7	.0000	.0017	.0064	.0173	.0701	.1611	.2365	.2201	.1721	.1107	.0158
	8	.0000	.0002	.0011	.0037	.0234	.0806	.1774	.2568	.2581	.2215	.0710
	9	.0000	.0000	.0001	.0005	.0052	.0269	.0887	.1998	.2581	.2953	.2131
	10	.0000	.0000	.0000	.0000	.0007	.0054	.0266	.0932	.1549	.2362	.3835
	11	.0000	.0000	.0000	.0000	.0000	.0005	.0036	.0198	.0422	.0859	.3138

TABLE VIII   **589**

							p					Table VIII (Cont.)
n	*k*	.1	.2	.25	.3	.4	.5	.6	.7	.75	.8	.9
12	0	.2824	.0687	.0317	.0138	.0022	.0002	.0000	.0000	.0000	.0000	.0000
	1	.3766	.2062	.1267	.0712	.0174	.0029	.0003	.0000	.0000	.0000	.0000
	2	.2301	.2835	.2323	.1678	.0639	.0161	.0025	.0002	.0000	.0000	.0000
	3	.0852	.2362	.2581	.2397	.1419	.0537	.0125	.0015	.0004	.0001	.0000
	4	.0213	.1329	.1936	.2311	.2128	.1208	.0420	.0078	.0024	.0005	.0000
	5	.0038	.0532	.1032	.1585	.2270	.1934	.1009	.0291	.0115	.0033	.0000
	6	.0005	.0155	.0401	.0792	.1766	.2256	.1766	.0792	.0401	.0155	.0005
	7	.0000	.0033	.0115	.0291	.1009	.1934	.2270	.1585	.1032	.0532	.0038
	8	.0000	.0005	.0024	.0078	.0420	.1208	.2128	.2311	.1936	.1329	.0213
	9	.0000	.0001	.0004	.0015	.0125	.0537	.1419	.2397	.2581	.2362	.0852
	10	.0000	.0000	.0000	.0002	.0025	.0161	.0639	.1678	.2323	.2835	.2301
	11	.0000	.0000	.0000	.0000	.0003	.0029	.0174	.0712	.1267	.2062	.3766
	12	.0000	.0000	.0000	.0000	.0000	.0002	.0022	.0138	.0317	.0687	.2824
13	0	.2542	.0550	.0238	.0097	.0013	.0001	.0000	.0000	.0000	.0000	.0000
	1	.3672	.1787	.1029	.0540	.0113	.0016	.0001	.0000	.0000	.0000	.0000
	2	.2448	.2680	.2059	.1388	.0453	.0095	.0012	.0001	.0000	.0000	.0000
	3	.0997	.2457	.2517	.2181	.1107	.0349	.0065	.0006	.0001	.0000	.0000
	4	.0277	.1535	.2097	.2337	.1845	.0873	.0243	.0034	.0009	.0001	.0000
	5	.0055	.0691	.1258	.1803	.2214	.1571	.0656	.0142	.0047	.0011	.0000
	6	.0008	.0230	.0559	.1030	.1968	.2095	.1312	.0442	.0186	.0058	.0001
	7	.0001	.0058	.0186	.0442	.1312	.2095	.1968	.1030	.0559	.0230	.0008
	8	.0000	.0011	.0047	.0142	.0656	.1571	.2214	.1803	.1258	.0691	.0055
	9	.0000	.0001	.0009	.0034	.0243	.0873	.1845	.2337	.2097	.1535	.0277
	10	.0000	.0000	.0001	.0006	.0065	.0349	.1107	.2181	.2517	.2457	.0997
	11	.0000	.0000	.0000	.0001	.0012	.0095	.0453	.1388	.2059	.2680	.2448
	12	.0000	.0000	.0000	.0000	.0001	.0016	.0113	.0540	.1029	.1787	.3672
	13	.0000	.0000	.0000	.0000	.0000	.0001	.0013	.0097	.0238	.0550	.2542
14	0	.2288	.0440	.0178	.0068	.0008	.0001	.0000	.0000	.0000	.0000	.0000
	1	.3559	.1539	.0832	.0407	.0073	.0009	.0001	.0000	.0000	.0000	.0000
	2	.2570	.2501	.1802	.1134	.0317	.0056	.0005	.0000	.0000	.0000	.0000
	3	.1142	.2501	.2402	.1943	.0845	.0222	.0033	.0002	.0000	.0000	.0000
	4	.0349	.1720	.2202	.2290	.1549	.0611	.0136	.0014	.0003	.0000	.0000
	5	.0078	.0860	.1468	.1963	.2066	.1222	.0408	.0066	.0018	.0003	.0000
	6	.0013	.0322	.0734	.1262	.2066	.1833	.0918	.0232	.0082	.0020	.0000
	7	.0002	.0092	.0280	.0618	.1574	.2095	.1574	.0618	.0280	.0092	.0002
	8	.0000	.0020	.0082	.0232	.0918	.1833	.2066	.1262	.0734	.0322	.0013
	9	.0000	.0003	.0018	.0066	.0408	.1222	.2066	.1963	.1468	.0860	.0078
	10	.0000	.0000	.0003	.0014	.0136	.0611	.1549	.2290	.2202	.1720	.0349
	11	.0000	.0000	.0000	.0002	.0033	.0222	.0845	.1943	.2402	.2501	.1142
	12	.0000	.0000	.0000	.0000	.0005	.0056	.0317	.1134	.1802	.2501	.2570
	13	.0000	.0000	.0000	.0000	.0001	.0009	.0073	.0407	.0832	.1539	.3559
	14	.0000	.0000	.0000	.0000	.0000	.0001	.0008	.0068	.0178	.0440	.2288

Table VIII (Cont.)

| | | | | | | | p | | | | | |
| | | | | | | | | | | | | |
n	k	.1	.2	.25	.3	.4	.5	.6	.7	.75	.8	.9
15	0	.2059	.0352	.0134	.0047	.0005	.0000	.0000	.0000	.0000	.0000	.0000
	1	.3432	.1319	.0668	.0305	.0047	.0005	.0000	.0000	.0000	.0000	.0000
	2	.2669	.2309	.1559	.0916	.0219	.0032	.0003	.0000	.0000	.0000	.0000
	3	.1285	.2501	.2252	.1700	.0634	.0139	.0016	.0001	.0000	.0000	.0000
	4	.0428	.1876	.2252	.2186	.1268	.0417	.0074	.0006	.0001	.0000	.0000
	5	.0105	.1032	.1651	.2061	.1859	.0916	.0245	.0030	.0007	.0001	.0000
	6	.0019	.0430	.0917	.1472	.2066	.1527	.0612	.0116	.0034	.0007	.0000
	7	.0003	.0138	.0393	.0811	.1771	.1964	.1181	.0348	.0131	.0035	.0000
	8	.0000	.0035	.0131	.0348	.1181	.1964	.1771	.0811	.0393	.0138	.0003
	9	.0000	.0007	.0034	.0116	.0612	.1527	.2066	.1472	.0917	.0430	.0019
	10	.0000	.0001	.0007	.0030	.0245	.0916	.1859	.2061	.1651	.1032	.0105
	11	.0000	.0000	.0001	.0006	.0074	.0417	.1268	.2186	.2252	.1876	.0428
	12	.0000	.0000	.0000	.0001	.0016	.0139	.0634	.1700	.2252	.2501	.1285
	13	.0000	.0000	.0000	.0000	.0003	.0032	.0219	.0916	.1559	.2309	.2669
	14	.0000	.0000	.0000	.0000	.0000	.0005	.0047	.0305	.0668	.1319	.3432
	15	.0000	.0000	.0000	.0000	.0000	.0000	.0005	.0047	.0134	.0352	.2059
20	0	.1216	.0115	.0032	.0008	.0000	.0000	.0000	.0000	.0000	.0000	.0000
	1	.2702	.0576	.0211	.0068	.0005	.0000	.0000	.0000	.0000	.0000	.0000
	2	.2852	.1369	.0669	.0278	.0031	.0002	.0000	.0000	.0000	.0000	.0000
	3	.1901	.2054	.1339	.0716	.0123	.0011	.0000	.0000	.0000	.0000	.0000
	4	.0898	.2182	.1897	.1304	.0350	.0046	.0003	.0000	.0000	.0000	.0000
	5	.0319	.1746	.2023	.1789	.0746	.0148	.0013	.0000	.0000	.0000	.0000
	6	.0089	.1091	.1686	.1916	.1244	.0370	.0049	.0002	.0000	.0000	.0000
	7	.0020	.0545	.1124	.1643	.1659	.0739	.0146	.0010	.0002	.0000	.0000
	8	.0004	.0222	.0609	.1144	.1797	.1201	.0355	.0039	.0008	.0001	.0000
	9	.0001	.0074	.0271	.0654	.1597	.1602	.0710	.0120	.0030	.0005	.0000
	10	.0000	.0020	.0099	.0308	.1171	.1762	.1171	.0308	.0099	.0020	.0000
	11	.0000	.0005	.0030	.0120	.0710	.1602	.1597	.0654	.0271	.0074	.0001
	12	.0000	.0001	.0008	.0039	.0355	.1201	.1797	.1144	.0609	.0222	.0004
	13	.0000	.0000	.0002	.0010	.0146	.0739	.1659	.1643	.1124	.0545	.0020
	14	.0000	.0000	.0000	.0002	.0049	.0370	.1244	.1916	.1686	.1091	.0089
	15	.0000	.0000	.0000	.0000	.0013	.0148	.0746	.1789	.2023	.1746	.0319
	16	.0000	.0000	.0000	.0000	.0003	.0046	.0350	.1304	.1897	.2182	.0898
	17	.0000	.0000	.0000	.0000	.0000	.0011	.0123	.0716	.1339	.2054	.1901
	18	.0000	.0000	.0000	.0000	.0000	.0002	.0031	.0278	.0669	.1369	.2852
	19	.0000	.0000	.0000	.0000	.0000	.0000	.0005	.0068	.0211	.0576	.2702
	20	.0000	.0000	.0000	.0000	.0000	.0000	.0000	.0008	.0032	.0115	.1216

TABLE VIII **591**

Table VIII (Cont.)

							p					
n	k	.1	.2	.25	.3	.4	.5	.6	.7	.75	.8	.9
25	0	.0718	.0038	.0008	.0001	.0000	.0000	.0000	.0000	.0000	.0000	.0000
	1	.1994	.0236	.0063	.0014	.0000	.0000	.0000	.0000	.0000	.0000	.0000
	2	.2659	.0708	.0251	.0074	.0004	.0000	.0000	.0000	.0000	.0000	.0000
	3	.2265	.1358	.0641	.0243	.0019	.0001	.0000	.0000	.0000	.0000	.0000
	4	.1384	.1867	.1175	.0572	.0071	.0004	.0000	.0000	.0000	.0000	.0000
	5	.0646	.1960	.1645	.1030	.0199	.0016	.0000	.0000	.0000	.0000	.0000
	6	.0239	.1633	.1828	.1472	.0442	.0053	.0002	.0000	.0000	.0000	.0000
	7	.0072	.1108	.1654	.1712	.0800	.0143	.0009	.0000	.0000	.0000	.0000
	8	.0018	.0623	.1241	.1651	.1200	.0322	.0031	.0001	.0000	.0000	.0000
	9	.0004	.0294	.0781	.1336	.1511	.0609	.0088	.0004	.0000	.0000	.0000
	10	.0001	.0118	.0417	.0916	.1612	.0974	.0212	.0013	.0002	.0000	.0000
	11	.0000	.0040	.0189	.0536	.1465	.1328	.0434	.0042	.0007	.0001	.0000
	12	.0000	.0012	.0074	.0268	.1140	.1550	.0760	.0115	.0025	.0003	.0000
	13	.0000	.0003	.0025	.0115	.0760	.1550	.1140	.0268	.0074	.0012	.0000
	14	.0000	.0001	.0007	.0042	.0434	.1328	.1465	.0536	.0189	.0040	.0000
	15	.0000	.0000	.0002	.0013	.0212	.0974	.1612	.0916	.0417	.0118	.0001
	16	.0000	.0000	.0000	.0004	.0088	.0609	.1511	.1336	.0781	.0294	.0004
	17	.0000	.0000	.0000	.0001	.0031	.0322	.1200	.1651	.1241	.0623	.0018
	18	.0000	.0000	.0000	.0000	.0009	.0143	.0800	.1712	.1654	.1108	.0072
	19	.0000	.0000	.0000	.0000	.0002	.0053	.0442	.1472	.1828	.1633	.0239
	20	.0000	.0000	.0000	.0000	.0000	.0016	.0199	.1030	.1645	.1960	.0646
	21	.0000	.0000	.0000	.0000	.0000	.0004	.0071	.0572	.1175	.1867	.1384
	22	.0000	.0000	.0000	.0000	.0000	.0001	.0019	.0243	.0641	.1358	.2265
	23	.0000	.0000	.0000	.0000	.0000	.0000	.0004	.0074	.0251	.0708	.2659
	24	.0000	.0000	.0000	.0000	.0000	.0000	.0000	.0014	.0063	.0236	.1994
	25	.0000	.0000	.0000	.0000	.0000	.0000	.0000	.0001	.0008	.0038	.0718

Table IX
Critical values and significance levels
for a one-tailed Wilcoxon signed-rank test

Sample size, n	Critical value, d	Significance level, α
3	0	.125
4	0	.062
	1	.125
5	0	.031
	1	.062
	2	.094
6	0	.016
	1	.031
	2	.047
	4	.109
7	0	.008
	1	.016
	2	.023
	3	.039
	4	.055
	6	.109
8	1	.008
	2	.012
	3	.020
	5	.039
	6	.055
	8	.098
9	3	.010
	5	.020
	8	.049
	11	.102
10	5	.010
	7	.019
	10	.042
	11	.053
	14	.097
11	7	.009
	10	.021
	14	.051
	18	.103
12	10	.010
	13	.021
	17	.046

Table IX (Cont.)

Sample size, n	Critical value, d	Significance level, α
12	18	.055
	22	.102
13	12	.009
	16	.020
	21	.047
	22	.055
	26	.095
14	16	.010
	20	.021
	25	.045
	26	.052
	31	.097
15	19	.009
	24	.021
	30	.047
	31	.053
	36	.094
16	23	.009
	28	.019
	35	.047
	36	.052
	42	.096
17	28	.010
	33	.020
	41	.049
	48	.095
18	33	.010
	38	.019
	47	.049
	55	.098
19	38	.010
	44	.020
	53	.048
	62	.098
20	43	.010
	50	.020
	60	.049
	69	.095

Adapted from D. B. Owen, *Handbook of Statistical Tables*. Courtesy of the Atomic Energy Commission. Reading, Mass.: Addison-Wesley, 1962.

Table X
Critical values for a two-tailed runs test with $\alpha = 0.05$

	5	6	7	8	9	10	11	12	13	14	15
2	*	*	*	*	*	*	*	2 6	2 6	2 6	2 6
3	*	2 8	2 8	2 8	2 8	2 8	2 8	2 8	2 8	2 8	3 8
4	2 9	2 9	2 10	3 10	3 10	3 10	3 10	3 10	3 10	3 10	3 10
5	2 10	3 10	3 11	3 11	3 12	3 12	4 12	4 12	4 12	4 12	4 12
6	3 10	3 11	3 12	3 12	4 13	4 13	4 13	4 13	5 14	5 14	5 14
7	3 11	3 12	3 13	4 13	4 14	5 14	5 14	5 14	5 15	5 15	6 15
8	3 11	3 12	4 13	4 14	5 14	5 15	5 15	6 16	6 16	6 16	6 16
9	3 12	4 13	4 14	5 14	5 15	5 16	6 16	6 16	6 17	7 17	7 18
10	3 12	4 13	5 14	5 15	5 16	6 16	6 17	7 17	7 18	7 18	7 18
11	4 12	4 13	5 14	5 15	6 16	6 17	7 17	7 18	7 19	8 19	8 19
12	4 12	4 13	5 14	6 16	6 16	7 17	7 18	7 19	8 19	8 20	8 20
13	4 12	5 14	5 15	6 16	6 17	7 18	7 19	8 19	8 20	9 20	9 21
14	4 12	5 14	5 15	6 16	7 17	7 18	8 19	8 20	9 20	9 21	9 22
15	4 12	5 14	6 15	6 16	7 18	7 18	8 19	8 20	9 21	9 22	10 22

Adapted with permission from D. B. Owen, *Handbook of Statistical Tables.* Courtesy of the Atomic Energy Commission. Reading, Mass.: Addison-Wesley, 1962.

Answers to Selected Exercises and Review Tests

Chapter 1

1. Inferential. The study draws conclusions about all American TV viewers (the population) from the sample responding to the survey.
3. Descriptive.   5. Descriptive.   7. a) Inferential   b) Descriptive   c) No

Review Test

1. Inferential   2. Descriptive   3. Descriptive   4. Inferential

---

Chapter 2

Section 2.1

1. a) Ordinal   b) Qualitative   c) Count
3. Metric   5. Metric (but it could also be considered count data)
7. a) Qualitative   b) Count

Section **2.2** 1. *Annual income of* 20 *families* (*to nearest* $1,000)

Class	Frequency	Relative frequency	Class mark
1–5	1	0.05	3
6–10	5	0.25	8
11–15	4	0.20	13
16–20	4	0.20	18
21–25	3	0.15	23
26–30	3	0.15	28
	20	1.00	

3. *Number of children at home*

Number of children	Frequency	Relative frequency
0	5	0.25
1	5	0.25
2	7	0.35
3	2	0.10
4	1	0.05
	20	1.00

5. *Heights of trees*

Height	Frequency	Relative frequency	Class mark
5–9	2	0.08	7
10–14	5	0.20	12
15–19	12	0.48	17
20–24	4	0.16	22
25–29	2	0.08	27
	25	1.00	

7. *Home runs in softball*

Home runs	Frequency	Relative frequency
0	4	0.20
1	2	0.10
2	3	0.15
3	3	0.15
4	2	0.10
5	2	0.10
6	1	0.05
7	1	0.05
8	1	0.05
9	0	0.00
10	1	0.05
	20	1.00

Section **2.3**   1. a) *Annual income in thousands of dollars*

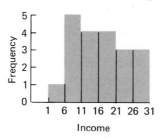

b)

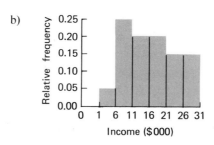

c)

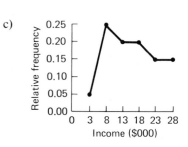

d)

Less than	Cumulative frequency	Cumulative percentage
1	0	0
6	1	5
11	6	30
16	10	50
21	14	70
26	17	85
31	20	100

e)

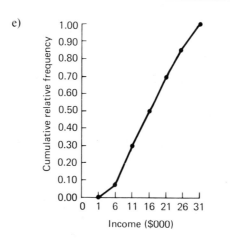

3. a)

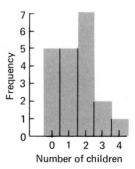

Frequency vs Number of children

b)

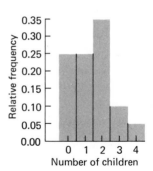

Relative frequency vs Number of children

c)

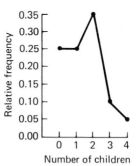

Relative frequency vs Number of children

<u>5.</u> a)

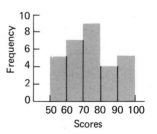

Frequency vs Scores

b)

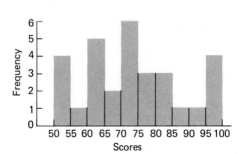

Frequency vs Scores

c) (b)

1. Stem-and-leaf diagram for average maximum temperature

Section **2.5**

```
4 | 7
5 | 9 8 9 8 5 9 9 7 4 9 5
6 | 9 4 0 7 3 8 5 6 3 4 0 4 8
7 | 7 0 1 1 0
8 | 5 3 3 0
```

3. Weights of 18-year-old males

```
12 | 4 6 0 8 7 9
13 | 6 5 1 2 5 6
14 | 1 0 7 8 3 2 2 3 4 0 0
15 | 4 6 3 8 8 2 5
```

Review Test   1. a) Qualitative   b) Count   2. a) Ordinal   b) Metric

3. *Number of family members*

Number	Frequency	Relative frequency
1	3	0.200
2	5	0.333
3	2	0.133
4	3	0.200
5	1	0.067
6	1	0.067
	15	1.000

4.

Class	Tallies	Frequency	Relative frequency	Class mark
900–999	||||	4	0.20	949.5
1000–1099	||||	4	0.20	1049.5
1100–1199	|||	3	0.15	1149.5
1200–1299	||	2	0.10	1249.5
1300–1399	||||	4	0.20	1349.5
1400–1499	||	2	0.10	1449.5
1500–1599	|	1	0.05	1549.5
		20	1.00	

5. a)

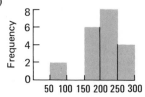

b)

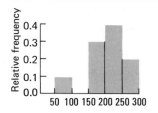

6. a)

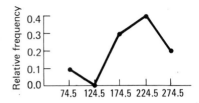

b)

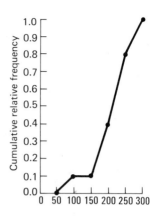

7.
2	7 8 4
3	2 5 3 1 6 9
4	7 3 2 5 7 1
5	5 2 3 2
6	1

Chapter **3**

Section **3.1**

1. Mean = 72.125°, median = 70°; modes are 61, 65, 70, 74, 85, 87

3. Mean = 74.48, median = 64, mode = 100

5. Mean = 288.3, median = 250, mode = 191

7. Answers may vary.

Section **3.2**

1. a) 27   b) 5.4   3. a) 563   b) 10   c) 56.3

5. a) 16
   b) $n = 6$
   c) 2.667
   e) 54

   d)
x	x^2
2	4
3	9
4	16
4	16
0	0
3	9

7. a) 47
   b) $n = 7$
   c) 6.714
   e) 443

   d)
x	x^2
11	121
3	9
8	64
4	16
6	36
14	196
1	1

Section **3.3** 1. $s \doteq 15.35$

3. a) Range $= 143 - 49 = 94$ b) $s^2 \doteq 604.2$ c) $s \doteq 24.6$

5.

Data set	Range	s^2	s
1	8	11.1	3.3
2	8	17.8	4.2
3	0	0	0
4	8	7.3	2.7

7. a) 6.82, 1.99
   b) 24, 7
   c) 4.6, 1.4

Section **3.4**   1.

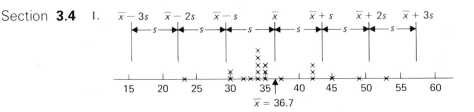

$\bar{x} = 36.7$

3.

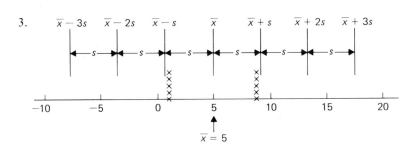

$\bar{x} = 5$

5.

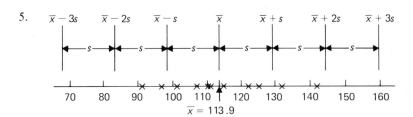

$\bar{x} = 113.9$

7. a) $1 - \dfrac{1}{k^2} = 1 - \dfrac{1}{9} = 89\%$;   89% of 20 = 17.8

At least 18 of the times should be within 3 standard deviations of $\bar{x}$, i.e., between 15.7 and 57.7.

b) In fact, 100% of the data lies between 15.7 and 57.7.

Section **3.5**

1.

x	f	xf	x^2	x^2f
4	32	128	16	512
3	44	132	9	396
2	14	28	4	56
1	6	6	1	6
0	4	0	0	0
	100	294		970

a) $\bar{x} = \dfrac{\Sigma xf}{n} = \dfrac{294}{100} = 2.94$

b) $s = \sqrt{\dfrac{100(970) - (294)^2}{100(99)}}$

$= \sqrt{1.07} = 1.03$

3.

x	f	xf	x^2	x^2f
65	2	130	4225	8450
66	0	0	4356	0
67	2	134	4489	8978
68	0	0	4624	0
69	3	207	4761	14283
70	6	420	4900	29400
71	4	284	5041	20164
72	3	216	5184	15552
73	4	292	5329	21316
74	1	74	5476	5476
	25	1757		123619

a) $\bar{x} = \dfrac{1757}{25} = 70.28$

b) $s = \sqrt{\dfrac{25(123619) - (1757)^2}{25(24)}}$

$= \sqrt{5.71} = 2.39$

5. a)

Grade class	Class mark	Frequency	xf	x^2f
0–9	4.5	0	0	0
10–19	14.5	0	0	0
20–29	24.5	0	0	0
30–39	34.5	2	69	2380.5
40–49	44.5	0	0	0
50–59	54.5	0	0	0
60–69	64.5	3	193.5	12480.75
70–79	74.5	3	223.5	16650.75
80–89	84.5	8	676	57122
90–99	94.5	3	283.5	26790.75
100–109	104.5	1	104.5	10920.25
		20	1550	126345

b) $\bar{x} = \dfrac{1550}{20} = 77.5$ c) $s = \sqrt{\dfrac{20(126345) - (1550)^2}{20(19)}} = \sqrt{327.4} = 18.1$

Section **3.6** 1.

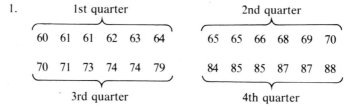

1st quarter

| 60 | 61 | 61 | 62 | 63 | 64 |

2nd quarter

| 65 | 65 | 66 | 68 | 69 | 70 |

| 70 | 71 | 73 | 74 | 74 | 79 |

3rd quarter

| 84 | 85 | 85 | 87 | 87 | 88 |

4th quarter

64.5 = 1st quartile;   70 = 2nd quartile;   81.5 = 3rd quartile

3. a) 1st quartile—191; 2nd quartile—250; 3rd quartile—365

b)

Decile	1	2	3	4	5	6	7	8	9
	138.5	190.5	204.5	229.5	250	277.5	328	389	527

5.

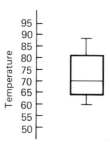

7.

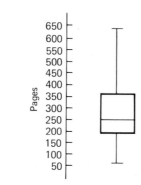

9. a) $\bar{x} = 36.7$, $s = 7$

b)

2	3
3	7 0 2 4 5 5 4 4 0 4 5 4 3
4	2 9 2 5 2
5	3

1. a) $\mu = 75$ b) $\sigma = 2.19$ 3. a) $\mu = 1.15$ b) $\sigma = 1.295$

1. Mean = 788.9;   median = 1000;   mode = 1000
2. Mean = 6;   median = 5.5;   modes are 3,5
3. $\bar{x} = 123.3$ 4. $\bar{x} = 10.62$ 5. $s = \sqrt{5.64} = 2.37$
6. (a) and (b)

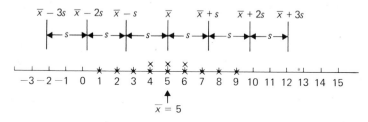

c) 75   d) 100%

7. $\bar{x} = 27.9$ 8. $s = \sqrt{1.39} = 1.18$
9. a) 1st quartile = 11;   2nd quartile = 16;   3rd quartile = 20.5
   b)

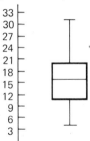

10. a) $\mu = 185$ b) $\sigma = 22.6$

---

1. a) $\frac{5}{36}$ b) $\frac{18}{36} = \frac{1}{2}$ c) $\frac{8}{36} = \frac{2}{9}$ d) $\frac{7}{36}$
3. a) $\frac{306}{3560} = 0.086$ b) $\frac{580}{3560} = 0.163$ c) 0.493
5. a) 0.359   b) 0.425
7. a) 0.778   b) 0.077   c) 0.145
9. a) 0.672   b) 0.896   c) 0.007   d) 0.769
11. a) 0.045   b) 0.285   c) 0.110

Section **4.2**

1.

x	P(x)
23	0.05
30	0.10
32	0.05
33	0.05
34	0.25
35	0.15
37	0.05
42	0.15
45	0.05
49	0.05
53	0.05
	1.00

3. a)

x	P(x)
4	0.32
3	0.44
2	0.14
1	0.06
0	0.04

b) 0.44
c) 0.32
d) 0.14

5. a)

x	P(x)
2	1/13
3	1/13
4	1/13
5	1/13
6	1/13
7	1/13
8	1/13
9	1/13
10	4/13
11	1/13

b) $P(x = 10) = 4/13$
c) $P(x = 6) = 1/13$

7. a)

Income class	Yearly income	P(x)
1	under 2000	0.021
2	2000–2999	0.024
3	3000–3999	0.034
4	4000–4999	0.041
5	5000–6999	0.083
6	7000–9999	0.128
7	10000–11999	0.089
8	12000–14999	0.134
9	15000–24999	0.304
10	25000 and over	0.141

b) 0.034
c) 0.304
d) 0.134

9. a)

x = mileage	f	P(x)
32	1	0.067
33	1	0.067
34	4	0.267
35	4	0.267
36	1	0.067
37	4	0.267

b) 0.067   c) 0.267   d) 0.067   e) 0.267

Section **4.3**

1. a) $\frac{3}{4}$ b) $\frac{1}{2}$ c) $\frac{7}{8}$ d) $\frac{7}{8}$

3. a) A is the event ($x = 3$ or $x = 4$); B is the event ($x \geq 5$) or, equivalently, ($x > 4$);
   C is the event ($x \leq 4$) or, equivalently, ($x < 5$)
   b) $\frac{7}{16}$ c) $\frac{3}{8}$ d) $\frac{5}{8}$

5. a) 0.17   b) 0.61   c) 0.07

7. a) $\frac{4}{13}$ b) $\frac{5}{13}$ c) $\frac{6}{13}$

9. a) 0.668   b) 0.334   c) 0.401

Section **4.4**

1. a) Not mutually exclusive.   b) Not mutually exclusive
   c) Not mutually exclusive.

3. a) Mutually exclusive. $P(A \text{ or } B) = \frac{13}{16}$.
   b) Not mutually exclusive.
   c) Mutually exclusive. $P(A \text{ or } D) = \frac{17}{32}$.
   d) Not mutually exclusive.   e) Not mutually exclusive.   f) Not mutually exclusive.

5. a) Mutually exclusive. $P(A \text{ or } B) = 0.78$
   b) Not mutually exclusive.   c) Mutually exclusive. $P(B \text{ or } C) = 0.68$.

7. a) $1 - \frac{1}{8} = \frac{7}{8}$ b) $1 - 0.05 = 0.95$

9. a) $P(\text{not } A) = \frac{18}{32} = \frac{9}{16}$ b) $P(\text{not } B) = \frac{20}{32} = \frac{5}{8}$ c) $P(\text{not } C) = \frac{18}{32} = \frac{9}{16}$

11. 0.90

13. $P((A \text{ or } B) \text{ or } C) = P(A \text{ or } B) + P(C) = P(A) + P(B) + P(C)$

Section **4.5**

1. a) $P(A) = \frac{11}{16}$ b) $P(B) = \frac{8}{16} = \frac{1}{2}$ c) (A & B) = number of heads is 2 or 4
   d) $P(A \text{ \& } B) = \frac{7}{16}$ e) $P(A \text{ or } B) = \frac{11}{16} + \frac{8}{16} - \frac{7}{16} = \frac{12}{16} = \frac{3}{4}$

3. a) $P(A) = 0.75$ b) $P(B) = 0.25$
   c) (A & B) means the expenditures were at least \$3000, but less than \$5000.
   d) $P(A \text{ \& } B) = 0.15$ e) $P(A \text{ or } B) = 0.75 + 0.25 - 0.15 = 0.85$

5. a) $P(A) = 0.780$ b) $P(B) = 0.933$ c) (A & B) = ($18 \leq \text{Age} \leq 64$)
   d) $P(A \text{ \& } B) = 0.713$ e) $P(A \text{ or } B) = 0.780 + 0.933 - 0.713 = 1.000$

7. a) $P(A) = 0.20$ b) $P(B) = 0.44$
   c) (A & B) = person is a Catholic Democrat.
   d) $P(A \text{ \& } B) = 0.11$ e) $P(A \text{ or } B) = 0.20 + 0.44 - 0.11 = 0.53$

9. a) No   b) $\frac{1}{12}$

Section **4.6**

1. a) $P(E|F) = \dfrac{P(F \ \& \ E)}{P(F)} = \dfrac{\frac{1}{6}}{\frac{1}{4}} = \dfrac{2}{3}$ b) $P(F|E) = \dfrac{P(E \ \& \ F)}{P(E)} = \dfrac{\frac{1}{6}}{\frac{1}{3}} = \dfrac{1}{2}$

3. a) $P(D|C) = \dfrac{P(C \ \& \ D)}{P(C)} = \dfrac{0.11}{0.20} = 0.55$ b) $P(J|R) = \dfrac{P(R \ \& \ J)}{P(R)} = \dfrac{0.02}{0.56} = 0.036$

   c) 55% of the Catholics are Democrats.   3.6% of the Republicans are Jewish.

5. a) $P(D|S) = \dfrac{P(S \ \& \ D)}{P(S)} = 0.56$ b) $P(H|D) = \dfrac{P(D \ \& \ H)}{P(D)} = 0.81$

   c) 56% of the Senators were Democrats.   81% of the Democrats were in the House.

7. a) $P(J|M) = \dfrac{P(M \ \& \ J)}{P(M)} = 0.286$ b) $P(F|J) = \dfrac{P(J \ \& \ F)}{P(J)} = 0.381$

   c) $P(J|F) = \dfrac{P(F \ \& \ J)}{P(F)} = 0.262$

9. a) $P(H) = \frac{1}{6}$ b) $P(K|H) = \frac{2}{5}$ c) $P(H \ \& \ K) = \frac{1}{6} \cdot \frac{2}{5} = \frac{1}{15}$

   d) $\frac{2}{6} \cdot \frac{1}{5} = \frac{1}{15}$ e) $\frac{3}{6} \cdot \frac{2}{5} = \frac{1}{5}$

Section **4.7**

1. $P(E|F) = \dfrac{P(F \ \& \ E)}{P(F)} = 0.42;$ $P(E) \doteq 0.44$
   Since $P(E|F) \neq P(E)$, E and F are dependent.

3. a) $\frac{4}{52} \cdot \frac{4}{52} = \frac{1}{169} \doteq 0.006;$ b) $\frac{4}{52} \cdot \frac{3}{51} = \frac{1}{221} \doteq 0.005$

5. a) $P(E \ \& \ F) = P(E) \cdot P(F) = \frac{1}{3} \cdot \frac{1}{4} = \frac{1}{12}$
   b) $P(E \text{ or } F) = P(E) + P(F) - P(E \ \& \ F) = \frac{1}{3} + \frac{1}{4} - \frac{1}{12} = \frac{1}{2}$

7. a) $P(M|W) = 0.86$ b) No. $P(M) = 0.58 \neq P(M|W)$.
   c) $P(M|H) = 0.458 \neq P(M)$. No.

9. $P(A \ \& \ B) = P(A) \cdot P(B)$, $P(A \ \& \ C) = P(A) \cdot P(C)$, $P(A \ \& \ D) = P(A) \cdot P(D)$,
   $P(B \ \& \ D) = P(B) \cdot P(D)$, $P(B \ \& \ C) = P(B) \cdot P(C)$, $P(C \ \& \ D) = P(C) \cdot P(D)$,
   $P(A \ \& \ B \ \& \ C) = P(A) \cdot P(B) \cdot P(C)$, $P(A \ \& \ B \ \& \ D) = P(A) \cdot P(B) \cdot P(D)$,
   $P(A \ \& \ C \ \& \ D) = P(A) \cdot P(C) \cdot P(D)$, $P(B \ \& \ C \ \& \ D) = P(B) \cdot P(C) \cdot P(D)$,
   and $P(A \ \& \ B \ \& \ C \ \& \ D) = P(A) \cdot P(B) \cdot P(C) \cdot P(D)$.

11. $P(A) = \frac{1}{2};$ $P(B) = \frac{1}{2};$ $P(C) = \frac{1}{4}$

   $P(A \ \& \ B) = \frac{1}{4},$ $P(A \ \& \ C) = \frac{1}{8},$ $P(B \ \& \ C) = \frac{1}{8}$
   $P(A \ \& \ B \ \& \ C) = \frac{1}{16}$
   Therefore, $P(A \ \& \ B) = P(A) \cdot P(B)$, $P(A \ \& \ C) = P(A) \cdot P(C)$, $P(B \ \& \ C) = P(B) \cdot P(C)$,
   and $P(A \ \& \ B \ \& \ C) = P(A) \cdot P(B) \cdot P(C)$. So, A, B, and C are independent.

1. a)

Outcome	Probability
sss	$\frac{1}{6}\cdot\frac{1}{6}\cdot\frac{1}{6}=\frac{1}{216}$
ssf	$\frac{1}{6}\cdot\frac{1}{6}\cdot\frac{5}{6}=\frac{5}{216}$
sfs	$\frac{1}{6}\cdot\frac{5}{6}\cdot\frac{1}{6}=\frac{5}{216}$
fss	$\frac{5}{6}\cdot\frac{1}{6}\cdot\frac{1}{6}=\frac{5}{216}$
sff	$\frac{1}{6}\cdot\frac{5}{6}\cdot\frac{5}{6}=\frac{25}{216}$
fsf	$\frac{5}{6}\cdot\frac{1}{6}\cdot\frac{5}{6}=\frac{25}{216}$
ffs	$\frac{5}{6}\cdot\frac{5}{6}\cdot\frac{1}{6}=\frac{25}{216}$
fff	$\frac{5}{6}\cdot\frac{5}{6}\cdot\frac{5}{6}=\frac{125}{216}$

b) $P(x = 1) = \frac{75}{216}$

c)

x	$P(x)$
0	$\frac{125}{216}$
1	$\frac{75}{216}$
2	$\frac{15}{216}$
3	$\frac{1}{216}$

Section **4.8**

3. $p = 0.49$, $n = 3$

   a) $P(x = 1) = \binom{3}{1}(0.49)^1(0.51)^2 = 0.382$

   b) $P(x \ge 1) = 1 - P(x = 0) = 1 - \binom{3}{0}(0.49)^0(0.51)^3 = 1 - 0.133 = 0.867$

   c) $P(x = 3) = \binom{3}{3}(0.49)^3(0.51)^0 = 0.118$

5. $p = 0.8$, $n = 2$.
   a) $P(x \ge 1) = P(x = 1) + P(x = 2) = 0.32 + 0.64 = 0.96$
   b) $P(x = 2) = 0.64$

7. $p = 0.9$, $n = 8$.
   $P(x \ge 6) = P(x = 6) + P(x = 7) + P(x = 8) \doteq 0.149 + 0.383 + 0.430 = 0.962$

1. $\mu = 3.5$, $\sigma = 1.71$　　　3. $\mu = 1.5$, $\sigma = 0.866$　　　5. $\mu = 7.31$, $\sigma = 2.91$　　Section **4.9**

7. $\mu = 7.36$, $\sigma = 1.43$　　　9. $\mu = 0.5$, $\sigma \doteq 0.645$　　11. 17

1. a) Observing whether or not a report of a burglary is received between 10:00 and　Section **4.10**
   11:00 P.M. on a weekday night.
   b) E is receiving a report of a burglary.　　c) $P(E) = 0.34$
   d) During a large number of weekday nights between 10:00 and 11:00 P.M., there will be
   a burglary reported about 34% of the time.

3. a) Observing whether or not a calculator is defective.
   b) E is the event that a calculator is defective.　　c) $P(E) = 0.003$
   d) About 0.3% of the calculators produced by this company are defective.

5. 2500　　　7. 0.837　　　9. 0.485

11. $P(|\hat{p} - \frac{1}{2}| \le \frac{1}{4}) = \frac{1}{2}$;　$n = 2$
    $= \frac{3}{4}$;　$n = 3$
    $= \frac{7}{8}$;　$n = 4$

1. a) 0.2　　b) 0.75　　c) 0.05　　　　　　　　　　　　　　　　Review Test

2. a)

x	18	19	20	21	22	23
$P(x)$	0.30	0.25	0.20	0.20	0.025	0.025

b) 0.70　　c) 0.95　　d) 0.65

3. a) No   b) Yes   c) $P(A$ or $C) = 1$ d) $x \geq 1$ e) $P(A$ or $B) = 0.815$
   f) $x \geq 2$ g) $P(A \& B) = 0.327$ h) $P(A$ or $B) = P(A) + P(B) - P(A \& B)$

4. a) $P(M) = 0.575$ b) $P(M|P) = 0.5$ c) $P(P)$ d) $P(M \& P) = 0.125$
   e) $P(C|F) \doteq 0.47$
   f) No   g) $P(C|F) = 0.47$; $P(C) = 0.5$; so $P(C|F) \neq P(C)$.

5. a) 0.0403   b) 0.1612   6. $\mu = 2.6$, $\sigma = 0.917$.

7. a) E is snowstorm on March 15. About 20% of the years there is a snowstorm on March 15 in this town.
   b) E is making a sale. About 36% of the visits lead to a sale for this salesman.

Chapter 5

Section **5.1**

1. a) Lengths (in days) of all strikes in U.S. in 1974.   b) $\bar{x} = 23.6$ days

3. a) College Board scores of 1979 high school seniors.   b) 488.1

5. a) Number of people applying for unemployment insurance in Colorado for each day in October, 1978.
   b) 3,375

7. $n = 2$.

Bullfrogs selected	1,2	1,3	1,4	1,5	1,6	2,3	2,4	2,5	2,6	3,4	3,5	3,6	4,5	4,6	5,6
$\bar{x}$	16.5	17	14	17.5	18	14.5	11.5	15	15.5	12	15.5	16	12.5	13	16.5

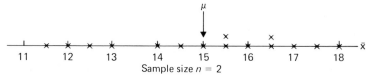

Sample size $n = 2$

9. $n = 4$.

Bullfrogs selected	1,2,3,4	1,2,3,5	1,2,3,6	1,3,4,5	1,3,4,6	1,3,5,6
$\bar{x}$	14.25	16	16.25	14.75	15	16.75

Bullfrogs selected	1,2,4,5	1,2,4,6	1,2,5,6	1,4,5,6	2,3,4,5	2,3,4,6
$\bar{x}$	14.5	14.75	16.5	15.25	13.5	13.75

Bullfrogs selected	2,4,5,6	2,3,5,6	3,4,5,6
$\bar{x}$	14	15.5	14.25

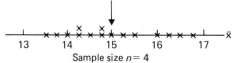

Sample size $n = 4$

1. a)

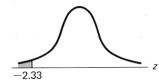

−2.33
Area = 0.0099

b)

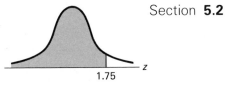

1.75
Area = 0.9599

c)

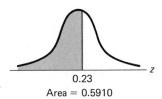

0.23
Area = 0.5910

d)

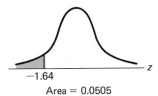

−1.64
Area = 0.0505

e)

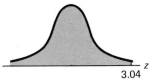

3.04
Area = 0.9988

f)

0
Area = 0.5

3. a)

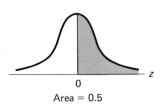

2.98
Area = 0.0014

b)

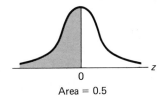

−1.64
Area = 0.9495

c)

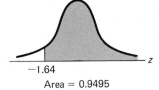

0
Area = 0.5

d)

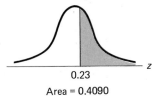

0.23
Area = 0.4090

5. a)

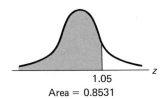

1.05
Area = 0.8531

b)

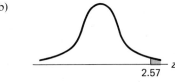

2.57
Area = 0.0051

c)

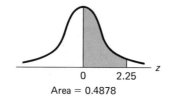

Area = 0.4878

d)

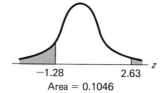

Area = 0.1046

Section **5.3**   1. a)

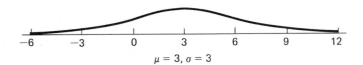

$\mu = 3, \sigma = 3$

b)

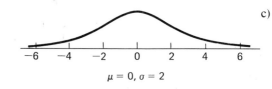

$\mu = 0, \sigma = 2$

c)

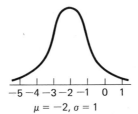

$\mu = -2, \sigma = 1$

3. a)

Normal curve
$(\mu = 1, \sigma = 2.5)$

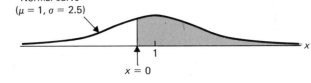

$x = 0$

$$z = \frac{x - \mu}{\sigma} = \frac{x - 1}{2.5}$$

Change to z

$$x = 0 \rightarrow z = \frac{0 - 1}{2.5} = -0.4$$

Area $= 1 - 0.3446 = 0.6554$

Standard
normal
curve

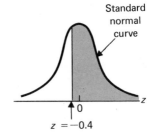

$z = -0.4$

b)

Normal curve
$(\mu = 1, \sigma = 2.5)$

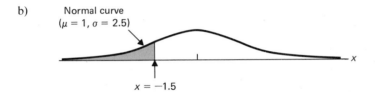

$x = -1.5$

$$z = \frac{x - \mu}{\sigma} = \frac{x - 1}{2.5}$$

Change to z
⟶

$$x = -1.5 \rightarrow z = \frac{-1.5 - 1}{2.5} = -1$$

Area = 0.1587

Standard
normal
curve

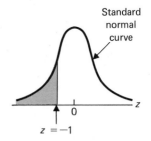

$z = -1$

c)

Normal curve
$(\mu = 1, \sigma = 2.5)$

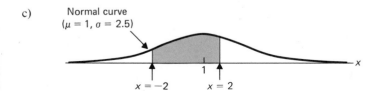

$x = -2$ $x = 2$

$$z = \frac{x - \mu}{\sigma} = \frac{x - 1}{2.5}$$

Change to z
⟶

$$x = -2 \rightarrow z = \frac{-2 - 1}{2.5} = -1.2$$

$$x = 2 \quad \rightarrow z = \frac{2 - 1}{2.5} = 0.4$$

Area = 0.6554 - 0.1151 = 0.5403

Standard
normal
curve

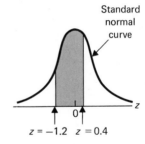

$z = -1.2$ $z = 0.4$

5. $\mu = 4$, $\qquad \sigma = 2$. $\qquad x = -2 \to z = \dfrac{-2-4}{2} = -3$; $\qquad x = 10 \to z = \dfrac{10-4}{2} = 3$;

Area $= 0.9987 - 0.0013 = 0.9974$

7. $x = \mu - 3\sigma \to z = \dfrac{(\mu - 3\sigma) - \mu}{\sigma} = -3$; $\qquad x = \mu + 3\sigma \to z = \dfrac{(\mu + 3\sigma) - \mu}{\sigma} = 3$;

Area $= 0.9974 \doteq 99.7\%$

Section 5.4

1. a) $P(72.5 < x < 75.5) = 0.55$; $\qquad$ Area $= 0.7704 - 0.2119 = 0.5585$

 b) $P(x > 75.5) = 0.23$; $\qquad$ Area $= 0.2296$

3. $\mu = 1750$, $\sigma = 185$

$$P(1500 < x < 2000) = P\left(\frac{1500 - 1750}{185} < z < \frac{2000 - 1750}{185}\right) = P(-1.35 < z < 1.35)$$

$$= 0.9115 - 0.0885 = 0.8230.$$

5. $\mu = 30$, $\sigma = 3.3$

$$P(25 < x < 35) = P\left(\frac{25 - 30}{3.3} < z < \frac{35 - 30}{3.3}\right) = P(-1.52 < z < 1.52)$$

$$= 0.9357 - 0.0643 = 0.8714.$$

7. $\mu = 3875$, $\sigma = 291$

$$P(3500 < x < 4000) = P\left(\frac{3500 - 3875}{291} < z < \frac{4000 - 3875}{291}\right) = P(-1.29 < z < 0.43)$$

$$= 0.6664 - 0.0985 = 0.5679$$

9. $\mu = 168$, $\sigma = 15$

$$P(x < 150) = P\left(z < \frac{150 - 168}{15}\right) = P(z < -1.2) = 0.1151$$

Section 5.5

1. a) 1 $\qquad$ b) -2 $\qquad$ 3. a) -3 $\qquad$ b) -0.5

5. a) 165 $\qquad$ b) 126.5 $\qquad$ c) 187 $\qquad$ d) 154

7. a) \$5.55, \$6.15 $\qquad$ b) \$5.25, \$6.45 $\qquad$ c) \$4.95, \$6.75

Section 5.6

1. a) 0.4512 $\qquad$ b) 0.8907 $\qquad$ c) 0.6231 $\qquad$ d) 0.377

3. $\mu = 15$, $\sigma \doteq 2.74$

5. $n = 50$, $p = 0.1$; $\quad np = 5$, $n(1 - p) = 45$; $\qquad \mu = 5$, $\sigma = 2.12$; $\qquad k = 5, 6, 7, 8, 9, 10$;

$$P(4.5 < x < 10.5) = P\left(\frac{4.5 - 5}{2.12} < z < \frac{10.5 - 5}{2.12}\right) = P(-0.24 < z < 2.59)$$

$$= 0.9952 - 0.4052 = 0.59$$

7. $n = 20, p = 0.7$; $np = 14, n(1 - p) = 6$; $\mu \doteq 14, \sigma = 2.05$; $k = 10, 11, 12, 13, 14, 15$;

$$P(9.5 < x < 15.5) = P\left(\frac{9.5 - 14}{2.05} < z < \frac{15.5 - 14}{2.05}\right) = P(-2.20 < z < 0.73)$$

$$= 0.7673 - 0.0139 = 0.7534.$$

9. $n = 20, p = 0.6$; $np = 12, n(1 - p) = 8$; $\mu = 12, \sigma = 2.19$;

$$P(x > 15.5) = P\left(z > \frac{15.5 - 12}{2.19}\right) = P(z > 1.60) = 0.0548$$

1. 0.1020   2. 0.0041   3. $0.9788 - 0.0322 = 0.9466$ Review Test
4. 0.9082   5. 0.6098   6. 0.6393
7. -2.5
8. \$1900, \$3500
9. $n = 100, p = 0.5$; $\mu = 50, \sigma = 5$

$$P(59.5 < x < 64.5) = P\left(\frac{59.5 - 50}{5} < z < \frac{64.5 - 50}{5}\right) = P(1.9 < z < 2.9)$$

$$= 0.9981 - 0.9713 = 0.0268$$

Chapter 6

Section 6.1

1. a) 1,2   1,3   1,4   2,3   2,4   3,4   b)1/6

3. a) 1,2,3   1,2,4   1,2,5   1,3,4   1,3,5   1,4,5   2,3,4   2,3,5   2,4,5   3,4,5
   b) 1/10
7. b) 1/6

Section 6.2

1. a) $\mu = \$250, \sigma = \$322.10$

b)

Prizes	A, B	A, C	A, D	B, C	B, D	C, D
Values	150, 50	150, 0	150, 800	50, 0	50, 800	0, 800

c) 100, 75, 475, 25, 425, 400.

d)

$\bar{x}$	25	75	100	400	425	475
$P(\bar{x})$	1/6	1/6	1/6	1/6	1/6	1/6

e) $\mu_{\bar{x}} = 250, \sigma_{\bar{x}} = 185.97$

3. a) $\mu = 2.2, \sigma = 1.72$ b) 2,1;   2,0;   2,5;   2,3;   1,0;   1,5;   1,3;   0,5;   0,3;   5,3
   c) 1.5, 1, 3.5, 2.5, 0.5, 3, 2, 2.5, 1.5, 4
   d)

$\bar{x}$	0.5	1	1.5	2	2.5	3	3.5	4.
$P(\bar{x})$	0.1	0.1	0.2	0.1	0.2	0.1	0.1	0.1

e) $\mu_{\bar{x}} = 2.2, \sigma_{\bar{x}} = 1.05$

5. a) $\mu_{\bar{x}} = 16, \sigma_{\bar{x}} = 0.117$ b) $\mu_{\bar{x}} = 16, \sigma_{\bar{x}} = 0.099$ c) $\mu_{\bar{x}} = 16, \sigma_{\bar{x}} = 0.082$

7. a) $\mu_{\bar{x}} = 138, \sigma_{\bar{x}} = 0.88$ b) $\mu_{\bar{x}} = 138, \sigma_{\bar{x}} = 0.625$ c) $\mu_{\bar{x}} = 138, \sigma_{\bar{x}} = 0.57$

Section **6.3**   1. a) 0.5403   b) 0.5403   c) 0.9544

3. a) 0.7458   b) 0.7404   c) 1 (to four decimal places)

5. 0.7876   7. 0.4833

9. The populations are normally distributed.

Review Test   1. a) Not random   b) Random

2. a) 3.5   b) $\sigma \geqslant \sigma_{\bar{x}}$

3. a) 50, 16.67   b) 50, 11.11   c) 50, 10

4. a) 50   b) 2.25   c) Normally distributed   d) 0.8707

Chapter **7**

Section **7.1**   1. a) 0.8414   b) 108, 112, 84%
c) She can be 84% confident that μ is between 108 and 112.

3. a) 0.9282   b) 9.95, 10.01, 93%
c) He can be 93% confident that μ is between 9.95 and 10.01.

5. a) 0.8860   b) 9.5, 10.5, 89%
c) She can be 89% confident that μ is between 9.5 and 10.5.

Section **7.2**   1. 0.51   3. 2.33   5. 2.05   7. 1.44   9. 0.67

11. 9.953 to 10.007

Section **7.3**   1. 3.462 to 3.538   3. 3.481 to 3.519   5. 109.9 to 114.7

7. 46.6 to 48.4   9. 98.44 to 102.72   11. 108.68 to 111.50   13. 96.56 to 100.40

Section **7.4**   1. 0.433 to 0.547   3. 0.943 to 0.977   5. 0.315 to 0.445

7. 0.673 to 0.767   9. 0.801 to 0.859

11. a) 0.24, 0.24, 0.25, No   b) 0.25   c) 1068, 2401, 9604
d) 0.511 to 0.609   e) 2401

Section **7.5**   1. t-curve with d.f. = 13   3. t-curve with d.f. = 19   5. t-curve with d.f. = 15

7.

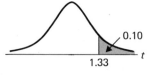

d.f. = 19, $t_{0.10} = 1.33$

9.

d.f. = 22, $t_{0.01} = 2.51$

11.

d.f. = 3, $t_{0.005} = 5.84$

13.

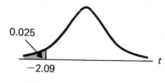

d.f. = 20, $-t_{0.025} = -2.09$

15.

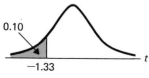

d.f. = 17, $-t_{0.10} = -1.33$

17. 0.90

19.

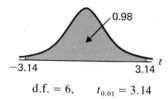

d.f. = 6, $t_{0.01} = 3.14$

21.

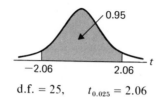

d.f. = 25, $t_{0.025} = 2.06$

23. 1.80   25. 3.01   27. 2.90

1. 16.63 to 19.37   3. 136.28 to 140.44   5. 2238.3 to 2313.7   Section **7.6**

7. 34.47 to 38.99   9. 17.60 to 18.18

11. a) 34.68 to 38.78

1. a) 2.05   b) 1.64   c) 0   d) −1.28   e) 1.64   Review Test

2. 69.4 to 70.6   3. $29,068 to $30,932   4. 0.66 to 0.74

5. a) 1.78   b) 2.49   c) 1.73   d) 2.10

6. 173.74 to 182.26

Chapter **8**

1. a) $P(\bar{x} \leq 35.67) = 0.2266$.   No.   b) $P(\bar{x} \leq 34.90) = 0.006$.   Yes.   Section **8.1**

3. a) $P(\bar{x} \leq 1957.4) = 0.0015$.   b) Yes.

5. a) The teacher would believe that the students study on the average more than six hours, when in fact they don't.

b) The teacher would believe that the students study 6 hours on the average, when in fact they study more.

c) The teacher would believe that the students study 6 hours on the average, when in fact they do, *or* the teacher would believe that the students study more than 6 hours on the average, when in fact they do.

Section **8.2**

1. a) $z = \dfrac{\bar{x} - 25}{s/\sqrt{50}}$ b) $c = -2.33$ c)   d) No.

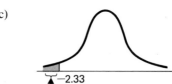

3. a) $z = \dfrac{\bar{x} - 9.5}{s/\sqrt{75}}$ b) $c = -1.28$ c)   d) Yes.

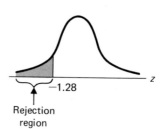

5. *Null hypothesis: $\mu = 12{,}500$.   Alternative hypothesis: $\mu < 12{,}500$.*

7. *Null hypothesis: $\mu = 1{,}388$.   Alternative hypothesis: $\mu < 1{,}388$.*

Here μ is the mean public school expenditure per pupil in California.

9. a) α b) 0.01   c) 0.10

Section **8.3**

1. a) *Null hypothesis: $\mu = 2144$*   b) *Alternative hypothesis: $\mu < 2144$*
c) $c = -1.64$

d)

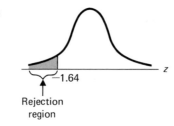

e) $z = -3.94$ f) Reject the null hypothesis and conclude that $\mu < 2144$.

3. a) *Null hypothesis:* $\mu = 18$ b) *Alternative hypothesis:* $\mu < 18$
Here μ is the mean weight for all boxes of this cereal.

c) $c = -1.28$

d)

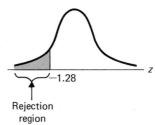

e) $z = -1.15$ f) The null hypothesis cannot be rejected.

5. a) *Null hypothesis:* $\mu = 14$ b) *Alternative hypothesis:* $\mu > 14$.   c) $c = 1.64$

d)

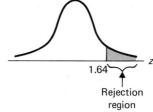

e) $z = 18.97$
f) Reject the null hypothesis. Evidently the advertising campaign is effective.

7. a) *Null hypothesis:* $\mu = 996$.   b) *Alternative hypothesis:* $\mu < 996$.
c) $c = -1.64$

d)

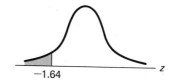

e) $z = -8.58$ f) Reject the null hypothesis. It appears that $\mu < 996$.

1. a) *Null hypothesis:* $\mu = 64$ b) *Alternative hypothesis:* $\mu \neq 64$ Section **8.4**
c) $c = 1.96$ and $-c = -1.96$

d)

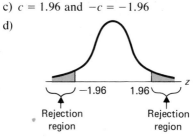

e) $z = -1.08$ f) The null hypothesis cannot be rejected.

3. a) *Null hypothesis:* $\mu = 18264$ b) *Alternative hypothesis:* $\mu \neq 18264$.

c) $c = 1.96$ and $-c = -1.96$

d)

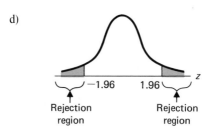

e) $z = 1.89$ f) The null hypothesis cannot be rejected.

5. a) *Null hypothesis:* $\mu = 454$.   b) *Alternative hypothesis:* $\mu \neq 454$.

c) $c = 1.64$ and $-c = -1.64$

d)

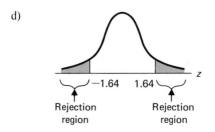

e) $z = -1.74$

f) Reject the null hypothesis, and conclude that the area's mean for personal savings accounts is different than the national mean.

7. a) *Null hypothesis:* $\mu = 1082$ b) *Alternative hypothesis:* $\mu \neq 1082$

c) $c = 2.58$ and $-c = -2.58$

d)

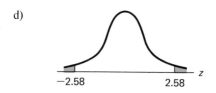

e) $z = -1.90$ f) The null hypothesis cannot be rejected.

9. a) 63.92 to 64.02   b) Yes   c) 1024.77 to 1073.23

d) No   e) (i) Accept; (ii) reject.

1. a) 319   b) 181   c) *Null hypothesis:* $p = 0.638$
   d) *Alternative hypothesis:* $p \neq 0.638$ e) $c = 1.96$ and $-c = -1.96$

f)

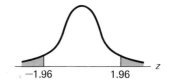

g) $z = -2.69$
h) Reject the null hypothesis, and conclude that the attitude on city street repair has changed.

3. a) 15.65   b) 484.35   c) *Null hypothesis:* $p = 0.0313$
   d) *Alternative hypothesis:* $p < 0.0313$ e) $c = -2.33$

f)

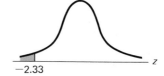

g) $z = -0.42$ h) The null hypothesis cannot be rejected.

5. a) 52.8   b) 67.2   c) *Null hypothesis:* $p = 0.44$
   d) *Alternative hypothesis:* $p > 0.44$ e) $c = 1.64$

f)

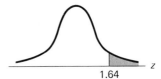

g) $z = 1.88$
h) Reject the null hypothesis. It appears that the percentage has increased.

7. a) 980   b) 20   c) *Null hypothesis:* $p = 0.98$
   d) *Alternative hypothesis:* $p < 0.98$ e) $c = -1.64$

f)

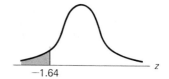

g) $z = -11.29$ h) Reject the null hypothesis. Apparently, the drug had become less effective.

9. a) 5.75   b) 394.25   c) *Null hypothesis:* $p = 0.0144$
   d) *Alternative hypothesis:* $p > 0.0144$ e) $c = 1.64$

f)

1.64

g) $z = 0.10$ h) The null hypothesis cannot be rejected.

Section 8.6   1. a) *Null hypothesis:* $\mu_1 = \mu_2$ b) *Alternative hypothesis:* $\mu_1 < \mu_2$

c)

$c = -1.64$

d) $z = -4.70$
e) Reject the null hypothesis. It appears that the mastery testing method is more effective.

3. a) *Null hypothesis:* $\mu_1 = \mu_2$ b) *Alternative hypothesis:* $\mu_1 < \mu_2$

c)

$c = -1.28$

d) $z = -1.52$
e) Reject the null hypothesis and conclude that the training program increases sales effectiveness.

5. a) *Null hypothesis:* $\mu_1 = \mu_2$ b) *Alternative hypothesis:* $\mu_1 < \mu_2$

c)

$c = -1.64$

d) $z = 2.63$ e) The null hypothesis cannot be rejected.

7. a) *Null hypothesis:* $\mu_1 = \mu_2$ b) *Alternative hypothesis:* $\mu_1 \neq \mu_2$
c)

$-c = -1.96$ $c = 1.96$

d) $z = -1.35$ e) The null hypothesis cannot be rejected.

<u>9.</u> a) -0.74 to 0.14 b) 0

1. a) $\mu = 1$ b) $\mu > 1$

c)

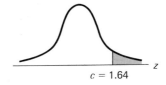

d) $z = 0.18$ e) The null hypothesis cannot be rejected.

3. a) $\mu = 0$ b) $\mu > 0$

c)

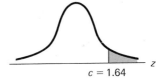

d) $z = 42.15$
e) Reject the null hypothesis. It appears that the training program will improve salaries of unskilled workers.

5. a) $\mu = 0$ b) $\mu < 0$

c)

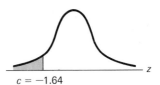

d) $z = -5.06$
e) Reject the null hypothesis. It appears that the running program reduces heart rates.

7. a) $\mu = 0$ b) $\mu < 0$

c)

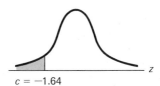

d) $z = -12.01$
e) Reject the null hypothesis. Evidently the new program is effective in reducing the cholesterol levels of children.

<u>9.</u> a) 1.70 to 5.06

Section **8.8**   1. 2.58   3. − 1.73   5. 2.31

Section **8.9**   1. a) $\mu = 36$ b) $\mu \neq 36$ c) ±2.09

d)
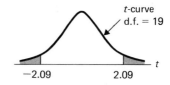

e) $t = -1.49$ f) The null hypothesis cannot be rejected.

3. a) $\mu = 2000$ b) $\mu > 2000$ c) $c = 1.71$

d)
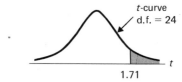

e) $t = 1.54$ f) The null hypothesis cannot be rejected.

5. a) $\mu = 33.8$ b) $\mu > 33.8$ c) $c = 1.73$

d)
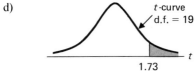

e) $t = 2.13$
f) Reject the null hypothesis. The new fertilizer appears to increase yield.

7. a) $\mu = 50$ b) $\mu \neq 50$ c) ±2.06

d)
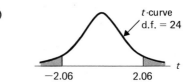

e) $t = -2.00$ f) The null hypothesis cannot be rejected.

9. a) The yields (in tons) of all one-acre plots of the sugar cane farmer using the new fertilizer.
   b) The weights of all ''50-pound'' bags of dog food produced by this manufacturer.

11. a) ±1.96   b) ±2.09   c) Less.

1. a) $\mu_1 = \mu_2$ b) $\mu_1 < \mu_2$ c) -1.70

d)

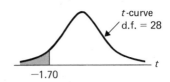

e) $t = -5.47$
f) Reject null hypothesis. It appears that the new paint lasts longer on the average.

3. a) $\mu_1 = \mu_2$ b) $\mu_1 \neq \mu_2$ c) ± 2.07

d)

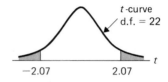

e) $t = 1.13$ f) The null hypothesis cannot be rejected.

5. a) $\mu_1 = \mu_2$ b) $\mu_1 \neq \mu_2$ c) ± 2.18

d)

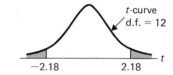

e) $t = 0.56$ f) The null hypothesis cannot be rejected.

7. $c = -1.64$ and $z = -1.60$. The null hypothesis cannot be rejected.

1. $c = -2.33$ and $z = -3.48$. Reject the null hypothesis and conclude $\mu < 7321$.

Review Test

2. $c = 1.64$ and $z = 10.04$. Reject the null hypothesis and conclude $\mu > \$59,000$.

3. a) $p < 0.8$ b) The null hypothesis cannot be rejected.

4. Critical values are ± 1.64. $z = -0.31$. The null hypothesis cannot be rejected.

5. $c = -1.64$ and $z = -2.36$. Reject the null hypothesis and conclude that $\mu < 20$.

6. d.f. $= 14$, $c = 2.62$ and $t = 6.45$. Reject the null hypothesis and conclude that $\mu > \$65$.

7. d.f. $= 14$; critical values ± 2.14; $t = 2.38$. Reject the null hypothesis.

Chapter **9**

Section **9.1**

1. a) $\sigma^2 = 0.01$ b) $\sigma^2 > 0.01$ ($\sigma^2 < 0.01$ would also be appropriate.)
   c) $s^2 = 0.008$ d) The null hypothesis cannot be rejected.

3. a) $\sigma^2 = 0.02$ b) $\sigma^2 > 0.02$ ($\sigma^2 < 0.02$ would also be appropriate.)
   c) $s^2 = 0.029$ d) Don't reject the null hypothesis.

5. a) $\sigma^2 = 0.5$ b) $\sigma^2 \neq 0.5$ c) $s^2 = 0.297$
   d) Reject the null hypothesis.

Section **9.2**

1.

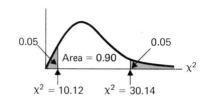

3.

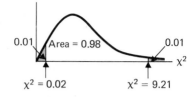

5. a) 23.21   b) 3.94   c) 3.25   d) 2.56

Section **9.3**

1. a) $\sigma^2 = 0.01$ b) $\sigma^2 > 0.01$ c) 36.42

d)

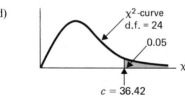

e) 19.2   f) The null hypothesis cannot be rejected.

3. a) $\sigma^2 = 0.01$ b) $\sigma^2 > 0.01$ c) 27.20

d)

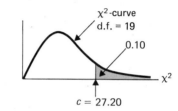

e) 36.10   f) Reject the null hypothesis. It appears that $\sigma^2 > 0.01$.

5. a) $\sigma^2 = 0.02$ b) $\sigma^2 > 0.02$ c) 25.00

d)

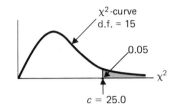

e) 30   f) Reject the null hypothesis and conclude $\sigma^2 > 0.02$.

7. a) $\sigma^2 = 50$ b) $\sigma^2 > 50$ c) 14.07

d)

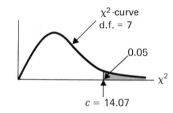

e) 8.02   f) The null hypothesis cannot be rejected.

9. *Null hypothesis:* $\sigma^2 = 0.02$ *Alternative hypothesis:* $\sigma^2 < 0.02$
d.f. $= 14$; $c = 6.57$; $\chi^2 = 5.25$
Reject the null hypothesis and conclude that $\sigma^2 < 0.02$.

11. a) $\chi^2_{0.99} = 10.86$, $\chi^2_{0.01} = 42.98$
b) Acceptance: $10.86 \leq \chi^2 \leq 42.98$ Rejection: $\chi^2 < 10.86$ or $\chi^2 > 42.98$.
c) $\chi^2 = 16.80$. The null hypothesis cannot be rejected.

1. 0.005 to 0.020	3. 0.006 to 0.016	5. 0.075 to 0.277	Section **9.4**
7. 0.45 to 3.19	9. 0.15 to 0.99		

1. 2.70	3. 4.64	5. 1.89	7. 4.54	Section **9.5**

9. a) $\sigma_1^2 = \sigma_2^2$ b) $\sigma_1^2 > \sigma_2^2$ c) $c = 2.33$

d)

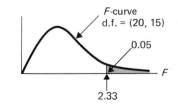

e) $F = 1.18$ f) The null hypothesis cannot be rejected.

11. a) $\sigma_1^2 = \sigma_2^2$ b) $\sigma_1^2 > \sigma_2^2$ c) $c = 2.91$

d)
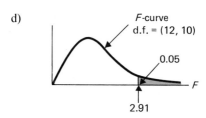

e) $F = 9.42$ f) Reject the null hypothesis and conclude that $\sigma_1^2 > \sigma_2^2$.

13. 0.41   15. 0.28

17. The critical values are 0.24 and 4.36. $F = 1.07$. The null hypothesis cannot be rejected.

Review Test   1. 20.09, 2.73, 15.51   2. 3.94, 18.31

3. a) $c = 31.41$. $\chi^2 = 22.75$. The null hypothesis cannot be rejected.   b) 0.060 to 0.317

4. a) 4.10   b) 4.50   c) 0.24   d) 0.22

5. a) $F = 3.36$, d.f. $= (20,16)$, $c = 2.28$. Reject the null hypothesis.
   b) *Null hypothesis:* $\sigma_1^2 = \sigma_2^2$ *Alternative hypothesis:* $\sigma_1^2 \neq \sigma_2^2$

Chapter **10**

Section 10.1   1. a) Distribution of number of persons per household in the midwest city is the same as the national distribution.
   b) The distributions are different.

c)

E	$O - E$	$(O - E)^2$	$(O - E)^2/E$
196	-11	121	0.62
306	10	100	0.33
174	-4	16	0.09
156	5	25	0.16
168	0	0	0.00

d) $\chi^2 = 1.20$ e) If $\chi^2 > 9.49$, reject the null hypothesis.
f) Don't reject the null hypothesis.

3. a) The die is "fair" (i.e., $p = 1/6$ for each number).   b) The die is "loaded."

c)

E	$O - E$	$(O - E)^2$	$(O - E)^2/E$
50	11	121	2.42
50	-8	64	1.28
50	6	36	0.72
50	8	64	1.28
50	-7	49	0.98
50	-10	100	2.00

d) $\chi^2 = 8.68$ e) If $\chi^2 > 11.07$, reject the null hypothesis.
f) Don't reject the null hypothesis.

1. The null hypothesis cannot be rejected.   Section **10.2**

3. a) The unified school district has the same distribution as the national distribution.
b) The distributions are different.

c)

E	$O - E$	$(O - E)^2$	$(O - E)^2/E$
27.9	-12.9	166.41	5.96
23.7	-3.7	13.69	0.58
25.7	9.3	86.49	3.37
22.7	7.3	53.29	2.35

$\chi^2 = 12.26$; $c = 7.81$.
Reject the null hypothesis.

5. a) Distributions for 1959 and 1969 are the same.
b) The 1959 and 1969 distributions are different.

c)

E	$(O - E)$	$(O - E)^2$	$(O - E)^2/E$
391.7	-7.7	59.29	0.15
529.8	10.2	104.04	0.20
92.8	-30.8	948.64	10.22
70.2	-20.2	408.04	5.81
36.2	8.2	67.24	1.86
317.0	29.0	841.00	2.65
436.9	38.1	1451.61	3.32
389.4	47.6	2265.76	5.82

$\chi^2 = 30.03$; $c = 14.07$.
Reject the null hypothesis.

7. a)

Number of hits	x	0	1	2	3	4
Probability	$P(x)$	0.130	0.346	0.346	0.154	0.026

b)

Number of hits	E	O	$(O - E)^2/E$
0	5.2	4	0.28
1	13.8	17	0.74
2	13.8	12	0.23
3	6.2	5	0.23
4	1.0	2	1.00

c) $\chi^2 = 2.48$, $c = 9.49$. The null hypothesis cannot be rejected.

Section 10.3

1. a) Annual income and education are independent.

b) Annual income and education are dependent.

c) Population: all white wage earners, 25 years and older.

e) 12 cells.

Annual income

		0–6,999	7,000–14,999	15,000–24,999	25,000+	Total
	0–8	34 / 15.8	41 / 29.5	10 / 28.9	4 / 14.8	89
Years of schooling	9–12	36 / 37.7	72 / 70.2	78 / 68.8	26 / 35.3	212
	over 12	10 / 26.5	36 / 49.3	58 / 48.3	45 / 24.8	149
	Total	80	149	146	75	450

h) Annual income and education are probably dependent.

3. a) Opinions on overall effectiveness and economic effectiveness of the President are independent.

b) Opinions on overall effectiveness and economic effectiveness of the President are dependent.

c) Population: Undergraduate engineering students.   e) 4 cells.

Question 1

	Yes	No	Total
Yes	17 · 7.7	18 · 27.3	35
No	5 · 14.3	60 · 50.7	65
Total	22	78	100

Question 2

h) The opinions appear to be dependent.

5. a) $d_r = \dfrac{17}{35} - \dfrac{5}{65} = 0.49 - 0.08 = 0.41$. Difference in proportions of "Yes" and "No" on Question 2.

b) $d_c = 0.77 - 0.23 = 0.54$. Difference in proportions of "Yes" and "No" on Question 1.

c) $\phi = 0.47$ d) Medium positive association.

Section **10.4**

1. a, b) See answer to Exercise 1, Secton 10.3.
   c) $\chi^2 = 81.75$, d.f. $= 6$, $c = 16.81$. Reject the null hypothesis.

3. a, b) See answer to Exercise 3, Section 10.3.
   c) $\chi^2 = 22.15$, d.f. $= 1$, $c = 3.84$. Reject the null hypothesis.

5. a) *Null hypothesis:*   Residence and income are independent.
   b) *Alternative hypothesis:*   Residence and income are dependent.

c)

	0–6,999	7,000–14,999	15,000–24,999	25,000 and over	Total
Central cities	23 · 22.2	42 · 43.2	38 · 39.0	18 · 16.5	121
Outside central cities	26 · 38.0	64 · 73.9	75 · 66.7	42 · 28.3	207
Outside metro areas	41 · 29.8	69 · 57.9	45 · 52.2	7 · 22.1	162
Total	90	175	158	67	490

d.f. $= 6$, $c = 16.81$, $\chi^2 = 30.65$.
Reject the null hypothesis.

7. a) $\chi^2 = 0.69$. The null hypothesis cannot be rejected.
   b) $\chi^2 = 1.37$. The null hypothesis cannot be rejected.
   c) $\chi^2 = 6.88$. Reject the null hypothesis.

Section **10.5**

1. a) *Null hypothesis:* The two geographical areas are homogeneous in their voting preference.
   b) *Alternative hypothesis:* The two geographical areas are not homogeneous.

   c)

	New England	S. Atlantic	Total
Democrat	274 / 283.6	612 / 602.4	886
Republican	243 / 233.4	486 / 495.6	729
Total	517	1098	1615

   d.f. $= 1$, $c = 3.84$, $\chi^2 = 1.06$.

   The null hypothesis cannot be rejected.

3. a) *Null hypothesis:* Divorced males and females are homogeneous with respect to age groups.
   b) *Alternative hypothesis:* Divorced males and females are not homogeneous with respect to age groups.

   c)

	25–34	35–44	45–54	Total
Male	67 / 67.2	56 / 56.5	56 / 55.3	179
Female	103 / 102.8	87 / 86.5	84 / 84.7	274
Total	170	143	140	453

   d.f. $= 2$, $c = 5.99$, $\chi^2 = 0.02$.

   The null hypothesis cannot be rejected.

5. a) *Null hypothesis:* Cities were homogeneous in viewing.
   b) *Alternative hypothesis:* Cities were not homogeneous in viewing.

   c)

	Roots	Graffiti	Marathon	Total
City 1	43 / 41	28 / 30.5	29 / 28.5	100
City 2	39 / 41	33 / 30.5	28 / 28.5	100
Total	82	61	57	200

   d.f. $= 2$, $c = 4.61$, $\chi^2 = 0.62$.

   The null hypothesis cannot be rejected.

1. a) *Null hypothesis:* The metropolitan area has the same distribution as the country
with respect to income.

   Alternative hypothesis: The distributions are different.

   b) i) d.f. = 4,   ii) $c = 9.49$

   c)

Class	O	E	$O - E$	$(O - E)^2/E$
Under $5,000	10	13	-3	0.69
$5,000–9,999	19	21	-2	0.19
$10,000–14,999	20	22	-2	0.18
$15,000–24,999	34	30	4	0.53
$25,000 +	17	14	3	0.64

   d) Yes   e) The null hypothesis cannot be rejected. $\chi^2 = 2.23$

2. a) *Null hypothesis:* Party affiliation and religion are independent.
   Alternative hypothesis: Party affiliation and religion are dependent.

   b) i) d.f. = 2   ii) $c = 9.21$ d) Yes

   e)

	Catholic	Protestant	Jewish	Total
Democrat	130 / 105.2	360 / 394.5	36 / 26.3	526
Republican	110 / 134.8	540 / 505.5	24 / 33.7	674
Total	240	900	60	1200

   f) $\chi^2 = 22.15$.   g) Reject the null hypothesis.

Chapter **11**

1. a) *Null hypothesis:* There is no difference between the mean effectiveness of the three
waxes.

   b) *Alternative hypothesis:* There is a difference between the mean effectiveness of the
three waxes.

   d) d.f. = (2, 12),   e) $c = 3.89$ f)

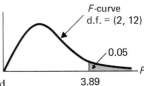

   g) MSW = 6.733,   h) MSB = 21.600

   i) $F^* = 3.21$.   j) The null hypothesis cannot be rejected.

3. a) *Null hypothesis:* The mean lives of the three kinds of calculators are the same.
   b) *Alternative hypothesis:* The mean lives are not all the same.

   d) d.f. = (2, 12),   e) $c = 3.89$ f)

   g) MSW = 0.097,   h) MSB = 0.060

   i) $F^* = 0.62$. The null hypothesis cannot be rejected.

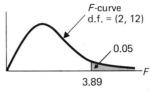

5. a) *Null hypothesis:* The mean weight gains produced by the three treatments are the same.
   b) *Alternative hypothesis:* The mean weight gains produced by the three treatments are not all the same.

   d) d.f. = (2, 9),   e) $c = 4.26$ f)

   g) MSW = 0.134,   h) MSB = 1.611

   i) $F^* = 12.02$.   j) Reject the null hypothesis.

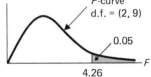

Section 11.2   1. SSB = 43.200, SSW = 80.800, $F^* = 3.21$ 3. SSB = 0.120, SSW = 1.162, $F^* = 0.62$

5. *Null hypothesis:* The advertising policies yield the same mean sales.
   Alternative hypothesis: The advertising policies do not yield the same mean sales.
   Step 1. $\alpha = 0.05$.
   Step 2. $c = F_{0.05} = 5.14$, for d.f. = $(k - 1, n - k) = (2, 6)$
   Step 3. MSB = 12.111, MSW = 9.778, $F^* = 1.24$
   Step 4. The null hypothesis cannot be rejected.

Review Test   1. *Null hypothesis:* There is no difference in mean sales level of the three designs.
   Alternative hypothesis: There is a difference in mean sales level of the three designs.
   $\alpha = 0.01$, d.f. = (2, 9), $c = 8.02$;   MSW = 43.389, MSB = 9.250, $F^* = 0.21$.
   The null hypothesis cannot be rejected.

2. d.f. = (3, 16), $\alpha = 0.05$, $c = 3.24$;   MSW = 0.220, MSB = 2.421, $F^* = 11.00$
   Reject the null hypothesis.

Chapter **12**

Section 12.1   1. a) $y = 50 + 5x$ b)

x	y
1	55
2	60
3	65

c)

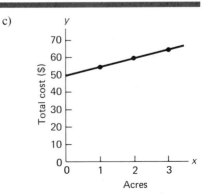

d) $70

3. a) $y = 100,000 - 10,000x$

c)

b)

x	y
1	90,000
4	60,000
8	20,000

d) $40,000

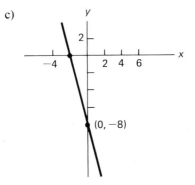

5. a) 2   b) −1

c)

7. a) −4   b) −8

c)

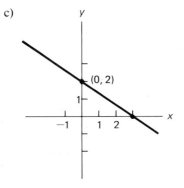

9. a) −2/3   b) 2

c)

11. $y = 5 + 2x$

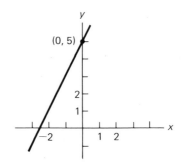

13. $y = -2 - 3x$

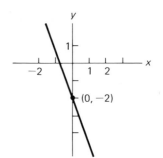

15. $y = -1 - \frac{3}{2}x$

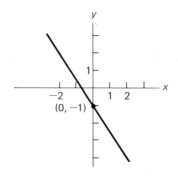

17. Increasing   19. Decreasing   21. Increasing

Section **12.2**   1. a)

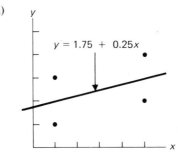

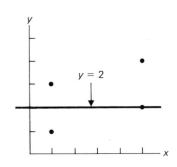

b)

			Line A			Line B	
x	y	$\hat{y}$	e	e^2	$\hat{y}$	e	e^2
1	1	2	-1	1	2	-1	1
1	3	2	1	1	2	1	1
5	2	3	-1	1	2	0	0
5	4	3	1	1	2	2	4
				4			6

c) $y = 1.75 + 0.25x$

3. $\hat{y} = 1.75 + 0.25x$

5. a) $\hat{y} = 12.86 + 1.21x$ b)

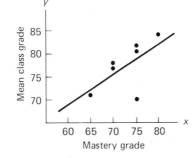

c) Food costs increase as income does.

7. a) $\hat{y} = 27.30 + 0.69x$ b)

c) Mean class grade increases as
   mastery level does.

9. a) $\hat{y} = 96.79 - 0.96x$ b)

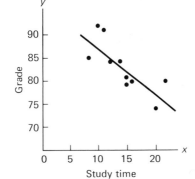

c) Grades decrease as study time increases.

<u>11.</u> b) Between 60 and 80   c) $x = -1$ d) Between 1 and 6

<u>13.</u> a) $\text{Cov}(x, y) = 1.33$ b) $\text{Cov}(x, y) = -4$

Section 12.3

1. a) $r = -0.447$ b) SST = 5, SSE = 4, SSR = 1   c) 0.20, 0.80   d) -0.447

3. a) $r = 0.74$ 5. $r = 0.63$ 7. $r = -0.77$

9. a) $r = 0$ b) No

c) Yes. In fact, the points all lie on the parabola $y = 9 - (x - 3)^2$.

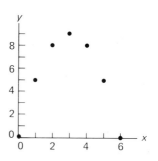

Section 12.4

1. a) All possible pairs of values (x, y), where x is family disposable income for middle income families of four, and y is food cost.

b) Assumptions I–III; regression line, normality, equal variances.

c) Similar to Fig. 12.15

d) $y = \beta_0 + \beta_1 x + \varepsilon$, where ε is normally distributed with mean 0 and variance σ^2.

3. a) All possible pairs of values (x, y), where x is the age of Ford pickups for sale in Denver in May 1979, and y is price.

b) Assumptions I–III.

c) Similar to Fig. 12.15

d) $y = \beta_0 + \beta_1 x + \varepsilon$, where ε is normally distributed with mean 0 and variance σ^2.

5. (b)   7. a) MSE = 2   b) 2

9. *Note:* Answers may vary due to roundoff error.

a)

x	y	$\hat{y}$	e	e^2
10	29	23.10	5.90	34.86
25	72	72.18	-0.18	0.03
19	40	52.55	-12.55	157.43
42	128	127.81	0.19	0.04
22	64	62.36	1.64	2.68
28	87	82.00	5.00	25.01

220.05 = SSE

MSE = 55.01

b) 55.01

Section 12.5

Note: Answers may vary due to roundoff error.

1. d.f. = 2, critical values ± 4.30, $t = 0.71$. The null hypothesis cannot be rejected.

3. d.f. = 4, critical values ± 4.60, t = 10.49. Reject the null hypothesis.

5. d.f. = 6, critical values ± 2.45, t = 19.12. Reject the null hypothesis.

7. 64.91 to 144.73   9. 3.19 to 5.01

<u>11.</u> a)

x	Prediction interval
1	1.08 to 3.07
2	2.14 to 4.03
3	3.19 to 5.01
4	4.21 to 6.00
5	5.23 to 7.01
6	6.22 to 8.04
7	7.19 to 9.09
8	8.15 to 10.15

b)

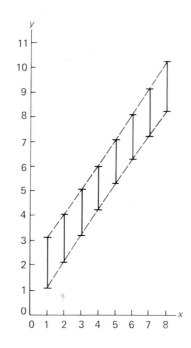

c)

x	Prediction interval
1	0.56 to 3.59
2	1.65 to 4.52
3	2.72 to 5.48
4	3.76 to 6.46
5	4.77 to 7.47
6	5.75 to 8.51
7	6.71 to 9.57
8	7.64 to 10.66

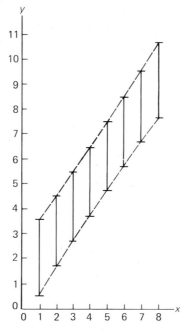

Section **12.6**

1. $\alpha = 0.05$, d.f. = 23, critical values ± 0.396, $r = 0.31$. Not significant.

3. $\alpha = 0.05$, d.f. = 4, critical values ± 0.811, $r = 0.98$. Significant.

5. $\alpha = 0.10$, d.f. = 7, critical values ± 0.582, $r = -0.95$. Significant.

<u>7.</u> a) i) 0.9   ii) 0.9   iii) 0.9   b) No.

Review Test

1. a) $\hat{y} = 1.30 + 3.30x$ b) $\hat{y} = 12.85$

2. $r = 0.9914$ 3. a) $r \doteq 0$ b) $r > 0$ c) $r < 0$

4. a) SSE = 1.9   b) SST = 110.8   c) SSR = 108.9   d) 98.29%   e) $r = 0.9914$

5. a) d.f. = 21, critical values ± 2.08, $t = 3.52$. Reject the null hypothesis.

   b) 127.60 to 182.48 pounds   c) Yes.

Chapter **13**

Section **13.2**

1. 0.1172   3. a) 0.5596   b) 0.3596

5. 0.0420   7. 0.9130   9. 0.8418

Section **13.3**

1. *Null hypothesis:* Md = 12.1   *Alternative hypothesis:* Md > 12.1

   $n = 8$, $\alpha = 0.0351$

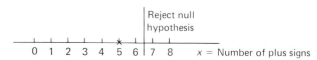

   $x = 5$. The null hypothesis cannot be rejected.

3. *Null hypothesis:* Md = 8618   *Alternative hypothesis:* Md < 8618

   $n = 7$, $\alpha = 0.0078$

   $x = 4$. The null hypothesis cannot be rejected.

5. a) *Null hypothesis:* Md = 5   *Alternative hypothesis:* Md < 5

   b) $n = 12$, $\alpha = 0.0729$.

   $x = 2$. Reject the null hypothesis.

<u>7.</u> b) $k = 1$ and $k = 8$

c)

d) $x = 8$. Reject the null hypothesis.

1. *Null hypothesis:* Md $= 50$ *Alternative hypothesis:* Md < 50
   $\alpha = 0.055, d = 6$

Sample values, x	Differences, $D = x - 50$	Positive differences, $\lvert D \rvert$	Rank	Signed rank, R
50.5	$+0.5$	0.5	3	$+3$
50.2	$+0.2$	0.2	2	$+2$
50.1	$+0.1$	0.1	1	$+1$
49.4	-0.6	0.6	4	-4
49.3	-0.7	0.7	5	-5
48.9	-1.1	1.1	6	-6
48.1	-1.9	1.9	7	-7
46.5	-3.5	3.5	8	-8

$R^+ = 6$. Reject the null hypothesis.

3. $\alpha = 0.008, d = 0$

Sample values, x	Differences, $D = x - 8618$	Positive differences, $\lvert D \rvert$	Rank	Signed rank, R
8610	-8	8	1	-1
9346	$+728$	728	2	$+2$
11428	$+2810$	2810	6	$+6$
17347	$+8729$	8729	7	$+7$
6092	-2526	2526	5	-5
7817	-801	801	4	-4
9385	$+767$	767	3	$+3$

$R^+ = 18$. The null hypothesis cannot be rejected.

5. *Null hypothesis:* Md = 23   *Alternative hypothesis:* Md > 23
   $\alpha = 0.053, d = 11$

Sample values, x	Differences, $D = x - 23$	Positive differences, $\|D\|$	Rank	Signed rank, R
35	+12	12	5.5	+5.5
60	+37	37	10	+10
28	+5	5	2.5	+2.5
22	−1	1	1	−1
31	+8	8	4	+4
18	−5	5	2.5	−2.5
53	+30	30	9	+9
42	+19	19	7	+7
35	+12	12	5.5	+5.5
47	+24	24	8	+8

$R^- = -3.5, \|R^-\| = 3.5$. Reject the null hypothesis.

7. a) $d = 50$ b) $R^+ = 103, \|R^-\| = 107$. The null hypothesis cannot be rejected.
   c) $d = 43$

9. a) $\mu = 52.50, \sigma \doteq 15.93$
   b) Reject null hypothesis if $\|R^-\| < 15.38$.   c) No.

Section **13.5**   1. $n = 5, k = 5, c = 4$.

	Mfr. I					Mfr. II				
Lifetimes	2.3	3.7	5.9	6.8	3.5	1.9	3.8	6.4	5.6	4.9
Rank	2	4	8	10	3	1	5	9	7	6

$S_1 = 27, T = 27 - \dfrac{5(5 + 1)}{2} = 12$

The null hypothesis cannot be rejected.

3. $n = 4, k = 6, c = 3$

	Group I				Group II					
Averages	81	62	53	58	75	91	76	83	64	49
Rank	8	4	2	3	6	10	7	9	5	1

$S_1 = 17, T = 17 - \dfrac{4(4 + 1)}{2} = 7$

The null hypothesis cannot be rejected.

5. $n = 5, k = 5, c = 4$

	Mfr. I					Mfr. II				
Stress	52	55	49	53	52	58	57	56	53	55
Rank	2.5	6.5	1	4.5	2.5	10	9	8	4.5	6.5

$$S_1 = 17, T = 17 - \frac{5(5 + 1)}{2} = 2$$

Reject the null hypothesis and conclude that Manufacturer I makes cord with less median strength than Manufacturer II.

7. a) $\mu = 75, \sigma \doteq 18.03$
   b) Using the normal approximation, $c = 45.43$. Using Table 13.11, $c = 44$.
   c) Using the normal approximation, $c = 72.96$. Using Table 13.11, $c = 72$.

1. $R = 7, c_1 = 4, c_2 = 13$. The null hypothesis (of randomness) cannot be rejected.   Section **13.6**

3. $R = 6, c_1 = 6, c_2 = 16$. Reject the null hypothesis (of randomness).

1. $n = 8, \alpha \doteq 0.0351$ Review Test
   $x = 3$. The null hypothesis cannot be rejected.

2. $\alpha = 0.039, d = 5$

Sample values, x	Differences, $D = x - 12.1$	Positive differences, $\lvert D \rvert$	Rank	Signed rank, R
12.0	-0.10	0.10	1.5	-1.5
12.2	$+0.10$	0.10	1.5	$+1.5$
11.9	-0.20	0.20	3.5	-3.5
11.7	-0.40	0.40	7.5	-7.5
12.3	$+0.20$	0.20	3.5	$+3.5$
12.4	$+0.30$	0.30	5	$+5$
12.5	$+0.40$	0.40	7.5	$+7.5$
12.45	$+0.35$	0.35	6	$+6$

$R^- = -12.5, \lvert R^- \rvert = 12.5$. The null hypothesis cannot be rejected.

3. $n = 5, k = 5, c = 4$

	Present model					New model				
Lifetime	2.93	2.74	2.81	3.02	2.98	3.07	3.12	2.85	2.94	3.05
Rank	4	1	2	7	6	9	10	3	5	8

$$S_1 = 20, \quad T = 20 - \frac{5(5 + 1)}{2} = 5$$

The null hypothesis cannot be rejected.

4. $R = 5, c_1 = 4, c_2 = 12$. The null hypothesis (of randomness) cannot be rejected.

Chapter **14**

Section **14.3**   1. Answers will vary.   3. Answers will vary.

5. a) Number the suites 1 to 48, and then use a table of random numbers to select four of the suites at random.
   b) Probably not, since friends are likely to have similar opinions.

   c)

Stratum	Percentage	Number to be selected
Freshmen	33.33%	16
Sophomores	29.17%	14
Juniors	25.00%	12
Seniors	12.50%	6
		48

Number the freshmen 1 to 128 and use a table of random numbers to select sixteen of these freshmen; number the sophomores 1 to 112 and use a table of random numbers to select fourteen of these sophomores; and so on.

Section **14.4**   1. a) By rejecting the null hypothesis $\mu = 100$, when it is actually true.
   b) 0.05
   c) By failing to reject the null hypothesis, when in fact $\mu > 100$.
   d) Decrease

3. a) $\sigma_{\bar{x}} \doteq 0.0021$ b) $d \doteq 0.9966$ c) 0.05

   d) A Type I error would occur if $\mu = 1$ and $\bar{x}$ turned out to be less than 0.9966. The manufacturer would not like to see a Type I error.

   e) A Type II error would occur if $\mu < 1$ and $\bar{x}$ turned out to be at least 0.9966. The regulatory agency would like to guard against a Type II error.

   f) 0.2296

g)

	True value of μ	
	$\mu = 1$	$\mu = 0.995$
Sample size	Probability of Type I error	Probability of Type II error
64	0.0951	0.2776
81	0.0694	0.2514
100	0.05 (0.0505)	0.2296
144	0.0244	0.1867

Note: To construct this table (and answer part (f)) we used $d = 0.996556$, which is the exact answer for (b). If you use $d = 0.9966$ you will obtain a slightly different table.

Index